Chinese Mathematics in the Thirteenth Century

M.I.T. East Asian Science Series
Nathan Sivin, general editor

Volume 1

Chinese Mathematics in the Thirteenth Century

The Shu-shu chiu-chang of Ch'in Chiu-shao

Ulrich Libbrecht

The MIT Press
Cambridge, Massachusetts, and London, England

This book was set in Monotype Baskerville,
printed on Mohawk Neotext Offset
by The Colonial Press Inc.,
and bound by The Colonial Press Inc.
in the United States of America.

Library of Congress Cataloging in Publication Data

Libbrecht, Ulrich.
    Chinese mathematics in the thirteenth century.

    (M.I.T. East Asian science series, v. 1)
    Bibliography: p.
    1. Mathematics, Chinese. 2. Algebra—Early works to 1800. 3. Ch'in, Chiu-
shao. I. Title. II. Series: Massachusetts Institute of Technology. M.I.T. East
Asian science series, v.1.
QA 27.C5L54   510'.951   72–2320
ISBN 0–262–12043–7

Dedicated to Nathan Sivin

Contents

The M.I.T. East Asian Science Series

One of the most interesting developments in historical scholarship over the past two decades has been a growing realization of the strength and importance of science and technology in ancient Asian culture. Joseph Needham's monumental exploratory survey, *Science and Civilisation in China,* has brought the Chinese tradition to the attention of educated people throughout the Occident. The level of our understanding is steadily deepening as new investigations are carried out in East Asia, Europe, and the United States.

The publication of general books and monographs in this field, because of its interdisciplinary character, presents special difficulties with which not every publisher is fully prepared to deal. The aim of the M.I.T. East Asian Science Series, under the general editorship of Nathan Sivin, is to identify and make available books which are based on original research in the Oriental sources, and which combine the high methodological standards of Asian studies with those of technical history. This series will also bring special editorial and production skills to bear on the problems which arise when scientific equations and Chinese characters must appear in close proximity, and when ideas from both worlds of discourse are interwoven. Most books in the series will deal with science and technology before modern times in China and related Far Eastern cultures, but manuscripts concerned with contemporary scientific developments or with the survival and adaptation of traditional techniques in China, Japan, and their neighbors today will also be welcomed.

Volumes in the Series

Ulrich Libbrecht, *Chinese Mathematics in the Thirteenth Century: The Shu-shu chiu-chang of Ch'in Chiu-shao*

Shigeru Nakayama and Nathan Sivin (eds.), *Chinese Science: Explorations of an Ancient Tradition*

Manfred Porkert, *The Theoretical Foundations of Chinese Medicine: Systems of Correspondence* (in preparation)

Sang-woon Jeon, *Science and Technology in Korea: Traditional Instruments and Techniques* (in preparation)

Foreword

When we first become aware of the existence of an ancient science and begin exploring it, we naturally proceed from one familiar landmark to another. We search out what is "scientific" by analogy with our contemporary knowledge, picking out precursors of today's techniques and theories from their background of what seem to be retrograde notions and wild guesses. This extensive exploration yields a general map, which is perhaps all we need so long as our interest remains concentrated upon landmarks in the anticipation of our own particular state of knowledge.

Sooner or later, as we seek to deepen our comprehension, we are forced to enter a new phase. Our initial view of the science as a gradual accumulation of isolated discoveries connected only by their common end is in the last analysis teleological. It can point to A as a step on the way to B, but it cannot lead to an understanding of how B evolved out of the inner necessity, historical experience, and social consequences of A. Evolution is, after all, a matter of ambience. A did not appear at a certain time because some final cause ordained its time had come, but rather because it was coherent with other ideas, attitudes, and prejudices of its time. The problem of this intensive phase of exploration thus becomes the recognition of crises and of points open to innovation within an integral system of scientific thought and practice, influenced by and influencing the wider social and intellectual climate of its time. Ideas which in the earlier phase of study were perceived merely as the outdated and misguided backdrop of "modern" anticipations now must be evaluated as seriously as the latter, for they played no less important a role in defining the ancient scientist's conception of the natural world—and thus the direction and style of his investigations.

It is not at all unusual to uncover methods or concepts which, despite negligible intrinsic interest to the physicist or chemist today, played crucial roles in early scientific change. Among examples which spring to mind are the medieval Western speculations about the time-variant fall of bodies which long after, when grounded by Galileo in measure and experiment, contributed to the foundations of mechanics. The Chinese concepts of *wu-hsing*—which used to be translated "five elements"—and *yin-yang*, until lately dismissed as superstitious nonsense or (at best) as notions of a pre-Socratic kind, are now turning out, under disciplined philological analysis, to be two interrelated systems for abstracting the phases of a cyclic process. Far from being archaic impediments to the advance of science, they reveal themselves increasingly as forms by which the Chinese structured what was most objective as well as most abstract in their thought.

It is perhaps not intuitively obvious that this distinction between extensive and intensive stages of exploration should be equally cogent for mathematics, which after all has been from the very start the art of number. There is nothing the ancient Chinese mathematicians were able to do which cannot be transcribed in modern notation. Enough of this exploratory labor of reconstruction and transcription has been accomplished to give us at least a sketchy idea of the gamut of problems and computational techniques, and to confirm certain generalizations about the tradition. Mikami's *History of Mathematics in China and Japan* is still in print after sixty years, and the mathematical volume of Joseph Needham's *Science and Civilisation in China* has been out for a decade. It is difficult to imagine that anyone aware the ancient Chinese did mathematics is not also conscious of the overwhelmingly practical orientation of their writing, their great early strength in numerical and algebraical procedures and corresponding lack of development in geometry, and their perennial reliance on computational devices (first the computing rods and then the abacus). But we still have to learn how these tendencies grew and hardened, and what embryonic alternatives were tried and rejected, and why.

Questions of this sort can be answered only by intensive investigations. As Dr. Libbrecht points out, modern transcription can obscure the point of a technique as well as reveal it. An algebraic equation is a completely general assertion of equality, implying no preference among possible values of its unknowns or among its various solutions. The Chinese problem which it purports to express, on the other hand, may or may not envision highly specific values and solutions. It was a consequence of Ch'in Chiu-shao's practical orientation, for instance, that (like his predecessors) he did not proceed beyond the simplest solution of indeterminate equations. To imply that, like the modern mathematician, he thought of the remainder theorem's solution as many-valued is to misunderstand him seriously.

It is equally easy, through lack of openness to the inner logic of a problem, to overlook attempted moves beyond the limits of practical application. One of the many *tours de force* of Dr. Libbrecht's book has to do with a problem which Ch'in solves using an equation of the tenth degree. This problem has been taken up by historians before, and it has generally been noted in passing that the solution is unnecessarily complicated. It is quite true that an equation of the third degree would have been adequate to the problem as stated. But Dr. Libbrecht perceives the crux for the first time. Ch'in Chiu-shao accepted the traditional requirement that problems be stated in practical terms. There was in fact no other language available in which they could have been stated. Given Ch'in's desire to explore equations of the tenth degree, it is hard to imagine any situation in his experience which unequivocally required them. It is also hardly realistic to expect a pioneer to be much concerned about identifying a situation which in *our* vastly different experience requires equations of such high degree. In other words, this is an example not of inelegant solution but of experimentation in a direction not easily accommodated by the traditional character of Chinese mathematics.

Ch'in's account of the general method for solving indeterminate problems is the first general mathematical formulation in the Chinese literature; it moves mathematical discourse onto a new level of abstraction. Again we can only regret that he had no posterity before the decline of the tradition had set in.

Chinese Mathematics in the Thirteenth Century is four closely related studies of Ch'in, one of the very greatest representatives of his tradition, and his work. The first brings together what is known of his life and of the history of his one extant work, the *Shu-shu chiu-chang*. The second (Parts II–IV) delineates the whole range of mathematical techniques and problems found in Ch'in's book. They are to a large extent typical of the Chinese art at its apogee. Although economy of space demands considerable use of modern transcription, there is ample translation and description of Ch'in's actual procedures to support many suggestive hypotheses about their underlying rationales. Even in general matters Dr. Libbrecht's combination of mathematical skill and intimate knowledge of his sources results in a new increment of clarity. There is no doubt in my mind, for instance, that his discussion of decimal place-value will be considered definitive. It is certainly badly needed.

There follows (Part V) a study in breadth and depth of what modern mathematicians still call the Chinese remainder theorem for the solution of indeterminate equations of the first degree. This was Ch'in's most original contribution to mathematics—so original that no one who had access to copies of Ch'in's manuscript could correctly explain his procedure until the beginning of the nineteenth century, and his book was not printed until 1842.

In fact Dr. Libbrecht demonstrates, through his historical survey of indeterminate equations in India, Islam, Europe, and China, how unprecedented Ch'in's technique was. We can see once and for all that, despite the claims of one historian of mathematics after another, the ancient Indian *kuṭṭaka* method was not identical with it. The *kuṭṭaka* anticipates Lagrange's approach to indeterminate equations, and Ch'in's procedure that of Euler and Gauss. We are provided not only with full translations and explanations of the ten indeterminate problems in Ch'in's book but with an analysis which reveals the single consistent algebraic pattern uniting them. Given this consistency, we are further encouraged to question the conventional wisdom about the Chinese mathematicians' inability to rise to the level of general rules. As Dr. Libbrecht remarks, a rule can be general without being deductive.

The chapter headed "General Evaluation of Ch'in Chiu-shao's *Ta-yen* Rule" has much wider implications than its title indicates. In actuality it deals with the question of how to determine rigorously which of two solutions to a given problem is superior—and under what conditions no judgment of superiority can be meaningfully made. Because we are always faced with the issue of whether a chronologically later solution is logically more advanced, Dr. Libbrecht's discussion is a most useful methodological foundation for valid comparisons of mathematical methods across gaps of space and time.

The sixth part will interest even those to whom mathematics is a more exotic language than Chinese. By bringing together information on artisanal, economic, administrative, and military affairs dispersed throughout Ch'in's problems, it provides a trove of sidelights on Chinese life a millennium ago. It is striking how consistently the authenticity of these data can be demonstrated from other sources of the period. Thus when we are confronted with the complete nomenclature of the components of the gate tower on a city wall, we can be confident that they represent the practice of the time.

This book comes at the beginning of Dr. Libbrecht's second career. It is his doctoral dissertation, written to the very high standards of a degree *cum laude* at Leiden. It shows him already to be without peer in the West as a historian of Chinese mathematics.

Ulrich Libbrecht's first career was as a teacher of mathematics. It was an avocational interest in languages which led him to an awareness of the mathematical riches of old Cathay. Although he has taken two degrees in classical Sinology, he is by necessity entirely self-taught in the lost terminology of the Chinese art of number. Only immense discipline and perseverance have made up for remoteness from a major collection of Chinese mathematical literature. I am confident that the reader will detect in this book a high devotion to learning of a kind which, however quickly it may be disappearing elsewhere under the onslaught of academic bureaucratization and fashionable anti-intellectualism, still blooms from time to time in the ancient towns of East Flanders.

Nathan Sivin
La Pomme d'Or
Oudenaarde, Belgium
15 June 1971

Preface

Since the publication of Joseph Needham's *Science and Civilisation in China,* interest in old Chinese mathematics has awakened again in the West, and my work has its roots in this important study. In order to get a true idea of Chinese mathematical knowledge, we need at present a series of monographs based on the original Chinese texts. I have tried to fill this need for Ch'in Chiu-shao's *Shu-shu chiu-chang.*

Chinese mathematics forms part of medieval mathematics, of the algorithmic phase we find in all civilized countries at that time. In reading Ch'in's text, I tried to place it within this algorithmic mathematical conception, which was the preamble to modern algebraic logistic. One can dispute the algebraical nature of this kind of mathematics and state that modern algebraic symbolism marks the beginning of true algebra, but we must not underestimate the important role of this algorithmic phase. In the first place, it was at this stage that material was gathered in the form of problems and valid methods were invented; through this dark period of trial and error, mathematicians succeeded in handing down some valuable procedures, which formed the foundation for the logical structures of modern algebra. Modern mathematics did not develop *in vacuo.* We see that the great mathematicians of the eighteenth century took the traditional problems of arithmetic textbooks as their starting points; they selected the right procedures and proved them or rejected the methods *ad hoc.* Although in modern times algebraical methods were developed from a limited set of axioms in a purely deductive way, it would be pointless to hold that modern algebra came into being from nothing. By arranging old procedures in a new deductive structure, modern algebra was born and could build up its marvelous set of relations.

Mathematics is also a part of our culture, and, where logic has not yet come into existence, it allows us to lift a corner of the veil and to see how far practical logic has developed. Like literature and fine arts, logical reasoning is an expression of culture. For that reason, mathematics on every level is worth investigation. The history of mathematics is concerned not only with mathematics but also with history, that is, with the development of human knowledge and culture. Even though mathematics in a certain period may have been on a low level, compared with the modern counterpart, it is always interesting to see how difficult it was for the pioneers. I hope that my work has succeeded in giving a true idea not only of the results of Chinese mathematics (these are in fact without importance for our time) but also of its nature, its peculiar procedures, and its unusual notation.

The first part of this study is devoted to the *Shu-shu chiu-chang* in general; the second part is in the form of a monograph on the Chinese remainder theorem, which is of interest, even in the context of our modern theory of numbers. As demonstrated in this work, the solution reached its first apogee in Ch'in Chiu-shao's *ta-yen* rule; but, as Chinese mathematics has been long neglected, I am convinced that an evaluation is possible only through a thorough comparative study. Secondary sources are not always reliable because they translate the texts into modern algebraic symbols, sometimes generalizing what is in fact not general. In all circumstances I took the original texts as starting point, copied, and translated them. This source material may be a mere distraction to some scholars who live in places farther from the Tower of Babel than is Belgium, but I hope they will understand that I hoped to render a service to the history of science by starting from original texts, which are always a sure and reliable base.

From a methodological point of view, the relation between the Chinese text and its mathematical content is another reliable base. But for a good understanding of the contents, we have to analyze—and if necessary, to imitate—the *mechanical* methods with which Chinese mathematicians solved their problems.

In all circumstances, logical patterns are indispensable in analysis—opinions and allegations are not scientific—and I have tried to apply them everywhere in this work. In most cases analysis is a matter of patience and accuracy, and sometimes we have to sacrifice the aesthetic dimension to exactitude.

But I believe that science can only be promoted by rigorously pursuing truth and for that reason I did not try to write another novel on Chinese mathematics.

Acknowledgments

I do not know how I can express my gratitude to Professor Nathan Sivin of the Massachusetts Institute of Technology (Cambridge, Massachusetts); there is no doubt that without his help this book would never have been completed, not only because of the difficult bibliographical circumstances in which I had to work and which he solved by providing me with copies of all the texts I needed, but also because of his constant encouragement, his assistance in solving countless difficulties, and his proofreading of the whole manuscript and aid in preparing it for publication. I can only express my heartfelt thanks by dedicating my humble work to him as much too poor a compensation for so great a friendship.

I am much indebted to the staff of the Sinologisch Instituut at Leiden, especially to Professor Dr. A. F. P. Hulsewé for his constant assistance and encouragement, for his critical reading of the manuscript, for correcting many passages and providing me with valuable information; to Professor E. Zürcher for reading the whole work; and to Messrs. D. R. Jonker and D. Van der Horst for help in the bibliographical field.

My warm thanks also to Professor Yang Lien-shêng of Harvard University for reading the chapter on "Socioeconomic Information" and providing me with valuable corrections and suggestions; to Mr. P. van der Loon of Cambridge University, England, for doing the same for the chapter on "The *Shu-shu chiu-chang*: History and Investigation"; and to Professor K. Yabuuchi of Kyoto for his very helpful assistance in the study of the astronomical problems.

I am also much obliged to Joseph Needham, F.R.S., of Cambridge University, England, who was willing to help me in all circumstances; I hope my work will be worthy of his gigantic *Science and Civilisation in China,* where it found its inspiration.

I owe a great deal to many scholars who helped and encouraged me in solving special problems, provided me with books and periodicals I needed, and gave me valuable information, namely, Professor H. Franke (Munich), Professor T. Pokora (Prague); Dr. Berezkina (Moscow); Professor K.

Vogel (Munich); Professor M. L. Righini-Bonelli (Florence); Professor D. J. Struik (Massachusetts Institute of Technology); Professor W. Franke (Hamburg); Dr. J. Deleu (Ghent).

I am much indebted to Professor J. J. Bouckaert, Prorector of the University of Ghent; to Professor W. Couvreur, Director, and Professor A. Scharpé, Professor D. Ellegiers, and Professor W. Acker, all of the Oriental Institute of the University of Ghent, for their support in obtaining the scholarships indispensable for research or for their encouragement.

For assistance in compiling the bibliographies of Indian works, I am indebted to Professor D. Pingree, University of Chicago, and for assistance with Arabian works to Professor R. E. Hall of Imperial University, London.

For help with special problems I am much obliged to Professor P. Dingens (Astronomical Observatory, University of Ghent), Professor P. Bultiau (Leuven), Dr. Shigeru Nakayama (Tokyo), Professor C. C. Grosjean and his assistant Drs. W. Bossaert (Computing Laboratory, University of Ghent), Mr. A. G. Velghe (Royal Observatory of Belgium), Drs. F. Van Ommeslaeghe (Brussels), Mr. P. Costabel (Académie Internationale d'Histoire des Sciences, Paris).

As I did not have easy access to materials, I am much indebted for their kind assistance to Professor K. G. Van Acker, Librarian of the University Library, Ghent; to Dr. M. I. Scott (University Library, Cambridge, England); to Mr. Chi Wang (Library of Congress, Washington, D.C.); to Mr. G. Bouckaert (University Library, Ghent).

I am extremely grateful to the staffs of the Sinologisch Instituut at Leiden and of the Chinese Department of the University Library, Cambridge, England; and for the assistance of the Harvard-Yenching Institute and the Widener Library, Cambridge, Massachusetts, for giving me permission to make use of their rich collections of Chinese and Western books. As my own country has no Chinese library which could provide me with the necessary works, my debt to these institutions is extremely great. I wish also to express my thanks to the Massachusetts Institute of Technology, which appointed me guest of the Institute during my stay in the United States.

Thanks to the scholarship awarded by the Government of the Netherlands, I was enabled to work at the Sinologisch Instituut at Leiden; I am grateful for enjoying its confidence.

It is certain that this work could not have been completed without the sabbatical year granted me by the Nationaal Fonds voor Wetenschappelijk Onderzoek (National Foundation for Scientific Research) of Belgium; I am much indebted for this great honor.

For technical assistance I thank Mrs. R. Lemarcq, Mrs. P. Polet, and Miss Susan McCorkendale for typing the manuscript, Miss Els Raes for proofreading the typed text, and Mr. G. Van Rysselberghe for making the diagrams.

And finally I would like to thank my wife and children for the inconvenience they had to endure during the long time I was working on this manuscript, at home and abroad.

Maximas vobis gratias agimus, majores etiam habemus.

Bibliographical Note

There is only one text edition of the *Shu-shu chiu-chang*, and that is the first printing in the *I-chia-t'ang ts'ung-shu* (1842); the edition in the *Ts'ung-shu chi-ch'êng* (1936) is a mere reprint.[1] Another text still extant is preserved in the *Ssŭ-k'u ch'üan-shu* (1782), but it is not available;[2] all we know about older texts is based on the emendations made by Shên Ch'in-p'ei and Sung Ching-ch'ang in the *Shu-shu chiu-chang cha-chi*. Their comparative notes were also published in the *I-chia-t'ang* edition.[3]

In the area of bibliographical works on Chinese mathematics, the extremely valuable bibliographies in the third part of Needham's *Science and Civilisation in China* (1959) leave little to be desired.[4] For Western books and articles they are almost complete; some supplementary listings, especially for Russian books, can be found in Yushkevitch (4), of which there is an English translation in press, as well as a German translation (1964). In order to compile a bibliography covering the last ten years, it was necessary to review the bibliographical notes in all available periodicals devoted to Chinese studies. However, the best information can be found in two periodicals that specialize in history of science: *Isis* and *Archeion*.

For information on Chinese works the older books include Wylie's *Notes on Chinese Literature* (1867), several pages of which are devoted to astronomy and mathematics. In Chinese there are the indispensable *Ssŭ-k'u ch'üan-shu tsung-mu t'i-yao* (1782),[5] containing much information on older mathematical works, and Mei Wên-ting's *Li-suan shu-mu* (1702), which is lacking in many respects.[6]

For all bibliographical references to mathematical works, there is the extremely rich work by Ting Fu-pao (1'); other

---

[1] On the history of the text, see Chapter 4.
[2] On the copies of this collection that have been preserved, see Hummel (1), p. 121.
[3] See Chapter 4.
[4] Needham's bibliography goes up to 1957.
[5] The Harvard-Yenching Index (no. 7) to this work appeared in 1932.
[6] Many ancient mathematical works had not been recovered at the time when it was written.

valuable notes are collected in Hu Yü-chin (1′).

Bibliographical problems in the history of Chinese mathematics have been almost entirely eliminated by the work of Li Yen and Yen Tun-chieh, which falls into two categories:

(1) *Bibliography of text editions and older studies on Chinese mathematics:* Li Yen wrote several historical surveys of Chinese mathematics, the greatest (but not the only) value of which is their bibliographical completeness.[7] Of particular importance are the list of Ch'ing mathematicians and their works in Li Yen (10′), pp. 178–423, and a discussion of some mathematical books containing bibliographical notes in the same volume, pp. 424–478. Li Yen (11′) is a catalogue of 448 mathematical books preserved in his own library, including some helpful annotations. Li Yen (1′) contains descriptive notes on rare Chinese mathematical books in the Oriental Library of the Commercial Press;[8] a short account of mathematical books available in *ts'ung-shu* (collectanea) or published separately is included in (23′).[9] On old Chinese mathematical books in Shanghai see Yen Tun-chieh (1′).

(2) *Modern studies on the history of Chinese mathematics:*[10] A bibliography of modern papers on the history of Chinese mathematics appears in Li Yen (10′), beginning on p. 23. Other compilations, in chronological order, are shown in the accompanying table.

These bibliographical lists give us almost all the publications on Chinese mathematics up to 1948. It is not easy, however, to compile bibliographies for the last twenty years. Of course the most important works are known in the West, but as for articles it is very difficult to keep oneself posted.

There are some specialized bibliographies on the works of single authors, such as Ogura and Ôya (1′) and Yajima (1) on

[7] For the T'ang period see (10′), pp. 26–99; for Sung and Yüan, see (3′); for Ming, see (10′), pp. 149–178.
[8] This work is lacking in Needham's bibliography.
[9] Published again in (10′), pp. 18 ff.
[10] These bibliographical notes are not complete; we restrict ourselves to works we had at our disposal.

Year of Publication	Author and Work	Period Covered
1928	Li Yen (19′)	1799–1927[11]
1932	Li Yen (2′)	1912–1931
1936	Têng Yen-lin and Li Yen (1′)[12]	All periods
1940	Li Yen (4′)	1912–1939
1944	Li Yen and Yen Tun-chieh (18′)	1938–1944
1947	Li Yen (5′)	1917–1947
1948	Li Yen (24′)	1937–1947
1948	Li Yen (6′), vol. 5, pp. 116 ff[13]	1912-1948
1953	Li Yen (6′), vol. 2, pp. 32 ff	All periods

Mikami's work[14] and Wong Ming (1) on Li Yen's work.[15]

Mathematical dictionaries are a problem, because all of them give only modern mathematical terminology. I have relied on Wang Chu-chi's *English-Chinese Dictionary of Mathematical Terms*, Taranzano's *Vocabulaire des sciences mathématiques, physiques et naturelles*, and Wylie's *Mathematical and Astronomical Terms* in Doolittle (1), vol. 2, pp. 354 ff.[16] Other works I used are Chao Liao-shih (1′)[17] and Nagasawa (1′). However, the only dictionary which contains the old terminology is one that I compiled for my own use, deriving the meaning of the terms from the mathematical context (see Glossary). Useful also are Huang Chieh (1′) and Needham (1), vol. 3.

Biographical dictionaries include first of all the *Chung-kuo*

[11] The year 1799 was the date of publication of the *Ch'ou-jên chuan* of Juan Yüan.
[12] For a discussion of this work, see Hummel (2), p. 179.
[13] This work comprises Li Yen (2′), (4′) and (18′).
[14] Yajima's bibliographical list is difficult to use, because titles are given only in translation.
[15] Yuan Tung-li's *Bibliography of Chinese Mathematics 1918–1960* is devoted mainly to modern mathematics.
[16] The recent *Chinese-English Glossary of the Mathematical Sciences* by J. de Francis contains only modern terminology. On this work, see Ho Pêng-yoke in *JAOS*, 85, 2 (1965), pp. 212 ff.
[17] The first section gives "The technical terms containing [sic] in elementary mathematics," the second section a "vocabulary in English and Chinese of terms containing [sic] in mathematics."

jên-ming ta-tz'ŭ-tien for general information. A very important reference work is Juan Yüan's *Ch'ou-jên chuan,* which contains biobibliographical data derived from older works; the sources are indicated. For the Ch'ing period there is a very useful work by Hummel, *Eminent Chinese of the Ch'ing Period,* and the thoroughly documented *Ch'ing-shih lieh-chuan,* containing also biographies of minor scholars. For the Ming, there is the *Ming-jên chuan-chi tzŭ-liao so-yin* (1965).

For historical background I used O. Franke (1), vol. 4, and H. Franke (3); at present there are the very important works of the "Sung Project," including Cochini and Seidel's *Chronique de la dynastie des Sung,* a companion volume to the *Sung-shih;* Wêng T'ung-wên's *Répertoire des dates des hommes célèbres des Song;* Chang Fu-jui's *Les fonctionaires des Song: index des titres;* and Lee Mei Ching-ying's *Index des noms propres dans les annales principales des Song.*

For geographical names I made use of the *Historical and Commercial Atlas of China,* by A. Herrmann; the *Chung-kuo ku-chin ti-ming ta-tz'ŭ-tien;* and F. S. Couvreur's *Géographie ancienne et moderne de la Chine.* For use in locating places, there is the small but valuable *Chung-kuo ti-ming tz'ŭ-tien,* a list of modern place names with their geographic coordinates.

For bibliographies of Western mathematical writings, apart from the very valuable lists in *Isis* and *Bibliotheca Mathematica* (covering older works), the main sources were Tropfke (1) and Hofmann (1); moreover, there are important bibliographical notes in Dickson (1) and Vogel (3). For Arabian works I consulted J. D. Pearson's *Index Islamicus,* C. Brockelmann's *Geschichte der arabischen Literatur,* and C. A. Storey's *Persian Literature;* but as I had no access to Arabian texts, it was impossible to pursue this research. For works in Western languages the bibliography in Yushkevitch (4) is the best I know.[18] For Indian works there was no general reference work available, and I compiled my bibliography with the help of notes to which Professor D. Pingree (Chicago) provided important addenda.

[18] I am much indebted to Professor R. E. Hall for his kind assistance in compiling a bibliography on Islamic mathematics.

Reference Table

References to problems in the *Shu-shu chiu-chang* will follow the division of the Chinese text in nine chapters, each containing nine problems. For example, VI, 3 refers to the third problem of Chapter 6.

The third and fourth columns of this table show the *chüan* and the pages where the problem appears in the *Ts'ung-shu chi-ch'êng* edition. The last column shows page references for this book, with numbers in italic for pages where problems are dealt with in some detail.

Shu-shu chiu-chang		Ts'ung-shu chi-chêng		Page(s) in this work
Chapter	Problem	Chüan	Page	
I	1	1	1	82, 86, 335, *388*
	2		10	37, 82, 83, 86, 336, 353, 368, *391*
	3		19	83, 336, *396*, 438, 442
	4		23	82, 336, *350*, *382*, 418, 419,420,421
	5	2	31	77, 82, 337, *399*, 433
	6		34	76, 83, 337, *401*
	7		38	64, 69, 83, 85, *402*
	8		42	82, 85, 334, *406*, 448
	9		50	82, 85, 337, *408*
II	1	3	55	37, 69, 73, 75, 86, 368, 467
	2		56	37, 368, *469*
	3		57	37, 65, 86, 272, 368, *409*
	4		79	64, 76, *470*
	5	4	93	*472*
	6		107	86, 96, *113*, 474
	7		108	96, *115*, 474
	8		109	86, 96, *119*, 474
	9		110	86, 193, 198, 202, 210, 474
III	1	5	117	64, 70, 85, *97*, *180*, 193, 195, 210
	2		127	64, *99*, 209
	3		128	64, 86, *100*, 209
	4		129	64, *101*, 209
	5		131	64, 86, *108*, 209
	6		132	64, 210
	7	6	137	64, 85, *104*, 194, 201, 209, 337
	8		139	64, 77, *105*, 199, 209, 211

Shu-shu chiu-chang		Ts'ung-shu chi-chêng		Page(s) in this work
Chapter	Problem	Chüan	Page	
			154	64, 84, 85, 193, 202, 209, 439
IV	1	7	161	64, 77, 84, 86, *123*
	2		166	64, 85, *126*, 194, 209, 463
	3		177	64, 65, 123, *130*
	4	8	181	64, 65, 71, *132*
	5		183	64, *134*, 193, 211
	6		190	64, 123, *140*, 211, 465, 466
	7		202	64, 65, *144*, 467
	8		204	*144*, 195, 210
	9		205	64, 65, 84, *145*
V	1	9	209	64, 77, 78, 88, 89, 442, 444
	2	10	241	64, 71, 77, 88, 438, 445
	3		242	64, 438, 445
	4		243	64, *162*, 445
	5		249	64, 446
	6		250	446
	7		254	64, 77, 442, 443, 447
	8		259	64, 65, 88, 442, 447
	9		265	64, 447
VI	1	11	269	64, 65, 423, 425, 426, 437
	2		287	36, 39, 64, 424, 425, 436
	3		287	64, 424, 425, 434, 437
	4	12	291	64, 78, 86, *110*, 111, 193, 199, 202, 210
	5		308	64, 71, 77, 78, 436
	6		309	64, 433, 436
	7		320	64, 88, 418
	8		321	64, 72, 74, 94, 429, 430,
	9		322	64, 79, 88
VII	1	13	325	64, 113, 424, 448, 449, 451, 463
	2		332	64, 449, 450, 457
	3		333	64, 113, 172, 440
	4		341	64, 75, 437, 441
	5	14	345	64, 69, 71, 77, 83, 111, 457, 458
	6		363	88, 94, 461
	7		363	64, 449, 461

Shu-shu chiu-chang		Ts'ung-shu chi-chêng		Page(s) in this work
Chapter	Problem	Chüan	Page	
	8		365	64, 94
	9		366	64, *173*
VIII	1	15	369	64, *107*, 196, 209, 463, 464
	2		376	64, 172, 464, 465
	3		383	64, *174*, 210, 465
	4	16	387	64, *175*, 465
	5		394	64, 65, 123, *147*, 465
	6		399	64
	7		401	64
	8		402	37, 39, 64, 466
	9		406	64, *164*, 466
IX	1	17	415	64, 65, 71, *153*, 432
	2		421	64, *161*, 425, 426, 427, 432
	3		439	64, 88, 89, 427, 428
	4		440	64, 90, 431
	5	18	443	64, *90*, 431
	6		447	64, 425, 426
	7		451	87, 429, 430
	8		459	37, 39, 64, 85, *94*, 429, 430
	9		463	64

I

Ch'in Chiu-shao and
His Shu-shu chiu-chang

1

General Characteristics of Chinese Mathematics in Sung and Yüan

The *Shu-shu chiu-chang* 數書九章 (Mathematical treatise in nine sections) was written by Ch'in Chiu-shao 秦九韶 and published in 1247. This later phase of the Sung marks both the apogee of the development of mathematics in China and its terminal point.[1]

"The achievement of Chinese mathematics up to the late Ming dynasty was certainly not inferior to that of any other contemporary civilized country. In fields such as algebra, China was even more advanced than some other countries."[2] Ch'in Chiu-shao was the first of four great mathematicians, all living in the same half-century. As one observes in the history of many peoples, a time of political decline is sometimes a period of scientific prosperity. During Ch'in's life, China had fallen into decay; the northern part was in the hands of the Tartar Chin dynasty (1115–1234), and the western part was occupied by the Tangut dynasty of the Hsi Hsia (990–1227). Around 1230 both parts were conquered by the Mongols, who were from that time on a constant menace to the Southern Sung (1127–1279), who had established their new capital at Hang-chou.[3] The empire was in a state of great unrest; nevertheless, on both sides of the demarcation line mathematics flourished. "The intellectual spirit of the period, of which neo-Confucianism represented but one aspect, is also characterized by rationalism and a tendency toward systematization."[4] It is

---

[1] There are many guesses as to the reason for this decline; as this matter falls outside the scope of this work, the reader is referred to Needham (2), pp. 322 and 325 f.

[2] Wang Ping (1), p. 777.

[3] For a concise description, see H. Franke (3), pp. 213 ff; a more detailed historical survey is in O. Franke (1), vol. 4.

[4] "Die Geistigkeit der Zeit, von der der Neokonfuzianismus nur einen Aspekt darstellte, ist auch geprägt durch Rationalismus und einen Hang zur Systematik." H. Franke (3), pp. 218 ff.

not impossible that the evolution of mathematics was due to this "rationalism"; the "tendency toward systematization" can be seen in the technological works of the Sung, such as the famous *Ying-tsao fa-shih* (Architectural standards) and the *Wu-ching tsung-yao* (Conspectus of essential military techniques). The influence of technology on mathematics should not be underestimated.[5] Town planning, digging of irrigation canals, and other achievements do not pose significant mathematical problems, but the mathematicians proved themselves able to plan these works by means of calculations.

What was a mathematician in those times? In the first place the starting point of mathematics was at a very high level: the *Chiu-chang suan-shu* (Nine chapters on mathematical techniques), compiled in the first century, "is a work of highest rank, and in its influence probably the most significant of all Chinese mathematical books. It is the oldest textbook on arithmetic in existence, and its 246 problems make it incomparably richer than any other collection of examples that has been preserved from ancient Egyptian and Babylonian texts. In fact, Greek collections of arithmetic problems are known to us...only from the later Hellenistic and Byzantine periods."[6]

The wealth of the mathematical contents of the *Chiu-chang suan-shu* paradoxically tended to hinder further exploration, for the book became a classic. "It dominated the practice of Chinese reckoning-clerks for more than a millennium. Yet in its social origins it was closely bound up with the bureaucratic government system, and devoted to the problems which the ruling officials had to solve."[7] Becoming a classic can be a great

[5] See Part VI.

[6] ". . .Ist ein Werk höchsten Ranges und in seinem Einfluss wohl das bedeutendste aller mathematischen chinesischen Bücher; es ist das älteste Lehrbuch der Rechentechnik überhaupt und mit seinen 246 Problemen als Aufgabensammlung ungleich reichhaltiger als andere aus der Antike, die sich in ägyptischen und babylonischen Texten erhalten haben. Griechische arithmetische Aufgabensammlungen...kennen wir sogar erst aus späthellenistischer und byzantinischer Zeit." K. Vogel (2), p. 1.

[7] Needham (2), p. 325; see also Gauchet (2), p. 537: "It is particularly important because of its overwhelming influence on writers in the centuries that followed; all of Chinese mathematics bears its imprint, as to both ideas

honor for a literary work, but it is a mixed blessing in the sciences, because evolution is the *conditio sine qua non* of all scientific thought. Indeed, we see that even in Ch'in Chiu-shao's mathematical work the negative influence of the *Chiu-chang suan-shu* is very great.[8] Ch'in, like other late mathematicians, tends to compose variations on the canonical problems, as men of letters make variations on ancient poems; this practice may give rise to valuable new creations, but in mathematics nothing of value is achieved by changing the figures in a problem. Moreover, if we examine the section on mathematics in the *Yü-hai* encyclopedia (1267), we find only the traditional methods of the *Chiu-chang suan-shu* and some other old works, and not the slightest word about the great mathematicians of the Sung.[9]

Mathematics of a very simple kind was one of the essential accomplishments of the post-Confucian gentleman, on the same level as propriety, music, archery, charioteering, and calligraphy. In the *Yen-shih chia-hsün* (Family instructions for the Yen clan), written by Yen Chih-t'ui in the sixth century,[10] there is an interesting text that clearly shows the high esteem in which mathematics was held: "Mathematics is an important subject in the six arts. Through the ages all scholars who have participated in discussions on astronomy and calendars have to master it. However, you may consider it as a minor occupation, not as a major one" (p. 205). In other words, although mathematics was not considered a suitable livelihood for a gentleman, it was among the foremost of the arts of which he was encouraged to become an amateur.

This relation to astronomy and the calendrical sciences was a typical feature of mathematics in China; mathematics was the servant of the more important sciences of the heaven.

---

and terminology." (Son importance vient surtout de l'influence prépondérante qu'il a exercée sur les écrivains des siècles suivants; toute la mathématique chinoise porte son empreinte, et comme idée et comme terminologie.)

[8] See Chapter 5.

[9] In the *T'ien-chung chi* encyclopedia of 1569 (mathematical section, ch. 41, pp. 14a–16b) there is not the slightest word on the mathematics of the Sung and Yüan.

[10] See the translation by Têng Ssŭ-yü (1).

As early as the Sui and T'ang periods, there was founded an Office of Mathematics (*Suan-kuan* 算館), where minor officials were trained, not in theoretical mathematics, but in practical mensuration. We know the names of the textbooks used in T'ang times to prepare students for the official examination.[11] Some prefects had at their disposal a few clerks who were able to solve problems on irrigation, taxes, trade, and so on. The number of students in the *Suan-kuan* was the lowest of all offices.[12] This organization of mathematical studies seems to have been a great hindrance to their further development. As we shall see later, the great mathematicians of the Sung and Yüan did not study in these offices, and they were not professors of mathematics in the formal sense.

As Nakayama[13] points out, "the other kind of mathematics traditionally taught in China, known as *li-suan* 曆算 (calendrical mathematics), was the concern, not of the Office of Mathematics, but of the Board of Astronomy." Ch'in Chiu-shao studied at this Board, and this was rather exceptional; it was also why he dealt with calendrical problems, and why the important Chinese remainder problem has been preserved.

We have reason to believe that the "independent mathematician" appears for the first time in the Sung, judging from what we can deduce from biographical data. Ch'in Chiu-shao was never an official mathematician, although we know that "somebody recommended him to the throne on account of his *calendrical* science."[14] Chu Shih-chieh was a wandering teacher,[15] Yang Hui a civil servant, and Li Yeh a recluse scholar. Needham is right when he says that "the greatest mathematical minds were now (with the exception of Shên Kua) mostly wandering plebeians or minor officials."[16] This was perhaps one impulse for the development of mathematics in the Sung;

[11] Des Rotours (1), pp. 139 ff. See also the interesting material gathered by Li Yen (6'), vol. 5, pp. 15 ff.
[12] See des Rotours (1), pp. 179 f; Nakayama (1), p. 15.
[13] Nakayama (1), p. 16.
[14] Chou Mi (1'), part C, p. 6b.
[15] Ho Pêng-yoke (4), p. 1.
[16] (1), vol. 3, p. 42. The word "plebeian" is perhaps confusing, since all the great mathematicians were members of the educated minority.

precisely because these men were not mathematics teachers, they "broke out into fields much wider than the traditional bureaucratic preoccupations. Intellectual curiosity could now be abundantly satisfied."[17]

However, for the very reason that they were not mathematics teachers, they could not found a school, and their period of activity was of short duration. This seems to have been the vicious circle of Chinese mathematics.

In the bureaucratic China of the Sung, the fact that these mathematicians were not in the service of the Office of Mathematics was not the only reason for the stagnation of official mathematics; however, it is true that their writings were never used for instruction by the Office, either in their time or in the Ming or Ch'ing periods. Moreover, their works were unknown to the people of the Ming.[18] Only Li Yeh's works were printed; fortunately Yang Hui's and Ch'in Chiu-shao's works were preserved in the *Yung-lo ta-tien*,[19] but Chu Shih-chieh's work was preserved only because it was printed in Korea and later in Japan.

I am convinced that the solution of the difficult problem of the social position of a Chinese mathematician in general could throw light upon the nature and the background of Chinese mathematics. However, the data at our disposal are very scanty. "The low opinion of mathematics held by the government from the time that Confucianism became the state orthodoxy"[20] is responsible for the low social position of mathematicians. "The divisions of law, calligraphy, and mathematics were reserved for the sons of petty officials and even for the common people, for these areas of study were held in low esteem."[21] In his preface Ch'in Chiu-shao complained of this

---

[17] Needham (2), p. 325.
[18] The attitude toward mathematics in the Sung and later is illustrated by the example of T'ai Tsu, who added mathematics and military arts to the usual classical learning; after a short time they were abolished.
[19] Yang Hui's works were preserved only in part.
[20] Nakayama (1), p. 15.
[21] Des Rotours (1), p. 38, and Nakayama (1), pp. 15 ff: "The status of a professor of mathematics was low, as was that of his students, who were

situation: "In later generations scholars were very proud of themselves and, considering [these arts] inferior, did not teach [or discuss] them. These studies were almost defunct [through neglect]"; and, speaking of "clerks who applied themselves to simple calculations," he says: "If those who did computations were only that sort of man, it was merely right that they should be disdained." This passage clearly shows the contempt he felt for official mathematics, and in his last paragraph he tried to convert the literati of his time: "Perhaps they [my mathematical problems] will serve as material for gentlemen of broad knowledge to peruse in their spare time, for although [mathematics is] a minor art it is worth pursuing. Thus I wish to offer this work to my colleagues. If they say that their skill [in the minor arts] is complete and that this is merely for people like astronomers [*ch'ou-jên*] and provincial clerks, and ask why this should deserve to be used throughout the empire, will that not show them to be benighted!"

We know that "students of mathematics were preparing for careers as technical specialists rather than administrators."[22] A Chinese mathematician was in the first place a technologist, who was able to solve a variety of practical problems in the fields of chronology and astronomy (if he studied in the Board of Astronomy), or in the fields of financial affairs, taxation, architecture, military problems, and so forth; and this determined his social role. We believe that these mathematicians tried to solve problems for those "artisans, no matter how greatly gifted, [who] remained upon the other side of an invisible wall which separated them from the scholars of literary training."[23] When Ch'in Chiu-shao solves mathematical problems for the building of a city wall, the problem (even from the standpoint of organization) is entirely solved as far as it concerns calculations. These are not simple problems involving

---

mostly descendents of minor officials and commoners, and had by birth been denied admission to more prestigious educational institutions."
[22] Nakayama (1), p. 15.
[23] Needham (2), p. 325.

only one mathematical procedure, such as those dealt with in the older mathematical treatises that were used as textbooks in the Office of Mathematics, but real plans for immediate practical use. In the *Chiu-chang suan-shu* we find the formula for calculating the volume of a city wall, but not the project for building a whole city wall. Ch'in himself says in his preface: "As for the details [of the mathematical problems], I set them out in the form of problems and answers meant for *practical use*." There is no doubt about the fact that the problems solved in Ch'in's book were of great interest to technicians in various fields. Whether they were ever used or not is another question, but we have to stress the fact that they were not the bureaucratic problems of the Office of Mathematics. In a splendid work on architecture, the *Ying-tsao fa-shih*, there is a full description of materials and constructions, but what is lacking is plans for carrying out the work: the calculation of the building materials, the number of workmen, the provisions and wages. All these we find in Ch'in Chiu-shao's work. He even does calculations concerning the assignment of tasks. For this reason some of the problems are extremely simple, and others are very intricate, but all are classified according to application and are obviously intended for practical use.

The *Chiu-chang suan-shu* was undoubtedly a textbook for minor officials that left the details to be set forth by a teacher (this would explain why the operations on the computing board are not included, although we are almost sure that the instrument was used). The same holds for the mathematical handbooks of the T'ang. Were the works of the Sung and Yüan also textbooks? In Ch'in's work all the basic operations (even the square-root extraction) are taken for granted. For a beginner's textbook its problems are much too complex; it would be useful only for advanced students. If we maintain that these books were textbooks, we must add that they were never used as such in official instruction, although it is not impossible that the authors themselves taught from them, as we have good reason to believe was true in the case of Chu Shih-chieh. It is possible that they were unsuccessful substitutes for older books written in a less advanced phase of mathematical knowledge. But as mentioned earlier, none of the authors was a mathematics teacher; even Li Yeh's work, which was printed in the thirteenth century, was never used as a textbook. We have the impression that these books were written for the use of technologists or for the mathematicians they employed.

However, this does not mean that Chinese mathematicians did not try to prove their ability in solving rather abstract and difficult mathematical equations. We shall see that Ch'in Chiu-shao sometimes constructed equations of a degree higher than necessary for solving his problems; the only explanation is that he wanted to prove that he was able to solve them. And here we meet the true mathematician as opposed to the technologist.[24] Thus these works were written partly for the sake of mathematics by men who had a natural turn for mathematics (as in the West); but the presentation of the problems was determined by Chinese tradition. They are practical problems; in the West mathematical problems are usually given in anecdotal form.

It is thus to be expected that a Chinese mathematical work should be a conglomeration of valuable mathematical techniques and very simple arithmetical operations. We have to read a great number of primary-school problems in order to find a few mathematical methods that are indeed on a very high level, considering the early period of their origin.

That Chinese mathematics is merely empirical is a frequently heard objection. Yushkevitch[25] draws attention to the following points:

[24] See the famous equation of the tenth degree (Chapter 13 of this volume). See also Yushkevitch (4), p. 87: "... that they did not confine themselves to tasks that were dictated by immediate practical considerations, but that they developed more abstract branches of mathematics deriving from such problems, which at that time had not yet found any applications of a nonmathematical nature"(... dass sie sich nicht auf Aufgaben beschränkten, die unmittelbar von der Praxis diktiert wurden, sondern dass sie von derartigen Problemen ausgehend abstraktere Teile der Mathematik entwickelten, die zu jener Zeit noch keinerlei Anwendungen aussermathematischer Art gefunden hatten).
[25] Yushkevitch (4), p. 5. ("Der Dogmatismus der Darstellung, das mecha-

1. Almost all the medieval mathematics books lack proofs.
2. Mathematics books were mainly intended for practical use.
("The dogmatism of the presentation, the mechanical memorizing of various rules, and the multiplicity and splintering of the latter were predicated on the idea that medieval textbooks were primarily for practical use by merchants, land surveyors, officials, builders, and so on. This type of reader required mechanical and brief rules for the solutions of a clearly defined and limited series of problems.")
3. "Many scientific results could not be obtained empirically but had to be based upon logical deductions." This is very likely, but at the same time it means an insurmountable obstruction when we wish to penetrate the logical background of Chinese algebra. In special cases such as the general rule for solving indeterminate equations, it is not difficult to prove that the rules are in fact not empirical, but this is only a negative proof (by exclusion). If we take a positive approach, nothing can be proved; that is, we cannot find out how Chinese mathematicians built up their methods.

If we are to isolate the general characteristics of Chinese mathematics, we must not forget that it was on the same level as mathematics everywhere in the world in medieval times. "Basically the oriental mathematics of the Middle Ages was a curriculum of constant magnitudes and invariable geometric figures. However, such a characterization is not sufficiently concrete. It was primarily a numerical mathematics, a body of algorithms for the solution of problems in arithmetic, algebra, and geometry, which at first were relatively simple but became more difficult later on."[26] Yushkevitch also draws attention to

---

nische Auswendiglernen verschiedener Regeln sowie die Vielfalt und Zersplitterung der letzteren waren dadurch bedingt, dass die mittelalterlichen Lehrbücher vor allem für Praktiker, wie Kaufleute, Landvermesser, Beamte, Bauleute usw. bestimmt waren. Solche Leser benötigten mechanische und nach Möglichkeit kurze Regeln zur Lösung eines scharf umrissenen und engen Problemkreises"; and "Viele wissenschaftliche Ergebnisse konnten überhaupt nicht empirisch gewonnen werden und mussten auf eine logische Deduktion stützen.")
[26] Ibid. p. 2. ("Die orientalische Mathematik des Mittelalters war im

a very important fact: "One must differentiate between the *type of representation,* which is primarily determined by the purpose of the book, and the method of the investigation."[27] We know almost nothing about the research methods Chinese mathematicians applied, but it is quite clear that difficult problems cannot be solved by trial and error; this means that some kind of logical structure is indispensable for constructing methods as intricate as the *ta-yen* rule. But we are completely in the dark as to how these mathematicians constructed their rules, and it is very likely that this problem will never be solved completely.

A last point must be emphasized here. In the West there is some belief that the problem of the theoretical background of Chinese mathematics is simply to be reduced to its special notation on the counting board, and that its discoveries are derived from the matrix notation used there. In fact the counting board does not solve problems, but merely provides a matrix notation by which one can keep track of operations performed in the mind. It is true that the board encouraged developments in some directions and discouraged innovations in other directions: matrix operations were easy, while equations of higher degrees in several unknowns with intermediate powers of each unknown were very inconvenient. This is equally true of any notational instrument.[28]

We may conclude that Chinese mathematics was "far removed from the ideals . . . formulated by the Greek classical writers. It did not succeed in deductively forming whole disciplines on the basis of a few premises or in developing a theory

---

Grunde genommen eine Lehre von konstanten Grössen und unveränderlichen geometrischen Figuren. Jedoch ist eine derartige Charakterisierung noch nicht konkret genug. Sie war vor allem eine numerische Mathematik, eine Gesamtheit von Algorithmen zur Lösung arithmetischer, algebraischer und geometrischer Aufgaben, die zunächst verhältnismässig einfach waren, später aber schwieriger wurden.")

27 Ibid., p. 5 ("Man muss aber zwischen der Art der Darstellung, die hauptsächlich durch den Zweck des Buches bestimmt ist, und der Untersuchungsmethode unterscheiden.")

28 I am much indebted to Nathan Sivin for this important suggestion.

of mathematical proof."[29] Chinese mathematics is a part of this algorithmic medieval mathematics; but modern algebra has its roots in these algorithms, and in this sense, "while we recognize the great achievements of modern mathematics, we must not underestimate the great work accomplished by the pioneers."[30]

[29] Yushkevitch (4), p. 87. (". . .von dem durch die griechischen Klassiker geschaffenen Ideal der Mathematik weit entfernt. Es gelang ihnen nicht auf der Grundlage weniger Voraussetzungen ganze Disziplinen deduktiv aufzubauen und eine Theorie der mathematischen Beweisführung zu entwickeln.") See also Needham (2), p. 235: "Of mathematics 'for the sake of mathematics' there was extremely little. This does not mean that Chinese scholars were not interested in truth, but it was not abstract systematized irrelevant truth after which sought the Greeks."
[30] Konantz (1), p. 310.

2

Mathematicians and Mathematical Methods in Sung and Yüan

1. Mathematical Methods

There is no reason to doubt that the last half of the thirteenth century was the culminating point of Chinese mathematics.[1] Although we have at our disposal much information about this period (mainly in Chinese), we must be careful not to judge wrongly by comparing Chinese medieval mathematics to the modern Western variety.[2] Chinese mathematics as a whole forms part of the algorithmic phase we find in all civilized countries at that time. "It is pointless, therefore, to subject the old Chinese contributions to the yardstick of modern mathematics. We have to put ourselves in the position of those who had to take the earliest steps and try to realize how difficult it was *for them*."[3] We can make a relevant judgment only if we are well informed about the state of mathematics elsewhere in the world at the same time.[4]

A general characteristic of Chinese mathematics is its *algebraic nature*. Algebra is a study of mathematical patterns, of structures: this implies a constellation of relations, in which the meaning of the symbols is irrelevant. What was the reason that the Chinese were satisfied to construct algebraic matrices that

---

[1] Sédillot, writing in 1869, expressed the opinion "that these people had never known what mathematics is; that in the thirteenth century they had not progressed beyond the right triangle" (que ce peuple n'avait jamais su ce que c'était les mathématiques; qu'au 13e siècle de notre ère, il ne s'était point encore élevé au delà du triangle rectiligne rectangle) [(3), p. 3]. Such statements may be forgiven in the case of scholars who lived at a time when there was much dispute and little information, but it is difficult to know what to think of J. F. Scott (1), who said in 1958: "The eleventh and twelfth centuries were barren. There was a little activity during the thirteenth; after that there was but slight development in mathematics on Chinese soil until more recent times" (p. 82).

[2] The value judgments of Van Hee and Loria might be mentioned here.

[3] Needham (2), p. 320.

[4] For that reason, when discussing the *ta-yen* rule, we shall investigate the

enabled them to solve practical problems? One of the main features of their world view was the idea that the whole cosmos was structured according to unaltered but dynamic relationships. We may suppose that this belief influenced their mathematical thinking, for, from a relational, logical point of view, a mathematical problem is solved if the relation between a set of data and the solutions is analyzed, that is, if there is a sure and general way of proceeding from data to solutions.[5] However, if we are content with such an algorithm, we can solve only a set of analogous problems: its practitioners took pleasure in solving endless series of identical problems, changing only the numerical values and the mode of application.[6] This being so, mathematics developed only very slowly; and if we take into account this practical orientation, we should not be surprised by what seems to be a real stagnation. Although Chinese mathematics was built up as a constellation of relations, the relations were unique, and there was no *general algebraic structure*. The main reason for this lack was the absence of general analysis, which could have lent structural insight. There are many beautiful bricks in Chinese mathematics, but there is no building.

We must realize that even in Sung mathematics a great part

---

evolution of the same rule in India, Islam, and Europe.

[5] Yushkevitch (4), p. 6, tries to explain from a Marxist point of view the development of geometrical mathematics in Greece and of algebraical mathematics in the Orient. It is possible that the structure of society contributed to this striking difference. Nevertheless, we can reduce the problem more plausibly to the difference in logical conception. In Greek philosophy, logic is attributive, that is, characteristics are attributed to a substance. A geometrical form is such a substance, of which we investigate the special properties. Algebra is a summation of relational patterns, and indeed Chinese thinking is relational (it takes the form of relational logic). Of course this is only a restatement of the problem, but it takes into account general conceptions of the structure of the cosmos.

[6] Of course, the fact that their writings were intended as textbooks encouraged this tendency, but the question here is that of evolution: what is the difference between the first and the last of the Ten Manuals as official textbooks? What is the difference between the solutions of the remainder problem in the *Sun Tzŭ suan-ching* and in Yang Hui's work?

of the procedures was on a level no higher than that of the *Chiu-chang suan-shu*, then a thousand years old. In the first place, many practical problems required the same mathematical procedures as those of ten centuries earlier. In addition the taste for analogy together with the canonical reputation of the *Chiu-chang suan-shu* were undoubtedly responsible for this slow evolution.

As for mathematical activity in those times, there was much more than the few names of great mathematicians would imply. Ch'in Chiu-shao informs us in his preface that there were more than thirty mathematical schools (or books?) at this time. Needham states that Ts'ai Ching (1046–1126) "encouraged the study of mathematics for the imperial examination (1108)."[7] Li Yen devotes a special chapter to instruction in mathematics in China,[8] where we see that the central source of instruction in Sung times was the *Suan-ching shih-shu* (The ten mathematical manuals), which first appeared in 656 and was printed in 1084. This collection, in which we find the *Chiu-chang suan-shu*,[9] was a kind of textbook for "official" arithmetic. It is obvious from Ch'in Chiu-shao's work that this book was the basis of his own mathematical studies as the disciple of a recluse scholar. If it was the major source used by Sung and Yüan mathematicians, they must have been very talented, because their works far surpass what was gathered in the Ten Manuals.

Another difference between the Ten Manuals and Ch'in Chiu-shao's work is that the *field of practical application* of the latter was much larger. Vogel[10] describes the everyday problems of the *Chiu-chang suan-shu;* they are restricted to such very simple applications as trade and barter, payment of wages, and taxes, whereas the *Shu-shu chiu-chang* contains problems on

[7] Needham (1), vol. 3, pp. 40 f. On Ts'ai Ching's reforms see H. Franke (3), pp. 205 f and Kuo Ping-wên (1), pp. 46 f.

[8] Li Yen (6′), pp. 238 ff; on the Sung, pp. 252 ff.

[9] For the history of this work, see Needham (1), vol. 3, p. 18 and Hummel (1), p. 697.

[10] K. Vogel (2), pp. 124 f.

chronology, meteorology, architecture, military problems, and so on. Moreover, Ch'in Chiu-shao's work is divided according to these fields of practical application; only the first chapter deals with a mathematical method, the *ta-yen* rule. The other chapters are on chronology, surveying, "trigonometry," levies of service, taxation, architecture, military calculations, and trade. The division of the *Chiu-chang suan-shu*, however, is mixed; for example, Book 6 treats of taxation, Book 9 of the right triangle. This comparison gives the impression that the mathematical works of Sung and Yüan were compiled as amplifications of the older works, and, as their purpose was to serve as arithmetic textbooks for professional use, they could only repeat all the ancient procedures.[11] This necessarily practical attitude was an impediment to the unfolding of the genius of some mathematicians; it is a striking fact that, as mentioned earlier, Ch'in Chiu-shao twists and turns to construct practical problems (which do not look practical at all) in order to get equations of a degree high enough to prove his ability in solving them. All this points to one of the main reasons for the final stagnation of Chinese mathematics. Indeed, it is, in the traditional Chinese mind, foolish to solve an equation of the tenth degree when there is no practical problem that requires it.

As for the progress of mathematics in Sung and Yüan, we find first of all that it consisted of the *combination* of older mathematical methods for solving more complex problems. In the *Chiu-chang suan-shu* the problems are very simple and in general each problem contains one single operation. In the Sung books sometimes more than twenty operations are to be performed before one gets the result; there are indeed some very complex problems.

It is a typical characteristic of Chinese mathematics that its development at any time is not as great in geometry as in algebra. In the field of areas and volumes there are some complex problems in the Sung book (and there are at least

---

[11] It would be interesting to make a comparison from this point of view between the *Chiu-chang suan-shu* and the *Shu-shu chiu-chang*.

several ingenious combinations in Ch'in's work), but there are no new formulae in the restricted geometrical sense.

In the field of "telemetry" (a kind of prototrigonometry, that makes use of similar right triangles), Ch'in Chiu-shao does not show great ability in making combinations. Again, it is typical that a large number of problems are reduced to algebraic equations.

New algebraic methods were the following:

1. *Indeterminate analysis* (the Chinese remainder problem), for which Ch'in Chiu-shao gives a fully elaborated method of solution, even for the case where the moduli are not relatively prime. It may have been derived from calendrical computations, but undeniably he was the first to see its full mathematical significance (as far as we know, of course!). This problem was not solved in Europe until the eighteenth century, with the work of Euler and Gauss.

2. *Numerical higher equations,* which occur for the first time in the work of Ch'in Chiu-shao. We find them also in Li Yeh's and Chu Shih-chieh's works. Such equations were solved in Europe in the beginning of the nineteenth century (Ruffini and Horner).

3. *The t'ien-yüan notation for nonlinear equations,* first used by Li Yeh and Chu Shih-chieh. It was peculiar to Chinese mathematics.

4. *The Pascal triangle,* first given by Yang Hui and Chu Shih-chieh. Its European date is the sixteenth century.

5. *Cubic interpolation formulae,* used by Kuo Shou-ching (1231–1316) and identical with the Newton-Sterling formulae (1711/1730).

6. A considerable development in the field of *series* beyond the studies of Shên Kua in the works of Ch'in Chiu-shao, Yang Hui, and particularly of Chu Shih-chieh.

7. Kuo Shou-ching's *prototrigonometry,* which developed out of the arc-sagitta method of Shên Kua.

It is important that none of these methods seems to have been invented by the mathematicians in whose books we find them. As for indeterminate analysis, we know only that the method was used in calendar reckoning, but beyond that fact we are entirely at sea. As for equations of higher degree, it is obvious that the method was much older than the mathematics of Sung and Yüan, as it was known to both Ch'in Chiu-shao and Li Yeh, who had no contact with each other.[12] Moreover, it seems to have been a natural evolution from the root-extraction methods in the *Chiu-chang suan-shu*.[13] Li Yeh, in whose work we find the *t'ien-yüan* notation for the first time, did not say that he himself was the originator of the method. According to his *Ching-chai ku-chin chu* he copied it from a certain P'êng Cha, of whom we know practically nothing. And in the epilogue to Chu Shih-chieh's *Ssŭ-yüan yü-chien* written by Tsu I-chi, several works older than Chu's are cited in which this method was mentioned.[14]

As for the Pascal triangle, Yang Hui states that it was derived from an older work written by Chia Hsien (fl. c. 1050).[15]

For more details the reader must consult general historical studies of Chinese mathematics. It is clear enough from what has been said, however, *that Chinese mathematics is largely anonymous,* and that the great mathematicians represent only the latest (and unfortunately final) phase of a long and slow evolution.

2. Works

From the preceding statement it is obvious that only a few names have been recorded, and that the greater part of Chinese mathematical works have been lost. Although this means that we cannot follow the evolution of mathematical procedures step by step, it is very likely that all the truly interesting metb-ods have been preserved in the few works which are still extant. Li Yen[16] compiled from bibliographies a list of all titles that can be attributed to this period,[17] thus showing that there was a wealth of mathematical literature in the Sung.

[12] See the second section of this chapter.
[13] Wang Ling and Needham (2).
[14] See Ch'ien Pao-tsung (8'), p. 179; Ho Pêng-yoke (4), p. 5.
[15] Ho Pêng-yoke (2), p. 7.
[16] Li Yen (6'), vol. 4, pp. 252 ff.
[17] Needham (1), vol. 3, p. 40: "Some of them were contained in bibli-

The same author has devoted a special study[18] to fragments of mathematical works of the thirteenth and fourteenth centuries. Some of these have been preserved in two volumes of the *Yung-lo ta-tien* devoted to mathematics[19] and others in the *Chu-chia suan-fa chi hsü-chi* 諸家算法及序記 (Records of mathematical methods and prefaces of all schools).[20]

The works of this period that have survived intact are
1. Ch'in Chiu-shao, *Shu-shu chiu-chang* (1247)
2. Li Yeh
a). Ts'ê-yüan hai-ching (Sea mirror of circle measurement, 1248)
b). I-ku yen-tuan (New steps in computation, 1259)
3. Yang Hui
a). Hsiang-chieh chiu-chang suan-fa tsuan lei (Compendium of analyzed mathematical methods in the 'Nine Chapters,' 1261)
b). Jih-yung suan-fa (Mathematical rules for daily use, 1262)[21]
c). Yang Hui suan-fa (Yang Hui's mathematical methods),[22] including
   (1). *Ch'êng-ch'u t'ung-pien pên-mo* (Complete "Mastery of metamorphoses" in multiplication and division, 1274)
   (2). *T'ien-mou pi-lei ch'êng-ch'u chieh-fa* (Easy rules of mathematics for surveying, 1275)
   (3). *Hsü-ku chai-ch'i suan-fa* (Continuation of ancient and curious mathematical methods, 1275)

---

ographies of the Sung which still survive, such as the library catalogue of Yu Mou (*Sui-ch'u t'ang shu-mu*), where ninety-five titles are listed in the mathematical section."
[18] Li Yen (3'). It must be said that most of these fragments are not of great interest.
[19] See Chapter 4. *Yung-lo ta-tien*, ch. 16,343–16,344.
[20] This work exists only in manuscript form; the compiler is unknown. It is supposed to be a remnant of the *Yung-lo ta-tien*. It was kept in the library of Mo Yu-chih (1811–1871) and found by Li Yen in 1912. See Ting Fu-pao (1'), Appendix, pp. 42b and following page, no. 165. Needham (1), vol. 3, p. 50, says that some of the treatises of Yen Kung are preserved in the *Yung-lo ta-tien*. This is true, but the parts containing indeterminate problems come from the *Chu-chia suan-fa chi hsü-chi*.
[21] Only fragments have been preserved.
[22] This is the title under which some of Yang Hui's works were later collected.

4. Chu Shih-chieh

a). Suan-hsüeh ch'i-mêng (Introduction to mathematical studies, 1299)

b). Ssŭ-yüan yü-chien (Jade mirror of the four unknowns, 1303)

5. Kuo Shou-ching, *Shou-shih li i ching* (Manual of explanations of the Shou-shih Calendar, 1280).[23]

There is additional bibliography available on the subjects covered in the first two sections of this chapter.[24]

3. Ch'in Chiu-shao and Other Mathematicians

In his preface Ch'in Chiu-shao says: "Only the *ta-yen* rule is not contained in the *Chiu-chang suan-shu*." From analysis of his work we are sure that the *Chiu-chang suan-shu* was his main source. A great part of his problems can be reduced to procedures given in this work; moreover, his methods[25] are mostly adopted from it. Ch'in does not treat the method for solving numerical equations of all degrees as a special case, probably because he considers it as only an extension of the root-extraction procedures of the *Chiu-chang suan-shu*.[26] About the *ta-yen* rule he says: "No one has yet been able to derive it [from other procedures]. Calendar-makers, in working out their methods, have made considerable use of it. Those who consider it as belonging to 'equations' (*fang-ch'êng*) are wrong." In all probability Ch'in learned the *ta-yen* rule at the Board of Astronomy (see his biography), but he was apparently the first who thoroughly understood the method.

[23] Preserved only in ch. 52–55 of the *Yüan-shih*. None of his original works are extant. See Needham (1), vol. 3, p. 48.

[24] For good monographs on Sung and Yüan mathematics see Ch'ien Pao-tsung (2') and Yabuuchi (3'). In a Western language there are the articles of Ho Pêng-yoke, some already cited. For some special fields see also Yen Tun-chieh (2'). Needham says in (1), vol. 3, p. 42 that this last work deals with "the background of the Sung algebraists"; this statement is somewhat misleading, because this article treats only of some special points in Sung algebra, such as the appearance of the zero, the *tuo-mên* (several methods) of which Shên Kua speaks, mathematical progressions, foreign influences, and the calendar.

[25] See Chapter 5.

[26] See Wang Ling and Needham (2).

Ch'in does not mention any of the names of his predecessors, and what he learned from his "recluse scholar," we do not know, although we are sure that the *Chiu-chang suan-shu* was among his textbooks.

As for his other great contemporaries, there is not the slightest indication that Li Yeh, Yang Hui, or Chu Shih-chieh ever saw Ch'in's work. Ch'ien Ta-hsin[27] was the first who called attention to the fact that Ch'in Chiu-shao and Li Yeh could never have met; included in his internal evidence are the following arguments:

1. The *t'ien-yüan-i* method[28] has an entirely different mathematical sense in both works.

2. Ch'in's work is dated 1247, Li Yeh's 1248; this makes it very likely that Li Yeh never saw Ch'in's work (which was not printed before modern times, even though printing was very widespread by the end of the Sung).

3. The two mathematicians were separated geographically by the war between northern and southern China that was carried on during their lifetimes. Until now, no trace of the influence of Ch'in Chiu-shao has been detected in the works of Yang Hui and Chu Shih-chieh. The *ta-yen* rule, a procedure that one would expect to be attractive to advanced mathematicians, is not given in these works. In Yang Hui's *Hsü-ku chai-ch'i suan-fa* the indeterminate problem of Sun Tzŭ is treated and even extended to other moduli, but this has nothing to do with Ch'in's procedure.

As for Li Yeh and Ch'in Chiu-shao, they make use of the same symbols, but although their systems for solutions of higher equations are generally the same, their terminology is different. All this gives the impression that Ch'in's influence was almost nil.

[27] Ch'ien Ta-hsin (1'), ch. 30, p. 1b and following page.
[28] See Chapter 17, section on "Interpretation of the Text."

3

Biography of Ch'in Chiu-shao

There is no biography of Ch'in Chiu-shao in the *Sung-shih* 宋史 (History of the Sung dynasty), but we have several fragmentary data which make it possible to reconstruct his curriculum vitae.[1] Reference to Figure 1 will aid in the location of place names mentioned in this chapter.

From an inscription in Fu-chou 涪州,[2] we know that Ch'in Chiu-shao's father was called Ch'in Chi-yu 秦季槱. The inscription says: "The prefect Li Yü (Kung-yü) 李琚公玉 and Ch'in Chi-yu (Hung-fu) 秦季槱宏父, the new prefect of T'ung-ch'uan 潼川[3] . . . Chiu-shao (Tao-ku) 九韶道古, the son of Chi-yu, and Chê-min 澤民, the son of Yü, in remembrance of their joint excursion to the 'Stone Fish.'[4] . . . Twelfth day of the first

---

[1] Ch'ien Ta-hsin (1728–1804), in (2′), ch. 14, p. 332, gives biographical data found in Ch'ên Chên-sun's *Chih-chai shu-lu chieh-t'i*, ch. 12, pp. 354 f. [Ch'ên Chên-sun (c. 1190–after 1249) was a contemporary of Ch'in Chiu-shao and owner of a copy of the *Shu-shu chiu-chang*; see Chapter 4.] Ch'ien Ta-hsin also used Chou Mi's *Kuei-hsin tsa-chih, hsü-chi,* (c. 1290) [see des Rotours (1), p. cxii]; Li Liu's *Mei-t'ing hsien-shêng ssŭ-liu piao-chun*, ch. 36, p. 7b [Li Liu or Li Mei-t'ing; on his work see *Ssŭ-k'u ch'üan-shu tsung-mu t'i-yao*, vol. 5, p. 3,402]; and *Ching-ting Chien-k'ang chih* (Gazetteer of Nanking in the Ching-ting period, i.e. 1260–1265), ch. 24, p. 14a and p. 34a. The same sources are given in Juan Yüan's *Ch'ou-jên chuan*, but we know that this work was written with the help of other scholars, among them Ch'ien Ta-hsin [see Hummel (1), p. 402]. Lu Hsin-yüan (1834–1894), in his *I-ku t'ang t'i-pa*, ch. 8, pp. 2a–3a and suppl., ch. 8, pp. 20b–21a [see Têng and Biggerstaff (1), p. 47; Hummel (1), p. 546] gives other sources, including Yao Chin-yüan's *Fu-chou Shih-yü wên-tzŭ so-chien lu*, C, p, 9a; Li Ts'êng-po's *K'o-chai hsü-kao hou*, ts'ê 21 (no. 425), ch. 6, p. 42a and following pages, and Liu K'ê-chuang's *Hou Ts'un hsien-shêng ta-ch'üan-chi*, ch. 81, pp. 1a–1b.

[2] Now Fuling 涪陵 in Szechwan.

[3] Now San-t'ai 三臺 in Szechwan.

[4] At the bottom of the Yang-tzŭ river near Fu-chou, there were two ancient stone carvings of fishes; when the water fell so low that these fishes became visible, it was considered to presage a splendid harvest. Whenever this happened, sightseers went to the spot and rendered their impressions, often cut in stone. Yao Chin-yüan collected all references to these visits, mostly in the form of inscriptions.

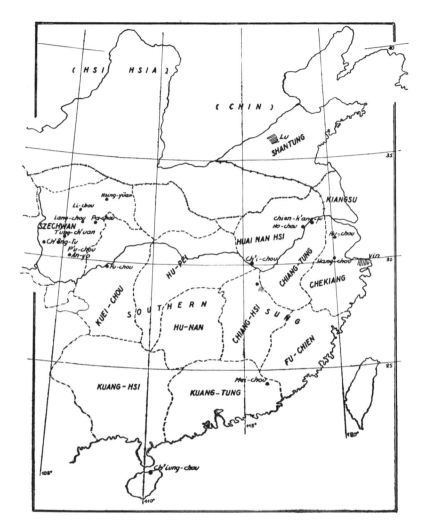

Figure 1. Biographical map (based on Herrmann, *An Historical and Commercial Atlas of China*)

month of the second year of the Pao-ch'ing period [1226]."[5]
From this text we know that the *tzŭ* (courtesy name) of Ch'in
Chiu-shao was Tao-ku.

Some data about his father are given in the historical records.
According to the *Nan-Sung kuan-ko hsü-lu* 南宋館閣續錄,[6] he was
a native of An-yüeh 安岳 in P'u-chou[7] 普州 and a *chin-shih* 進士
(third-degree graduate) of 1193.[8]

In the *Sung-shih*[9] under the year 1219, third month, twelfth
sexagenary day, we read: "Ch'üan Hsing 權興, a soldier from
Hsing-yüan 興元,[10] and others made rebellion and invaded Pa-
chou 巴州,[11] and the prefect Ch'in Chi-yu left the town."

In the ninth month of 1224 he was appointed to a position in
the capital[12] as assistant librarian (*pi-shu shao-chien* 秘書少監).[13]
In 1225 he was prefect of T'ung-ch'uan.[14] As he was appointed
in the sixth month, he was in the capital for only ten months.[15]

In the preface to the *Shu-shu chiu-chang* we read: "Ch'in
Chiu-shao, from the Lu 魯 district." Long before that time, Lu
(in Shantung) had fallen into the hands of the Chin. Consequently, this cannot be Ch'in's place of origin. Attention was
drawn to this fact in the commentary of the *Ssŭ-k'u ch'üan-shu
tsung-mu t'i-yao,*[16] where we read: "In his own preface he gives
as his native place Lu commandery. But the preface [is dated]
1247 and long before [that time] Lu commandery had already
fallen into the hands of the Yüan. Chiu-shao may be stating the
native place of his ancestors." This explanation is now generally

---

[5] Yao Chin-yüan (1'), C, p. 9a.
[6] Ch. 7, p. 11a.
[7] Now An-yüeh in Szechwan.
[8] According to this text, he passed the examination at the same time as the
philosopher Ch'ên Liang 陳亮 (see Fung Yu-lan (1), vol. 2, p. 556).
[9] Ch. 40, p. 3b.
[10] Now Han-chung 漢中 in Shensi.
[11] Now Pa-chung 巴中 in Szechwan.
[12] *Nan-Sung kuan-ko hsü-lu*, ch. 7, p. 11a; the capital of the Southern Sung
was Hang-chou 杭州.
[13] Ibid., ch. 9, p. 17a and p. 28a.
[14] Ibid., ch. 7, p. 11a.
[15] See the first part of this chapter. See also Wei Liao-wêng (1'), ch. 4, pp.
10a–10b.
[16] Ch. 104, p. 2,206.

accepted. Ch'ên Chên-sun[17] calls him "a man from Lu commandery" and also "a man from Shu 蜀, or Szechwan." On the
other hand, Chou Mi[18] says that he is a man from Ch'in-fêng 秦
鳳. Lu Hsin-yüan[19] criticizes this opinion and calls it laughable;
Ch'ien Pao-tsung agrees with Lu in his later publications.[20]
Ch'in-fêng is a part of Shensi. As Ch'in Chiu-shao's father was
from An-yüeh in Szechwan, Ch'in himself cannot be "a man
from Ch'in-fêng."[21]

According to Ch'ien Pao-tsung,[22] we can reconstruct Ch'in
Chiu-shao's birth year. In the *Sung-shih*,[23] year 1219, third
month, sixtieth sexagenary day, we read: "Chang Fu 張福,
Mo Chien 莫簡, and others, soldiers of [the garrison] of Hsing-
yüan 興元,[24] revolted. They took red caps as their emblem.[25]
They conquered several cities in Szechwan, among them Li-
chou 利州,[26] Lang-chou 閬州,[27] Kuo-chou 果州,[28] Sui-ning 遂
寧,[29] and P'u-chou 普州,[30] the place of origin of Ch'in Chi-yu.
Chang Wei 張威, military commander of Hsieh-chou 瀉州,
suppressed the revolt, and in the seventh month Chang Fu was
killed and Mo Chien committed suicide, and so order was
restored."[31] In Chou Mi[32] we read: "When [Ch'in Chiu-shao]

---

[17] Ch'ên Chên-sun (1'), ch. 12, pp. 354 f.
[18] Chou Mi (1'), C, p. 6a.
[19] Lu Hsin-yüan (1'), suppl., ch. 8, p. 21a.
[20] (2'), p. 60; but not in his earlier works, for example, (7'), p. 125.
[21] Chou Mi contains the text "*Ch'in Chiu-shao tzŭ Tao-ku Ch'in Fêng chien
jên* 秦九韶字道古秦鳳間人." Ch'ien Pao-tsung (2') reads: "*Ch'in, Fêng
chien jên*" (p. 60); Ch'ien Ta-hsin (2'), ch. 14, p. 332 quotes Chou Mi,
and Hu Yü-chin (1'), quoting Ch'ien Ta-hsin, reads: "*Chiu-shao, Ch'in,
Fêng chien jên*." (p. 208). Ch'in is an abbreviation of Shênsi.
[22] (2'), p. 60.
[23] Ch. 40, p. 3b.
[24] Now Han-chung 漢中 in Shênsi.
[25] See Cochini and Seidel (1), p. 190.
[26] Now Kuang-yüan 廣元.
[27] Now Lang-chung 閬中.
[28] Now Nan-ch'ung 南充.
[29] Now Sui-ning 遂寧.
[30] Now An-yüeh 安岳.
[31] *Sung-shih*, ch. 40, p. 4a; Cochini and Seidel (1), p. 191.
[32] (1'), C, p. 6a.

was eighteen years old, he was commander of the volunteers in his native place." This was in all probability P'u-chou (An-yüeh),[33] the birthplace of his father, which was indeed occupied by the "red caps." At this time, his father was no longer in Pa-chou; he left the town on the twelfth sexagenary day, and the rebellion of the "red caps" began on the sixtieth day. If this historical reconstruction is right, Ch'in must have been born in 1202.

Chou Mi says: "He was commander of the volunteers in his native place; he was brave and vigorous [but] unrestrained. He went with his father when he [the latter] took up his prefec-ture."[34] His father was appointed in the capital between the ninth month of 1224 and the sixth month of 1225.[35] Ch'in Chiu-shao writes in his preface: "In my youth I was living in the capital,[36] so that I was enabled to study in the Board of Astronomy;[37] subsequently I was instructed in mathematics by a recluse scholar." Analysis of his work shows us that these mathematical studies consisted mainly of the old *Chiu-chang suan-shu*.[38]

The inscription quoted earlier tells us that in 1226 he ac-companied his father, who was by then prefect of T'ung-ch'uan, on an excursion to the "Stone Fish" at Fu-chou.[39] This means that he was living with his father at T'ung-ch'uan in 1226.

About 1233, Li Liu 李劉 (literary name: Mei-t'ing 梅亭) was fiscal intendant[40] in Ch'êng-tu, the capital of Szechwan. Ac-cording to Chou Mi,[41] Li Liu taught *p'ien-li* 駢儷 prose[42] and

[33] This is not entirely certain because *hsiang-li* 鄉里 indeed means "native place," but it is somewhat strange that Chou Mi should give two different native places in the same line.
[34] (1'), C, p. 6a.
[35] See Chapter 5.
[36] The capital of the Southern Sung, now Hang-chou.
[37] On the Board of Astronomy, see Nakayama (1), pp. 14 ff; Needham (1), vol. 3, p. 191 and p. 421.
[38] The *Chiu-chang suan-shu* was printed in 1084.
[39] See the first part of this chapter.
[40] Translations of titles are from Chang Fu-jui (1).
[41] (1'), C, p. 6b.
[42] The parallel or antithetical prose style in four or six characters, "a

poetry to Ch'in Chiu-shao. In his *Mei-t'ing hsien-shêng ssŭ-liu piao-chun* (Standards for the 'four-six' prose of Mr. [Li] Mei-t'ing),[43] Li Liu mentions Ch'in Chiu-shao[44] in a short note, entitled "Reply to the sheriff Ch'in Chiu-shao, who accepted an appointment to do collation." This text proves that Ch'in was appointed as a military official, but we do not know where. It may be noted here that Ch'in, after the preface of his work, gives a kind of poetic description of the contents of the several chapters, written in this *p'ien-li* style, each phrase containing four characters.

In 1234 the Chin empire was decisively conquered by the Mongols, who became the new and much more dangerous enemies of the Southern Sung.[45] In 1235 the Mongols began their conquest of Szechwan and destroyed many cities, among them Hsing-yüan 興元 and Ch'êng-tu 成都, the capital of Szechwan.[46] It was a very troubled time for the frontier provinces. In Ch'in's preface we read: "At the time of the troubles with the barbarians, I spent some years at the distant frontier; without care for my safety among the arrows and stone missiles, I endured danger and unhappiness for ten years."[47]

According to Li Liu,[48] Ch'in was a military official in 1233. He wrote his mathematical work in 1247, ten years after the invasion of the Mongols in Szechwan. In his preface Ch'in

---

highly mannered writing style, structured according to antithetical sentence elements, which derived from the prose-poems of Han times" (eine höchst verkünstelte und in antithetisch gebauten Satzgliedern strukturierte Schreibweise, die sich von den Prosagedichten der Han-Zeit herleitete). H. Franke (3), p. 146. See also J. R. Hightower (1).

[43] "Four-six" was another name for the *p'ien-li* prose style.

[44] See note 1.

[45] See H. Franke (3), p. 225.

[46] See Cochini and Seidel (1), pp. 219 ff; *Yüan-shu*, ch. 2, p. 3b and following page; *Sung-shu*, ch. 89, p. 6a.

[47] Mikami (1) gives a totally different interpretation of this text: "He survived the dangers of stones and arrows, but he was caught with a disease from which he suffered for ten long years, when, his heart sinking within him, he found himself an utterly disappointed man (p. 64)." This seems to be impossible, as we know that he held various official appointments during this period (1236–1247).

[48] See note 1.

states: "My heart was withered, and my vital power fell away. [But] I knew truly that none of these things was without its 'number,' and I let loose my imagination among these numbers." Thus he tells how he returned to the mathematical studies begun in his youth, after what he considered a dark time in his life.

Following the long period of distress caused by military events, he left his home and escaped calamity by going to the southeast.[49] According to Ch'ien Ta-hsin[50] it was there that he was in contact with Ch'ên Chên-sun, who devoted a note to his book. In Liu K'ê-chuang's 劉克莊 *Hou Ts'un hsien-shêng ta-ch'üan-chi*[51] we find a petition to the emperor concerning Ch'in Chiu-shao's evil deeds[52] where we learn that he was appointed subprefect in Ch'i-chou 蘄州 and that he behaved badly, whereupon some exasperated soldiers revolted. Later he was appointed prefect in Ho-chou 和州, where he was responsible for the salt trade and sold salt illegally to the people.[53]

Chou Mi says that he left the southeast as a rich man, and that he dwelt in a house at the eastern gate of Hu-chou 湖州[54] in Chekiang.[55]

In the eighth month of 1244 he was appointed vice-administrator of the prefecture of Chien-k'ang-fu 建康府,[56] but already in the eleventh month of that year he was given leave because of his mother's death,[57] and returned to Hu-chou. From the

[49] Chou Mi (1′), C, p. 6a and following page.
[50] (2′), ch. 14, p. 332.
[51] Ch. 81, pp. 1a–2b.
[52] See Chapter 3, Appendix.
[53] See also Lu Hsin-yüan (1′), suppl., ch. 8, p. 20b. Ch'i-chou is now Ch'i-ch'un 蘄春 in Hupeh. Ho-chou is now Ho-hsien 和縣 in Anhwei.
[54] Chou Mi gives a detailed description of this house in (1′), C, p. 6b. See Chapter 3, Appendix.
[55] According to Lu Hsin-yüan (1′), suppl., ch. 8, p. 20b, Chou Mi and Ch'in Chiu-shao were living at the same time in Hu-chou. Ch'in Chiu-shao arrived at Hu-chou about 1240; but Chou Mi was born in 1232, and was only a young boy at that time. Moreover, Chou Mi attributes his information to a certain Ch'ên Shêng-kuan. Lu Hsin-yüan supposes that Chou Mi and Ch'in were personal enemies, but this is hardly possible.
[56] Now Nanking 南京 in Kiangsu.
[57] According to the *Ching-ting Chien-k'ang chih*, ch. 24, p. 14a.

preface of the *Shu-shu chiu-chang* we know that his mathematical work appeared in the ninth month of 1247. As the customary mourning period was three years, in all probability his mathematical treatise was written at Hu-chou during this period. Indeed, one of his calendrical problems (I, 2) deals with the year 1246. In Chou Mi we read: "Someone recommended him to the throne on account of his calendrical science, and he was allowed to take part in the examination."[58] This occurred between 1247 and 1254.

According to a list of officials in the *Ching-ting Chien-k'ang chih*,[59] during the period 1253–1259 Ch'in Chiu-shao was again in Chien-k'ang-fu, where he was appointed advisor to the Directorate of Military Affairs (*Chih-chih-ssŭ* 制置司). However, after a short time he resigned and went back to his native home. He paid a visit to Chia Ssŭ-tao 賈似道 (1213–1275), who was an influential chancellor at that time,[60] and he got an appointment as prefect of Ch'iung-chou 瓊州.[61]

After a few months, he had to leave Ch'iung-chou because he was impeached for corruption and exploitation of the people.[62] According to Chou Mi,[63] he followed his friend Wu Ch'ien 吳潛 to the district of Yin 鄞,[64] and in 1259 he was appointed assistant in the agricultural office. Wu Ch'ien subsequently became a minister, but "Wu thereupon was disgraced. When Chia [Ssŭ-tao] had become chancellor, he gradually collected data about Ch'in, and put him away in Mei-chou 梅州;[65] in Mei-chou he was very active in administration and finally

---

[58] (1'), C, p. 6b.

[59] Ch. 25, p. 34a.

[60] On the political role of Chia Ssŭ-tao, see H. Franke (3), pp. 225 f.

[61] Chou Mi (1'), C, p. 7a; Li Ts'êng-po (1'), ch. 6, p. 42a and following page. Ch'iung-chou is now Ch'iung-shan-hsien 瓊山縣 in Hainan.

[62] Chou Mi (1'), C, p. 7b; Liu K'ê-chang (1'), ch. 81, p. 1b and following page.

[63] (1'), C, p. 7b.

[64] The district where Ningpo 寧波 is situated, in Chekiang. Wu Ch'ien was himself an officer of naval affairs there.

[65] *Ts'uan-chê* 竄謫 means "to transfer as a punitive measure." The same expression is used in Shên Kua's *Mêng-ch'i pi-t'an*, vol. 2, p. 721, para. 396. Mei-chou: now Mei-hsien 梅縣 in Kuangtung.

died there."[66] This probably happened in 1261, for in 1262 Wu Ch'ien and his clique were banned from the civil service.[67] In all probability Ch'in Chiu-shao died at this post, in 1261 at the age of sixty years.

[66] Chou Mi (1'), C, p. 7b.
[67] *Sung-shih,* ch. 45, p. 6a.

Appendix to Chapter 3

From Chou Mi's Kuei-hsin tsa-chih, hsü-chi 癸辛雜識續集 (Miscellaneous Notes from Kuei-hsin Street, in Hangchow, First Addendum)

This appendix is not an attempt to analyze Ch'in's character; it will do no more than convey some data that have been recorded. Ho Pêng-yoke provides a good description of his nature: "Ch'in has often been judged an intriguing and unprincipled character, reminding one of the sixteenth-century mathematician Hieronimus Cardano. In love affairs he had a reputation similar to Avicenna."[1]

Liu Kê-chuang says that he was "violent as a tiger or a wolf and poisonous as a viper or a scorpion."[2]

The biography earlier in this chapter mentioned some of his evil deeds. This appendix will complete the description by giving a translation of Chou Mi's note on Ch'in Chiu-shao.[3]

"Ch'in Chiu-shao, with courtesy name Tao-ku, was a man from Ch'in-fêng.[4] When he was eighteen years old he was head of the volunteers in his native place. He was brave and vigorous [but] unrestrained. He followed his father when he [the latter] took up his prefecture. Once when his father was giving a feast, a crossbow pellet suddenly flew out from behind his father. All the guests were amazed and nobody knew where it came from. After a short time they inquired into the matter. It turned out that Chiu-shao and a singing girl had been playing together, and at the time had also taken places at the feast. It was from their direction that the pellet had come.

"When he went to the southeast, he generally consorted with the grand and rich [families] there. By nature he was extremely

[1] Ho Pêng-yoke (1), p. 1.
[2] Liu K'ê-chuang (1'), ch. 81, p. 1b.
[3] Chou Mi (1'), ch. 2, p. 5b and pages following.
[4] See, however, p. 29 f.

ingenious; all such things as astronomy, harmonics, mathematics, and even architecture he investigated thoroughly. Later, from Li Mei-t'ing he learned the antithetical prose style and regular and irregular poetry. As for sports—polo playing, archery, and swordplay—there was nothing that he could not learn.

"His character was extravagant and boastful. He was obsessed with his own advancement and took only himself into consideration. Somebody recommended him to the throne on account of his calendrical science and he was allowed to take part in the examinations. His memorials are extant, as well as the *Shu-hsüeh ta-lüeh*, which he wrote.

"He was on especially familiar terms with Wu Li-chai [Wu Ch'ien]. Wu had a plot of land outside the western gate of Hu-chou. The name of this plot was Ts'êng-shang.[5]

"It was situated just so that the T'iao River flowed through it before entering the city wall. Its situation was magnificent. And then with a trick Ch'in got hold of it. Immediately he had a hall built on it. This hall was extremely spacious: one of its chambers was 70 feet wide. He sought out wonderful materials from overseas [literally, of oversea-rafts] to construct the lintels at the front. The emplacement was entirely planned by himself. All the ridges, the eaves [?], and the flanks of the roof were made of tile. The central hall was composed of seven rooms. On the back he made a series of rooms for lodging beautiful female musicians and singers. All compositions and songs which he wrote were in the most fastidious taste. The expense was incalculable.

"He met [a monk] who begged at the doors leading to the women's apartments, and [thus] the fact that his elder brother's son, whom he had adopted, had illicit intercourse with his own son's concubine, leaked out; he had the concubine locked up and had her starved until she died. He further ordered a servant to set out on a journey with this son; he gave the servant

---

[5] Uncertain translation. One can also translate as "the fame of this area had been very high."

poison and a sword, and said: 'Take him to an uninhabited region. First persuade him to swallow the poison; if you are unable to do so, make him commit suicide. If again you are unable to do so, push him into the water. ' The servant falsely promised to obey, and took the son to O-islet,[6] where his own brother was living. He came back and reported that the task was performed. But later [Ch'in] gradually learned the truth. The servant was afraid and fled. Ch'in put a price on both their heads and, spending all he had saved, set out personally to find his son and the servant, for the purpose of gratifying his thirst for revenge. He told people: 'I am going to set out for Yang-[chou] equipped with 100,000 in cash, and only an autumn ditch will lodge me.' Once he arrived, he paid visits everywhere to those in high places. Hung Ju-chai[7]————————[8] and congratulated him, saying: 'Recently there was a rumor that your son had met his death in an irregular way. That you now are looking for him means that the rumor was wrong. Is that not an occasion for congratulations?' Ch'in did not [reply].[9]

Chia [Ssŭ-tao] arranged matters for him, and he obtained [the prefecture of] Ch'iung-chou. Even before he arrived, from anger at the fact that those who were to receive him were not on time, he captured the *yamen*-runner and had him executed. A few months after he had arrived in the prefecture he was relieved [of his post] and returned. And he took great riches with him.

"In the year *chi-wei* [1269], [when the Mongols] crossed [the Yangtzŭ], Ch'in seemed pleased. Since he had no savings [?], he said: 'My livelihood has been stolen by others.'[10]

"In those times Wu Li-chai [Wu Ch'ien] was staying in Yin, and [Ch'in] hastened there to take refuge with him. At this time Wu was about to become minister and he made him

---

[6] O-islet: in the Hsikiang river in Hupeh, Wu-ch'ang district.
[7] Unidentified.
[8] Obscure sentence.
[9] Text mutilated; presumably "reply" or some such word.
[10] The sense of this passage is not clear.

[Ch'in] leave first, saying: 'I shall have to consider what measures to take.' Nevertheless, Ch'in continued to follow him. After a short while, Wu fell out of favor. When Chia became chancellor, he gradually gathered information about Ch'in, and put him away in Mei-chou. In Mei-chou he was very active in his administration and finally died there. On the day some time before when he had been banished to Mei-chou and had left his home, the central lintel in front of the main hall broke in two; people said that this was an inauspicious sign.

"When Ch'in had died, his adoptive son came back and stayed there together with his youngest brother.

"I heard this from Yang Shou-chai [Yang Tsan]: On the day I went to Hu-chou as prefect, Ch'in was at home. During a summer night, he was intimate with one of his concubines in the moonlight. By chance there was a servant who came to draw water below the hall. He [Ch'in] believed that he [the servant] had been watching him. The next day he had him accused of theft and sent him under escort to the prefecture. Moreover he went himself and reported the matter to the prefect with the purpose of having him tatooed. Mr. Yang quite understood this affair, and as its punishment was not as heavy as that [demanded by Ch'in], he sentenced [the servant] to the bastinado. Although Ch'in was highly dissatisfied, he concealed his anger and continued to consort with him [Yang]. Yang knew that he hated him. He always watched for [Ch'in] to be absent [from his house] and then went to call on him. Only when he was about to be replaced did he go to take leave of him. [Ch'in] then invited him into the inner hall and did his utmost to detain him; but Yang strongly refused. Next he offered Yang a cup: [the liquid] was entirely the color of ink. Yang was highly afraid and left without drinking it. For when Ch'in was in Kuangtung he collected many poisons. Persons he did not like were sure to come into contact with his poisonous tricks: it may be understood how dangerous he was. Ch'ên Shêng-kuan wrote this."[11]

[11] Unidentified.

4

The Shu-shu chiu-chang: History and Investigation

The authenticity of Ch'in Chiu-shao's work is beyond all doubt. Not only is there a continuous chain of bibliographical evidence; it is also possible to date the work from internal information.

Loria[1] expresses his doubt about the authenticity of Chinese texts when he speaks of "the extreme uncertainty which the historian always feels as to the date and authenticity of any Chinese manuscript." This, however, is true only for very ancient texts; there is seldom any doubt as to the authenticity of works from the Sung and Yüan periods. D. E. Smith[2] writes: "Of all the works needing critical study those of the thirteenth and fourteenth centuries are the most prominent. This is not so much because of doubts as to their authenticity, for they are late enough to make changes in the text more readily detected, but because of the obscurity of certain statements in the works of three mathematicians, Ch'in Chiu-shao, Li Yeh and Chu Shih-chieh." This kind of study is a question of mathematical analysis, which this work will attempt.

As for "changes in the text," the history of the *Shu-shu chiu-chang* (see the diagram in Figure 2) gives evidence to the effect that the problems as a whole, if not the figures and computations, have been handed down almost intact. Moreover, since the publication of the extant volumes of the *Yung-lo ta-tien* 永樂大典,[3] in which three of Ch'in's problems are preserved, it has been possible to make a restricted textual comparison (see Chapter 5). As regards Problem VI, 2 (this system of notation is explained in the Reference Table preceding the text) there

---

[1] Loria (1), p. 517.

[2] Smith (3), p. 248.

[3] This encyclopedia was compiled in 1403–1408. Only a few volumes remain; these were published in the form of photographic reproductions by the Chung-hua Press in 1960.

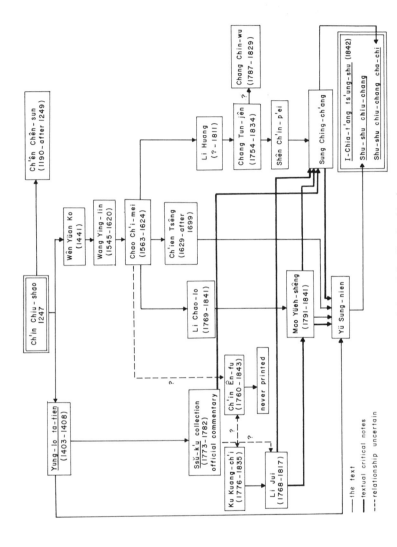

Figure 2. Diagram of the manuscript tradition of the *Shu-shu chiu-chang*

are only a few unimportant differences; only the words "17 *chieh hui-tzŭ*" (seventeenth conversion of paper money)[4] are not in the *Yung-lo ta-tien*. In Problem VIII, 8, some mathematical changes have been made. It is true that the text contained in the *Yung-lo ta-tien* is not without mistakes, but these are mathematical in nature. In his *Shu-shu chiu-chang cha-chi*, Sung Ching-ch'ang corrects some mistakes that are not in the *Yung-lo ta-tien;* this means they were in the text he was working on. Problem IX, 8 agrees entirely with the *Yung-lo ta-tien*. From these facts alone, it is clear that the text is corrupt in many places, as I shall show in detail in the present chapter. But it is also clear, as we may perhaps deduce from these three problems, that the text as a whole is authentic. The mistakes in it result partly from the fact that nobody in the Ming period was able to understand these texts and partly from the copyists' perennial carelessness with figures.

As for internal means of dating, we have very strong arguments at our disposal. One of the chronological problems (I, 2) refers to the year 1246, the year before the completion of the *Shu-shu chiu-chang* (1247).[5] In II, 1 Ch'in gives a chronological problem for the year 1230; in II, 2 for the year 1204; and in II, 3 he mentions the *K'ai-hsi* 開禧 calendar (1205–1207).

In addition, all the socioeconomic information we find in his work is undoubtedly Southern Sung.[6] In the monetary problems there is a reference to the seventeenth *chieh* (year of convertibility of paper money); this corresponds to 1240.[7]

Ch'ên Chên-sun 陳振孫 (c. 1190–after 1249),[8] who was a contemporary of Ch'in Chiu-shao, mentions the latter's work in his *Chih-chai shu-lu chieh-t'i* 直齋書錄解題 (Catalogue of the books of Chih-chai and explanation of their contents).[9] It is an im-

---

[4] On "17 *chieh hui-tzŭ*," see Part VI.
[5] Ch'ien Ta-hsin (1′), ch. 30, p. 1b drew attention to this fact.
[6] See Part VI.
[7] See Part VI.
[8] These dates are from Wêng T'ung-wên (1), p. 99. Ch'ên's dates are uncertain, but we know that he was appointed inspector of the salt trade (*t'i-chü* 提舉) in Chê-hsi 浙西 in the period 1234–1237.
[9] Ch'ên Chên-sun (1′), part 3, ch. 12, p. 355. Chih-chai is the *hao* (literary

portant fact that this work is "an annotated catalogue of the works preserved by the compiler"[10] and that Ch'ên Chên-sun copied several books from other libraries.[11] No doubt he owned one of the first copies of the *Shu-shu chiu-chang*. From the descriptive note on this work we learn that "the name of this book is *Shu-shu* 數術"; this must have been the original title. Ch'ên Chên-sun also refers to it as the *Shu-shu ta-lüeh* 數術大略 (Outline of mathematical methods).[12]

Chou Mi 周密 (1232–after 1308),[13] in his *Kuei-hsin tsa-chih hsü-chi* (Miscellaneous notes from Kuei-hsin street, in Hangchow, first addendum) (c. 1297),[14] gives the title "*Shu-hsüeh ta-lüeh* 數學大略"[15] (Outline of mathematics).

Although the original manuscript has long been lost, the work is preserved because it was included in the *Yung-lo ta-tien*. Of this immense encyclopedia in 11,095 volumes, divided into 22,211 *chüan,* and compiled in the years 1403–1408, only 370 volumes, scattered all over the world, are extant. These volumes were photographically reproduced and published in 1959–1960. However, the original catalogue,[16] issued by order of the Ming Emperor Ch'êng-tsu 成祖 (r. 1403) has been preserved in the National Library of Peking[17] and published in the *Lien-yün-i ts'ung-shu* 連筠簃叢書 (1848).[18] From this we learn that *chüan* 16,329 to 16,365 were devoted to mathematics (*suan* 算). Of these, only *chüan* 16,343 and 16,344 still exist.[19] These are in

---

name) of Ch'ên Chên-sun. For a description of the work, see Têng and Biggerstaff (1), pp. 21 f.

[10] Ibid.

[11] Ibid.

[12] As it was later known as the *Shu-shu chiu-chang*, Sarton (1), supposed that these were different works (vol. 1, part 2, p. 626). Ch'ien Ta-hsin (1728–1804) was the first who drew attention to the different titles in (3'), ch. 1, p. 23b.

[13] These dates are from Wêng T'ung-wên (1), p. 105.

[14] For a description of this work, see des Rotours (1), pp. cxii ff.

[15] Part C, p. 5b.

[16] *Yung-lo ta-tien mu-lu* 永樂大典目錄.

[17] See Ting Fu-pao (1), p. 42b.

[18] The mathematical chapters are in ch. 42, pp. 18a–20a.

[19] See Yüan T'ung-li (1'), p. 135. This article lists all the extant volumes of the *Yung-lo ta-tien*.

the Cambridge University Library[20] (see Figure 3). From the catalogue and the surviving mathematical texts it is obvious that the *Yung-lo ta-tien* did not reproduce the various mathematical works as a whole, but that the part dealing with mathematics was subdivided according to the different mathematical methods[21] and illustrated by extracts from several works.[22] These works are listed in Li Yen (8'), and among them we find the *Shu-shu chiu-chang*. *Chüan* 16,343 includes Problems VI, 2; VIII, 8; and IX, 8 from Ch'in Chiu-shao's work,[23] all of which make use of the *su-mi* 粟米 method.[24]

These three problems are provided as illustrations for a mathematical procedure called *i-ch'êng t'ung-ch'u* 異乘同除. It is obvious that this is the rule given in the *Chiu-chang suan-shu* under the name *su-mi* (literally, millet and rice), a general procedure for calculating proportions, here described as "multiplying by different [numbers], dividing by the same [number]." It is very likely that the whole *Shu-shu chiu-chang* was included in this way in the *Yung-lo ta-tien*,[25] and this is perhaps the meaning of Lu Hsin-yüan's words, "It was collated from the *Yung-lo ta-tien;* the division in *chüan* is not the same, the arrangement is also different. . . ."[26]

The *Shu-shu chiu-chang* is also listed in the *Wên-yüan-ko shu-mu* 文淵閣書目, compiled by Yang Shih-ch'i 楊士奇 and others in 1441.[27] "This is a catalogue of the Ming Imperial Library,

[20] See Needham (1), vol. 3, pp. 31 f.

[21] In the Chinese sense, of course.

[22] See Li Yen (8'), vol. 2, p. 295 for a quotation from Ch'êng Ta-wei's *Suan-fa t'ung-tsung* (1593), which may be an allusion to this aspect of the structure of the encyclopedia. In the same work on pp. 296 f, Li Yen reproduces this part of the catalogue. See also Li Yen (6'), vol. 2, pp. 47–53.

[23] Ch. 16,343, pp. 14a–15b, and 17a–18b. As usual, the problems that have been preserved are not those of greatest interest.

[24] This method was first given in the *Chiu-chang suan-shu*, ch. 2, pp. 23–36. See Needham (1), vol. 3, pp. 25 f; Vogel (1), pp. 17–27.

[25] A description of the mathematical works in the *Yung-lo ta-tien* is also given in Kuo Pai-kung (1'), p. 213.

[26] Lu Hsin-yüan (1'), ch. 8, p. 2a.

[27] Without author's name, as usual in this catalogue. See Têng and Biggerstaff (1), p. 22.

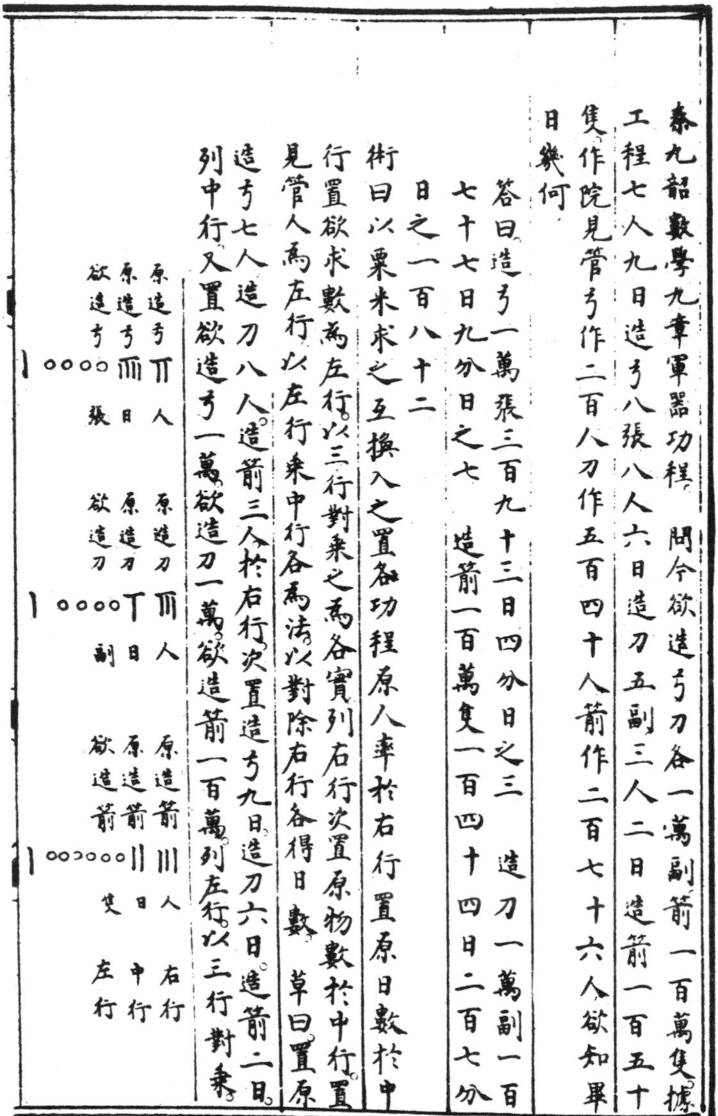

秦九韶數學九章軍器功程　問今欲造弓刀各一萬副箭一百萬隻據
工程七人九日造弓八張八人六日造刀五副三人二日造箭一百五十
隻作院見管弓作二百八刀作五百四十人箭作二百七十六人欲知畢
日幾何

答曰造弓一萬張三百九十三日四分日之三　造刀一萬副一百
七十七日九分日之七　造箭一百萬隻一百四十四日二百七十分
日之一百八十二

術曰以粟米求之互換入之置各功程原人率於右行置原日數於中
行置欲求數為左行以三行對乘之為各實列右行次置原物數於中行置
見管人為左行乘以右行乘中行各為法以對除右行各得日數　草曰置原
造弓七人造箭三人於右行次置造弓六日造箭二日
列中行又置欲造弓一萬欲造箭一百萬列左行以三行對乘

原造弓　Ⅲ人
原造弓　｜日
欲造弓　|〇〇〇〇張

原造刀　Ⅲ人
原造刀　丁日
欲造刀　|〇〇〇〇副

原造箭　Ⅲ人
原造箭　‖日
欲造箭　|〇〇〇〇〇隻

右行
中行
左行

Figure 3. A page from the *Shu-shu chiu-chang* as copied into the great Ming manu-script encyclopedia *Yung-lo ta-tien* 永樂大典, ch. 16, 343, p. 14a. Reproduced from a photographic reprint (Peking, 1960) of the original fascicule preserved in the Cambridge University Library.

compiled early in the dynasty. In 1421 the books were moved from Nanking to Peking and housed . . . in the *Wên-yüan-ko*."[28] In the catalogue we find in *chüan* 14, under the sub-division "*Suan-fa* 算法" (Arithmetical methods), the *Shu-hsüeh chiu-chang* 數學九章,[29] proving that the work belonged to this library. In this catalogue the work is indicated as "*ch'üeh* 闕" (no longer extant), but, as Van der Loon pointed out,[30] the words "complete, incomplete, or missing" were probably added in the sixteenth century. It is very likely that this was the copy used for compiling the *Yung-lo ta-tien*.[31]

However, a copy made from the book preserved in the *Wên-yüan-ko* came into the possession of Wang Ying-lin 王應遴 (1545–1620).[32] Chao Ch'i-mei 趙琦美 (1563–1624) borrowed it from him and made a copy of the text (1616).[33] In his preface to the *Shu-shu chiu-chang* he says: "This book at first was a manuscript of the [*Wên-yüan*]-*ko*. Wang Yün-lai (Ying-lin) 王雲來 (應遴) from Kuei-chi made a copy of it. I borrowed it and copied it. In the original volume they cut off the title after *Shu-shu*[34]; the two characters *chiu-chang* were added by Wang."[35] The preface is dated "the last day of the seventh month, 1616." In the postface Chao Ch'i-mei adds: "Originally there was no table of contents. I made it and added it."[36] The postface is dated "The fifth day of the first month, 1617."

Ch'in Chiu-shao's work is mentioned in the *Ch'ien-ch'ing t'ang shu-mu* 千頃堂書目, a catalogue of Ming works compiled

---

[28] Ibid., p. 22; Lu Hsin-yüan (1'), ch. 8, p. 2a.

[29] In *Ts'ung-shu chi-ch'êng*, vol. 30, p. 180.

[30] Van der Loon (1), p. 385n.

[31] In the *Yung-lo ta-tien*, the title is also given as *Shu-hsüeh chiu-chang*.

[32] For a short note on him, see *Ming-jên chuan-chi tzŭ-liao so-yin*, vol. 1, p. 76.

[33] See the biographical note in *Ming-jên chuan-chi tzŭ-liao so-yin*, vol. 2, p. 736.

[34] Compare with Ch'ên Chên-sun's work.

[35] These two characters were added as early as the *Wên-yüan ko shu-mu* (1441); however, the title was *Shu-hsüeh chiu-chang*, whereas the title in Chao Ch'i-mei's text is *Shu-shu chiu-chang*.

[36] Ch'in Chiu-shao (1'), p. 437.

by Huang Yü-chi 黃虞稷 (1629–1691).[37] "None of the Sung works listed appear in the bibliographical section of the *Sung-shih....*"[38] This is also true of Ch'in's work. The title is given as *Ch'in Chiu-shao shu-hsüeh chiu-chang, 9 chüan.*

Ch'ien Tsêng 錢曾 (1629–after 1699),[39] "beginning in 1669, and continuing for fifteen years . . . labored on another bibliographical work, the *Tu-shu min-ch'iu chi* 讀書敏求記 in 4 *chüan*. This work, first printed by Chao Mêng-shêng 趙孟升 in 1726, lists 601 Sung and Yüan books and manuscripts with detailed comparative annotations concerning editions." "An edition, carefully collated from various texts by Chang Yü 章鈺 (1865–1937), appeared in 1926 with comments and annotations, under the title *Ch'ien Tsun-wang, Tu-shu min-ch'iu chi chiao-chêng* 錢遵王讀書敏求記校證."[40] The information given in Chang's edition is the same as that in Chao Ch'i-mei's preface, from which it may have been derived. Here we find for the first time the title *Shu-shu chiu-chang* 數書九章. Ch'ien Tsêng's copy fell into the hands of Yü Sung-nien, the editor of the *I-chia t'ang* collection of 1842.[41]

In the years 1773–1782 the *Ssŭ-k'u ch'üan-shu* 四庫全書 was compiled, and several works from the *Yung-lo ta-tien* were copied, among them the *Shu-shu chiu-chang.*[42] Three officials were appointed to compile the mathematical and astronomical part of the collection.[43] Ni T'ing-mei 倪廷梅 was the author of the

---

[37] See Têng and Biggerstaff (1), p. 25; Hummel (1), pp. 355 f.
[38] Ibid., p. 25.
[39] See the biography in Hummel (1), p. 157.
[40] Ibid., p. 157. The text on Ch'in Chiu-shao is in ch. 1, C, pp. 27b–28a. The text, together with a part of Chang Yü's commentary, is reproduced in Ting Fu-pao (1), p. 573b; however, there is no indication of its source.
[41] According to Chang Yü (ch. 1, C, p. 27b); on Yü Sung-nien, see p. 47.
[42] There were only seven copies of this encyclopedia; some of them still exist in Chinese libraries (see Hummel (1), pp. 121 f.), but only a small part of the material they contain has been published. As we know from the *Ssŭ-k'u ch'üan-shu tsung-mu t'i-yao* that critical notes were added, we are in want of a valuable source of information. For general information on the mathematical works in the *Ssŭ-k'u ch'üan-shu*, see Li Yen (8'), pp. 464 ff.
[43] See *Ssŭ-k'u ch'üan-shu tsung-mu t'i-yao, Pan-li ssŭ-k'u ch'üan-shu tsai-shih chu ch'ên chih-ming*, p. 10, and Kuo Pai-kung (1'), p. 61.

critical and descriptive notes on mathematical works, published in the *Ssŭ-k'u ch'üan-shu tsung-mu t'i-yao* 四庫全書總目提要 and compiled in 1781.[44] The author says: "The present work is taken from the *Yung-lo ta-tien;* I corrected the mistakes in it and pointed out the errors; I arranged in good order what was confused [literally, upside-down] and added critical footnotes, in order to reveal the mistakes, which can be investigated by mathematicians." Some of the textual emendations are quoted in Sung Ching-ch'ang's commentary, discussed later in this chapter. The author of the descriptive notes tries to explain some of Ch'in's methods, with the result that he proves himself incapable of understanding much of the text; and the greater part of his note is useless talk.[45] This shows clearly that the mathematical meaning of Ch'in's work had been entirely forgotten, and in checking the other great encyclopedia of the Ch'ing period, the *Ku-chin t'u-shu chi-ch'êng* 古今圖書集成 (1725), one is not surprised to find that it omits any reference to the great mathematicians of the Sung and Yüan.[46] "None of these mathematicians of the Ming . . . was a master of the Sung and Yüan algebra. It fell completely out of use, and was not revived until long after the introduction of algebra from Europe by the Jesuits and others. . . ."[47] In 1782 a table of contents of the works copied in the *Ssŭ-k'u* collection, together with short notes about the contents, was drawn up and published under the title *Ssŭ-k'u ch'üan-shu chien-ming mu-lu* 四庫全書簡明目錄.[48] The information given in it is not very pertinent, even from the biblio-

---

[44] See Hummel (1), p. 121. The descriptive note on the *Shu-shu chiu-chang* is in ch. 106 (p. 2,206 in the *Kuo-hsüeh chi-pên ts'ung-shu* edition, 1968).

[45] All he says about the *ta-yen* rule is: "He wished to replace the old divination method of the *I-ching* of the Chou by a new method, [but] he completely changed the old meaning."

[46] The section on arithmetic (ch. 109–128) reproduces Ch'êng Ta-wei's *Suan-fa t'ung-tsung,* which is very inferior to the works of the Sung and Yüan mathematicians. For a description of the mathematical part of the *Ku-chin t'u-shu chi-ch'êng,* see Li Yen (8′), pp. 461 ff.

[47] Needham (1), vol. 3, p. 51. Note that Chao Ch'i-mei's copy is dated 1616. Matteo Ricci arrived in Peking in 1601; the first six volumes of Euclid were translated in 1603–1607. See Mikami (1), p. 113.

[48] See Hummel (1), p. 121 and Têng and Biggerstaff (1), p. 32.

graphical point of view, but its goal is rather restricted—to provide a quick, handy idea of the contents and the value of each book.[49]

After a long period in which the mathematics of the Sung and Yüan were entirely forgotten, the study of the culminating points in Chinese algebra started again. Mei Wên-ting 梅文鼎 (1633–1721) and his grandson Mei Ku-ch'êng 梅瑴成 (d. 1763) are the forerunners of the "mathematical renaissance" in China.[50] "The actual contributions of Mei Wên-ting to the sciences of mathematics and astronomy are not of great significance, but his labors served to popularize these subjects in China and to revive an interest in older Chinese mathematical discoveries."[51] "Mei Ku-ch'êng is important in the history of Chinese mathematics, not so much for his writings, as for the encouragement he gave to the study of older Chinese works on mathematics and to the recovery of works believed to have been lost."[52] The most important scholars who recovered old works of the Sung and Yüan were Tai Chên 戴震 (1724–1777)[53] and Lo Shih-Lin 羅士琳 (d. 1853).[54] Ch'ien Ta-hsin 錢大昕 (1728–1804), who was acquainted with Tai Chên, gathered valuable biographical and bibliographical notes on Ch'in Chiu-shao.[55] They were published in his *Ch'ien-yen t'ang wên-chi* (Collection of studies from Ch'ien-yen studio),[56] where Ch'ien discusses the internal dating of the work[57] and the difference in the meaning of the term *t'ien-yüan-i* 天元一 (Celestial element unity) as Ch'in Chiu-shao and Li Yeh used it to refer to their algebraic meth-

[49] On Ch'in Chiu-shao, see ch. 11, p. 13a.
[50] On Mei Wên-ting, see Hummel (1), pp. 570 f; on Mei Ku-ch'êng, ibid., p. 569; on both, Mikami (1), pp. 120 ff.
[51] Hummel (1), p. 571.
[52] Ibid., p. 569.
[53] Ibid., pp. 695 ff.
[54] Ibid., pp. 538 ff.
[55] On Ch'ien Ta-hsin, see *Ch'ing-shih lieh-chuan*, ch. 68, p. 42, and Hummel (1), pp. 152 ff.
[56] Part 8, ch. 30, pp. 1b–2a. First published in 1806.
[57] See Chapter 3. He attracts aettntion to the fact that one of the chronological problems (I, 2) deals with the year 1246.

od.[58] He also mentions the fact that these mathematicians were unacquainted with each other's work. Another note is published in his *Shih-chia chai yang-hsin lu* (Record of cultivating self-renewal from the Shih-chia studio) (c. 1780),[59] in which he collected several bibliographical data concerning Ch'in Chiu-shao's work from Sung and Yüan sources.[60] In a third note, published in his *Chu-t'ing hsien-shêng jih-chi ch'ao* (Diary of Mr. Chu-t'ing),[61] he gives some information he learned from Li Jui,[62] namely, the dating of the work, the table of contents, and some other data already given in his other works.

Mathematical analysis of the *Shu-shu chiu-chang* was done for the first time by Li Jui 李鋭 (1768–1817),[63] who was a disciple of Ch'ien Ta-hsin.[64] In the biographical note in the *Ch'ing-shih lieh-chuan*[65] we read: "In the house of his fellow-townsman Ku Kuang-ch'i 顧廣圻[66] he got the *Chiu-chang suan-ching*[67] of Ch'in Chiu-shao, and then he made a thorough investigation of the *t'ien-yüan-i* method. . . ."[68] This was in 1796.[69] In the preface of the *Shu-shu chiu-chang cha-chi* 數書九章札記,[70] written by Yü Sung-nien 郁松年, we learn that Li Jui was owner of a copy of the *Shu-shu chiu-chang.* Ch'ien Pao-tsung[71] writes: "Li Jui got a handwritten copy of the *Shu-shu chiu-chang* from the *Ssŭ-k'u*

---

[58] On the *t'ien-yüan-i,* see Chapter 17.
[59] Ch. 14, p. 332. The text is also published in Hu Yü-chin (1), pp. 830 ff.
[60] Namely, Ch'ên Chên-sun's *Chih-chai shu-lu chieh-t'i;* Chou Mi's *Kuei-hsin tsa-chih;* Li-Liu's *Mei-t'ing hsien-shêng ssŭ-liu piao-chun,* and the *Ching-ting Chien-k'ang chih.*
[61] Chu-t'ing is the literary name (*hao*) ôf Ch'ien Ta-hsin.
[62] See next paragraph.
[63] There is some doubt about these dates. Hummel, in (1), p. 144, gives 1765–1814; Needham, in (1), vol. 3, p. 731, and Li Yen, in (10'), p. 250, give 1768–1817.
[64] Hummel (1), p. 153.
[65] A biography of Li Jui is provided in *Ch'ing-shih lieh-chuan,* ch. 69, p. 24b and following pages.
[66] On Ku Kuang-ch'i, see p. 48.
[67] This title is of course a mistake. See Li Yen (8'), p. 478.
[68] Ch. 69, p. 25a.
[69] Ch'ien Pao-tsung (8'), p. 291.
[70] See p. 49.
[71] (2'), p. 65.

encyclopedia. He put it in order, made additions, revised, and commented on it." In Sung Ching-ch'ang's 宋景昌 preface[72] no mention is made of this fact. The text says: "Mr. Mao also took along that which the first-degree graduate Li [Li Jui] of Yüan-ho had collated, from the latter's private library. As for the *Ssŭ-k'u kuan-pên*, I also rely on the comparative study made by Mr. Sung [Sung Ching-ch'ang]." In this edition there is a blank space between these sentences; we find the same in Ch'ien T'ai-chi's 錢泰吉 *P'u-shu tsa-chi* (1838),[73] where Sung Ching-ch'ang is quoted, showing clearly that Li Jui is not connected with the *Ssŭ-k'u* collection. However, in Ting Fu-pao's 丁福保 edition of Ch'ien T'ai-chi's text, the punctuation is changed,[74] giving the impression that Li Jui collated the text from the *Ssŭ-k'u* collection. On the other hand, it is very likely that the copy owned in Ku Kuang-ch'i's library was a reproduction of the *Ssŭ-k'u* text. In any case, the text collated by Li Jui and his critical notes were never published. According to Pai Shang-shu 白尚恕[75] a copy of this text was made by Wang Hsüan-ling 王萱齡[76] and must be still extant.[77] The critical notes were used by Shên Ch'in-p'ei 沈欽裴 and Sung Ching-ch'ang for the textual criticism in the *Shu-shu chiu-chang cha-chi*.[78]

Chang Tun-jên 張敦仁 (1754–1834), a friend of Li Jui and Ku Kuang-ch'i,[79] writes in the preface to his *Ch'iu-i suan-shu* 求一算術: "During my official stay in Chiang-yu I visited commissioner Li Yün-mên 李雲門[80] who lent me for copying all the books of Ch'in and Li he owned. Thus I was able to investi-

[72] Sung Ching-ch'ang (1′), preface.

[73] Ch. C, p. 39b.

[74] Ting Fu-pao (1), p. 575a and Appendix, p. 3.

[75] In Ch'ien Pao-tsung (2′), pp. 290 f in several notes.

[76] There is a biographical note in the *Ch'ing-shih lieh-chuan*, ch. 69, pp. 36b–37a.

[77] As this manuscript was never published, we have no access to its contents and cannot provide further information about it.

[78] This is the only information we have about Li Jui's critical notes.

[79] See Hummel (1), p. 417.

[80] This is Li Huang, whose *tzŭ* was Yün-mên. In spite of a diligent search we are unable to identify the post held by Li Huang (*wên-hsüeh shih* 文學使).

gate thoroughly the secrets of the *Li t'ien-yüan-i* 立天元一[81] and the *Ch'iu-i* 求一.[82]

"I also went to Wu-mên 吳門,[83] where Li Shang-chih (Jui) 李尚之(鋭), a graduate of the first degree from Yüan-ho 元和, was living."[84] Li Huang 李潢 (?–1811)[85] must have owned Ch'ao Ch'i-mei's copy,[86] as Yü Sung-nien 有松年 (mid-nineteenth century) says in the preface to the *Shu-shu chiu-chang cha-chi*: "Mr. Shên 沈 [Shên Ch'in-p'ei 沈欽裴] of the College of Literature (*kuang-wên* 廣文) from Yüan-ho got the manuscript of Ch'ao Ch'i-mei, a man of the Ming, in the house of the prefect Chang [Tun-jên] of Yang-ch'êng." Chang Tun-jên says in his own preface about Li Jui: "Together we discussed day and night and investigated thoroughly the secret [of the text]";[87] Chang Tun-jên then wrote his *Ch'iu-i suan-shu* (1803),[88] a study of the indeterminate analysis of Ch'in Chiu-shao.[89]

In 1799 Juan Yüan 阮元 (1764–1849) published his well-known *Ch'ou-jên chuan* (Biographies of mathematicians and astronomers).[90] In his article on Ch'in Chiu-shao[91] he gives a short biography,[92] a discussion of the contents of the several chapters, a summary of Ch'in's preface, and a personal note.[93]

At this time many copies of the *Shu-shu chiu-chang* were in the hands of various scholars. According to Ch'ien T'ai-chi[94] and

---

[81] That is, a method for solving equations of all degrees.
[82] That is, indeterminate analysis.
[83] Now Wu-hsien 吳縣 in Kiangsu.
[84] Preface, p. 2a.
[85] See Ting Ch'ü-chung (2'), preface, p. 2a; Li Yen (10'), p. 249.
[86] See p. 41.
[87] *Ch'iu-i suan-shu*, p. 2a.
[88] It was printed in 1831. The preface is dated 1803.
[89] For a discussion of this work, see Chapter 16.
[90] On this work, see Vissière (1); Van Hee (3); W. Franke (1) and Hummel (1), p. 402.
[91] Part 2, pp. 277 ff.
[92] See Chapter 3, note 1.
[93] For further information, see Chapter 16.
[94] (1), C, p. 39b.

Yü Sung-nien,[95] Li Chao-lo 李兆洛 (1769–1841)[96] also had a copy, which came into the possession of Yü Sung-nien.[97] In 1826 Chang Chin-wu 張金吾 (1787–1829)[98] in his *Ai-jih-ching-lu ts'ang-shu chih* (1826)[99] simply reproduced Ch'in Chiu-shao's preface and Chao Ch'i-mei's postface. He added: "An old manuscript. A book from the *Mo-wang* library." This last was Chao Ch'i-mei's library, and Chang's copy was in all probability a reproduction from the text owned by Chang Tun-jên. Another valuable note on the *Shu-shu chiu-chang* was written by Chou Chung-fu 周中孚(1768–1831) in his *Chêng-t'ang tu-shu chi*,[100] including a short account of the *Ssŭ-k'u* edition and a discussion of the titles given to the work and its contents. As he quotes Juan Yüan, this note must have been written after 1799. Ku Kuang-ch'i 顧廣圻 (1776–1835)[101] wrote several prefaces and postfaces in his *Ssŭ-shih chai-chi*, among them a preface for the *Shu-shu chiu-chang*.[102] It was written for Hsia Wên-tao 夏文燾.[103] Ku Kuang-ch'i said that Ch'in Ên-fu 秦恩復 (1760–1843)[104] collated his own text, that they began to make blocks for printing, and that they relied on the check of the calculations made by Hsia Wên-tao. He wrote that there were many discrepancies among the problems, the methods, and the procedures, and that there were mistakes in the calculations.[105] As for this printed

---

[95] In his preface to the *Shu-shu chiu-chang cha-chi*.
[96] On Li Chao-lo, see Hummel (1), pp. 448 ff. He met Juan Yüan in Kuangtung in 1820.
[97] See note 95.
[98] See Hummel (1), p. 33; Têng and Biggerstaff (1), p. 44.
[99] Ch. 23, pp. 5b–9a; the preface was written by Ku Kuang-ch'i.
[100] Ch. 45, p. 856 f. On this work, see Têng and Biggerstaff (1), p. 32.
[101] On Ku Kuang-ch'i, see Hummel (1), pp. 417 ff.
[102] Ch. 9, p. 1a and pages following.
[103] Literary name: Hsia Fang-kuang 夏方光. The greater part of Ku Kuang-ch'i's works were written for other scholars who supported him financially. See Hummel (1), p. 418.
[104] On Ch'in Ên-fu, see Hummel (1), p. 417: "In the following year [1805] he [Ku Kuang-ch'i] was invited to Yangchow by Chang Tun-jên. . . There Ku became acquainted with a famous bibliophile, Ch'in Ên-fu. . . ."
[105] This was of course owing to the fact that the book was copied several times without comprehension. Even today the whole text has not been reconstructed.

version, nobody ever saw a copy, and it is very likely that it was never completed.[106] In his article, Ku Kuang-ch'i devoted special attention to the *ta-yen* rule, quoting Li Jui and Li Liu.

At the beginning of the nineteenth century, Shên Ch'in-p'ei began to collate the text of the *Shu-shu chiu-chang*. He found the manuscript of Chao Ch'i-mei in the house of Chang Tun-jên at Yang-ch'êng,[107] and he spent several years correcting its mistakes. As he became old and ill, he was not able to accomplish this task. His disciple Sung Ching-ch'ang 宋景昌[108] continued his studies and wrote the important collection of reading notes, *Shu-shu chiu-chang cha-chi*, relying largely on Shên Ch'in-p'ei's manuscript from the library of Chao Ch'i-mei, but also making use of Li Jui's studies.[109] In his work he points out the textual errors; he quotes the official comments from the *Ssǔ-k'u* encyclopedia (*kuan-an* 館案), his master Shên Ch'in-p'ei, Li Jui, and Mao Yüeh-shêng 毛嶽生. For some of the problems he gives a mathematical reco⸴struction or correction, but as a whole this work is a textual collation. However, it is very useful for the study of Ch'in's writings.[110]

Mao Yüeh-shêng (1791–1841) played an important part in the preparation of the first printed edition of the *Shu-shu chiu-chang*.[111] Yü Sung-nien, in his introduction to the *Shu-shu chiu-chang cha-chi*, says that he relied on Mao Yüeh-shêng in studying the original text, and Sung Ching-ch'ang quotes some of Mao's criticisms. Mao Yüeh-shêng also found in Li Jui's house the

---

[106] See Ch'ien Pao-tsung (2'), p. 65.

[107] According to Yü Sung-nien in his preface to the *Shu-shu chiu-chang cha-chi*.

[108] On Shên Ch'in-p'ei, see Li Yen (8'), p. 478. On Sung Ching-ch'ang, see the preface to Tsou An-ch'ang (1'); Miu Ch'üan-sun (1'), ch. 20, p. 38b and following page; Liu Shêng-mu (1'), ch. 9, p. 2a; Ch'ien T'ai-chi (1'), C, pp. 39b and page following.

[109] Ting Fu-pao (1'), Appendix, p. 3, no. 7.

[110] On Sung Ching-ch'ang, see Ch'ien T'ai-chi (1'), C, pp. 39b–40a (mainly a quotation from Yü Sung-nien's preface); Miu Ch'üan-sun (1'), ch. 20, pp. 38b–39a; Liu Shêng-mu (1'), ch. 9, p. 2a. According to these last two sources, Sung Ching-ch'ang was a disciple of Li Chao-lo.

[111] There is a biographical note in *Ch'ing-shih lieh-chuan*, Part 10, p. 29b.

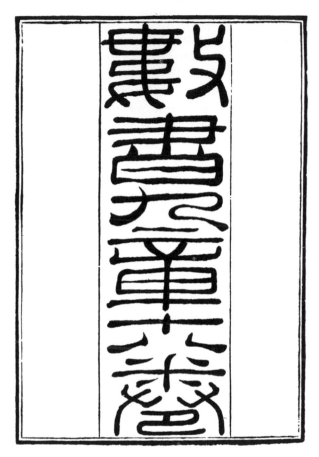

Figure 4. Title page of the *Shu-shu chiu-chang* in the *I-chia t'ang* 宜稼堂 collectanea of 1842.

text collated by this latter, studied it, and made a new collation.[112]

This was the final step to publication. In 1842 Yü Sung-nien
published the *I-chia t'ang ts'ung-shu* (*I-chia t'ang* collection)[113]
(see Figure 4). Volumes 43–48 comprise the *Shu-shu chiu-chang*
and volumes 49–50 the *Shu-shu chiu-chang cha-chi*. He also wrote
a postface to Ch'in's work in the form of a biographical note[114]
and an introduction to the *Cha-chi*.[115] In preparing this edition,
Yü Sung-nien made use of several versions.[116] According to
Lao Ch'üan 勞權 [117] he obtained Ch'ien Tsêng's copy[118] and
compared it with the collation made by Li Jui and with the
Yung-lo ta-tien. He also made use of the work done by Shên
Ch'in-p'ei and Sung Ching-ch'ang, and found also Li Chao-lo's
copy. After investigating these versions, he concluded that Chao
Ch'i-mei's text was the most nearly perfect, and he took it as
the basic text for printing. He said that all these texts were
different from each other, and that all had copyists' errors and
mathematical mistakes.[119]

Lu Hsin-yüan 陸心源 (1834–1894) gathered some biographi-
cal material in his *I-ku t'ang t'i-pa*.[120] Ch'ü Yung 瞿鏞 (first half

[112] Li Yen (9′), p. 142.
[113] The *I-chia* library in Shanghai, which was owned by Yü Sung-nien.
See Hummel (1), pp. 545 f.
[114] See Chapter 5.
[115] See Ting Fu-pao (1′), p. 573b. In the *Ts'ung-shu chi-ch'êng* edition of
1936, the postface to the *Shu-shu chiu-chang* and the preface to the *cha-chi*
are retained. In the *I-chia t'ang* edition both appear with the *cha-chi* (vol.
49). This means that the postface is not signed. Furthermore, the prefaces
in the *I-chia* edition are not in good order (at least in the copy at Harvard-
Yenching), because the postface is mixed up with the biographical note of
Chou Mi. The signature of the preface refers also to the postface.
[116] The *I-chia t'ang shu-mu* 宜稼堂書目 reads: "*Shu-shu chiu-chang*, a manuscript."
[117] There is a biographical note in the *Pei-chuan chi-pu*, ch. 50. This quote
is not from the original text, but from Chang Yü's notes to Ch'ien Tsêng
(1), ch. 1, C, p. 27b.
[118] See pp. 42 f.
[119] See his preface to the *Shu-shu chiu-chang cha-chi*.
[120] On this work, see Têng and Biggerstaff (1), p. 47, and Hummel (1),
p. 546. Lu Hsin-yüan's notes on Ch'in Chiu-shao are in ch. 8, pp. 2a–3a;
in the supplementary volumes, in ch. 8, pp. 20b–21a. See also Chapter
3 of this work.

of the nineteenth century), the son of Chang Chin-wu,[121] provides a bibliographical note in his *T'ieh-ch'in-t'ung-chien lou
ts'ang-shu mu-lu*.[122]

Ch'ien Pao-tsung[123] says that the *Shu-shu chiu-chang* was also
included in the *Ku-chin suan-hsüeh ts'ung-shu* 古今算學叢書,
published in Shanghai in 1898 by Liu To 劉鐸. The work on
this *ts'ung-shu* was never completed. The *Shu-shu chiu-chang* is
listed in the table of contents, but according to Li Yen the last
parts, among them Ch'in's work, were never printed.[124] Ting
Fu-pao[125] gives the table of contents of the whole work in his
Ssŭ-pu tsung-lu suan-fa pien, but after the *Ssŭ-yüan chieh* he
writes: "The rest was never published." Tsou An-ch'ang 鄒安
鬯 prepared the text for this edition,[126] but it was never printed.

About 1879 Fêng Ch'êng 馮澂 wrote his *Suan-hsüeh k'ao
ch'u-pien* 算學考初編 in which he provided a short note on
Ch'in Chiu-shao.[127] Chang Yü 章鈺 (1865–1937) published in
1926 Ch'ien Tsêng's *Tu-shu min-ch'iu chi chiao-chêng*,[128] which
contains a note on Ch'in Chiu-shao with annotations by Chang
Yü.[129]

In 1899 Ting Fu-pao 丁福保 (1874–1952) wrote his *Suan-
hsüeh shu-mu t'i-yao*, in which he gathered valuable notes on

[121] Hummel (1), p. 34; Têng and Biggerstaff (1), p. 45.
[122] Ch. 15, pp. 4b–5a.
[123] (2'), p. 65.
[124] (10'), p. 375. Li Yen (11'), in a survey of the mathematical books in
his personal library, ends his "contents" of the *Ku-chin suan-hsüey ts'ung-shu*
with the *Ssŭ-yüan chieh*. See (11'), p. 1,551, item 438. Nathan Sivin, who
studied the Kyoto copy, writes: "*Shu-shu chiu-chang* appears in the table
of contents of the Kyoto copy of *Ku-chin suan-hsüeh ts'ung-shu*, in conformity
with the list in *Ssŭ-pu tsung-lu suan-fa pien* (by Ting Fu-pao), but the actual
contents of the series goes only up through *Ssŭ-yüan chieh*. Perhaps it was
never printed" (personal communication).
[125] Ting Fu-pao (1'), pp. 16b–18b.
[126] He was acquainted with Sung Ching-ch'ang, with whom he published
a small work entitled *K'ai-fang chih fên huan-yüan shu* 開方之分還原術 (see
the preface to this work).
[127] I have not seen the work, and I rely on Ting Fu-pao, Appendix, p. 3a.
According to the introduction to this latter work (p. 11), the National
Library of Peking owns the original manuscript. It was printed in 1897 in
the *Ch'iang-tzŭ-li chai-chi* 強自力齋集.
[128] See p. 42.
[129] In his annotations he quotes Lao Ch'üan and Ch'ien Ta-hsin.

mathematical books.[130] It was incorporated in the most valuable bibliographical work on Chinese mathematics, the *Ssŭ-pu tsung-lu suan-fa pien* (General catalogue of the four departments of literature, section on mathematics), prepared by Ting Fu-pao and Chou Yün-ch'ing 周雲青 and published by the Commercial Press in 1957. The work consists of three parts, each containing bibliographical notes on Ch'in Chiu-shao.[131]

Other bibliographical notes are included in Miu Ch'üansun and Ch'ên Ssŭ (1)[132] and Liu Shêng-mu(1).[133] In 1936 a new edition of the *Shu-shu chiu-chang* was included in the *Ts'ung-shu chi-ch'êng* 叢書集成 collection, published by Wang Yün-wu 王雲五. It is a typeset reprint of the *I-chia t'ang* edition of 1842.

Additional notes to the bibliographical works mentioned here are provided in Hu Yü-chin 胡玉縉 (1), printed in 1964.[134] According to Ting Fu-pao,[135] an article by Mêng Sên 孟森, entitled "Additional Investigations on the *Shu-shu chiu-chang* of the *Ssŭ-k'u t'i-yao* and the Author Ch'in Chiu-shao," was published in 1940.[136]

[130] I have not located the work, but his notes are included in the *Ssŭ-pu tsung-lu suan-fa pien*. The note on Ch'in Chiu-shao is on pp. 574b–575a.
[131] Published in 1926 by Ting Fu-pao and his disciple Chou Yün-ch'ing, the *Ssŭ-pu tsung-lu* (General catalogue of the four departments of literature) was a compilation of bibliographical notes in four parts. The fourth section was called *Tzŭ-pu* 子部 (Miscellaneous notes), and part of it (*Suan-fa pien* 算法編) is devoted to mathematics. In the 1957 edition, the pagination of the first edition is preserved (pp. 560a–597b). See Li Yen's introduction, p. 7a. An appendix (*Pu-i* 補遺) is included (pp. 1a–53a); it was prepared by Chou Yün-ch'ing and covers 198 subjects, of which no. 7 is the *Shu-shu chiu-chang*. A third part is derived from two sources: the *Ku-chin suan-hsüeh shu-lu* 古今算學書錄 (complete title: *Jo-shui chai* 若水齋 *ku-chin suan-hsüeh shu-lu*, Jo-shui studio bibliography of old and new mathematical books, published by Liu To 劉鐸 in 1898; and the *Suan-hsüeh k'ao ch'u-pien*, written by Fang Chêng in 1897. The pagination and numbering of the subjects are continuous to part 2.
[132] Ch. 20, p. 38b and following page.
[133] Ch. 9, p. 2a. This is a note on Sung Ching-ch'ang.
[134] Hu Yü-chin died in 1940. The work was completed by Wang Hsin-fu 王欣夫 and published in 1964 (second edition: Taipeh, 1967). In connection with Ch'in Chiu-shao there are notes by Ch'ien Ta-hsin, Lu Hsin-yüan and Ku Kuang-ch'i. See pp. 208 f (in the 1964 edition, pp. 830–834).
[135] Appendix, p. 3a.
[136] In the "weekly literary supplement" of the *T'ien-chin I shih pao* 天津益世報, no. 79. I have not seen this article.

•

5

General Structure of the Shu-shu chiu-chang

The modern edition of the *Shu-shu chiu-chang* contains the following parts:
A. Preface by Ch'in Chiu-shao
1. The preface as such
2. The nine "poetic descriptions" of the nine chapters
B. Short preface by Chao Ch'i-mei, dated 1616
C. Index, divided into eighteen *chüan*
D. The nine chapters
1. *Ta-yen lei* 大衍類 (Indeterminate analysis)
2. *T'ien-shih lei* 天時類 ("Heavenly phenomena")[1]
3. *T'ien-yü lei* 田域類 ("Boundaries of fields," *that is,* surveying)
4. *Ts'ê-wang lei* 測望類 ("Telemetry")[2]
5. *Fu-i lei* 賦役類 (Taxes and levies of service)
6. *Ch'ien-ku lei* 錢穀類 (Taxes)[3]
7. *Ying-chien lei* 營建類 (Fortifications and buildings)[4]
8. *Chün-lü lei* 軍旅類 (Military affairs)
9. *Shih-wu lei* 市物類 (Commercial affairs)
E. Postface by the editor Yü Sung-nien
F. Appendix: bibliographical note from Chou Mi's
Kuei-hsin tsa-chih, hsü-chi (1308)
G. A short note by Chao Ch'i-mei[5]

Ch'in Chiu-shao's Preface
In this section, the preface is quoted in its entirety. Passages

---

[1] Of these problems on heavenly phenomena, five deal with chronological questions, the four others with meteorological questions (measurements of rainfall and snowfall).
[2] Or "measuring at a distance," a kind of proto-trigonometry.
[3] The literal translation of *Ch'ien-ku* is "money and grain."
[4] This is a chapter on various architectural problems.
[5] To the modern edition is added the *Shu-shu chiu-chang cha-chi*, containing the textual emendations and some critical notes by Sung Ching-ch'ang. For more details about this work and the role of Chao Ch'i-mei in recovering the text, the reader is referred to Chapter 4.

from the text are enclosed in quotation marks and are followed by annotations.

"The Six Arts of the teaching of the Chou were truly made complete by mathematics.[6] It is something that scholars and great officers have always esteemed."[7]

The dates of the Chou dynasty are 1122?–255 B.C. The six arts are mentioned in the *Chou-li* (Rites of Chou).[8] They were propriety, music, archery, charioteering, writing, and "mathematics"; they were not sciences, but arts. This shows clearly that mathematics was considered an ancient technique. Moreover, it is very likely that *shu* 數 (mathematics) meant both arithmetic and fate calculation.[9]

"Their application was based on the idea that 'the Great Void generates the One, and it oscillates without ending.' "

The "Great Void" (*T'ai-hsü* 太虛) is a Taostic term. The One is not a number, but it generates all numbers. See B. S. Solomon (1), pp. 257 ff, and Fung Yu-lan (1), vol. 2, p. 182.

"If we aim at the great, we can be in touch with the spiritual powers and thus live conformably with our destinies; if we aim at the small, we can settle the affairs of this life, and by classification deal with the myriad phenomena. How could this be possible merely by peering at the shallow and familiar? Thus in ancient times they arranged the stalks [for prognostication] in order to calculate the [auspicious days] and fixed the pitch pipes so as to determine the *ch'i* 氣."

This is an allusion to the prognostication methods of the *I-ching* (Book of Changes),[10] which were mathematical. It could

---

[6] The meaning seems to be that "mathematics" was the last in the series, when the six arts were enumerated.

[7] Compare the text in the Family Instructions of the Yen Clan, quoted in Chapter 1.

[8] The *Chou-li* has been traced by internal evidence to the fourth century B.C., although tradition ascribed it to the Duke of Chou (c. 1000 B.C); see B. Karlgren (1) and Ch'ien Mu (1') and (2').

[9] See Fung Yu-lan (1), vol. 1, p. 26.

[10] See Chapter 22.

also be an allusion to the *ta-yen* rule, by means of which the prognostication method of the *I-ching* can be rationalized.[11] The pitch pipes are the twelve musical tones.[12]

"By means of the plumb [*pi* 髀][13] and the carpenter's square they cleared their rivers and with the gnomon shadow template[14] they measured [the length of] the shadow."

The length of the shadow was measured for fixing the winter and summer solstices.

"The greatness of heaven and earth was enclosed by these methods and could not exceed their scope—much less the multiplicity of human life between [heaven and earth]. After [the coming of] the *Ho-t'u* diagram and the *Lo-shu* diagram, they [the ancients] unriddled and explained these secrets."

On these diagrams, which are magic squares, see Needham (1), vol. 3, pp. 56 ff. They were by tradition considered the origin of Chinese mathematics.[15] From Yang Hui's work we know that magic squares were still formed in Ch'in's time.

"With the eight diagrams [*pa-kua* 八卦] and the nine *ch'ou* 疇 [the nine categories], they intricately pieced together an understanding of the essentials and subtleties [of heaven and earth]."

The eight diagrams are the so-called trigrams, combinations of three broken or unbroken lines as given in the *I-ching*.[16] For information on the nine categories, see Fung Yu-lan (1), vol. 1, pp. 163 ff; the first of these categories is called the "five elements," and it is very likely that those are meant here.

"They reached their apogee in the application of [these methods to] the *Ta-yen* and *Huang-chi* calendars."

11 Ibid.
12 For the method of prognostication by means of pitch pipes, see Bodde (1).
13 The later meaning was probably *gnomon*. See Needham (1), vol. 3, p. 284; Maspero (1), pp. 217 ff.
14 See Needham (1), vol. 3, p. 286; Maspero (1), p. 217.
15 For a full explanation, see Chêng Chin-tê (1), pp. 499 ff and Mikami (1), p. 3. There are bibliographical notes in Needham (1), vol. 3, pp. 56 ff. See also Camman (1).
16 See Fung Yu lan (1), vol. 1, p. 379; Barde (1); Cassien-Bernard (1).

The *Ta-yen* calendar was drawn up by the Buddhist monk I-hsing in 724;[17] the *Huang-chi* calendar was compiled by Liu Ch'o about 600;[18] both were considered a leap forward in calendrical computations. It is believed that the first was the origin of the *ta-yen* rule, the other of the "Method of Finite Differences."[19]

"As for changes in the affairs of men, there was nothing that could not be accounted for. As for the dispositions of the spirits, there was nothing that could be hidden. The sages comprehended it wondrously; they talked about it and handed down the crude outlines; the common man found it obscure; and for that reason none of them were aware of it. If you look at essential aspects of their meanings, the numbers and the *Tao* do not derive from two [different] bases."

If we interpret the *Tao* as the dynamic structure of the cosmos, and if "mathematics" gives the relations that express the general rules of nature, then this is a very modern idea, even if the application of the principle was not at all scientific.[20]

"The Han period was not far remote from antiquity. There were people like Chang Ts'ang 張蒼, Hsü Shang 許商, Ch'êng-ma Yen-nien 乘馬延年, Kêng Shou-chang 耿壽昌, Chêng Hsüan 鄭[元]＝鄭玄, Chang Hêng 張衡, and Liu Hung 劉洪."

Chang Ts'ang (active 165–152 B.C.) is believed to have been the first compiler of the *Chiu-chang suan-shu*. See Needham (1), vol. 3, p. 24; he is also mentioned in Liu Hui's preface to the *Chiu-chang suan shu*.[21] The work of Hsü Shang (early Han, first century B.C.) is lost; see Needham, ibid., p. 28. Ch'êng-ma

---

[17] It is supposed that he made use of indeterminate analysis. See Chapter 15.
[18] See Yabuuchi (3), p. 458.
[19] See Needham (1), vol. 3, p. 123.
[20] Chinese thinking was deeply concerned with the fundamental order of nature. This order was not static, but dynamic; hence the principle of order is *Tao*, the Way of all events in the cosmos.
[21] On Chang Ts'ang as statesman, see Hulsewé (1), p. 380, n. 170.

Yen-nien is mentioned several times in the *Han-shu*.[22] Kêng Shou-chang was active 75–49 B.C.; according to Liu Hui's preface, p. 1, he was one of the commentators of the *Chiu-chang suan-shu* [see Needham, ibid, p. 24, and Mikami (1), p. 9].[23] On Chang Hêng, see Needham, ibid., p. 20 and note *d*. Liu Hung (active 178–183 A.D.) was the compiler of the *Ch'ien-hsiang* calendar. See Yabuuchi (3), pp. 452 ff, and Needham, ibid., p. 20. For a general discussion of the Han mathematicians, see Li Yen (10'), pp. 5 ff.

"Some of them investigated the *Tao* of heaven . . . "

The *Tao* of heaven refers to mathematical astronomy as well as astrology.

" . . . and their methods were handed down to posterity; some of them calculated the allocation of labor . . . "

That is, they solved practical, mundane problems for the government.

" . . . and they arrived at effective results in their time. In later generations scholars were very proud of themselves and, considering [these arts] inferior, did not teach [or discuss] them. Those studies were almost defunct [through neglect]. Only calendar-makers and mathematicians [*ch'ou-jên* 疇人] . . ."

Compare the title of Juan Yüan's *Ch'ou-jên chuan*. "The word *ch'ou* came to mean surveyor, and then later on was applied to all computers (*ch'ou-jên*), especially those surveyors of the heavens, the astronomers" [Needham (1), vol.3, p. 3].

" . . . were able to manage multiplication and division, but they could not comprehend square-root extraction or indeterminate analysis."

Ch'in seems to mean that "mathematics" was no longer studied by the literati, but only by individual mathematicians. This

[22] Ch. 29, p. 15b, where it is said that he was brilliant in computing, and that he was able to determine "tasks" and benefits.
[23] See also Swann (1), p. 192 ff.

passage recalls Needham's statement, quoted in Chapter 1, about the position of mathematicians in the Sung. Square-root extraction was already known to the compilers of the *Chiu-chang suan-shu;* perhaps Ch'in refers to the general method for solving numerical equations of all degrees (Horner's method).

"In case there were calculations to be performed in the government offices, one or two of the clerks might participate . . ."

See the story from the *T'ang ch'üeh-shih* 唐闕史 (855), repeated in Needham (1), vol. 3, p. 116; see also Nakayama (1), p. 16.

". . . but the position of the mathematicians was never held in esteem; their superiors left things to them and let them do as they pleased; [but] if those who did computations were only that sort of man, it was merely right that they should be disdained."

"In the field of music, there are conductors who can only arrange the sounds of the bells and sounding stones, but is it permissible to say that 'to produce complete harmony with heaven and earth' [which the Great Music is said to do] merely consists in this?"[24]

This is perhaps an allusion to a kind of "musical cosmology," as described in the *Yüeh-wei* 樂緯 (Apocryphal treatise on

[24] The first part of this sentence *(yüeh yu chih shih, chin chi k'êng ch'iang* 樂有制氏, 僅記鏗鏘*)* is directly derived from two almost identical passages in the History of the Former Han Dynasty, namely, *HS* 22.12, the treatise on Ritual and Music (no parallel in the *Shih-chi*) and *HS* 30.14b, the treatise on Bibliography; an almost identical passage was included in the *Fêng-su t'ung-i* of the second century A.D. In these texts, the decay of music is deplored; the musical tradition, they say, has foundered to such an extent that, "when the Han arose. the conductors, who for generations had been in the Office for Music because of [their knowledge of] the correct sounds of the Solemn Music, merely [in *HS* 22; the other two versions read "slightly"] knew how to arrange [*chi* 紀; Ch'in mistakenly has *chi* 記, "to record"] the sounds of the bells and sounding stones and the drumming and dancing, but they were incapable of explaining their significance." The second half of Ch'in's question contains a reference to the canonical text of the Book of Rites, the *Li-chi*, ch. 37, p. 6a, which says that the Great Music is in full harmony with Heaven and Earth *(ta yüeh yü T'ien Ti t'ung ho* 大樂與天地同和*)*.

music):[25] "Some [through their music] harmonized the Yin and Yang, some the five elements, some the growth and decay [of the seasons], some the pitch pipes and the calendar [equated with these pitch pipes], and some the five notes [of the scale]." Quoted from Fung Yu-lan (1), vol. 2, pp. 127 f.

"As for mathematical books, today there remain the works of more than thirty schools [or authors?]. [This part of mathematics], where heavenly configurations and celestial motions are calculated, is called 'the technique of threading' [*chui-shu* 綴術]."

Chui-shu is the title of a work compiled by Tsu Ch'ung-chih (430–501), now lost; the real meaning of the term *chui-shu* in this work is no longer known. For the meaning of *chui-shu* in Shên Kua's work, see Sivin (3), pp. 72 ff, and Needham (1), vol. 3, p. 97.

"Divination and fate calculation are referred to as *san-shih* 三式; all of these are called 'esoteric mathematics' (*nei-suan* 內算), to emphasize their esoteric nature."

The *san-shih* (three patterns) are three different methods of divination. See Needham (1), vol. 3, p. 141.

"What is contained in the Nine Chapters are the nine computations of the *Chou-li*."

The "nine computations" are in all probability nothing else than the multiplication table (although there is no proof of this); but what is interesting for our purposes is the fact that Liu Hui, in his preface to his edition of the *Chiu-chang suan-shu*, identifies these nine computations with the nine chapters of this work. The matter is discussed in Needham (1), vol. 3, p. 25. Ch'in Chiu-shao's preface shows that he also subscribed to this idea.

"What is related to squares and circles is 'geometry' (*chuan-shu* 裏術)."

[25] Probably written in the first century B.C.

For the term *chuan-shu*, see Needham (1), vol. 3, p. 97 and note *c*.

"These are all called 'exoteric mathematics' (*wai-suan* 外算), in contrast to 'esoteric mathematics.' Their applications are connected with each other, and we cannot divide them into two [different arts]."

He seems to be emphasizing the identity of astronomical and mensurational mathematics.

"Only the *ta-yen* rule is not contained in the Nine Chapters, for no one has yet been able to derive it [from other procedures]. Calendar-makers, in working out their methods, have made considerable use of it. Those who consider it as belonging to 'equations' (*fang-ch'êng* 方程) are wrong."

Fang-ch'êng is the title of the eighth book of the *Chiu-chang suan-shu;* the literal meaning is "square table method," being an allusion to the matrix calculation performed on the counting board. This whole eighth chapter is devoted to simultaneous linear equations. Ch'in Chiu-shao means that indeterminate equations cannot be solved in the same way as determinate equations; or, in other words, that the *ta-yen* rule is an innovation, not a mere development out of one chapter of the *Chiu-chang suan-shu*.

"Moreover, the things of the world are great in number, and people of ancient times made their calculations before things happened. When their prognostications were calculated, they acted on them. They looked up [to heaven] and looked down [to earth] . . ."

"Looking up" seems to be an allusion to Fu Hsi, who found the eight trigrams by looking up to the heaven; "looking down" seems to mean "examining the configuration of the land," as done by geomancers.

". . . the plans of men and the plans of spirits, there was nothing they did not draw attention to. By so doing they avoided faults in the results. This we can ascertain in every chapter of the historical records. But later generations, in their under-

takings and in their enterprises, were but rarely capable of reflection and contemplation.

"Gradually laws of heaven and affairs of men have become
vague and imperfectly known. Should we not search out the
reasons for this?

"I, Chiu-shao, am stupid and uneducated and not versed in
the arts. But in my youth I was living in the capital, so that I
was enabled to study in the Board of Astronomy; subsequently
I was instructed in mathematics by a recluse scholar. At the
time of the troubles with the barbarians, I spent some years at
the distant frontier; without care for my safety among the
arrows and stone missiles, I endured danger and unhappiness
for ten years. My heart was withered, and my vital power fell
away. [But] I knew truly that none of these things was without
its 'number,' and I let loose my imagination among these
numbers."

The biographical data are explained in the biographical
section. The meaning of the last sentence seems to be: my fate
has its mathematical sense in the order of nature, and I started
thinking about this mathematical sense of all things.

"I made inquiries among well-versed and capable [persons]
and investigated mysterious and vague matters. If I attained
some crude understanding, it was by what is called 'being in
touch with the spiritual powers and living conformably with
destiny.' " [corrupt sentence omitted.]

"As for the details [of the mathematical problems], I set them
out in the form of problems and answers meant for practical
use. I collected many of them and, not wanting to cast them
aside, I selected eighty-one problems and divided them in
nine classes; I drew up their methods and their solutions and
elucidated them by means of diagrams.

"Perhaps they will serve as material for gentlemen of broad
knowledge to peruse in their spare time, for although [mathe-
matics is] a minor art it is worth pursuing. Thus I wish to offer
this work to my colleagues. If they say that their skill [in the
minor arts] is complete and that this is merely for people like
astronomers [*ch'ou-jên*] and provincial clerks, and ask why this

should deserve to be used throughout the empire, will that not show them to be benighted!

"In the ninth month of the seventh year of *Shun-yu* (1241–1253), Ch'in Chiu-shao of Lu prefecture wrote this."

Organization of the Shu-shu chiu-chang

Although the *Shu-shu chiu-chang* is divided into nine chapters, this arrangement has nothing to do with the old *Chiu-chang suan-shu*. However, the influence of the latter on Ch'in's work is very important, as we shall see in the next paragraph.

The titles of the chapters in the *Chiu-chang suan-shu* are as follows: [26]

1. *Fang-t'ien* 方田 (Surveying of land)
2. *Su-mi* 粟米 (Millet and rice)
3. *Ts'ui-fên* 衰分 (Proportional division)
4. *Shao-kuang* 少廣 (Diminishing the breadth[27])
5. *Shang-kung* 商功 (Evaluation of labor[28])
6. *Chün-shu* 均輸 (Impartial taxation)
7. *Ying-nu* 盈朒 (Excess and deficiency)
8. *Fang-ch'êng* 方程 (Calculation by tabulation[29])
9. *Kou-ku* 勾股 (Right triangle)

These chapters were considered in Ch'in's work as an indication of general procedures. In each problem of the *Shu-shu chiu-chang* the method to be used is designated by the words: "*I . . . ch'iu chih* 以 . . . 求之 (solve it by means of . . .)".

Of the ten methods referred to in the *Shu-shu chiu-chang*, one, the *ta-yen* 大衍 rule (method for solving indeterminate equations), is not mentioned in the *Chiu-chang suan-shu*. The remaining nine are listed in Table 1, which displays them as they appear in the two works.[30]

---

[26] For more details about the contents, the reader is referred to Needham (1), vol. 3, pp. 25 ff; Vogel (2); Yushkevitch (4), pp. 24 ff.

[27] This term is difficult to translate. The chapter deals with the computation of one side of a figure when area or volume is given. Vogel (2) transcribes "Kleinere und grössere Breite" (greater and lesser widths).

[28] *Kung* is the usual term for "task," also used as unit of labor. See Part VI.

[29] Solving of simultaneous linear equations.

[30] For the sake of convenience in this comparison, the *Chiu-chang suan-shu* will be referred to as *CCSS*, the *Shu-shu chiu-chang* as *SSCC*.

Table 1. Methods in the *Chiu-chang suan-shu* and the *Shu-shu chiu-chang*

Method	Chapter in CCSS	Problems in SSCC
1. *Fang-t'ien*	1	III, 8; VII, 8; IX, 6
2. *Su-mi*	2	V, 4, 5, 7; VI, 2, 3, 9; VIII, 8; IX, 3, 4, 5, 6, 8
3. *Ts'ui-fên*	3	V, 1, 2, 5, 7, 9; VI, 7, 9; IX, 2, 9
4. *Shao-kuang*	4	III, 1–8; IV, 2; VI, 4; VII, 7; VIII, 1, 2
5. *Shang-kung*	5	III, 9; V, 3; VI, 4, 5, 6; VII, 1, 2, 3, 4, 5, 9; VIII, 3, 4
6. *Chün-shu*	6	I, 7; V, 8; VI, 1; VII, 5; VIII, 6, 7
7. *Ying-nu*	7	VI, 8; VIII, 9
8. *Fang-ch'êng*	8	II, 4; IX, 1, 2
9. *Kou-ku*	9	III, 7; IV, 1, 2, 3, 4, 5, 6, 7, 9; VIII, 5

From this comparison it is obvious that a large part of the *Shu-shu chiu-chang* was a reorganization of the problems derived from the *Chiu-chang suan-shu* that applied them to other practical problems, and usually to more difficult ones.

Each problem is divided into four parts:

1. *Wên* 問 : the question, always given in practical form
2. *Ta* 答 : the answer
3. *Shu* 術 : the method, given in a general form, without figures. We find the same "general rules" in the *CCSS*.
4. *Ts'ao* 草 : the solution, which shows the numerical working, usually with diagrams derived from the counting board.

The Methods (shu 術) in the Shu-shu chiu-chang

The "methods" given by Ch'in Chiu-shao are not mathematical methods in our sense.[31] At first sight they seem to be general and vague, and it is hard to believe that they were considered as algebraical methods (in the modern sense) by the Chinese mathematicians. They seem to have been used only as references to the chapters of the *Chiu-chang suan-shu*. If we take, for

---

[31] Of course the methods for solving the problems are real mathematical methods.

instance, the method *chün-shu*, we find it only as the title of Chapter 6 of the Nine Chapters, where twenty-eight problems are solved. For each separate problem the method is given. For instance, Problem V, 8 of the *Shu-shu chiu-chang* is, from the mathematical point of view, identical with Problem VI, 1 in the *Chiu-chang suan-shu*. Moreover, the purely technical parts of both methods are almost identical, even to the letter; in the *SSCC* we find: "*Chieh ju . . . êrh i ; . . . fu ping wei fa; i . . . ch'êng wei ping . . .; ko tê wei shih* 皆如 · · · 而一; · · · 副併爲法; 以 · · · 乘未併 · · ·; 各得爲實"; in the *CCSS* there is exactly the same wording.[32] From this statement it is quite clear that Ch'in refers to one of the methods of one of the chapters of the *Chiu-chang suan-shu*.[33]

Ch'in's work sets forth some other methods which are not in the *Chiu-chang suan-shu*. The first of these is the *ta-yen* rule. In II, 3 he speaks of *li-fa* 曆法 (calendrical method), which must have been the method used in the *K'ai-hsi* calendar (1205–1207), with which he is dealing. These are perhaps the methods he learned in the Board of Astronomy.

There are also some methods which do not refer explicitly to chapters in the *Chiu-chang suan-shu*, and which are given only in the procedures. For example, the rule on positive and negative numbers (*SSCC*, IX, 1) is literally the same as in the *CCSS*, VIII, 3. The *ch'ung-ch'a* 重差 method is used in the *SSCC*, IV, 3, 4, 7, 9 and VIII, 5; it is only another name for Chapter 9 in the *CCSS*.

Some of these methods are combined with other procedures. The usual formula is: "*I x ch'iu chih*;*y ju chih* 以 *x* 求之;*y* 入之" (solve by means of *x*; introduce *y*).

Finally, some of the old methods are extended to new ones

[32] It should be pointed out that this part of the text is the mathematical framework of the solution. This study has lead to conclusion that Ch'in's methods are derived from the *Chiu-chang suan-shu*. However, this does not mean that they represent no improvement on the source.

[33] It would be interesting to make a detailed comparison between the methods of both works. As this would require a thorough analysis of the *Chiu-chang suan-shu*, it cannot be done here.

which go far beyond the level of the *Chiu-chang suan-shu,* for example, the method for solving numerical equations of higher degree.[34]

[34] It is somewhat strange that this method has no special name.

II

**Mathematics in the Shu-shu chiu-chang:
Numerical Notation and Terminology**

6

Notation and Terminology

Numbers

For the usual notation in literary texts, including the literary statement of problems in scientific texts, the decimal system is used, but each power of ten is indicated by the everyday words for "ten," "hundred," "thousand," and so on. For example, the Chinese 三萬五千六百七十二 means "3 ten thousands 5 thousands 6 hundreds 7 tens 2" or 35,672.[1] A special condensed mathematical notation was developed under the influence of the counting board, and is often found in the parts of scientific texts devoted to the working out of problems.

The reader is referred to Needham[2] for a general survey of these *counting-rod forms*. Here we restrict ourselves to the forms used in the *Shu-shu chiu-chang*.

A counting-rod number has the general form . . . ABABAB. In Diagram 1 the forms are given in columns *A* and *B*. Column *B* is used for units, hundreds. . . , *A* for tens, thousands. . . .

	A		*B*	
1	—		I	
2	=		II	
3	≡		III	
4	≣	×	IIII	×
5	≣	ó	IIIII	ō
6	⊥		T	
7	⊥		TT	
8	⊥		TTT	
9	⊥	×	TTTT	⊠
0	O		O	

Diagram 1

[1] An expression such as "*i êrh san ssŭ* 一二三四" means: 1, 2, 3, 4.
[2] Needham (1), vol. 3, pp. 5 ff (Table 22); see also van Gulik (1); Fang Kuo-yü (1'); D. E. Smith (1), vol. 2, p. 40; Mikami (1), p. 73.

For example, the number 35,672 is written ‖‖ ≡ T ⊥ ‖ . Other examples taken from Ch'in's work are

⊤⊥×	864
∣ ○ ‖‖ × ×	15,344
‖‖‖=‖‖‖⊥○⊥⊤=⊤	424,657,627
—⊤○×≡⊤	175,938
‖‖‖× ‖‖≡○	49,380

THE ZERO

It is in the *Shu-shu chiu-chang* that we find, for the first time in China, the zero in print.[3] The nine digits, together with the zero, form *a complete decimal place-value system*.[4] One strange use of the zero in Ch'in's work must be mentioned here. In Problem I, 7 (p. 41) we find (in the method of solution of the problem), 一萬二千○○一 (12,001), wherever the usual form should be 一萬二千一. In the commentary we read: "In all old books we find empty places; none of them uses a circle. Elsewhere this book does the same. . . ." Perhaps this is a gradual change in the direction of a modern place-value notation: 一二○○一.

Sometimes the character *k'ung* 空 (empty) is used for zero in the written language (for example, in II, 1).

LARGE NUMBERS

Only the characters *i* 億 (10^8) and *wan* 萬 (10^4) are used. For example, in VII, 5, 4億3千2百9十4萬9千8百2十5 = 4 *i* 3,294 *wan* 9,825 = 432,949,825. We also find such forms as 28,674 *i* instead of 2 *chao* 兆 8,674 *i*.[5]

NEGATIVE NUMBERS

Positive numbers (*chêng* 正) were probably represented by red counting rods, and negative numbers (*fu* 負) by black ones.

---

[3] See also the very important article by Yen Tun-chieh (2′).
[4] See Wang Ling (1). For information on the transmission of this system from China to India, the reader is referred to Needham (1), vol. 3, pp. 10 ff. There is also a short paper by Van Hee (4).
[5] On large numbers, see Needham (1), vol. 3, p. 87, and Ch'ien Pao-tsung (7′), p. 77.

This is clearly stated in the *Chiu-chang suan-shu*.[6] However, we are not entirely sure of the representation of positive and negative numbers on the counting board as Ch'in Chiu-shao uses them. As for equations, the positive or negative sign of the coefficients is indicated before (to the right of) the representation of the equation. For instance in III, 1 (p. 120) we find:

0	*shang* 商	*shang* is positive
40,642,560,000	*shih* 實	*shih* is negative
	fang 方[7]	
763,200	*ts'ung-shang-lien* 從上廉	*ts'ung* is positive
	hsia-lien 下廉[7]	
1	*i-yü* 益隅	*i* is negative

This is the representation of $-x^4 + 763,200x^2 - 40,642,560,000 = 0$. Sometimes the signs are given in the text. For example, on page 426 we find: 2,353,200 *kuan* 貫 *chêng* and 525 *liang* 兩 *fu*,[8] which means $+2,353,200$ and -525. The characters *chêng* and *fu* also appear in the diagrams, in front of the numbers. There is an example on p. 427: *fu* $525 = -525$; *chêng* $52,104 = +52,104$. This last representation is very modern, in the sense that the two characters serve the function of plus and minus signs.

Fractions

The name of fractions is *fên* 分; the numerator is called *tzŭ* 子 (son) and the denominator *mu* 母 (mother). Simple fractions such as 1/2, 1/3, 1/4, 3/4, and 2/3 have special names.

1/2 = *chung-pan* 中半 (literally, middle half)
1/3 = *shao-pan* 少半 (literally, diminished half)
2/3 = *t'ai-pan* 太半 (literally, greater half)
1/4 = *jo-pan* 弱半 (literally, weak half)
3/4 = *ch'iang-pan* 强半 (literally, strong half)

These names are probably very old. Some of them (1/2, 1/3, 2/3) are used in the *Chiu-chang suan-shu*;[9] *t'ai-pan* and *shao-pan*

---

[6] Ch. 8, p. 129.
[7] The absence of these terms is indicated by the character *hsü* 虛, empty.
[8] *Kuan* and *liang* are units of measure.
[9] See Yushkevitch (4), p. 19.

occur even in Han bronze inscriptions.[10] The use of these special names was very common; we find them also in the *Sun Tzŭ suan-ching*, the *Chang Ch'iu-chien suan-ching* and in many non-mathematical texts.[11] Ch'in Chiu-shao (p. 349) himself defines the meaning of *shao-pan*, *chung-pan*, and *t'ai-pan*.[12] These special fractions are given in IV, 4; V, 2; VI, 5; VII, 5.

The general representation of fractions is by the expression: "*y fên chih x, y* 分之 *x*" for $\frac{x}{y}$.[13] Often the "specification" is also given in the fraction; for example, in IX, 1, "3,056 *t'ao* 套 4 *fen t'ao chih* 1"[14] means $3,056\frac{1}{4}$ packages. In the diagrams the fractions are indicated as in Diagram 2.

子				numerator	1	
母	‖‖			denominator	4	

$$\boxed{\begin{array}{cc} 子 & | \\ 母 & ‖‖ \end{array}} = \boxed{\begin{array}{cc} \text{numerator} & 1 \\ \text{denominator} & 4 \end{array}} = \frac{1}{4}$$

Diagram 2

The operations on fractions are the traditional ones. There is information on special terminology available to the reader.[15] In some respects there seems to have been no progress since the time of the *Chiu-chang suan-shu*; for instance, the common denominator is still the product of the denominators (no use is made of the lowest common denominator.)

Decimal Fractions

Decimal numbers are called *shou-shu* 收數 in the *Shu-shu chiu-chang*, where the following definition is given: "those the tail positions [*wei-wei* 尾位] of which contain *fên-li* 分釐." *Fên-li* is a general indication for the decimal fractions (*fên* = 1/10 *ts'un* 寸[16]; *li* = 1/10 *fên*); the "tail positions" are the decimals

[10] See L. S. Yang (1), p. 78, notes 13–15.
[11] Li Yen (6'), vol. 1, p. 19 ff and p. 19, note 2. Li Yen gives a full account of all these special names for fractions. We have to restrict ourselves to those that are given in the *Shu-shu chiu-chang*. See also Li Yen (17'), p. 54.
[12] See also Sung Ching-ch'ang's commentary, p. 119.
[13] The literal meaning is "*x* of *y* parts."
[14] *T'ao* means a package.
[15] See Needham (1), vol. 3, pp. 31 f; Yushkevitch (4), pp. 13 ff; and the study in Mikami (3).
[16] A *ts'un* or inch is 1/10 foot.

(or numbers after the decimal point) as shown in Diagram 3 (*SSCC*, p. 89).

○○○○○＝‖‖⊥丅	
收　　　　小　小	＝0.00002366
數　　　　分　秒	

Diagram 3

The origin of the decimal fractions can be found in the metrological system. In Ch'in Chiu-shao's work (VI, 8) this system is given as follows:

fên 分	0.1
li 釐	0.01
hao 毫	0.001
ssŭ 絲	0.0001
hu 忽	0.00001
wei 微	0.000001
ch'ên 塵	0.0000001
sha 沙	0.00000001
miao 渺	0.000000001
mang 莽	0.0000000001
ch'ing 輕	0.00000000001
ch'ing 清	0.000000000001
yen 煙	0.0000000000001

In VI, 8 an amount of 24,706,279.3484670703125 *wên* is given as follows: 24,706 *kuan* 279 *wên*[17] 3 *fên* 4 *li* 8 *hao* 4 *ssŭ* 6 *hu* 7 *wei* 0 *ch'ên* 7 *sha* 0 *miao* 3 *mang* 1 *ch'ing* 2 *ch'ing* 5 *yen*.[18] This method of writing decimal fractions differs from our modern system by the names given to the different decimal places. However, this was so only in the written language; in the notational diagrams

---

[17] 1 *kuan* = 1,000 *wên* (money).

[18] This seems to be the most extensive list of metrological decimals given in old Chinese mathematical books. For the development of these decimal fractions, see Needham (1), vol. 3, pp. 82 ff and Wang Ling (1). This list of decimals is also given in Li Yen (17'), p. 54. The best study in Chinese is Ch'ien Pao-tsung (2'), pp. 14 ff (on Ch'in Chiu-shao, see p. 16). The same system up to *sha* 沙 (but with *hsien* 纖 instead of *ch'ên* 塵) is given in C. W. Mateer (1'), B, p. 59, showing that it was used well into modern times.

representing the different decimal positions on a counting board, the decimals are not named.[19] The decimal point itself was unknown to Chinese mathematicians, but as all their problems were practical ones, it was not necessary. Above or below the unit place the measure is given, and this replaces our point. For example,

| | | | | | | | | | | | | means 1.1446154 days (p. 67)
日

○Ⅲ≡ⅢⅠ○○⊥‖⊥Ⅲ⊥ means 0.9340062736 inch
寸 (p. 97)

≡|≡|≡ⅢⅠ means 91.3134 degrees
度 (p. 97)

In our own decimal notation system we usually write, for example, 2.31 m. In the Chinese counting-rod system the notation is of the form 2m31. This seems to be the only difference between the two systems. To look in Chinese mathematics for a number like 2.31 would be futile, because all Chinese numbers are concrete numbers.

In some cases a centesimal system is used in metrology. The names of the centesimals in Ch'in's work are

fên 分	0.01
miao 秒	0.0001
hsiao-fên 小分	0.000001
hsiao-miao 小秒	0.00000001
wei-fên 微分	0.0000000001
wei-miao 微秒	0.000000000001.

This system is used for the division of time and degrees, as in II, 1:[20]

39 days 92 *k'ê* 刻 45 *fên* = 39.9245 days.
11 days 38 *k'ê* 20 *fên* 81 *miao* 80 *hsiao-fên* = 11.38208180 days.

[19] In our own modern arithmetic, one can read 2.34 either as "two point thirty-four" or as "2 units 3 tenths 4 hundredths." The difference is that our names for decimals denote place-value, whereas the Chinese do not. They are mostly words indicating small quantities or small things (dust, sand, smoke, and so on).
[20] A day is divided into 100 *k'ê*, a *k'ê* into 100 *fên*, and so on.

This metrological centesimal system becomes decimal when written in the counting-rod form. The counting-rod notation for 11 days 44 *k'ê* 61 *fên* 54 *miao* is

一 | 三IIII⊥ | 三IIII   or 11.446154 days.
日

The use of the zero in this system is important. In the written form, a missing decimal is indicated in several ways:

1. By means of the character *k'ung* 空 (empty): *k'ung ts'un* 寸

9,664 *fên* 40 *miao* = 0.96644 inch (in counting rods: 〇≡丅⊥
IIII三).

2. By means of the character *wu* 無 (none. . .) (VI, 8): . . . 7 *wei wu ch'ên* 7 *sha wu miao* . . . = 0 . . . 7070 . . . .

3. In most cases the decimal places are omitted: 193,440 *fên* 26 *hsiao-fên* = 193,440.0026 (in the counting-rod system: 一Ⅲ≡

IIII三〇〇〇二丅).

This shows clearly that the idea of decimal fractions was fully developed in the counting-rod system. Indeed, it would have been impossible to make calculations on the counting board without this fully developed conception of place-value.[21] Often in the diagrams the units are given only once, but the numbers are arranged in such a way that the units are in the same row, as in Diagrams 4 and 5 (*SSCC*, pp. 85 and 81).

度 ≡	一 Ⅲ ⊥ Ⅲ 三 Ⅲ ≡ ‖ 一 Ⅲ 日 | = Ⅲ 度 〇 = Ⅲ ≡ Ⅱ 〇〇〇〇 Ⅲ 一 Ⅲ ≡ ‖ 一 Ⅲ 秒	31.1373433213 128 0.2397 0.0003133213

Diagram 4

[21] Sometimes all the decimals (or centesimals) are named in the diagrams. On p. 83 we find:

‖ ≣ 度	3.9
—丁≣ 日	16.9
╎ ═○≣‖‖	120.95
‖ ≣ 度	3.9
—丁≣ 日	16.9
═○≣‖‖○‖‖‖≣	2044.055

Diagram 5

As for operations with decimal numbers, Ch'in Chiu-shao designed his procedures to yield the decimal fraction wanted, even in the cases of root extraction and the solving of numerical equations.

In VII, 4 the root of the equation $16x^2 + 192x - 1863.2 = 0$ is given as $x = 6.35$.[22] The conversion of a fraction into a decimal fraction was also known, as can be seen in II, 3:

$$365 \tfrac{79}{325} = 365.2431.$$

Metrology
"The history of Chinese decimal measures has not yet been thoroughly investigated."[23] There are some general studies available to the reader,[24] but this chapter is restricted to the information found in Ch'in's work.

DIVISION OF TIME
The division of a day is centesimal in II, 1:

○ ○ ○ — ╎ ═ ‖ ⊥ ‖
度　分　秒　小　小
　　　　　分　秒

However, as a rule, only the units that replace the decimal point are indicated.

[22] See Chapter 13.
[23] Yushkevitch (4), p. 21.
[24] The best is Wu Ch'êng-lo (1'). Material in this chapter was also drawn from Yang K'uan (1'), and from the appendix on metrology in Vogel (2), which was useful for comparison.

1 day = 100 *k'ê* 刻

1 *k'ê* = 100 *fên* 分.[25]

In I, 6 we find the division of a day into twelve hours (*shih-ch'ên* 時辰), which are designated by the Twelve Terrestrial Branches (*chih* 支).[26] The beginning of an hour is indicated by the character "*ch'u* 初," the middle by "*chêng* 正," and the end by "*mo* 末." For example,

shên-mo 申末 = 5:00 P.M. (*shên* = 3:00–5:00 P.M.)

yu-ch'u 酉初 = 5:00 P.M. (*yu* = 5:00–7:00 P.M.)

wei-chêng 未正 = 2:00 P.M. (*wei* = 1:00–3:00 P.M.).[27]

For the meaning of time in general these cycles are used: the year cycle, the monthly cycle, and the sexagenary cycle.[28] Their incommensurability forms the base for the calendrical problem.[29]

DEGREES (*TU* 度)

The subdivision of degrees is centesimal. In II, 4 we find

1 degree (*tu*) = 100 *fên* 分

1 *fên* = 100 *miao* 秒

1 *miao* = 100 *hsiao-fên* 小分

1 *hsiao-fên* = 100 *hsiao-miao* 小秒

1 *hsiao-miao* = 100 *wei-fên* 微分

1 *wei-fên* = 100 *wei-miao* 微秒.

A circle was divided into $365 \frac{1}{4}^{\circ}$.

LINEAR MEASURES

There are two kinds of linear measure:

1. Distance measures:

[25] The division of a *fên* is also centesimal.

[26] For a table of them, see Couvreur's *Dictionnaire classique de la langue chinoise,* p. 1,071.

[27] This system was very old in China. See Needham (1), vol. 3, pp. 396 ff, for additional bibliography.

[28] The last was an artificial cycle, like our week. See Chatley (1) and (2).

[29] See Part VI.

1 *li* 里 (mile) = 360 *pu* 步 (paces)[30]
1 *pu* = 5 *ch'ih* 尺 (feet)[31]

2. The linear measures for rolls of cloth:
In V, 7 we see that

1 *p'i* 疋 (roll) = 4 *chang* 丈
1 *chang* = 10 *ch'ih* 尺 (feet)
1 *ch'ih* = 10 *ts'un* 寸 (inch)[32]
1 *ts'un* = 10 *fên* 分, and so on, the usual decimal division.[33] The
relationship between these systems is made clear in III, 8 and
IV, 1:

1 *pu* = 50 *ts'un;*
hence 1 *li* = 360 *pu* = 18,000 *ts'un*.[34]

AREA MEASURES
In V, 1 and 2, we see that
1 *ch'ing* 頃 = 100 *mou* 畝
1 *mou* = 240 *pu* 步 (square), or 4 *chüeh* 角
1 *chüeh* = 60 *pu*.

MEASURES OF CAPACITY AND WEIGHT
A *hu* 斛 is a volume measure for grain (I, 5; VI, 5). The official
weight measure for rice (*shih* 石) had a volume capacity of 83
shêng 升 (I,5), whereas in An-chi the equivalent volume is
given as 110 *shêng* and in P'ing-chiang as 135 shêng (I,5). In
VII, 5 the grain measure is called *tou* 斗; the offical *tou* was
equal in capacity to 83 *ko* 合, and others were equivalent to
110, 135, 115, 120, and 118 *ko*. Thus, 1 *hu* = 10 *tou* because
1 *shêng* = 10 *ko* (measures of capacity).

[30] In the *Chiu-chang suan-shu* a *li* was 300 *pu*. See Vogel (2), p. 139.
[31] This was not a general rule; in I, 3 the *pu*-ratio is 5.8 feet, whereas in
IV, 7 and V, 3 it is 5 feet. These ratios are indicated in the problem.
[32] Same division as in the *Chiu-chang suan-shu*. See Vogel (2), p. 139.
[33] For a special study of the foot measure in China, consult Yang K'uan
(1') (Sung dynasty, pp. 80 ff). The length of the standard foot was more
than 0.3 meter, according to excavated foot measures, but the length was
not always the same [see Yang K'uan (1'), pp. 81 f].
[34] As 1 *pu* = 5 *ch'ih* and 1 *ch'ih* = 10 *ts'un*, this agrees with 1 *pu* = 50 *ts'un*.

According to VI,4 a *tou* (peck) has the form of a truncated pyramid with a square base of side 7 *ts'un* 寸 (inches), an upper side of 9.6 *ts'un*, and a height of 4 *ts'un*. The volume is computed as follows:

$$V = \frac{1}{3} \times 4\,(7^2 + 9.6^2 + \sqrt{7^2 \times 9.6^2}\,)$$

$$= \frac{1}{3} \times 4\,(49 + 92.16 + 67.2).$$

1 *tou* = 277.813 *ts'un*³;

1 *hu* = 10 *tou* = 2,778.13 *ts'un*³

$\qquad\qquad$ = 2.77813 *ch'ih*³ (ft³).

This agrees approximately with VI,5, where 1 *hu* is equal to 2.5 *ch'ih*³.[35] It is striking that there is no clear distinction between measures of capacity and measures of weight. In fact they are very close to each other, as it is necessary only to multiply the volume by the specific gravity to get the weight. Even in the *Chiu-chang suan-shu*, the nature of the *hu* (grain measure) is not very clear; it is at the same time a measure of capacity and a measure of weight. The reason is clearly explained by Vogel.[36] A *hu* for rice has a capacity of 1.62 *ch'ih*³,[37] and a *hu* for corn 2.7 *ch'ih*³;[38] their weight is the same. This means that the specific gravities of rice and corn are in the inverse proportion 1.62 to 2.7; or $\gamma_r \times 1.62 = \gamma_c \times 2.7$ (γ_r = specific gravity of rice; γ_c of corn). Consequently, the *hu* is a specific measure with constant weight and variable volume.[39]

The same confusion exists about the nature of the *shih* 石, which is both a dry measure for grain (a volume) and a weight. In the *Shu-shu chiu-chang*, V, 1, the decimal subdivisions of the *shih* are given as follows:

[35] In Han times the standard *hu* for rice was 1.62 *ch'ih*³, according to the *Chia-liang hu* 嘉量斛 standard measure (literally, good measure *hu*) of Liu Hsin. See Needham (1), vol. 3, p. 100; Vogel (2), p. 140; Ch'ien Pao-tsung (7'). p. 38.

[36] Vogel (2), p. 141.

[37] This is the capacity of Liu Hsin's standard *hu*, already mentioned.

[38] This agrees with the capacity of the *hu* as stated in VI, 4.

[39] Vogel's interpretation of the *hu* (p. 140) seems somewhat confusing. The *hu* was not a constant measure of capacity.

1 *shih* = 10 *tou*
1 *tou* = 10 *shêng*
1 *shêng* = 10 *ko*
1 *ko* = 10 *shao* 勺
1 *shao* = 10 *miao*
1 *miao* = 10 *ts'o* 撮
1 *ts'o* = 10 *kuei* 圭.[40]

The ambivalent nature of the *shih* is obvious from Problem VI, 9. There it is given that we have 1,534 *shih* of rice, but when checking a sample, we find that some other grain (apparently millet) is mixed among the rice. We take a sample of 254 grains, among which we find 28 grains of other grain. All the grains are measured by the rice ratio, which is 300 grains per *shao*. How much rice and other grain is there? How much rice do we need to replace the other grain?[41]

If rice and the other grain had the same specific gravity, we would have:

$$\frac{28 \times 1,534}{254} = 169 \text{ } shih \text{ of other grain;}$$

$$\frac{226 \times 1,534}{254} = 1,364 \text{ } shih \text{ of rice.}$$

Next the other grain is reduced by its density ratio, 50 (50 percent of rice):

169 *shih* of other grain × 0.5 = 84.5 *shih* of rice
$$+ \text{ } 1,364$$

1,448.5 *shih* of rice.

When all the grain was measured by the rice *shih*, we got 1,534 *shih*; the 169 *shih* of "rice-*shih*" grain have the same

[40] These subdivisions, beginning with the *ko*, were used as early as the *Sun Tzŭ suan-ching*, but the *miao* and the *ts'o* were inverted. See Needham (1), vol. 3, p. 85. At the time of the *Chiu-chang suan-shu* decimal subdivisions did not exist. See Vogel (2), p. 140.
[41] This is a shortened version of the problem, and the computations have been simplified.

weight as 84.5 of "rice-*shih*" rice. From this it is obvious that
a *shih* is a *volume*. In the *Shu-shu chiu-chang* the *shih* is used only
for cereals: rice, wheat, glutinous rice, glutinous corn, fermented rice, and for pulses and linseed. This confirms the
statement that the *shih* was a volume. On the other hand, as
any single *shih* standard was used only for one kind of cereal,
it can also be considered as a weight, as in any one case there
is an immediate relationship between volume and weight.

True weight was measured in *liang* 兩; in Ch'in's work it
is used for silk (*mien* 綿)[42] and raw cotton (*hsü* 絮), for gold and
silver, and in addition for frankincense and kneaded leaven.

The subdivision of weights is

1 *chin* 斤 = 16 *liang*[43]
1 *liang*   = 10 *ch'ien* 錢
1 *ch'ien* = 10 *fên* 分.

The subdivision of the *chin* (pound) into 16 *liang* occurs in
the *Chiu-chang suan-shu*;[44] the decimal subdivision of the *liang*
originated much later.

MONETARY UNITS

These are decimal: 1 *kuan* 貫 ("string of cash") is equal to
1,000 *wên* 文 ("cash"). Sometimes one finds notations of the
type "1 *kuan* 200."

---

[42] This *mien* is a kind of silk floss, and thus unwoven; woven silk was measured in rolls (*p'i*), explained under "Linear Measures."
[43] Only in V, 6 and VIII, 9.
[44] Vogel (2), p. 140.

III

Elementary Mathematical Methods

7

Arithmetic

This chapter on arithmetical methods will not explain the common arithmetical operations. The reader will find an explanation and discussion of terminology in Needham.[1] Ch'in never works out the fundamental operations step by step; even the process of square-root extraction is usually omitted.

Classification of Numbers

On pages 1–2, Ch'in Chiu-shao gives an elementary classification of numbers. He distinguishes among four kinds:[2]

1. *Yüan-shu* 元數[3] or whole numbers. The definition given is: "they are the numbers the 'tail positions'[4] of which are zero. The 'tail positions' are the decimal fractions."[5] After that, Ch'in provides some examples in which he uses this kind of number:[6] I,1 (moduli 1, 2, 3, 4); I,4 (moduli 12, 11, 10, 9, 8, 7, 6); I,5 (moduli 0.83, 1.10, 1.35, reduced to 83, 110, 135); I,8 (moduli 130, 110, 120, 60, 25, 100, 50, 20); I,9 (moduli 19, 17, 12).

2. *Shou-shu* 收數[7] or decimal fractions, "those the tail position of which contains *fên-li* 分釐."[8] The example given is I,2, where $365\frac{1}{4}$ is expressed as 365.25. However, in the problem itself, only $365\frac{1}{4}$ is used.[9]

---

[1] Needham (1), vol. 3, pp. 62 ff.

[2] The classification is given in connection with the *ta-yen* rule. Ch'in says: Set up all the *wên-shu* [problem numbers]. As for the names of the classes there are four ["*lei ming yu ssŭ* 類名有四...."].

[3] Modern: *chêng-shu* 整數.

[4] *Wei-wei* 尾位. One can also translate this as "tail figures," since *wei* is used many times in this sense in the *Shu-shu chiu-chang*.

[5] As an operational definition this makes sense.

[6] Chao Ch'i-mei 趙琦美, who edited the work in 1616, added the table of contents (see his postscript to the *Shu-shu chiu-chang*, p. 473). The names of the examples given by Ch'in do not agree with the names in the table of contents.

[7] Modern: *hsiao-shu* 小數 [see Li Yen (6), p. 129].

[8] See Needham (1), vol. 3, p. 86.

[9] Ch'ien Pao-tsung (2), p. 69, also reckons I, 5 (moduli 0.83, 1.10, 1.35)

3. *T'ung-shu* 通數 or fractions, that is, "all the numbers that have numerator and denominator"; for example, I,2 (moduli $365\frac{1}{4}$; $29\frac{439}{940}$).

4. *Fu-shu* 復數 or "multiples of 10^n," "are those the tail figures of which are 10, 100, 1,000, and so on." Examples are I,3 (moduli: 138,600, 146,300, 192,500, 184,800); I,7 (moduli: 300, 250, 200); I,6 (moduli: 300, 240, 180).

Some Notes on Fundamental Operations

For *multiplication* three different terms are used: *ch'êng* 乘,[10] *yin* 因, and *shêng* 生; the last is used only to express the multiplication of coefficients in equations when applying the so-called Horner method. In the counting-rod system *hsiang-ch'êng* 相乘 means mutual multiplication of the numbers in the same column $(a \times b \times c \times d)$; *hu-ch'êng* 互乘 is a kind of cross-multiplication, one number in the left column being multiplied by all the numbers in the right column except the one immediately opposite.

$a \times b' \times c' \times d'$
$a' \times b \times c' \times d',$

and so on.

As an example, Problem VII,5 shows columns as in Diagram 6.

5	2
7	3
5	3
1	1
tzŭ 子	*mu* 母

Diagram 6

The text says: "with the *mu* of the right column 'cross-multiply' [*hu-ch'êng*] the *tzŭ* of the left column. For the first you get 45 [that is, $5 \times 3 \times 3 \times 1$], for the second 42 [that is, $7 \times 2 \times 3 \times$

among the *shou-shu*, and although he is correct, this is not given as an example in the Chinese text.

[10] For a possible explanation of this term, see Needham (1), vol. 3, p. 63.

1], for the third 30 [that is, $5 \times 2 \times 3 \times 1$], for the last 18 [that is, $1 \times 2 \times 3 \times 3$]."

Tzŭ and *mu*, usually translated as "numerator" and "denominator," have the general meaning "factor";[11] other terms for factor are *ch'êng-lü* 乘率 and *yin-lü* 因率.

There are two kinds of division. The first, indicated by the term *ch'u* 除, is the common form of division;[12] the other one is indicated by *ju . . . êrh i* 如 . . . 而一. In the *Shu-shu chiu-chang*, for a division with decimal fractions in the quotient the term *ch'u* is used; divisions leaving no remainder and divisions the remainder of which is added to the quotient in the form of a fraction (in fact also leaving no remainder) are indicated by *ju . . . êrh i*. For example:

(III,9) $617 : 6 \qquad = 102\frac{5}{6}$

(IV,1) $1,476,720 : 4 \qquad = 369,180$

(IV,1) $462,213,360 : 920 \qquad = 502,405\frac{19}{29}$

(IV,9) $224,400 : 300 \qquad = 748.$

In all these divisions *ju . . . êrh i* is used.[13] Sometimes, where we would use division, but where only the remainder is to be used subsequently, a subtraction is indicated. The expression is "*man x ch'ü chih*" 滿x去之, "subtract the whole *x*'s from it (that is, subtract *x* as many times as possible)." The remainder is called *pu-man* 不滿, the incomplete part.

The remainder of a division is occasionally given as a fraction of the quotient.

---

[11] They are the factors of a division, and hence the special meaning.

[12] The terminology is the same as in the *Chiu-chang suan-shu*. See Needham (1), vol. 3, p. 65, and Vogel (2), p. 111; the usual expression is: *i a ch'u b*, 以 *a* 除 *b*, "by *a* divide *b*." In fact *ch'u* means "to subtract." Successive subtraction of the same subtrahend is the same operation as division.

[13] Vogel, in (2), p. 111, gives the right explanation: "*Shih ju fa êrh i* 實如法 而一 means, the dividend joins the divisor and makes one; the quotient is thus an entity, identical with a fraction." (Der Dividend kommt zum Divisor und (bildet) 1. Der Quotient ist also eine zusammengehörende Einheit, die mit einem Bruch identisch ist.) In II, 1 we find the expression, "*a ju b ch'u chih, a* 如 *b* 除之 "(divide *a* by *b*); the expression *ju. . . . êrh i* seems to be a shortened formula for *a ju b ch'u chih êrh i, a* 如 *b* 除之而一, "divide *a* by *b*, but make it [the quotient] one [unite it]."

(III,7)   3,200 : 11

3,200 $= 11 \times 29 + 1$, or $3,200 = 11 \times 29\frac{1}{11}$.

(III,9)   617 $= 6 \times 102\frac{5}{6}$.

The method for division by a decimal number can be illustrated by an example (IX,8): 48,249,600 : 402.08 = 120,000.

The problem is set up so that the first figure of the divisor is below the first figure of the dividend; this is indicated in the quotient by means of zeros.

D	48,249,600	D	48,249,600
d	402.08	d	40,208
Q		Q	00,000

The division is performed as follows:

D		R	8,041,600
d		d	40,208
Q	100,000		

The divisor is moved back one rank:

R	8,041,600
d	40,208
Q	120,000

In our system the divisor is multiplied by 10^n so that there is no longer a decimal part:

48,249,600 : 402.08 = 4,824,960,000 : 40,208.

These systems are essentially the same.

Powers are indicated by means of the following terms:

1. The square of a number is *tzǔ-ch'êng* 自乘 (literally, "multiplied by itself").[14] Sometimes the term *tzǔ-chih* 自之, short for *tzǔ-ch'êng-chih*, is used (for example, in II,8). *Mi* 冪, superficial measure, is also used to indicate squares of a number (II,7; II,9; III,1; IV,2).

----

[14] Needham (1), vol. 3, p. 65, says that *ch'êng-fang* 乘方 is the term current in Sung times, whereas *tzǔ-ch'êng* 自乘 is a modern term; however, the last is the usual term in the *Shu-shu chiu-chang*.

2. The cube is indicated by *li-fang* 立方[15] or by *tsai-tzŭ-ch'êng* 再自乘. The latter term is somewhat confusing; literally it means "multiplied by itself twice": $a \times a \times a = a^3$ (III,3).

Another term is *"liang-tu tzŭ-ch'êng"* 兩度自乘 or "two times multiplied by itself" (see III, 5).

3. Other powers are indicated in the same way; for instance, "multiplied by itself three times" (*san-ch'êng-fang* 三乘方) means to the fourth power.[16]
For root extraction as the solution of an equation, see the chapter on "Numerical Equations of Higher Degree."

Auxiliary Operations
Doubling and halving, although not considered as special operations, are used several times in Ch'in's work. Doubling (*pei* 倍) appears in IV,1 and in the expression *tuo i pei x* 多一倍 x "one more than double *x*" (I,2);[17] *pan* 半, or *chê-pan* 折半 (I,2) means "to halve"; the expression *sun ch'i pan pei* 損其半倍 means "diminish by half" (I,1).

Rounding off a number is indicated by *chiu-chin* 就近 ("to approximate"), as in the expression *chiu-chin shou miao wei i fên* 就近收秒爲一分, "round off the decimal *miao* and make 1 *fên*" (II,3).[18] Another expression is *ho sun i* 合損益 ("to join a diminution or an augmentation") (VI,4). To cast out decimals is indicated by the term *ch'i* 棄 (II,1), reducing, by the term *yüeh* 約. For reduction of decimals there is the expression *t'ung-wei* 通爲 (II,8; II,9) as in *t'ung-wei fên* 通爲分, "reducing to *fên*," and *chan-wei* 展爲, as in II,6: 20 *ts'un chan-wei* 2 *ch'ih*, 20寸展爲2尺, "reduce 20 inches to 2 feet."

The greatest common denominator is called *têng* 等, *têng-shu* 等數, or sometimes *tsung-têng* 總等 (I,2). The process for finding the greatest common denominator is not indicated; in

---

[15] Literally, an upright square. This is also the geometrical term for the cube.
[16] And not the third power, as one might expect.
[17] The literal meaning of the characters is "twice as much as *x*," but the interpretation given is the only one that makes mathematical sense. The original text may have read *"pei x tuo i* 倍 x 多一" or something similar.
[18] For the meaning of *miao* and *fên*, see Chapter 6.

all probability it was the process of continued division, because this was already known in the time of the *Chiu-chang suan-shu*.[19] Mutually divisible numbers are called *chi-ou t'ung-lei* 奇偶同類 and mutually indivisible numbers *chi-ou pu-t'ung-lei* 奇偶不同類. *Chi* means odd, *ou* means even. As we shall discuss later, the meaning of the phrase is generalized to "numbers having a common divisor." Ch'in does not use the lowest common denominator. For the common denominator in fractions the product of the denominators is used, as was done earlier in the *Chiu-chang suan-shu*.[20]

Proportions

In IX,7 three kinds of proportions are indicated by special names:[21]

1. *Fan-chui ch'a* 反錐差 (inverted-wedge difference). In the *Shu-shu chiu-chang*, it is explained that "if we have a continual decrease of 1, 2, 3 it is like an upright wedge" (p. 152). An inverted wedge has the proportions 3, 2, 1 (Figure 5). In IX, 7 it is said that taxes of 1 percent, 2.5 percent, and 3 percent are to be paid in the proportion of an "inverted wedge";

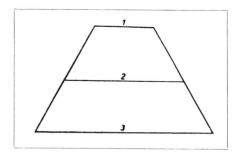

Figure 5

[19] See Vogel (2), p. 8, and Yushkevitch (4), p. 20. This method is the same as the Euclidean algorithm. Ch'in Chiu-shao makes use of this algorithm for solving the congruences in his *ta-yen* rule. See Chapter 17.
[20] See Vogel (2), p. 8.
[21] See Li Yen (17'), p. 54.

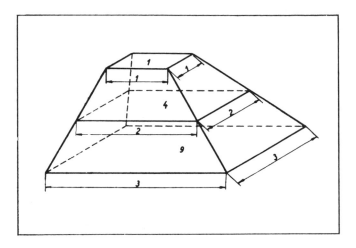

Figure 6

this means that for 3 parts there is a 1 percent tax, for 2 parts, 2.5 percent, and for 1 part, 3 percent.

2. *Fang-chui ch'a* 方錐差 (squared-wedge difference). The wedge differences are 1, 2, 3; the squares are 1, 4, 9 (Figure 6).

3. *Chi-li ch'a* 蒺藜差 (*chi-li* difference). In his *Botanicon Sinicum*, E. Bretschneider gives the following description of the *chi-li*: "This plant creeps on the ground. It has small leaves. The fruit is provided with *three* prickly horns."[22] From the last characteristic the proportion 1, 3, 6 may have been derived.

Other special indications of proportions, from V, 1, are

10 *fên-wai ch'a* 1 十分外差一, which means $\dfrac{a}{b} = \dfrac{10+1}{10} = \dfrac{11}{10}\left(=\dfrac{99}{90}\right)$

10 *fên-nei ch'a* 1 十分內差一, which means $\dfrac{a}{b} = \dfrac{10}{10-1} = \dfrac{10}{9}\left(=\dfrac{100}{90}\right)$

Many problems in the *Shu-shu chiu-chang* are solved by means of proportions. All the methods for direct and inverse proportions are known (V,8; VI,9; IX, 3, and so on); proportional divisions are used in VI,7 and VII, 6. In V,2 simultaneous equations are used:

[22] Bretschneider (1), vol. 2, p. 60. The *chi-li* is the *Tribulus terrestris* of the genus *Zygophyllacae*.

$$\frac{x}{y} = \frac{3}{4}$$

$$\frac{y}{z} = \frac{1}{3}$$

$x + y + z = 72,516,255.$

In V,1 the proportions $a/b = 11/10$ and $b/c = 11/10$ are given; the proportion $a/c = (11 \times 11)/(10 \times 10)$ is calculated. For solving complex proportions on the counting board a special configuration was used by the Chinese; it was called *yen-ch'ih* 雁翅, "wing of the wild goose." If the proportions $a/b = n_1/m_1$, $b/c = n_2/m_2$, $c/d = n_3/m_3$, $d/e = n_4/m_4$ are given, and the value of e is known, what is the value of a?

The configuration on the counting board is

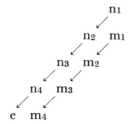

$$a = e \times \frac{n_1 \times n_2 \times n_3 \times n_4}{m_1 \times m_2 \times m_3 \times m_4}.$$

For instance:

$$\frac{a}{b} = \frac{3}{13}; \quad \frac{b}{c} = \frac{2}{84}; \quad \frac{c}{d} = \frac{15}{3.5}; \quad \frac{d}{e} = \frac{6}{7.2};$$

$e = 91,728.$

What is the value of a (IX, 3)?

The disposition on the counting board is

$$\frac{a}{e} = \frac{a}{b} \times \frac{b}{c} \times \frac{c}{d} \times \frac{d}{e} = \frac{3 \times 2 \times 15 \times 6}{13 \times 84 \times 3.5 \times 72}$$

$$a = \frac{3 \times 2 \times 15 \times 6}{13 \times 84 \times 3.5 \times 72} \times e = \frac{3 \times 2 \times 15 \times 6}{13 \times 84 \times 3.5 \times 72} \times 91,728 = 180.$$

The same method is used in IX,4.

Problem IX,5 will provide a more detailed explanation.

"According to the rates of a granary, from 7 *shih* of glutinous grain [in the husk] (*a*), they produce 3 *shih* of [unhusked] glutinous rice (*b*). 1 *tou* of glutinous rice (*b*) is exchanged for 1 *tou* 7 *shêng* wheat (*c*); 5 *shêng* wheat (*c*) are exchanged for 2 *chin* 4 *liang* kneaded leaven (*d*); 11 *chin* kneaded leaven (*d*) are exchanged for 1 *tou* 3 *shêng* fermented glutinous rice (*e*). Now they have [a total quantity of] 1,759 *shih* 3 *tou* 8 *shêng* of glutinous rice (*a+e*). As for the producing grain (*a*), the produced rice (*b*), the exchanged wheat (*c*), the kneaded leaven (*d*), and also the rice from the remaining grain after fermentation (*e*), we have to determine exactly the corresponding quantities."

Diagram of the Original Numbers					Diagram of the Converted Numbers			
	D	C	B	A				
1				7a				7a
2			10b	3b			10b	3b
3		5c	17c			20c	17c	
4	11d	2.25d			33d	9d		
5	13e				91a			

Diagram 7   Diagram 8

In Diagrams 7 and 8,

Column (*hang* 行)   A=*yu* 右 (right) ; B=*fu* 副 (second);
C=*tz'ŭ* 次 (next) ; D=*tso* 左 (left).

Row (*wei* 位)   1=*shang* 上 (above) ; 2=*fu* (second);
3=*chung* 中 ; 4=*tz'ŭ* (next) ; 5=
hsia 下 (below).

Arithmetic

The procedure says: "Put $7a$ and $3b$ in column A, ranks *1* and *2*; next put $10b$ and $17c$ in column B, ranks *2* and *3*; next put $5c$ and 2 *chin* 4 *liang* in column C, ranks *3* and *4*; next put $11d$ and $13e$ in column D, ranks *4* and *5* [see Diagram 7]. After arranging them according to their kind in a 'goosewing' configuration, examine the 2 *chin* 4 *liang* of column C. As for this 1/4 *chin,* reduce both ranks of column C to denominator 4 and add numerator 1 to C,4. Rank *3* of this [column] becomes 20, and rank *4* becomes 9.[23] Next investigate D,5. This *e* is of a different kind from *a;* we bring it into line and change it into *a;* we change it according to the $7a=3b$ of the opening sentence in the problem. Multiply $13e$ by 7; the answer is 91 in D,5. This is *a;*[24] but multiply $11d$ by $3b$, and it becomes 33 *chin d* in D. We get the numbers of the Diagram of Converted Numbers [see Diagram 8]." Next the following multiplications are made:

$33 \times 20 =$ *660*
$\qquad 660 \times 10 = 6,600$
$\qquad\qquad 6,600 \times 7 = 46,200.$

Put these numbers in the original places.

$3 \times 17 = 51$
$\qquad 51 \times 9 = 459$
$\qquad\qquad 459 \times 91 = 41,769.$

	D	C	B	A
Diagram of the Reconciled Numbers (*Ho-t'u* 合圖)				
1				46,200
2			6,600	3
3		660	51	
4	33	459		
5	41,769			

Diagram 9

[23] Or 2 *chin* 4 *liang* = 2 1/4 *chin* (1 *chin* = 16 *liang*) = 9/4; 5 = 20/4. As these numbers are only proportional values, the denominators may be omitted.

[24] The symbol *e* is the same unit as *b;* $7a = 3b$, hence $13b = (7a \times 13)/3 = 91a/3$. In *D, 4,* 11 is converted to 33/3; as these are proportional numbers, the denominators are omitted.

Arrange them as the numbers of the *Ho-t'u* [Diagram of Reconciled Numbers; see Diagram 9].

"Then investigate the four columns of the *Ho-t'u*. Multiply the opposite numbers by each other to bring them into agreement (*ho* 合). The entries A, *1* and D, *5* have no paired [number]. Treat them directly as rates[25] and put them in column A: *1* is 46,200, the rate of *a; 2* is 19,800, the rate of *b; 3* is 33,660, the rate of *c; 4* is 15,147, the rate of *d; 5* is 41,769, the rate of *a'*. Add the rates of *1* and *5* to get 87,969. This is the divisor (*fa-lü* 法率). Now find the greatest common divisor of all the

Diagram of the Rates		
	A	
	46,200	*a 1*
	19,800	*b 2*
175,938*	87,969	33,660 *c 3*
(*shêng*)	(*fa-lü*)	15,147 *d 4*
		41,769 *a' 5*

Diagram 10

*number of the problem (*wên-shu* 問數)

$$19,800 = 6,600 \times 3$$
$$33,660 = 660 \times 51$$
$$15,147 = 33 \times 459$$
$$87,969 = 46,200 + 41,769.$$

924.00 *shih a*	8,128,335,600
396.00 *shih b*	3,483,572,400
673.20 *shih c*	5,922,073,080
302.94 *chin d*	2,664,932,886
835.38 *shih a'*	7,348,754,322
3	
7	87,969 (divisor)
250,614	
(358.02 *shih e*)	

Diagram 11

[25] The text says *Ming chih* 命之, "name them...."

rates; you get 1; reduce them and you have only the original numbers.[26] This is the Diagram of the Rates [Diagrams 10 and 11]."

$$\frac{46,200 \times 175,938}{87,969} = 92,400$$

$$\frac{19,800 \times 175,938}{87,969} = 39,600$$

$$\frac{33,600 \times 175,938}{87,969} = 67,320$$

$$\frac{15,147 \times 175,938}{87,969} = 30,294$$

$$\frac{41,769 \times 175,938}{87,969} = 83,538.$$

The quantity a' is again converted into b. As $a/b = 3/7$, we have $(835.38 \times 3)/7 = 358.02$ *shih*.

Rationale: The proportions of the values (for the same quantity) are

$$\frac{a}{b} = \frac{3}{7}$$

$$\frac{b}{c} = \frac{17}{10}$$

$$\frac{c}{d} = \frac{9}{20}$$

$$\frac{d}{a'} = \frac{91}{33}$$

Thus

$$\frac{a}{a'} = \frac{3 \times 17 \times 9 \times 91}{7 \times 10 \times 20 \times 33} = \frac{46,200}{41,769},$$

and $a + a' = 175,938$.

Hence

$$a = \frac{175,938 \times 46,200}{87,969} = 92,400,$$

$$a' = \frac{175,938 \times 41,769}{87,969} = 83,538.$$

[26] This means that this reduction was a prescription of the general algorithm.

The *rule of three* is used in VII,8: "Seven men make 8 bows in 9 days. In how many days do 225 men make 10,000 bows?" This rule was already known in the time of the *Chiu-chang suan-shu*. In VII,6 we read: "Thirty feet of a foundation are built by 7 men in 2 days. How many men are required for building 3,570 feet in 3 days?"

Calculation of Interest

In Problem IX, 8 we have: Principal K + Interest I = 160,832 *wên* (K'). The interest is calculated at a rate of 2.2 *fên* 分 monthly for a time t of 464 days. Find the original capital. If $K' = K + I$, the formula used by Ch'in is

$$K = \frac{300\,K'}{pt + 300},$$

where p = 2.2 *wên*/100 *wên* 文 = 0.022 *wên/wên* and

$$t = \frac{\text{number of days}}{30} = \frac{464}{30} = \text{number of months.}$$

This formula can be derived from

$$K' = K \times \frac{p}{10} \times \frac{t}{30} + K,$$

where $p/10$ = the rate given in *wên*: in the problem the rate is given in *fên* (1 *wên* = 10 *fên*), and

$$p = 2.2\ \textit{wên}/100;$$

$$K' = K \times \frac{pt}{300} + 1 = K \times \frac{pt + 300}{300}.$$

Hence

$$K = \frac{300K'}{pt + 300}.$$

In Problem VI,8, the initial capital amounts to 500,000 *kuan* and is lent out at a rate of 6.5 percent monthly; the monthly interest is added to the capital, but 100,000 *kuan* is paid back every month. How many payments will be required to discharge the debt, and how much is the last payment?

The problem is solved simply by calculating the interest and adding it to the capital. After that the amount is diminished by amounts of 100,000 *kuan*. Thus,

$$500,000 + \frac{500,000 \times 6.5}{100} = 532,500$$

$$532,500 - 100,000 = 432,500$$

$$432,500 + \frac{432,500 \times 6.5}{100} = 460,612.5$$

$460,612.5 - 100,000 = 360,612.5$, and so on.

Mikami[27] says: "In solving this problem Ch'in merely calculates the remainders after each payment; he does not use any method of general treatment, although he employs equations of higher degrees in other places. He was probably unaware of the general formula that gives the sum of numbers in a geometrical progression."

Yen Tun-chieh[28] gives the following general solution to the problem: if $P=$ the capital, $r=$ the percentage, and $Q=$ the monthly repayment, then

$$P(1+r)^6 - Q[(1+r)^5 + (1+r)^4 + (1+r)^3 + (1+r)^2 + (1+r)] = k,$$

and the last repayment $L = k(1+r)$. Ch'in Chiu-shao, who did not know geometrical progressions, was unaware of the method for calculating $\sum_{n=5}^{1}(1+r)^n$.

[27] Mikami (1), pp. 72 f.
[28] Yen Tun-chieh (2'), pp. 110 f.

8

Geometry

Chinese geometry cannot be compared with Greek geometry, because the Chinese did not have the slightest conception of deductive systems. All we find in their mathematical handbooks are some practical geometrical problems concerning plane areas and solid figures. The more theoretical geometry had not developed beyond its embryonic form in the propositions of the Mohist Canon (*Mo-ching*, fourth century B.C.); and, as Needham pointed out, "their deductive geometry remained the mystery of a particular school and had little or no influence on the main current of Chinese mathematics."[1] The mathematical genius of the Chinese was pragmatic in nature, and for this reason their geometry was restricted to problems of land surveying and the capacity of various vessels. Moreover, Chinese geometry was algebraic in approach, and many problems were reduced to algebraic equations;[2] this chapter, however, will emphasize the purely geometrical aspects of the problems.

The problems in the *Shu-shu chiu-chang* display these specific features:

1. *The Pythagorean theorem,* known in China from early times, is used in several places in Ch'in's work.[3]

2. *Similar triangles* provide a basis for the solution of several geometrical problems, and in fact for the solution of all the so-called trigonometrical problems, as in II,6; II,7; II,8; and others.

3. *Plane figures:* Needham lists all the figures known by Chinese mathematicians.[4] Here we shall treat only the more special

[1] Needham (1), vol. 3, p. 94, where a very good evaluation of the *Mo-ching* geometry is given. The recent publication by Kao Hêng (1') is a revised version of a draft completed in 1944.

[2] A monograph on Chinese geometry has been published by Hsü Ch'un-fang (2').

[3] For a general account, see Needham (1), vol. 3, pp. 95 ff, and Hsü Ch'un-fang (2'), pp. 1 ff.

[4] Needham (1), vol. 3, p. 98. This list is interesting also from the termino-

plane areas dealt with in Ch'in's work. The third chapter is devoted to *"t'ien-yü* 田域,*"* or surveying problems.

4. *The value of* π is expressed as

a) the "old value" (*ku-lü* 古率): $\pi = 3$

b) the "precise value" (*mi-lü* 密率): $\pi = 22/7$

c) $\pi = \sqrt{10}$ (usually in the form $\pi^2 = 10$).[5]

In Ch'in's work, the value $\sqrt{10}$ occurs in Chinese mathematics for the first time since Chang Hêng 張衡 (A.D. 78–139). Although very precise values, such as

$$3.1415926 < \pi < 3.1415927,$$

had been worked out by Tsu Ch'ung-chih 祖沖之 (fifth century), they had been entirely forgotten by Ch'in Chiu-shao's time.[6]

Plane Figures

This section will examine in detail several problems from the *Shu-shu chiu-chang* that deal with areas and sides.

PROBLEM III, 1. THE AREA OF A "POINTED FIELD" (*CHIEN-T'IEN* 尖田)

Ch'in gives the formula thus: if x is the area (see Figure 7), x is a solution of[7] $-x^4 + 2(A+B)x^2 - (B-A)^2 = 0$, where

$$A = [b^2 - (c/2)^2] \times (c/2)^2;$$
$$B = [a^2 - (c/2)^2] \times (c/2)^2. \qquad (1)$$

We can prove the correctness of the formula as follows:

---

logical point of view.

[5] None of these values is very accurate. The value $\pi = 3$ was used as early as the time of the *Chou-pei suan-ching* and the *Chiu-chang suan-shu*. The value 3.14 was first used by Liu Hui 劉徽 (c. 250). For a history of the evaluation of π, the reader is referred to Needham (1), vol. 3, pp. 99 ff; Mikami (1), pp. 46 ff and pp. 135 ff; Mao I-shêng (1'); Chang Yung-li (1'); and Mikami (4). For the π value in Ch'in's work, see Mikami (1), p. 70; Mikami (4), pp. 197 f and Mao I-shêng (1'), p. 421.

[6] There is a special article in English on Tsu Ch'ung-chih by Li Yen (2).

[7] The formula is also given in Mikami (1), p. 71; Li Yen (8'), p. 202; Loria (3), vol. 1, pp. 158 f; Ch'ien Pao-tsung (2'), pp. 81 and 83.

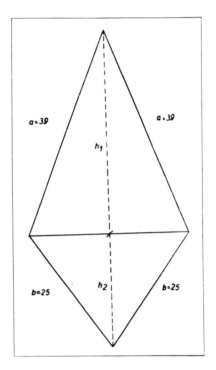

Figure 7

$h_1 = \sqrt{a^2 - (c/2)^2}; \quad h_2 = \sqrt{b^2 - (c/2)^2};$

$S = \dfrac{1}{2} c \times [\sqrt{a^2 - (c/2)^2} + \sqrt{b^2 - (c/2)^2}\,];$

or, making use of (1):

$$S = \frac{1}{2} c \times \left[\frac{\sqrt{A}}{c/2} + \frac{\sqrt{B}}{c/2} \right] = \sqrt{A} + \sqrt{B}. \tag{2}$$

The solution of Ch'in's equation, equalizing $x^2 = X$, is:

$$X = (B+A) \pm \sqrt{(B+A)^2 - (B-A)^2}$$
$$= (B+A) \pm 2\sqrt{(AB)}$$
$$x = \sqrt{(B+A) \pm 2\sqrt{(AB)}}. \tag{3}$$

$(2) = (3)$; indeed,

$$\sqrt{(B+A)+2\sqrt{(AB)}}=\sqrt{A}+\sqrt{B}$$
$$(B+A)+2\sqrt{(AB)}=(\sqrt{A}+\sqrt{B})^2$$
$$B+A+2\sqrt{(AB)}=B+A+2\sqrt{(AB)}.^8$$

As Loria says, the method is very complicated for such a simple problem. But the *"enigma cinese"* is not that the method used is indirect but the question of how the Chinese mathematicians were able to construct this kind of solution, which is indeed far too intricate for men who, as Loria and others assure us, did not know deductive logic. The probability that these algebraical constructions could have produced the correct formulae by trial and error alone is almost nil. Another important question is: What was the aim of these mathematicians, to find the area of the "field" or to construct equations in order to demonstrate their skill in solving them? In the pragmatic minds of Chinese mathematicians, it must have seemed senseless to construct equations that were not derived from everyday experience.[9]

PROBLEM III, 2. AREA OF A TRIANGLE

Hsü Ch'un-fang says that "for calculating the area of a triangle, ancient Chinese mathematicians originally knew only the method of taking half the product of base and height, but after Ch'in Chiu-shao's 'Area of a Triangle' appeared, they knew a new method of calculating the area from the three sides."[10]

The formula

$$S=\sqrt{s(s-a)(s-b)(s-c)}$$

first appeared in the *Geodaisia* of Heron of Alexandria (c. A.D. 50 or possibly A.D. 200).[11] We find this formula also in Al-

---

[8] As Ch'in considers only this possibility, the proof is restricted to positive values.
[9] The algebraical solution of Ch'in's equation will be treated in the chapter on equations of higher degree (Chapter 13).
[10] Hsü Ch'un-fang (2'), p. 40.
[11] Smith (1), p. 125; Struik (2), p. 68.

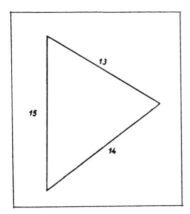

Figure 8

Karkhî's *Kâfî fîl Hisâb* (c. 1010–1016)[12] and in the work of Brahmagupta (c. 625).[13]

In Problem III,2 (see Figure 8) this formula has the form[14]

$$S = \sqrt{\frac{1}{4}\left[c^2a^2 - \left(\frac{c^2+a^2-b^2}{2}\right)^2\right]}.$$

PROBLEM III, 3. AREA OF A QUADRANGLE[15]

In Figure 9, the values of *a, b, c, d,* and *h* are known.

$$\triangle ABC = \frac{bh}{2};$$

$$\overline{DC}^2 = a^2 - h^2, \text{ and } DC = \sqrt{a^2 - h^2}.$$

$$\overline{BD} = b - \sqrt{a^2 - h^2};$$

$$\overline{AB}^2 = h^2 + \overline{BD}^2 = h^2 + \left(b - \sqrt{a^2 - h^2}\right)^2.$$

According to the formula given in III,2, we get

$$S = \sqrt{\frac{1}{4}\left[\overline{AB}^2 \times c^2 - \left(\frac{c^2 + \overline{AB}^2 - d^2}{2}\right)^2\right]} + \frac{bh}{2}.$$

[12] Ed. and trans. A. Hochheim, *Kâfi fil Hisâb des Abu Bekr Muhammed Ben Alhussein Alkarkhi* (Halle, 1878–1880). For the date, see Smith (1), vol. 1, p. 283.
[13] Yushkevitch (4), p. 155.
[14] Mikami (1), p. 70; Li Yen (8'), p. 202; Ch'ien Pao-tsung (2'), p. 83.
[15] See Ch'ien Pao-tsung (2'), p. 84.

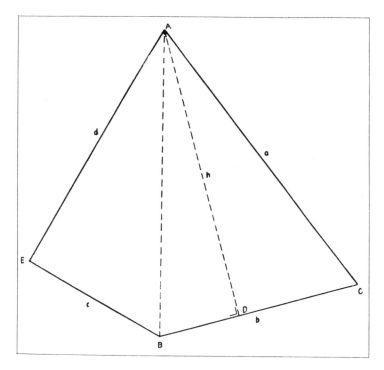

Figure 9

In this problem the explanation of the method is clearly stated and suggests that Chinese mathematicians were able to deduce their formulae.

PROBLEM III, 4

This is the same problem as III,3. In Figure 10,

$$\triangle ACD = \frac{ab}{2}$$

$$x^2 = a^2 + b^2$$

$$\triangle ABC = \sqrt{\frac{1}{4}\left[c^2 \times x^2 - \left(\frac{x^2+c^2-d^2}{2}\right)^2\right]}.$$

PROBLEM III, 6. EQUAL DIVISION OF A RUDDER-SHAPED FIELD

"A family possesses a field [Figure 11] which has the shape of a rudder; the south side is smaller and measures 34 *pu*; the

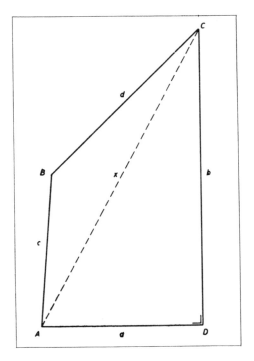

Figure 10

north side is larger and measures 52 *pu*; the perpendicular length [on the sides] is 150 *pu*. It belongs jointly to three brothers, who divide this field equally. On the side there is a footpath for those who wish to go in and out. Since the field is difficult to divide, an official is asked to divide it, assigning the southern parts to *A* and *B* and the northern one to *C*. Find the total area of the field, the equal part each man receives, and for each part the normal length and the larger and the smaller side."

The total area $= \dfrac{a+b}{2}\,h$.

Each part $S = \dfrac{1}{6}\,(a+b)h$.

$\triangle ADE \backsim \triangle ABC : DE/BC = x/h$, or

$$\dfrac{DF-a}{b-a} = \dfrac{x}{h},$$

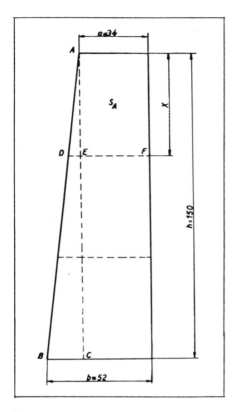

Figure 11

and

$$DF = \frac{b-a}{h} x + a.$$

Now

$$S_A = \frac{DF + a}{2} x = \frac{\left(\dfrac{b-a}{h} x + a \right) + a}{2} x,$$

$$S_A = \left(\frac{1}{2} \frac{b-a}{h} x + a \right) x,$$

from which

$$\frac{1}{2} (b-a)x^2 + ahx - S_A \times h = 0;$$

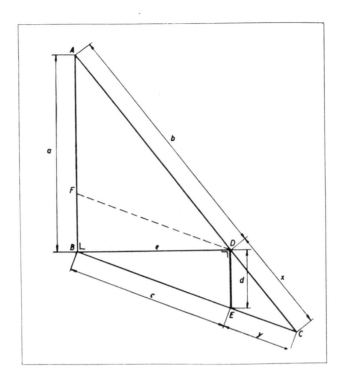

Figure 12

$$\frac{1}{2}(b-a)x^2+ahx-\frac{1}{6}(a+b)h^2=0 \text{ (Ch'in's formula).}[16]$$

PROBLEM III, 7

In Figure 12,[17]

AB $=a=16$ DE $=d=5$
AD $=b=20$ BD $=e$
BE $=c=13$ DC $=x$ and EC $=y$.

Draw DF parallel to BE; BF $=d$ and FD $=c$.

$\triangle$AFD$\backsim$$\triangle$ABC,

[16] See Mikami (1), p. 71, and Li Yen (8'), p. 203; a full explanation of the formula is given in Ch'ien Pao-tsung (2'), p. 80. According to Mikami, "This is the earliest instance of various problems of the same nature that attracted much attention among later Chinese and Japanese mathematicians."

[17] See Li Yen (8'), p. 200.

and $\dfrac{a}{a-d} = \dfrac{b+x}{b}$;

$$x = \dfrac{ab}{a-d} - b. \tag{1}$$

Thus $\dfrac{y+c}{c} = \dfrac{a}{a-d}$, or

$$y+c = \dfrac{ac}{a-d}. \tag{2}$$

Ch'in adopts a very intricate method, which proves clearly that he was not always sure of some simple geometrical properties. For equation (2) he gives:

$$(a-d)^2 \times e^2 \times (y+c)^2 = (d^2+e^2) \times a^2 \times e^2;$$

as $d^2+e^2=c^2$, we can reduce this to

$$y+c = \sqrt{\dfrac{a^2c^2}{(a-d)^2}} = \dfrac{ac}{a-d}.$$

After that, the areas are computed, but for the triangle DEC the formula $1/2\ dy$ is used. Ch'in knows the formula for finding the area of a triangle from its sides; what is the reason for this error? In the answer the area is given as $12\frac{8}{11}\ pu$, in the procedure as $14\frac{17}{22}\ pu$. This shows that Ch'in's text is mutilated here.

PROBLEM III, 8

In this problem all the formulae for circles are given:
1. The circumference C from the diameter D:

$C = \pi D$

(In fact, Ch'in's formula is $C = \sqrt{(\pi^2 D^2)}$. The reason for this complication is that here Ch'in uses $\pi^2 = 10$.)

2. The area A from the circumference C:

$A = \sqrt{D^2 C^2/16}$.

3. The diameter D from the circumference C:

$D = \sqrt{C^2/\pi^2}$ (here $\pi^2 = 10$).

4. The area of an annulus ("ring field," *huan-t'ien* 環田):
Ch'in gives the following equation, where x is the area of an annulus (see Figure 13).

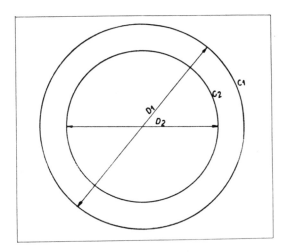

Figure 13

$$-x^4 + 2(A+B) \times x^2 - (A-B)^2 = 0.$$

$$A = \frac{C_1{}^2 \times C_1{}^2 / \pi^2}{16}; \quad B = \frac{D_2{}^2 \times D_2{}^2 \times \pi^2}{16}.$$

According to this equation,

$$x^2 = (A+B) - \sqrt{(A+B)^2 - (A-B)^2} = (A+B) - \sqrt{4AB}.$$

$$A = \frac{\pi^2 \times D_1{}^4}{16}; \quad B = \frac{\pi^2 \times D_2{}^4}{16}.$$

$$x^2 = \left[\left(\frac{\pi}{4} \times D_1{}^2\right)^2 + \left(\frac{\pi}{4} \times D_2{}^2\right)^2\right] - \sqrt{4 \times \frac{\pi^2 \times D_1{}^4}{16} \times \frac{\pi^2 \times D_2{}^4}{16}}$$

$$= \left(\frac{\pi}{4} \times D_1{}^2\right)^2 + \left(\frac{\pi}{4} \times D_2{}^2\right)^2 - 2 \times \frac{\pi D_1{}^2}{4} \times \frac{\pi D_2{}^2}{4}$$

$$x^2 = \left(\frac{\pi}{4} \times D_1{}^2 - \frac{\pi}{4} \times D_2{}^2\right)^2;$$

$$x = \frac{\pi}{4} D_1{}^2 - \frac{\pi}{4} D_2{}^2 = \frac{\pi}{4}(D_1{}^2 - D_2{}^2).$$

This is a much too intricate way of computing so simple a thing as the area of an annulus, which was known to the writers of the Nine Chapters.[18] Was Ch'in's purpose nothing

[18] See Needham (1), vol. 3, p. 25; and Vogel (2), pp. 15 ff.

more than to give a formula for this area? We are left with the impression that the geometrical aspect was only the basis for constructing algebraical equations.

PROBLEM VIII, 1

Find the side of a square camp in which 99 companies are encamped. Each company has at its disposal a square area with a side of 90 feet; between any two companies there must be a distance equal to the side of the same square (see Figure 14). We take the area for each company as being 4 times the occupied area; we put the rectangle BGFE below the rectangle DHJI. The width of the rectangle ABKE' is x and the length $x + 2$, giving an area $x(x + 2)$; the area is also equal to $4 \times 99 + 3$. This gives the equation

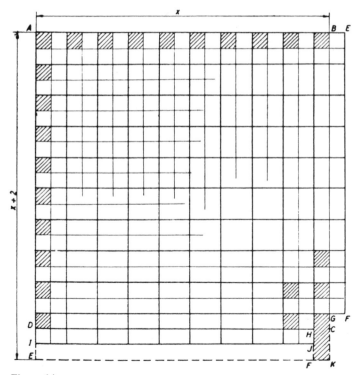

Figure 14

$x(x+2)=4\times99+3,$

or

$x^2+2x-399=0,$

giving the positive solution $x=19$.

PROBLEM III, 5. BANANA-LEAF-SHAPED FIELD[19] (SEGMENT FORMULA)

The "banana leaf" is formed by two secant circles of the same radius, as in Figure 15. According to Ch'in, the area A can be found by solving the equation

$$x^2+\left[\left(\frac{c}{2}\right)^2-\left(\frac{b}{2}\right)^2\right]x-10(b+c)^3=0,$$

where $x=2A$ and $10=\pi^2$. This formula cannot be correct. This can easily be proved by taking the extreme case where $b=c$. The intersection is a circle with diameter c, the area of which

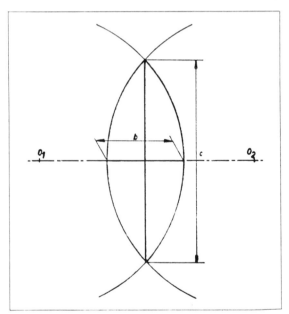

Figure 15

<hr />

[19] The formula is given in Mikami (1), p. 71, and Li Yen (8'), p. 202, but without comment. For a critical treatment, see Loria (3), vol. 1, pp. 161 f, and Ch'ien Pao-tsung (2'), pp. 84 f.

is $\pi(c/2)^2$. Taking b as equal to c in the above equation, we get $x^2=10(b+c)^3$ or $x^2=\pi^2(b+c)^3=\pi^2(2c)^3$.
Since $x=2A$, we get $4A^2=\pi^2(2c)^3$, or

$$A=\frac{1}{2}\pi\sqrt{(2c)^3}.$$

In order to make this formula correct, the constant part of the formula must be[20]

$$\frac{\pi^2}{4}\left(\frac{b+c}{2}\right)^4.$$

In the *Chiu-chang suan-shu* the segment formula is

$1/2(sp+p^2)$,
where $s=$chord, $p=$sagitta.

This formula is valid only for a segment equal to half a circle.[21] Chu Shih-chieh, in his *Ssŭ-yüan yü-chien* (1303), makes use of the formula[22]

$$A=\frac{1}{2}[bc+b^2+(\pi-3)(c/2)^2].$$

The second formula is more exact, and for $b=R$ and $c=2R$ we get the correct extreme value $\frac{1}{2}\pi R^2$.

In Ch'in's problem, the segment is very small ($b=34$ *pu*; $c=576$ *pu*); perhaps he was concerned merely with an approximate formula for small segments, since the formula in the *Chiu-chang suan-shu* gives very large relative errors. However, Ch'in's formula is less exact still, and no simpler.

The exact area of the segment is 6,512 *pu*^{2}. The formula in the *Chiu-chang suan-shu* gives the result 5,540.5 *pu*^{2}, or an error of approximately 15 percent. Chu Shih-chieh's formula gives the answer 5,673 *pu*^{2}. Ch'in's formula gives the less exact value of 5,435.5 *pu*^{2}.

I am not able to give the rationale of his formula.[23]

[20] Indeed $(2A)^2=\pi^2/4\times c^4$; thus $2A=\pi c^2/2$ and $A=\pi c^2/4$.
[21] See Vogel (2), p. 15; Mikami (1), p. 11. Yushkevitch (4), pp. 55 ff discusses the formula and gives the relative errors.
[22] See Hsü Ch'un-fang (2'), pp. 38 ff.
[23] For information on the evolution of the segment formula in China, the reader is referred to Li Yen (6'), vol. 3, pp. 254 ff.

Volumes

Ch'in gives formulae for several three-dimensional figures.

FRUSTUM OF A PYRAMID WITH A SQUARE BASE(VI,4)

The volume (Figure 16) is given as

$$I = \frac{1}{3} h \ (a^2 + ab + b^2).$$

This formula is stated in the *Chiu-chang suan-shu*, V,10. When b, h, and I are given, a is calculated by the following equation, in which $x = a$: $hx^2 + hbx + hb^2 - 3I = 0$.

In VI,4, $h = 16$, $b = 12$, and $3I = 4,167.2$. The equation becomes[24] $16x^2 + 192x - 1,863.2 = 0$. When a, b, and I are given, the formula for h is

$$h = \frac{3I}{a^2 + ab + b^2}.$$

All the formulae are correct.

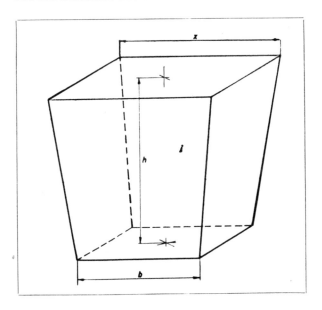

Figure 16

[24] For the solution, see Chapter 13.

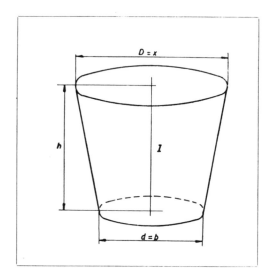

Figure 17

FRUSTUM OF A CONE

In VI,4 the following formula is used (see Figure 17):
$I = \frac{1}{4} h (D^2 + d^2 + D \times d)$.

Since $\pi = 3$, this is the same as

$I = \frac{1}{3} \pi h \times \frac{1}{4} (D^2 + d^2 + D \times d)$,

or $I = \frac{1}{3} \pi h (R^2 + r^2 + R \times r)$. In the same problem, h, d, and I are given. We find x [$=d$] by solving the equation $3hx^2 + 3hbx + 3b^2h - 4 \times 3I = 0$, or $\pi hx^2 + \pi hbx + \pi b^2h - 4 \times 3I = 0$. If $h = 12$, $b = 10$, and $3I = 4{,}167.2$, the equation becomes $36x^2 + 360x - 13{,}068.8 = 0$, from which $x = 14.7$.

TRUNCATED PYRAMID WITH RECTANGULAR BASE

In VII,5 (see Figure 18), the following formula is used:

$$I = \frac{[(2a+b) \times c + (2b+a) \times d] \times h}{6},$$

which can also be written

$$I = \frac{1}{3} h \left(ac + bd + \frac{bc + ad}{2} \right).$$

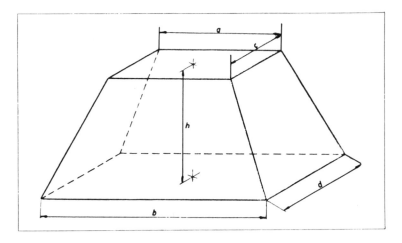

Figure 18

This formula is wrong; it should be

$$I = \frac{1}{3} h \left(ac + bd + \sqrt{ac \times bd} \right).$$

Ch'in's formula is valid only if the base is a square.[25]

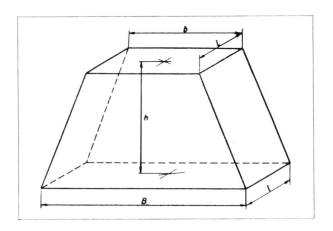

Figure 19

[25] See Hsü Ch'un-fang (3′), p. 76 and Vogel (2), p. 50.

PRISM WITH TRAPEZOIDAL BASE

In VII,1 and VII, 3 (see Figure 19), there is the formula

$$I = h\,\frac{(B+b)}{2} \times 1.$$

In VII,4 we find the derived formula:

$$h = \frac{I}{[(B+b)\times 1]/2}.$$

WEDGE WITH TRAPEZOIDAL BASE AND ONE SLOPING SIDE

The formula (see Figure 20) is

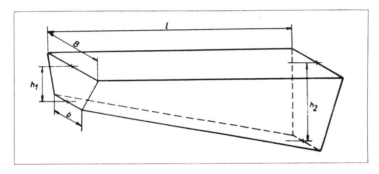

Figure 20

$$I = \frac{B+b}{2} \times \frac{h_1+h_2}{2} \times 1,$$

which is derived from the preceding one.

Here some interesting applications are given: In Problem II,6, ABCD is a conical vessel, filled with water to a height h. If this water is poured into a cylindrical vessel with diameter AB, what will be the height? In Figure 21,
h = the level of the water in the conical vessel, and
h' = the level of the same amount of water in the cylindrical vessel.

Ch'in gives the formula:

$$h' = \frac{h\,\{Hb\,[Hb+(a-b)\times h]+H^2b^2+[Hb+(a-b)\times h]^2\}}{3(aH)^2}$$

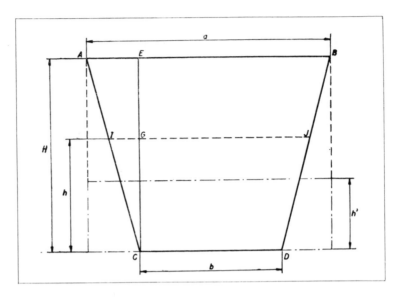

Figure 21

In this problem, $a=2.8$ feet; $b=1.2$ feet; $H=1.8$ feet; $h=0.9$ feet. Ch'in gives the correct solution: $h'=0.3$ feet. We can prove the correctness of the formula as follows: If $I_1=$ the water in the truncated cone, and $I_2=$ the water in the cylinder,

then

$$I_1 = \frac{1}{3}\pi h \left[\frac{c^2}{4} + \frac{b^2}{4} + \frac{c}{2} \times \frac{b}{2} \right],$$

and

$$I_2 = \pi \times \frac{a^2}{4} \times h'.$$

Since $I_1=I_2$, we have

$$\frac{1}{3}\pi h \times \frac{c^2+b^2+cb}{4} = \pi \times \frac{a^2}{4} \times h',$$

from which

$$h' = \frac{h(c^2+b^2+cb)}{3a^2}. \tag{1}$$

We eliminate c.

As $\triangle AEC \backsim \triangle IGC$, we get:

$$\frac{IG}{AE} = \frac{CG}{EC},$$

where

$$AE = \frac{AB - CD}{2},$$

and

$$IG = \frac{CG}{EC} \times \frac{AB - CD}{2}.$$

$$IJ = 2IG + b,$$

or

$$IJ = 2\frac{CG}{EC} \times \frac{AB - CD}{2} + b,$$

or

$$c = \frac{h}{H}(a-b) + b = \frac{h(a-b) + bH}{H}.$$

Substitution of c in (1) gives

$$h' = \frac{h\left\{\left[\frac{h(a-b)+bH}{H}\right]^2 + b^2 + \left[\frac{h(a-b)+bH}{H}\right] \times b\right\}}{3a^2}$$

$$= \frac{h\left\{\frac{[h(a-b)+bH]^2}{H^2} + \frac{b^2H^2}{H^2} + \frac{[h(a-b)+bH] \times bH}{H^2}\right\}}{3a^2}$$

$$= \frac{h\{[h(a-b)+bH]^2 + (bH)^2 + [h(a-b)+bH] \times bH\}}{3(aH)^2}$$

$$h' = \frac{h\{Hb[Hb+(a-b)\times h] + H^2b^2 + [Hb+(a-b)\times h]^2\}}{3(aH)^2}.$$

In Problem II,7, "A rain gauge has the form of a jar [*ying* 罌] [Figure 22]. The water level is at 1.2 feet. We use the precise value [*mi lü*] of π, or 22/7."

Mikami[26] gives a modern transcription of the problem, as follows: The diameters of mouth, middle section, and base are

[26] Mikami (1), p. 69.

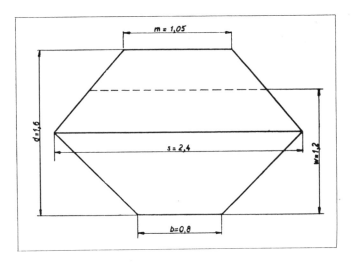

Figure 22

respectively m, s, and b. The depth of the jar is d, and the depth of the water w. To quote Mikami:

$$\text{``A}=(bs+b^2+s^2)\times \frac{d}{2}\times 11; \qquad\qquad A'=42$$

$$B = \frac{d}{2}\times m+\left[\left(\frac{d}{2}+w\right)-d\right]\times(s-m); \quad C=\frac{d}{2}\times s$$

$$D = (B^2+C^2+BC)\times 11;^{27} \qquad\qquad D'=\left(\frac{d}{2}\right)^2\times 42.$$

Hence:

$$(\text{lower half volume})=\frac{A}{A'}$$

$$(\text{diameter of water surface})=\frac{B}{d/2}$$

$$(\text{upper half volume})=\frac{D}{D'}$$

and the water contained would fill a cylindrical form with the section equal to the mouth of the keg and with the depth

$$\frac{(d/2)^2\times A+D}{m^2\times D'}\text{ ,,}$$

[27] The factor 11 is omitted in Mikami's text.

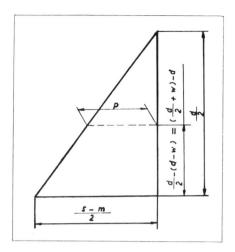

Figure 23

Mikami does not analyze the method, nor does he prove its correctness. If we represent the volume of the lower half by I_A, and the upper half by I_B (see Figure 23),

$$I_A = \frac{1}{3} \times \frac{22}{7} \times \frac{d}{2} \times \frac{1}{4}(b^2+s^2+bs) = \frac{11}{42} \times \frac{d}{2}(b^2+s^2+bs); \quad (1)$$

$$\frac{p}{d/2-\{[(d/2)+w]-d\}} = \frac{(s-m)/2}{d/2};$$

$$\frac{p}{d-w} = \frac{s-m}{d} \longrightarrow p = \frac{(d-w)(s-m)}{d}.$$

Here Ch'in makes a mistake, taking $[(d/2)+w]-d$ instead of $d/2-\{[(d/2)+w]-d\}$. But, as $[(d/2)+w]-d$ is the same as $d-w$, if $w=3d/4$ as in this case, the numerical value is not affected. However, the theoretical reasoning is wrong. If $y=$ the diameter of the water surface,

$$y=m+2p=m+\frac{2(d-w)(s-m)}{d}.$$

According to Ch'in we have

$$y = \frac{d/2 \times m + \{[(d/2)+w]-d\} \times (s-m)}{d/2}$$

$$= \frac{d/2 \times m}{d/2} + \frac{\{[(d/2)+w]-d\} \times (s-m)}{d/2}.$$

If we replace $\{[(d/2)+w]-d\}$ by $d-w$, we get

$$y=m+\frac{2(d-w)\ (s-m)}{d};$$

$$I_B=\frac{1}{3}\times\frac{22}{7}\times(d-w)\ (y^2+s^2+ys).$$

Making the same mistake as before, we get $d-w=d/4$, and

$$I_B=\frac{11}{42}\times\frac{d}{4}\times(y^2+s^2+ys).$$

Mikami gives the equation[28] $I_B=D/D'$, or

$$I_B=\frac{(B^2+C^2+BC)\times11}{(d/2)^2\times42}.$$

Since $B=y\times d/2$, we get

$$I_B=\frac{[(dy/2)^2+(ds/2)^2+(dy/2)\ (ds/2)]\times11}{(d/2)^2\times42}=\frac{(y^2+s^2+ys)\times11}{42}.$$

This formula is wrong, because there is no multiplication by the height $d/4$.

The correct solution of the problem is

$$x=\frac{I_A+I_B}{\frac{22}{7}\times\left(\frac{m}{2}\right)^2}=\frac{\frac{11}{42}\times\frac{d}{2}(b^2+s^2+bs)+\frac{11}{42}\times\frac{d}{4}(y^2+s^2+ys)}{\frac{22}{7}\times\frac{m^2}{4}}$$

$$=\frac{\frac{d}{6}\left[(b^2+s^2+bs)\times\frac{1}{2}(y^2+s^2+ys)\right]}{m^2}.$$

If we correct the formula for I_B, we get for x:

$$x=\frac{\left(\frac{d}{2}\right)^2\times A+D}{m^2\times D'}$$

$$=\frac{\left(\frac{d}{2}\right)^2\times(bs+b^2+s^2)\times\frac{d}{2}\times11+\left(\frac{d}{2}\right)^2(y^2+s^2+ys)\times11\times\frac{d}{4}}{m^2\times\left(\frac{d}{2}\right)^2\times42}$$

[28] Not in the text.

$$= \frac{\frac{11}{42} \times \frac{d}{2} \left[(b^2 + s^2 + bs) + \frac{1}{2} (y^2 + s^2 + ys) \right]}{m^2}.$$

The mistake is that the divisor m^2 should be $(22/7) \times (m/2)^2$. This does not seem to be the result of corruptions in the text, because the method (shu) fully agrees with the procedure (ts'ao).

Problem II, 8 says "We investigate the snow in order to prognosticate for the coming year. A wall is 12 feet high. A tree leaning against it is 5 feet away. The tip of the tree and the wall

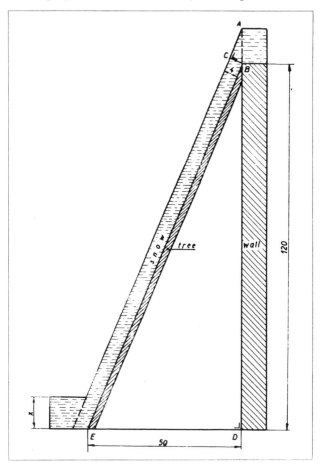

Figure 24

meet each other. The snow heaped up on the trunk is 4 inches deep. [See Figures 24 and 25.] The high layer is shallow; the layer on the soil is deep.[29] Find the depth of the snow on the soil.''

Sung Ching-ch'ang[30] uses the diagram shown in Figures 24 and 25,

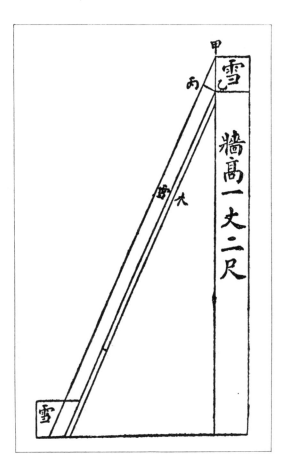

Figure 25. Problem II, 8 as reconstructed in Sung Ching-ch'ang (1'), p. 80

[29] That is, the snow on the trunk (measured perpendicularly to the trunk) is not as deep as the snow on the ground.
[30] (1'), p. 80.

where $\triangle ACB \backsim \triangle BDE$, and $AB/BE = BC/DE$.

$$AB = \frac{BC \times BE}{DE}, \quad BE = \sqrt{\overline{BD}^2 + \overline{DE}^2},$$

$$AB = \frac{4 \times \sqrt{120^2 + 50^2}}{50} = 10.4 \text{ inch or 1 foot } 4 \text{ } fên \text{ } 分.$$

Ch'in's formula is

$$AB = \sqrt{\frac{\overline{BC}^2 (\overline{BD}^2 + \overline{DE}^2)}{\overline{DE}^2}} = \sqrt{\frac{16(120^2 + 50^2)}{2,500}} = 10.4.$$

9

"Trigonometry"

Introductory Note

The title of this chapter must be qualified, because trigonometry was unknown in China.[1] However, Chinese mathematicians made use of a kind of primitive trigonometry in a method called *ch'ung-ch'a* 重差. Mikami says of this term that it "was originally intended to mean double or repeated applications of the consideration as met with in the chapter on the right triangle in the 'Nine Sections,' i.e., the double application of proportions."[2] Indeed, the Chinese *ch'ung-ch'a* concerns the properties of similar right triangles.[3] It is principally a kind of telemetry,[4] and

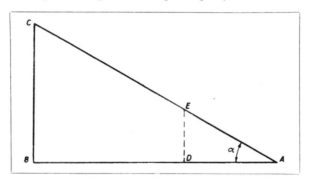

Figure 26

[1] For a general survey, see Needham (1), vol. 3, pp. 108 ff; Hsü Ch'un-fang (2'), pp. 14 ff; Li Yen (6'), vol. 3, pp. 191 ff. The term "practical trigonometry" is used in Wylie (1), pp. 173 and 179, and (2), p. 92. Although modern Western authors agree that this terminology is inaccurate [for example, Van Hee (6), p. 267; Needham (1), vol. 3, p. 31], some Chinese authors [among them Hsü Ch'un-fang (3'), p. 17] say that the *ch'ung-ch'a* method is the beginning (conceptually, not historically) of modern trigonometry.
[2] Mikami (1), p. 35. The last chapter of the *Chiu-chang suan-shu* is called *Kou-ku* 句股 (right triangles); the term *ch'ung-ch'a* was also used. See Vogel (2), pp. 90 ff.
[3] The *ch'ung-ch'a* is mentioned in the *Shu-shu chiu-chang* in IV, 3, 4, 7, 9 and VIII, 5.
[4] *Wang-yüan* 望遠, that is, "observing what is distant," is the usual term for this type of problem. See Needham (1), vol. 3, p. 104.

it is not unlike the methods applied in modern surveying, which, indeed, all rely on the use of properties of right triangles, expressed as a function of angles. Introducing specific names for these proportions (sine, cosine. . .) is not an improvement from the mathematical point of view, for "sine" and b/a are operationally equivalent. Goniometrical functions are nothing but proportions between the sides of a right triangle and can all be stated in this way. In Figure 26,

$BC = AB \times \tan a$
$BC = AB \times (DE/DA)$.

These formulae are equivalent.

Only from this point of view are the methods applied by the Chinese trigonometrical. The measurement of angles which characterizes modern trigonometry is entirely missing. The best definition of the *ch'ung-ch'a* is given by Needham: "it was a kind of empirical substitute for trigonometric functions."[5]

One is surprised by the large number of errors in Ch'in's "trigonometrical" problems. Of the problems translated in this section the greater part use the wrong methods; many of his mistakes are very gross, for example, in IV,3; IV,6; and VIII,5. These inaccuracies give the impression that Ch'in was not very familiar with this kind of problem and that he took his problems wholesale from the *Hai-tao suan-ching* 海島算經 (Sea island mathematical manual) of the third century,[6] or from a similar work now lost, without attempting to analyze them. Several common procedures are applied successively, without real insight into possible simplifications.

Problem IV, 1. Height and Distance of a Far-off Mountain

"There is a great mountain far from a town; we do not know its height or distance. On a plain outside the town there is a tree whose height is 23 feet. Suppose this is the first gnomon. Then

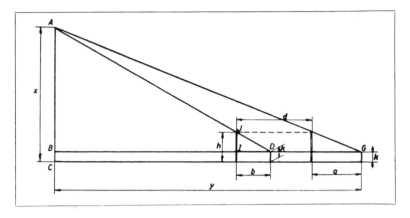

Figure 27

we set up the rear gnomon, of the same height as the tree. They
are 160 *pu* from each other. First we move back from the front
gnomon 30.9 feet; then we move back from the rear gnomon
31.3 feet. We look obliquely to the top of the mountain, and in
both cases it coincides with the end of the gnomon. The eye of
the observer is at a height of 5 feet [see Figure 27]. One *li* is 360
pu; one *pu* is 5 feet. Find the height and distance of the mountain."[7] According to Ch'in:

$$x = \frac{[(a-b)+d](h-k)}{a-b} = \frac{[(313-309)+8,200](230-50)}{313-309}$$

$$= 369,180 \ ts'un = 20 \ li \ 183 \tfrac{3}{5} \ pu.$$

$$y = \frac{[(a-b)+d](h-k)a}{(a-b)h} = 502,405 \tfrac{19}{23} \ ts'un = 27 \ li \ 328 \tfrac{67}{575} \ pu.$$

As already pointed out by Li Jui,[8] Ch'in is mistaken: his result
is not *x*, but *AB*.

The correct formula is

$$x = \frac{[(a-b)+d](h-k)}{a-b} + k.$$

[7] The symbols are those used by Pai Shang-shu (1′) in Ch'ien Pao-tsung
(2′), p. 291.
[8] Ibid., p. 291, n.2.

If we suppose that x is the same as AB, we can prove the formula of x as follows:

$$\frac{x}{h-k} = \frac{BD}{b} \longrightarrow BD = \frac{bx}{h-k} \qquad (1)$$

$$\frac{x}{h-k} = \frac{BG}{a} \text{ and } BG = BD + [d + (a-b)]$$

$$\longrightarrow BD = \frac{ax}{h-k} - [d + (a-b)] \qquad (2)$$

$$\frac{bx}{h-k} = \frac{ax}{h-k} - [d + (a-b)]$$

$$x = \frac{(h-k)[d + (a-b)]}{a-b}.$$

For the second formula:

$$\frac{y}{x} = \frac{a}{h} \longrightarrow y = \left(\frac{a}{h}\right)x = \frac{a(h-k)(d+a-b)}{h(a-b)}. \qquad (3)$$

Li Jui[9] and Shên Ch'in-p'ei[10] pointed out that the correct formula is

$$y = \frac{a(d+a-b)}{a-b}.$$

We get it by replacing h by $h-k$ in (3).

This problem first appears in the *Hai-tao suan-ching* (A.D.263).[11] The last chapter ("*Kou-ku*" 句股) of the *Chiu-chang suan-shu* is also known as "*Ch'ung-ch'a*" 重差 (double difference). Its subject is applying the properties of similar right triangles. As Needham points out, the *Hai-tao suan-ching* "seems indeed to have been intended as an extension of the last chapter of the *Chiu-chang.*"[12] Ch'in Chiu-shao's method (*shu* 術) is: "With the *kou-ku* method we solve it. With the *ch'ung-ch'a* we set it up." (*I kou-ku ch'iu chih. Ch'ung-ch'a ju chih* 以勾股求之. 重差入之.) From this it is obvious that Ch'in is applying a traditional method.

[9] Sung Ching-ch'ang (1′), p. 101.
[10] Ibid.
[11] There is a general description in Needham (1), vol. 3, p. 31; in Van Hee (6) and in Mikami (1), pp. 33 ff.
[12] Needham (1), vol. 3, p. 31.

In the first problem of the *Hai-tao suan-ching* the observer is supposed to have his eye in contact with the ground.[13] It is very likely that Ch'in wanted to make an obvious practical improvement, but he forgot to take into account this improvement when computing the height.[14]

Problem IV, 2. Measuring the Water in the Neighborhood of a Town

Question: "The tower of a city wall, close to water, has a perpendicular height of 3 *chang* [30 feet] (*c*).[15] They build a watchtower on it.[16] The foundation below slants to the base [and the difference in breadth from the upper part] is 2 feet. Below the rampart they set up a marking post in the sand, 1 *chang* 2 feet from the foundation (*d*); this exterior beacon protrudes from the ground 5 feet, and [its end] is on the same level as the base of the foundation. Now, the water has receded to an unknown distance. A man sticks a bamboo pole through the latticework of the railing at the top of the tower. Looking downward obliquely, he sees the end of a pole 4.15 feet long even with the edge of the water (*a*). The man's eye is at a height of 5 feet. [See Figures 28 and 29.] Find the depth to which the water has receded (*y*) and the slant distance along the uncovered bank between the foundation of the tower and the edge of the water."[17] Solution:

$$x^2 = \frac{c^2(\text{acd})^2[(\text{acd})^2 + (\text{ace})^2]}{[(\text{acd})(\text{bcd}) - (\text{acd})(\text{ace})]^2} \tag{a}$$

$$y^2 = \frac{e^2 c^2 (\text{acd})^2 [(\text{acd})^2 + (\text{ace})^2]}{(d^2 + e^2)[(\text{acd})(\text{bcd}) - (\text{acd})(\text{ace})]^2}. \tag{b}$$

The formulae Ch'in uses are extremely intricate. We can prove their correctness by reducing them to a simpler form:

---

[13] See Mikami (1), p. 35; Van Hee (6), p. 269.
[14] Ch'in's error is also discussed in Ch'ien Pao-tsung (2'), p. 98.
[15] The tower is a massive projection of the city wall.
[16] See Mao Yüan-i (1'), ch. 112, p. 13a.
[17] This account follows Pai Shang-shu's reconstruction, in Ch'ien Pao-tsung (2'), p. 292.

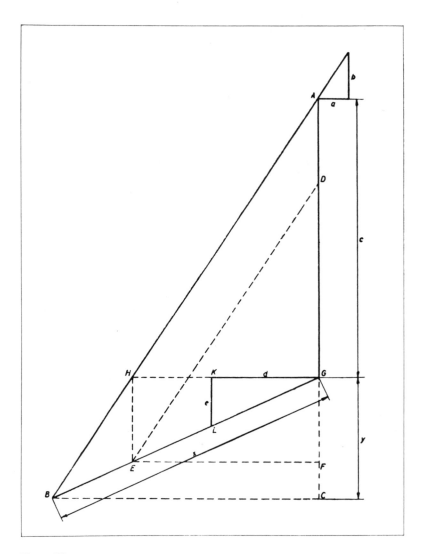

Figure 28

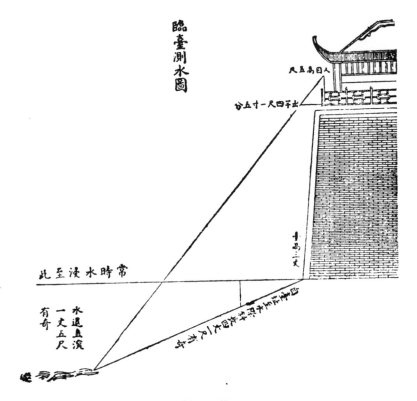

臨臺測水圖

尺五為目人

出平尺四寸一尺五分

此至浸水時常

高三丈

有奇

一丈五尺

水退直浸

退漲淺退法為水圖浸表目準

Figure 29. Illustration from the *SSCC*, p. 167

$$\frac{y}{BC} = \frac{e}{d}, \qquad dy = e \times BC;$$

$$\frac{AC}{BC} = \frac{b}{a}, \qquad b \times BC = a \times AC.$$

Or

$bdy = ae \times AC.$

Also,

$AC = y + c,$

$bdy = ae(y + c),$

$bdy = aey + aec,$

$(bd - ae)y = aec,$

or

$$y = \frac{aec}{bd-ae} \, . \tag{1}$$

As $x/y = GL/e$ we have

$$x = \frac{GL}{e} \times y = \frac{GL}{e} \times \frac{aec}{bd-ae} \, , \tag{2}$$

and $GL = \sqrt{d^2+e^2}$; thus

$$x = \frac{ac\sqrt{d^2+e^2}}{bd-ae} \, .$$

We can change formulae (a) and (b) as follows:

(a) becomes

$$x = \frac{c(acd)\sqrt{(acd)^2+(ace)^2}}{(acd)(bcd)-(acd)(ace)}$$

$$= \frac{\sqrt{(ac)^2(d^2+e^2)}}{bd-ae} = \frac{ac\sqrt{d^2+e^2}}{bd-ae} \, ; \tag{2}$$

(b) becomes

$$y = \frac{eac}{bd-ae} \, . \tag{1}$$

A reconstruction of the way these formulae may have been derived is given by Pai Shang-shu:

$$EF = \frac{ac}{b} \text{ (because } EF = HG)$$

$$FG = \frac{EF \times e}{d}$$

$$EG = \sqrt{\overline{EF^2 + FG^2}} \, .$$

$$\triangle GBC \infty \triangle GEF \longrightarrow \frac{GC}{GB} = \frac{FG}{EG} \text{ or } \frac{y}{x} = \frac{FG}{EG}$$

and $y = \dfrac{FG \times x}{EG}$. $\tag{1}$

$$\triangle ABC \infty \triangle DEF \longrightarrow \frac{BC}{AC} = \frac{EF}{FD} \text{ or } \frac{\sqrt{x^2-y^2}}{y+c} = \frac{EF}{FD}$$

$$FD = AG = c$$

$$\frac{\sqrt{x^2-y^2}}{y+c} = \frac{EF}{c} \longrightarrow \frac{\sqrt{x^2-y^2}}{y+c} = \frac{ac}{bc}$$

$$bc\sqrt{x^2-y^2} = ac(y+c). \tag{2}$$

From (1) and (2), we derive

$$[(acd)(bcd)-(acd)(ace)]x = c(acd)\sqrt{(acd)^2+(ace)^2}.$$

To avoid the square root, we square and get

$$x^2 = \frac{c^2(acd)^2[(acd)^2+(ace)^2]}{[(acd)(bcd)-(acd)(ace)]^2}.$$ (3)

Now

$$\triangle GBC \backsim \triangle GLK \longrightarrow y = \frac{e}{\sqrt{d^2+e^2}} \times x,$$

so that

$$y^2 = \frac{e^2}{d^2+e^2} \times x^2 = \frac{e^2c^2(acd)^2[(acd)^2+(ace)^2]}{(d^2+e^2)[(acd)(bcd)-(acd)(ace)]^2}.$$ (4)

The formulae for x^2 and y^2 as stated here give a wrong impression of the logical background of the method. The problem was solved step by step, making use only of similar right triangles and the Pythagorean theorem, as the preceding reconstruction shows. Formulae (3) and (4) give the impression that auxiliary values were not substituted, but from the reconstruction it is obvious that they were. However, Ch'in Chiu-shao was unable to see the method in a general way; in consequence he could not reduce the procedure to a simpler one. The main limiting factor in his thinking was lack of mathematical logic.

Problem IV, 3. Measuring a Stream from a Steeply Slanting Bank

"A division on the march comes upon a river. They have to calculate the length of a bamboo hawser needed to construct a pontoon bridge.[18] Now they hang down a rope to measure the slanting bank: its height is 30 feet. A man stands at the top and wishes to measure the width at the water's surface; he makes use of a bamboo pole 6 feet long as a perpendicular. He holds it horizontally 0.5 feet below his eye so that the end of the measuring rod touches his chin. When he looks out toward the other bank of the water, it coincides with the end of the rod. After

[18] There is a picture of a pontoon bridge in *San-ts'ai t'u-hui* (1607), *Kung-shih* 宮室, p. 36b [see Figure 64].

that, he looks toward the sandy border of this bank, which intersects the measuring rod 3.4 feet from its end. The eye of the observer is at a height of 5 feet [see Figures 30 and 31]. What is the width at the water's surface?"

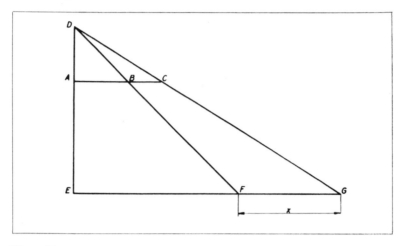

Figure 30

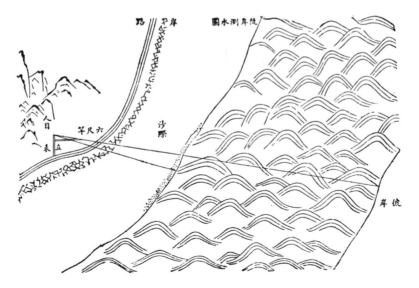

Figure 31. Illustration from the *SSCC*, p. 178

Solution: Ch'in's method is

$$x = \frac{(DE - DA) \times BC}{AD}$$

$$= \frac{34.5 \times 3.4}{5}.$$

Proof:[19]

In $\triangle DEG$, $\left.\begin{array}{l}\dfrac{DE}{DA} = \dfrac{EG}{AC}\\[2mm] \end{array}\right\}$ or $\dfrac{EG}{AC} = \dfrac{EF}{AB}$,

In $\triangle DEF$, $\left.\begin{array}{l}\dfrac{DE}{DA} = \dfrac{EF}{AB}\end{array}\right\}$

from which

$$\frac{EG - EF}{AC - AB} = \frac{EG}{AC} = \frac{DE}{DA}$$

$$\frac{x}{BC} = \frac{DE}{DA}$$

$$\frac{x}{3.4} = \frac{35}{0.5}.$$

Ch'in makes a mistake very common in his work when he states that $AE/AD = EC/AC$ . . . , giving the solution $x = 234.6$ feet instead of 238 feet.

Problem IV, 4. Observing a Square Town by Means of Gnomons

"We know neither the breadth nor the distance of an enemy town. We observe it from a wood at the foot of a mountain to the south of the town. At the edge of the wood there are two trees 160 feet apart, in line from north to south with the eastern edge of the town. To the east of the two trees we place two gnomons opposite each other, so that the gnomons and the trees form a square. [On the level] of the eye of the observer we connect them by a cord. From the rear gnomon at the east, [the observer] moves west 10 paces and observes the northeast corner of the town, which is 15 paces in from the forward gnomon at the east. Next he observes the southeast corner of the

[19] Pai Shang-shu (1'), p. 294.

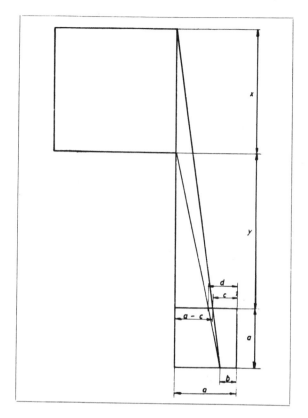

Figure 32

town, which is $48\frac{3}{4}$ paces[20] in from the forward gnomon at the east. [See Figures 32 and 33.] The *li* ratio is 360 *pu*. Find the side of the square and the distance."

Ch'in makes use of the formulae

$$x = \frac{a(a-c)}{c-b} \; ; \qquad y = \frac{a(a-d)}{c-b} .$$

As Li Jui[21] has pointed out, these formulae are incorrect. Sung Ching-ch'ang[22] makes the following corrections:

[20] Ch'in used the old expression *ch'iang-pan* 強半 for 3/4.
[21] See Pai Shang-shu (1'), p. 295.
[22] Sung Ching-ch'ang (1'), pp. 106 f. See also Pai Shang-shu (1'), p. 296.

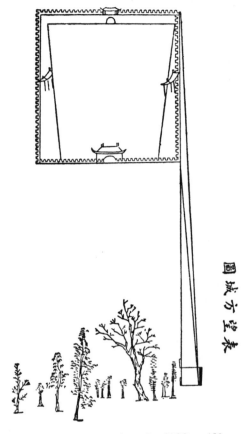

Figure 33. Illustration from the *SSCC*, p. 182

$$x = \frac{a(a-d)}{d-b} \, ;$$
$$y = \frac{a(a-c)(d-b) - a(a-d)(c-b)}{(c-b)(d-b)} \, .$$

Problem IV, 5. Measuring a Round Town from a Distance

"There is a round town of which we do not know the circumference and the diameter. There are four gates [in the wall]. Three *li* outside the northern [gate] there is a high tree. When

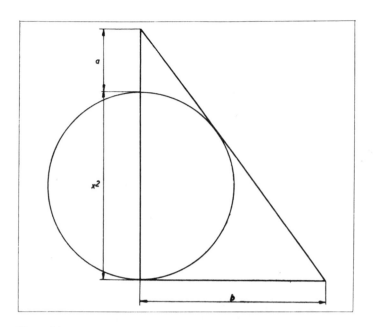

Figure 34

we go outside the southern gate and turn east, we must cover 9 *li* before we see the tree. [See Figures 34 and 35.] Find the circumference and the diameter [x^2] of the town ($\pi = 3$)."[23]

This explanation follows Pai Shang-shu.[24]

Ch'in says that x is the root of the equation:[25]

$$x^{10} + 5ax^8 + 8a^2x^6 - 4a(b^2 - a^2)x^4 - 16a^2b^2x^2 - 16a^3b^2 = 0. \qquad (1)$$

As $a = 3$ and $b = 9$, we get

$$x^{10} + 15x^8 + 72x^6 - 864x^4 - 11{,}664x^2 - 34{,}992 = 0.$$

[23] This problem is included in Mikami (1), p. 71; he quotes the famous equation of the tenth degree, but with the incorrect solution $x = 9$, instead of $x = 3$. The diameter is equal to x^2, not x. Loria, whose attitude toward Chinese mathematics was far from friendly, has a great deal to say about this problem in (3). His commentary, however, is worthless, since his only access to the text was Mikami.

[24] Pai Shang-shu (1′), pp. 296 ff.

[25] There are mistakes in the equation in Mikami's version (1), p. 72; he uses $7ax^8$ instead of $5ax^8$; $-4(b^2 - a^2)x^2$ instead of $-4a(b^2 - a^2)x^4$; and $-16a^2b^3$ instead of $-16a^3b^2$. All these mistakes are faithfully reproduced by Loria (3), p. 159. Yushkevitch (4), p. 72, gives the right form.

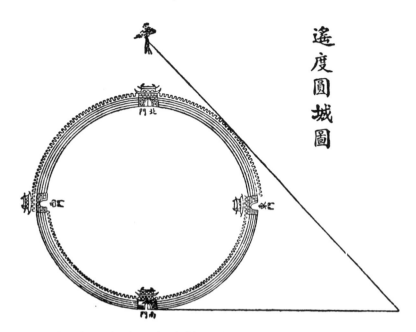

Figure 35. Illustration from the *SSCC*, p. 184

Ch'in gives only the root $x=3$, and for the diameter of the town $x^2=9$. If we take y as the diameter then we can reduce the equation of the tenth degree to one of the fifth degree:

$$y^5+15y^4+72y^3-864y^2-11{,}664y-34{,}992=0.$$

Li Jui[26] says: "In the Yüan period, of the 170 problems of Li Yeh 李治 in his *Ts'ê-yüan hai-ching* 測圓海鏡 (1248) there is only one that goes up to the sixth degree. As for myself I think this is still too intricate. This problem is not so difficult that it should go up to the tenth degree; there is no need for it. Therefore, I drew up another method . . . ." The method is given in Sung Ching-ch'ang.[27] The explanation is as follows:

Suppose that x is the diameter of the circle (see Figure 36).

[26] See Sung Ching-ch'ang (1'), p. 108.
[27] Ibid. pp. 108 f.

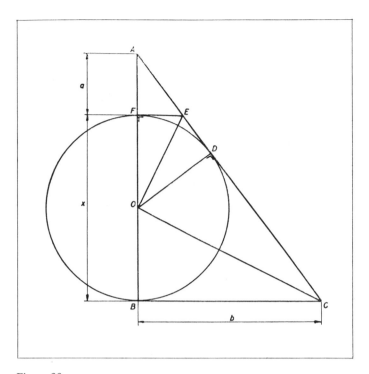

Figure 36

From $\triangle ABC \backsim \triangle ADO$ and $AC/AB=AO/AD$, we obtain

$$\frac{\sqrt{(x+a)^2+b^2}}{x+a}=\frac{(x/2)+a}{\sqrt{a(x+a)}}.$$

This gives the equation

$$x^4+2ax^3+a^2x^2-4ab^2x-4a^2b^2=0$$

or

$$x^4+6x^3+9x^2-972x-2,916=0$$

giving the solution $x=9$.

Li Jui's equation is of lower degree than Ch'in Chiu-shao's, but it can be reduced further. As $x=-a$ is impossible, we may divide the equation by $(x+a)$; we get

$$x^3+ax^2-4ab^2=0. \tag{2}$$

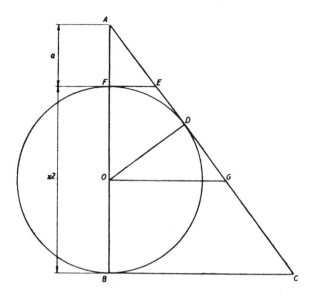

Figure 37

Shên Ch'in-p'ei,[28] taking the diameter$=x^2$, gives the equation $x^4(x^2+a)=4ab^2$. This is equation (2), replacing x by x^2.
Shên multiplies it by $(x^2+2a)^2$; the result is

$$x^{10}+15x^8+72x^6-864x^4-11,664x^2-34,992=0,$$

or Ch'in's equation, which is equivalent to Li Jui's.

Sung Ching-ch'ang says:[29] "This method of my master Shên
Ch'in-p'ei is based on the fourth problem of the *Pien-ku* 邊股
chapter and the *Ti-kou* 底句 chapter of the *Ts'ê-yüan hai-
ching*."[30] Problem 4 of Chapter 4 in the *Ts'ê-yüan hai-ching* gives
the solution[31] $x^3-ax^2-4ab^2=0$.

Ch'in's equation is $(x^2+2a)^2(x^3-ax^2-4ab^2)=0$.

[28] Ibid. pp. 109 f.
[29] Ibid. p. 298.
[30] *Pien-ku* is chapter 3; *Ti-kou* is chapter 4. In the *Ts'ung-shu chi-ch'êng*
edition of the *Ts'ê-yüan hai-ching* of Li Yeh, they are in part 1, p. 62 and p.
73.
[31] The problem is: "*B* goes outside the southern gate and walks straight
135 *pu. A* goes out through the northern gate and, going 200 *pu* eastward,
sees *B*." The problem in Chapter 3 is identical.

Pai Shang-shu, relying on Sung Ching-ch'ang, provides the following explanation: Sung says that, in order to derive the equation of the tenth degree, we have to take $x^2(x^2+2a)$ as *ching-lü* 徑率 and $b(x^2+2a)$ as *kou-lü* 勾率.

In Figure 37, $\triangle ADO \backsim \triangle AOG$; $AO/OG=AD/OD$, giving

$$\frac{(x^2/2)+a}{OG}=\frac{\sqrt{[(x^2/2)+a]^2-[x^2/2]^2}}{x^2/2}$$

or

$$x^2(x^2+2a)=4\sqrt{a(x^2+a)}\times OG \ (\textit{ching-lü}). \tag{1}$$

Also, $\triangle AOG \backsim \triangle ABC$; thus $AO/OG=AB/BC$, or

$$b(x^2+2a)=2(x^2+a)\times OG \ (\textit{kou-lü}). \tag{2}$$

Eliminating OG from (1) and (2), we get

$$\frac{x^2(x^2+2a)}{4\sqrt{a(x^2+a)}}=\frac{b(x^2+2a)}{2(x^2+a)} \tag{3}$$

or

$$\frac{x^4(x^2+2a)^2}{16a(x^2+a)}=\frac{b^2(x^2+2a)^2}{4(x^2+a)^2}.$$

If we reduce the denominator by $4(x^2+a)$, we get

$$\frac{x^4(x^2+2a)^2}{4a}=\frac{b^2(x^2+2a)^2}{(x^2+a)}.$$

This gives Ch'in Chiu-shao's equation.

Pai Shang-shu says: "Ch'in Chiu-shao does not take x as the diameter of the town, but notwithstanding the fact that it is more difficult represents the diameter by x^2. Or was his intention to construct an equation of higher degree to set a record? If so, he should not have reduced the denominators in (3), and he would have gotten an equation of the twelfth degree. Moreover he reduces only the denominators and not the numerators, which is difficult to explain. There is certainly some discrepancy between Sung's explanation and Ch'in's original method."[32]

[32] Pai Shang-shu (1') p. 299.

Pai Shang-shu proposes the following reconstruction of Ch'in's equation: $\triangle AOG \backsim \triangle AFE$, $AO/OG = AF/FE$ or

$$\frac{ab}{x^2+a}(x^2+2a) = 2a \times OG. \tag{a}$$

Multiplying (a) by (2), we get

$$b(x^2+2a)\left[\frac{ab}{x^2+a}(x^2+2a)\right] = 4a(x^2+a) \times \overline{OG}^2.$$

Subtracting the square of one-half of (1):

$$b(x^2+2a)\left[\frac{ab}{x^2+a}(x^2+2a)\right] - \left[\frac{x^2(x^2+2a)}{2}\right]^2 = 0,$$

which is equal to[33]

$$x^{10} + 5ax^8 + 8a^2x^6 - 4a(b^2-a^2)x^4 - 16a^2b^2x^2 - 16a^3b^2 = 0.$$

Problem IV, 6. Observing a Circular Camp of the Enemy[34]

"Near a river the enemy builds a circular camp of an unknown size. From the southern bank to a certain place the distance is 7 li (a). On this place we erect two gnomons, 2 pu apart (b). Keeping the western beacon and the north-south [diameter] of the enemy camp in line, the observer moves back from the western beacon 12 pu (c). He sights the eastern gnomon just on line with the edge of the enemy camp. For π we shall use the precise value $\left[\frac{22}{7}\right]$. The li-ratio is 360 pu. Find the circumference of the camp and the diameter."

According to Ch'in, if x is the diameter, the solution is obtained by

$$-\left[\frac{(c^2-b^2)}{4}\right]^2 \times x^4 + \left[a^2b^2 \times \frac{(b^2+c^2)}{2}\right]x^2 - (a^2b^2)^2 = 0, \tag{1}$$

or

$$-x^4 + \frac{8a^2b^2(b^2+c)^2}{(c^2-b^2)^2} \times x^2 - \frac{16(a^2b^2)^2}{(c^2-b^2)^2},$$

---

[33] For the solution of this equation, the reader is referred to Chapter 13.
[34] This explanation follows Pai Shang-shu (1'), pp. 299 f.

or

$$-x^4 + 1{,}534{,}464x^2 - 526{,}727{,}577{,}600 = 0.$$

The solution is given as $x = 2\ li = 720\ pu$. This agrees with the equation, but it does not agree with the geometrical problem, as already pointed out by Sung Ching-ch'ang.[35] He gives the correct solution as follows (see Figure 38): $\triangle ABC \backsim \triangle AB'C'$ and $AC/BC = AC'/B'C'$; or, as $\overline{AC^2} = AD \times AF = (a+c)(a+c-x)$, we get

$$\frac{\sqrt{(a+c)\ (a+c-x)}}{x/2} = \frac{c}{b},$$

or

$$x^2 + \frac{4b^2(a+c)}{c^2} \times x - \frac{4b^2(a+c)^2}{c^2} = 0.$$

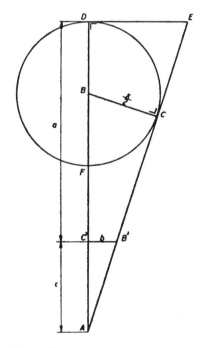

Figure 38

[35] Sung Ching-ch'ang (1'), p. 111.

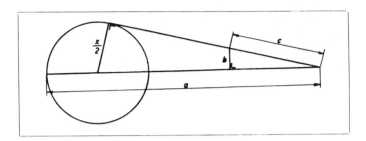

Figure 39

The solution is $x = 714 \frac{5004}{5131} \, pu.$

However, the correct solution does not remove the difficulties. In order to get Ch'in's solution, $x=720$, we have to change the problem as follows (see Figure 39):

$$\frac{x/2}{a-(x/2)} = \frac{b}{c}$$

$$x = \frac{2ab}{b+c} = 720.$$

This is a simple equation of the first degree.

If we accept this reconstruction, we can, according to Pai Shang-shu, arrive at Ch'in's equation as follows:
Since $\triangle ADE \backsim \triangle AC'B'$,

$$DE = \frac{ab}{\sqrt{c^2-b^2}};$$

and since $AE = \sqrt{\overline{AD}^2 + \overline{DE}^2}$, we get

$$AE = \sqrt{a^2 + \left(\frac{ab}{\sqrt{c^2-b^2}}\right)^2}.$$

From $\triangle ADE \backsim \triangle ACB$ we get $DE/AE = BC/AB$ and thus

$$\frac{(ab)/\sqrt{c^2-b^2}}{\sqrt{a^2 + [(ab)/\sqrt{c^2-b^2}]^2}} = \frac{x/2}{a-(x/2)},$$

which is equivalent to

$$-\left[\frac{(c^2-b^2)}{4}\right]^2 x^4 + a^2b^2 \frac{(b^2+c^2)}{2} x^2 - (a^2b^2)^2 = 0,$$

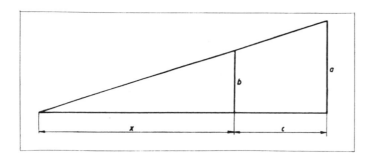

Figure 40

Figure 41. Illustration from the *SSCC*, p. 203

the only positive solution of which is

$$x = \frac{2ab}{b+c} .^{36}$$

[36] For the solution of this equation, see Chapter 13.

Problem IV, 7. Observing the Distance of the Enemy[37]
"An enemy army is at the foot of the North Mountain. We do not know its distance. On the flat plain we erect a gnomon 4 feet high. A man withdraws from the gnomon 900 paces [1 pace = 5 feet]. When he sights the foot of the mountain, it coincides with the top of the gnomon. The eye of the observer is 4.8 feet high. [See Figures 40 and 41.] Find the distance of the enemy army."

$$\frac{x}{x+c}=\frac{b}{a} \longrightarrow \frac{x}{x+c-x}=\frac{b}{a-b} \longrightarrow \frac{x}{c}=\frac{b}{a-b} \longrightarrow x=\frac{bc}{a-b}.$$

Problem IV, 8. Computing the Original Dimensions of an Ancient Pond
"There is an ancient pond in the shape of a circle within a square, overgrown and in ruins; only one corner is left. From the corner of the exterior square, measuring obliquely to the edge of the inner circle, the distance is 7.6 feet. [See Figure 42.] Using the old foundations, we wish to repair the pond. We wish to know the [size] of the circle and the square (x) and the diagonal of the square (y)."

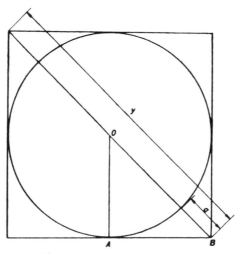

Figure 42

[37] See Ch'ien Pao-tsung (2'), pp. 95 f; Pai Shang-shu (1'), p. 301.

Ch'in gives the solution $\frac{1}{2}x^2-2ax-2a^2=0$. We can arrive at this equation by

$$\overline{OA}^2+\overline{AB}^2=\overline{OB}^2$$
$$\left(\frac{x}{2}\right)^2+\left(\frac{x}{2}\right)^2=\left(\frac{x}{2}+a\right)^2$$
$$x=2a(1+\sqrt{2})$$
$$y=x+2a=2a(2+\sqrt{2}).^{38}$$

Problem IV, 9. Observing a Pagoda by Means of a Gnomon

"There is a pagoda which leans to one side. We wish to change the central beam, but we do not know its height. At a distance of 60 feet from the pagoda there is a staff for banners, the height of which is also unknown. On the staff, 9.2 feet above the ground, a row of metal rings begins.[39] There are 14 rings, each 0.5 feet long. The lower edges of the rings are 2.5 feet from each other. We take the staff as a gnomon. The observer steps back from the staff 30 feet. He sights the top of the pagoda precisely in line with the top of the staff. After that he aims at the lower edge of the first of the nine horizontal wheels [just above the roof], and the line of sight coincides with the upper edge of the seventh iron ring on the staff. The eye of the observer is 4.8 feet above the ground. The central pillar rises 3 feet [above the roof] and terminates in a pinnacle. [See Figures 43 and 44.] Find the height of the pagoda, the height

---

[38] See Ch'ien Pao-tsung (2'), p. 80; Pai Shang-shu (1'), pp. 301 f.

[39] The text uses the very rare character "鈎," not listed in any of the great dictionaries. The author is much indebted to Professor A. Hulsewé (Leiden) for his suggestion about the meaning of this character, as well as for his kind assistance in elucidating the whole problem. These "iron rings" can be seen on a picture of a temple in Melchers (1), p. 58. Alternatively, Professor Nathan Sivin (Massachusetts Institute of Technology) suggests that the character in question is merely a local variant of *kou*, "hook," which it resembles closely in both form and sound (in view of its phonetic). Thus it would refer to the row of hooks rather than rings to which the banner was attached. It would be more natural to give the dimension of hooks as "length" and their lower edges as (literally) "haunches."

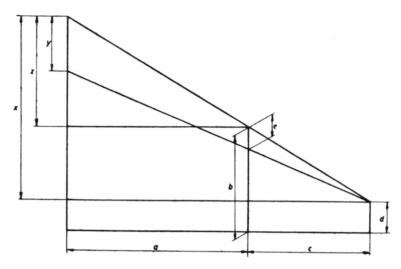

Figure 43

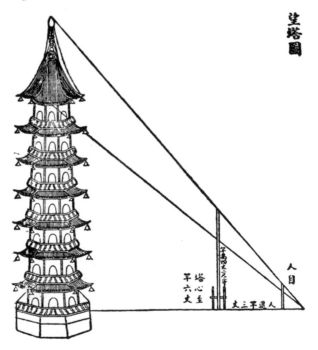

Figure 44. Illustration from the *SSCC*, p. 206

of the horizontal wheel, and the proper length of the central pillar of the pagoda."

This problem was solved as early as in the *Hai-tao suan-ching* 海島算經 by Liu Hui 劉徽 (third century).[40] This explanation follows Pai Shang-shu.[41] Ch'in gives the solution $y=ae/c$, which is incorrect; a should be replaced by $a+c$, giving $y=(a+c)e/c$. Indeed,

$$\frac{c}{b-d}=\frac{a+c}{x-d}\ ;\ c(x-d)=(a+c)(b-d);$$

$$\frac{c}{a+c}=\frac{(b-d)-e}{(x-d)-y};\ c[(x-d)-y]=(a+c)[(b-d)-e].$$

From which $c(x-d)-cy=(a+c)(b-d)-(a+c)e$ and

$$cy=(a+c)e$$
$$y=\frac{(a+c)e}{c}.$$

Problem VIII, 5. Detecting the Numerical Strength of the Enemy at a Distance

"The enemy pitches a circular camp on a sandy plain north of a river. We do not know the number of men. Spies report that the space occupied by each soldier in this camp is a square with a side of 8 feet. Our army is south of the river at the foot of a hill. Below the hill we set up a gnomon 80 feet high, its top level with an indentation on the hillside. From the top of the gnomon we stretch a cord. The distance through the air horizontally to the foot of the observer is 30 paces. The observer, who stands at this point, sights the northern border of the camp in line with the top of the gnomon. Next he sights the southern border of the camp 8 feet below the top of the gnomon. The eye of the observer is 4.8 feet above the ground. [See Figures 45 and 46.] We make use of the precise value of π [$=22/7$] and apply the *ch'ung-ch'a* 重差 method. Find the numerical strength of the enemy."

[40] See Hsü Ch'un-fang (3'), pp. 15–23; Mikami (1), p. 33; Needham (1), vol. 3, p. 30; there is a translation by Van Hee in (1) and (6).
[41] Pai Shang-shu (1'), pp 302 f.

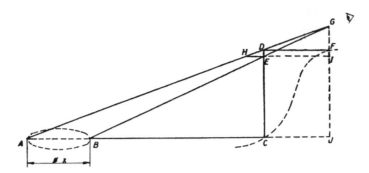

Figure 45

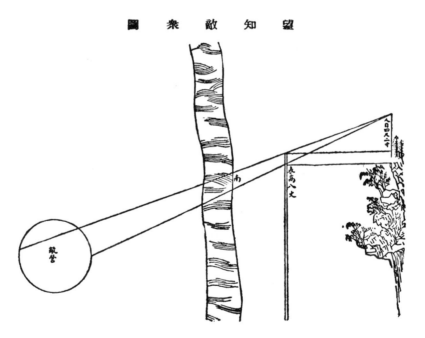

圖　棗　歟　知　望

Figure 46. Illustration from the *SSCC*, p. 395

In the figures, GF = 48 *ts'un;* DF = 30 *pu* = 1500 *ts'un;* DC = 8 *ch'ang* = 80 feet = 800 *ts'un;* DE = 8 feet = 80 *ts'un.*

Ch'in Ch'iu-shao uses the method $x = (\text{DF} \times \text{DE})/\text{GF}$. This is entirely incorrect.[42] In this computation x is equal to HE, because HE/DE=DF/GF. In fact, the solution is found from

$$\triangle GHE \backsim \triangle GAB, \frac{AB}{HE}=\frac{GB}{GE};$$

or

$$\triangle GJB \backsim \triangle GIE, \frac{GB}{GE}=\frac{GJ}{GI}=\frac{GF+DC}{GF+DE};$$

Thus,

$$AB=HE \times \frac{GF+DC}{GF+DE}= 2,500 \times \frac{48+800}{48+\ 80}= 16,562.5 \ ts'un.$$

The error is so gross that one wonders how Ch'in failed to see it.

[42] See Ch'ien Pao-tsung (2'), p. 97; Sung Ching-ch'ang (1'), pp. 147 ff.

IV

Algebra

10

Simultaneous Linear Equations

" . . . The first definite trace that we have of simultaneous linear equations is found in China."[1] In the *Chiu-chang suan-shu*, a full explanation of the "method by combination" is given. "The *fang-ch'êng* method represents the high point of the achievement of Chinese scholars in the area of linear problems."[2] Indeed, the procedure in the *Chiu-chang suan-shu* is almost perfect,[3] and the *Sun Tzŭ suan-ching* represents only a slight improvement.[4] In the *Chiu-chang* there are no diagrams,[5] although they have been reconstructed from the text.[6] For this reason there has been some question as to whether the simultaneous linear equations are meant to be solved on the counting board.[7] But that they are is obvious from the historical data, as explained in an article by Ch'ien Pao-tsung.[8]

If we compare the method as it is explained in the old *Chiu-chang* with Ch'in Chiu-shao's rule, the similarity is striking. We have good reason to think that Ch'in had the old mathematical classic close at hand when writing his work.[9] The applied

[1] Smith (1), vol. 2, p. 433; Smith gives an outline of simultaneous linear equations in Greece, China, India, and Europe. See also Vogel (2), p. 130.
[2] "Die Methode *fang-chang*...stellt den Gipfel des von den chinesischen Gelehrten auf dem Gebiet der linearen Probleme Erreichten dar." Yushkevitch (4), p. 32.
[3] A description is given in Vogel (2), pp. 130 ff; Mikami (1), pp. 16 ff; Yushkevitch (4), pp. 32 ff; Hsü Ch'un-fang (3'), pp. 3 ff.
[4] Mikami (1), p. 32; Smith (1), vol. 2, p. 433.
[5] The edition used here is the one in the *Ts'ung-shu chi-ch'êng*, abbreviated hereafter *TSCC*. Book 6 deals with *fang-ch'êng*.
[6] Vogel (2), pp. 80 ff; Hsü Ch'un-fang (3'), pp. 3 ff. (The second author always provides very clear and simple explanations, without pedantry. His works are well suited for nonmathematicians.)
[7] Mikami (1), p. 17; "If I rightly conjecture, the solution appears to have been practiced with the calculating pieces."
[8] Ch'ien Pao-tsung (3'). See also Needham (1), vol. 3, p. 126, note f.
[9] See Chapter 5.

methods *fang-ch'êng* 方程 and *chêng-fu* 正負 came from the *Chiu-chang*.[10] The rule of elimination that Ch'in Chiu-shao gives in his explanation of the *chêng-fu*[11] procedure is literally the same as the one in the *Chiu-chang*.[12] Of special interest in Ch'in's work is the fact that he uses diagrams, and from this fact we are sure that the simultaneous equations were solved on the counting board.

In the absence of the original texts, a full translation of one of these problems will show the cumbersome procedures of Chinese algebra, the influence of mechanical aids, the terminology in use, and the general pattern of the methods. It will demonstrate that Chinese mathematics is more different from our own than the use of modern algebraic symbolism would imply.

Problem IX, 1. "Find the Values of the Goods"

Question: "Three times we close a bargain; each time we pay for the goods an amount of precisely 1,470,000 *kuan*. The first time they send 3,500 bundles of garu wood, 2,200 *chin* 斤[13] of tortoiseshell, and 375 cases of frankincense. The next time they send 2,970 bundles of garu wood, 2,130 *chin* of tortoiseshell, and $3,056\frac{1}{4}$ cases of frankincense. The last time they send 3,200 bundles of garu wood, 1,500 *chin* of tortoiseshell, and 3,750 cases of frankincense. We wish to know the value of each bundle, *chin*, and case of the garu wood, the tortoiseshell, and the frankincense."[14]

---

[10] *TSCC* edition, pp. 125 ff.

[11] Literally, "positive-negative."

[12] Ch'in's rule says *"wu jên* 無人*"* instead of *"wu ju* 無入*"*, which is obviously a copyist's error. According to Ch'ien Pao-tsung (2′), pp. 91 f, this misprint appears in the Sung edition; Ch'in Chiu-shao did not correct it, but Yang Hui 楊輝 rectified it in his *Hsiang-chieh chiu-chang suan-fa* 詳解九章算法 (1261) (*TSCC*, vol. 2, p. 35). Compare with the passage in the *Chiu-chang suan-shu, TSCC*, p. 130.

[13] See Chapter 6.

[14] The problem is worked out in modern transcription in Ch'ien Pao-tsung (2′), pp. 89 ff.

Answer:

The value of the garu wood is 300 *kuan wên* 貫文 a bundle.

The value of the frankincense is 64 *kuan wên* a case.

The value of the tortoiseshell is 180 *kuan wên* a chin.

Method: "Solve it by the method of equations [*fang-ch'êng* 方程]; introduce the method of 'positive and negative' [*chêng-fu* 正負].[15] Arrange the total amount [the constant term] and the numbers of the goods below. Place the numbers of the columns in such a way that each is opposite to its own kind. Reduce those which have fractions, reduce those which can be reduced.[16] These are the definite numbers [*ting-lü-chi*定率積]. Arrange these numbers and multiply by each other all the lowest terms. Examine these numbers and subtract the smallest from the largest. Apply the 'positive-negative' rule, according to the constant terms, to the lowest numbers [of the goods]. If they have the same signs [*t'ung-ming* 同名], mutually subtract. If they have different signs [*i-ming* 異名], mutually add. A positive coefficient to which nothing is added is made negative; a negative coefficient to which nothing is added is made positive. If they have the same sign, add; if different signs, subtract. A positive coefficient to which nothing is added is taken as positive; a negative coefficient to which nothing is added is taken as negative."

According to Ch'ien Pao-tsung's explanation,[17] if $a > b$, the meaning is

(Subtraction)

$$\pm a - (\pm b) = \pm(a - b)$$
$$\pm a - (\mp b) = \pm(a + b)$$

(Addition)

$$\pm a + (\pm b) = \pm(a + b)$$
$$\pm a + (\mp b) = \pm(a - b)$$

---

[15] *Fang-ch'êng* is the title of the eighth chapter of the *Chiu-chang suan-shu* [see Needham (1), vol. 3, p. 26]. On the origin of the meaning, see ibid., p. 63. Needham gives the translation, "The way of calculating by tabulation" (ibid. p. 26). *Fang* seems to indicate the counting board; *ch'êng* means a pattern, a rule.

[16] The term for reducing a mixed number to an improper fraction is *t'ung* 通; that for reducing a fraction or an equation to its simplest form is *yüeh* 約.

[17] Ch'ien Pao-tsung (3'), p. 4; the same rule appears in Problem IX, 2.

$$0-(+b)=-b \qquad\qquad 0+(+b)=+b$$
$$0-(-b)=+b. \qquad\qquad 0+(-b)=-b.[18]$$

This is done by changing the red rods for black ones[19] as indicated in the *Chiu-chang*: "Positive numbers are red, negative numbers black."[20] So reads the first "law of signs" in mathematics.

The text continues: "Arrange the numbers so that as coefficients of the lowest terms [that is, of the lowest rank], you get [only] one coefficient; this is the divisor. The constant term is the dividend. Divide the dividend exactly by the divisor [so that no remainder is left]. With the exception of the result, subtract or add everywhere all the amounts and you get all the [other] divisors and dividends. Divide them. The remaining [amounts] are treated in the same way."

PROCEDURE

The procedure is illustrated in Diagrams 12–23.[21] In Diagram

A

	1		2		3
a	1,470,000		1,470,000		1,470,000
x	3,200	*x*	2,970	*x*	3,500
y	1,500	*y*	2,130	*y*	2,200
z	3,750	*z*	$3,056\frac{1}{4}$	*z*	375

	1		2		3
b	1,470,000		5,880,000		1,470,000
x	3,200	*x*	11,880	*x*	3,500
y	1,500	*y*	8,520	*y*	2,200
z	3,750	*z*	12,225	*z*	375

	1		2		3
c	29,400		392,000		58,800
x	64	*x*	792	*x*	140
y	30	*y*	568	*y*	88
z	75	*z*	815	*z*	15

Diagram 12

[18] See Vogel (2), p. 131, n. 1.
[19] See Needham (1), vol. 3, p. 116.
[20] *TSCC*, p. 129; see Needham (1), vol. 3, p. 90.
[21] The constant terms are given in ten *kuan* (*shih kuan* 十貫). In place of *x, y, z* the diagrams show the first character for each of the three kinds of goods: *ch'ên* 沈, *tai* 瑇, *ju* 乳.

12, row *a* shows the initial disposition on the counting board; row *b* shows the reduction of equation (2) by multiplying by 4; row *c* shows the reduction of *1* by the G.C.D. 50, of *2* by the G.C.D. 15, of *3* by the G.C.D. 25, or the definite disposition on the counting board (*ting-lü* 定率). The reason for the repetition of the numbers in *a* and *b* (*1* and *3*) in Diagram 12 is that the successive dispositions on the whole counting board are reproduced.

A

1a			*2*			*3a*	
	441,000			392,000			4,410,000
x	960		*x*	792		*x*	10,500
y	450		*y*	568		*y*	6,600
z	1,125		*z*	815		*z*	1,125

Diagram 13

The original values of *1*, *2*, and *3* have been given; we can remove them and pick them up again. *1a* and *3a* in Diagram 13 show the disposition after equalization of the coefficients of *z* in *1* and *3*, or,[22]

$$
\left\{
\begin{array}{ll}
64x+ 30y+ 75z= 29{,}400 & (1) \\
792x+568y+815z=392{,}000 & (2) \\
140x+ 88y+ 15z= 58{,}800 & (3)
\end{array}
\right.
\quad
\begin{array}{l}
\times 15; \\
\\
\times 75.
\end{array}
$$

$$
\left\{
\begin{array}{ll}
960x+ 450y+1{,}125z= 441{,}000; & (1a) \\
10{,}500x+6{,}600y+1{,}125z=4{,}410{,}000. & (3a)
\end{array}
\right.
$$

A

1			*2*			*A*	
	29,400			392,000			3,969,000
x	64		*x*	792		*x*	9,540
y	30		*y*	568		*y*	6,150
z	75		*z*	815		*z*	0

Diagram 14

[22] These factors are unnecessarily large, but the result is the same.

In Diagram 14, A is $3a - 1a$. The left space of the board is now empty and we can put equation (1) there.

$$
\begin{array}{ll}
10,500x + 6,600y + 1,125z = 4,410,000 & (3a) \\
\underline{-960x - 450y - 1,125z = -441,000} & (-1a) \\
9,540x + 6,150y + 0z = 3,969,000 & (A).
\end{array}
$$

A

1b		2b	A	
	23,961,000	29,400,000	3,919,000	
x	52,160	59,400	9,540	
y	24,450	42,600	6,150	
z	61,125	61,125	0	

Diagram 15

In Diagram 15, $1b$ and $2b$ are derived from 1 and 2 after equalization of the coefficients of z.

$$
\begin{array}{lll}
64x + 30y + 75z = 29,400 & \times 815. & (1) \\
792x + 568y + 815z = 392,000 & \times 75; & (2) \\
52,160x + 24,450y + 61,125z = 23,961,000, & & (1b) \\
59,400x + 42,600y + 61,125z = 29,400,000, & & (2b)
\end{array}
$$

A

1		B	A	
	29,400	5,439,000	3,969,000	
x	64	7,240	9,540	
y	30	18,150	6,150	
z	75	0	0	.

Diagram 16

In Diagram 16, $B = 2b - 1b$.

$$
\begin{array}{ll}
59,400x + 42,600y + 61,125z = 29,400,000 & (2b) \\
\underline{-52,160x - 24,450y - 61,125z = -23,961,000} & (1b) \\
7,240x + 18,150y + 0z = 5,439,000 & (B)
\end{array}
$$

The left side of the counting board is now free; we pick up 1 again (Diagram 17). The figures in A' are the reduced form of

A

1	B'	A'
29,400	543,900	132,300
64	724	318
30	1,815	205
75	0	0

Diagram 17

A (:30); those in B', of B (:10).

Until now, all computations have been performed on one counting board. In order to distinguish the boards from each other, they are given names. This board is called *kan-t'u* 干圖 (labeled **A**).

B

1		B''	A''
	29,400	111,499,500	240,124,500
x	64	148,420	577,170
y	30	372,075	372,075
z	75	0	0

Diagram 18

In Diagram 18, B'' and A'' are the reduced forms of A' and B' after equalization of the coefficient of y.

$$
\begin{array}{lll}
318x + \quad 205y = 132,300 & \times \quad 205 & (A') \\
724x + \quad 1,815y = 543,900 & \times 1,815 & (B') \\
577,170x + 372,075y = 240,124,500 & & (A'') \\
148,420x + 372,075y = 111,499,500 & & (B'')
\end{array}
$$

This last computation is performed on another counting board, called *kung-t'u* 宮圖 (labeled **B**). This is necessary because otherwise B' would disappear.

A

1		B'	C
	29,400	543,900	128,625,000
x	64	724	428,750
y	30	1,815	0
z	75	0	0

Diagram 19

$$577,170x + 372,075y = 240,124,500 \quad\quad (A'')$$
$$-148,420x - 372,075y = -111,499,500 \quad\quad (B'')$$
$$428,420x + 0y = 128,625,000 \quad\quad (C)$$

A

1		B'		x	
29,400		543,900		300	
64		724		1	
30		1,815		0	
75		0		0	

Diagram 20

$$x = \frac{128,625,000}{428,750} = 300.$$

C

$\beta'y + \gamma'z$		βy		x	
10,200		326,700		300	
0		0		1	
30		1,815		0	
75		0		0	

Diagram 21

$$(B') \quad = \alpha x + \beta y = p \longrightarrow \beta y = p - \alpha x$$
$$= 724x + 1,815y \quad\quad = 543,900$$
$$724 \times 300 + 1,815y = 543,900$$
$$217,200 + 1,815y = 543,900$$
$$1,815y = 543,900 - 217,200$$
$$1,815y = 326,700.$$

This computation must be done on a separate counting board, called *chih-t'u* 支圖 (labeled **C**), because in order to find the value of β (here, 1,815) it is necessary to compute αx, $p - \alpha x$, and $\beta'y + \gamma'z$.

On board A we still have equation (1).

$$(1) = a'x + \beta'y + \gamma'z = q \longrightarrow \beta'y + \gamma'z = q - a'x$$
$$(1) = 64x + 30y + 75z = 29,400$$

$$19,200+30y+75z=29,400$$
$$30y+75z=29,400-19,200=10,200.$$

This computation is done on a separate counting board, called *jun-t'u* 閏圖 (labeled **D**). (Of course, the functions of boards **B**, **C**, and **D** can all be performed on the same board.)

D

$\beta'y+\gamma'z$ 10,200	y 180	x 300
0	0	1
30	1	0
75	0	0

Diagram 22

After finding the value of y, we substitute in $\beta'y+\gamma'z$ and get

$$30\times180+75z=10,200$$
$$5,400+75z=10,200$$
$$75z=10,200-5,400=4,800$$
$$z=64.$$

This gives the final disposition shown in Diagram 23, from which

$$z=\frac{4,800}{75}=64.$$

D

$\gamma'z$ 4,800	y 180	x 300
0	0	1
0	1	0
75	0	0

Diagram 23

From this reproduction of the procedure on the counting boards, it can be seen that the apparently cumbersome method used by the Chinese is more logical than it first appears. We must keep in mind the fact that mechanical methods have a different structure from the usual written methods.[23] Some of the numbers vanish during the operations, but the general pattern requires that the numbers necessary for subsequent computations be preserved. [24] In general, this system is beautiful and efficient, but it is inherently limited, and one is left with the impression that Chinese mathematicians exploited it to the utmost limits of its possibilities.

A similar procedure is used in Problem IX,2, where

$$\begin{cases} 200x+ 40y+ \quad 0z+ \ 0w=106{,}000 \\ 0x+264y+ \ 800z+ \ 0w=106{,}000 \\ 0x+ \ 0y+1{,}670z+15w=106{,}000 \\ 58\tfrac{1}{3}x+ \ 0y+ \quad 0z+52w=106{,}000. \end{cases}$$

After reduction:

$$\begin{cases} 5x+ \quad y+ \qquad 0z + \quad 0w= \quad\quad 2{,}650 \quad (1) \\ 0x+ 33y+ \quad 100z + \quad 0w= \quad\; 13{,}250 \quad (2) \\ 0x+ \ 0y+ \quad 334z + \quad 3w= \quad\; 21{,}200 \quad (3) \times 156 \\ 175x+ \ 0y+ \qquad 0z +156w= \quad 318{,}000 \quad (4) \times 3 \end{cases}$$

$$\begin{aligned} 0x+ \ 0y+ \ 52{,}104z +468w= \ 3{,}307{,}200 \quad &(3') \\ (-)525x+ \ 0y+ \qquad 0z +468w= \quad 954{,}000 \quad &(4') \\ \hline -525x+ \ 0y+ \ 52{,}104z + \ 0w= \ 2{,}353{,}200 \quad &(A) \\ 5x+ \quad y+ \qquad 0z + \ 0w= \quad\quad 2{,}650 \quad &(1) \times 105 \\ \hline -525x+ \ 0y+ \ 52{,}104z + \ 0w= \ 2{,}353{,}200 \quad &(A) \\ 525x+105y+ \qquad 0z + \ 0w= \quad 278{,}250 \quad &(1') \\ \hline 105y+ \ 52{,}104z \qquad\quad = \ 2{,}631{,}450 \quad &(B) \times 11 \\ 0x+ 33y+ \quad 100z + \ 0w= \quad\; 13{,}250 \quad &(2) \times 35 \\ \hline 1{,}155y+573{,}144z \qquad\quad =28{,}945{,}950 \quad &(C) \\ (-)1{,}155y+ \ 3{,}500z \qquad\quad = \quad 463{,}750 \quad &(2') \\ \hline 569{,}644z \qquad\quad =28{,}482{,}200 \quad &(D) \end{aligned}$$

$$z \qquad =\frac{28{,}482{,}200}{569{,}644}=50.$$

---

[23] The counting board methods are comparable in this respect with those used in an electronic computer.
[24] For a general discussion on this matter, see Chapter 1.

Substituting z in (2) → $y=250$;
substituting y in (1) → $x=480$;
substituting x in (4) → $w=1,500$.

In Problem V,4, we see this computation:[25]

$$\begin{cases} 5a+3b=39,586 \\ a+\ b=\ 9,782 \end{cases}$$

$$\begin{array}{rcr} 5a+3b= & & 39,586 \\ -3a-3b= & & -29,346 \\ \hline 2a\ \ \ \ = & & 10,240 \\ a\ \ \ \ = & & 5,120 \end{array}$$

$b=9,782-5,120=4,662.$

[25] For the problem, see Part VI.

11

Determinants

"The Chinese method of representing the coefficients of the unknowns of several linear equations by means of rods on a calculating board naturally led to the discovery of simple methods of elimination. The arrangement of the rods was precisely that of the numbers in a determinant."[1] The idea of solving simultaneous linear equations by means of determinants was developed by Seki Kôwa (Seki Takakusu) in Japan in his work *Kai fukudai no ho* (Methods of solving problems by determinants) in 1683.[2] "The only surprising thing about the discovery is that it had not been stated much earlier, for example by the Sung algebraists."[3] One wonders if it is possible to find some trace of an elementary idea of determinants in Chinese works. "It is," as Mikami[4] says, "undeniable that the Japanese mathematics of the 17th century had been influenced by the Chinese science," and the tabulating of the coefficients of the equations by means of rods on the counting board may have stimulated the invention of determinants. However, in Ch'in Chiu-shao's work there is one problem solved by a method strongly resembling the procedure invented by G. Cramer (1704–1752), which makes use of determinants.[5]

The general solution of

$$\begin{cases} a\ x + b\ y = c \\ a'x + b'y = c' \end{cases}$$

is

---

[1] Smith (1), vol. 2, p. 475; Yushkevitch (4), p. 35.
[2] Smith (1), vol. 2, p. 475; Needham (1), vol. 3, p. 117; Mikami (1), p. 191, gives a full explanation of Seki's method. See Smith and Mikami (1).
[3] Smith and Mikami (1). In Europe, the theory of determinants began with Leibniz (1693); it was further developed by Vandermonde (1771), Laplace (1772), Lagrange (1773), and Gauss (1801).
[4] Mikami (1), p. 163.
[5] Now generally known as Cramer's rule.

$$x = \frac{D_x}{D} \text{ and } y = \frac{D_y}{D},$$

where

$$D = \begin{vmatrix} a & b \\ a' & b' \end{vmatrix}; D_x = \begin{vmatrix} c & b \\ c' & b' \end{vmatrix}; D_y = \begin{vmatrix} a & c \\ a' & c' \end{vmatrix}.$$

Problem VIII, 9 reads as follows:[6] "A storehouse has three kinds of stuff: cotton, floss silk, and raw silk. They take inventory of the materials and wish to cut out and make garments for the army. As for the cotton, if we use 8 rolls for 6 men, we have a shortage of 160 rolls; if we use 9 rolls for 7 men, there is a surplus of 560 rolls. As for the silk, if we use 150 *liang* for 8 men, there is a remainder of 16,500 *liang;* if we use 170 *liang* for 9 men, there is a remainder of 14,400 *liang.* As for the raw silk, if we use 13 *chin* for 4 men, there is a shortage of 6,804 *chin;* if we use 14 *chin* for 5 men it is exactly enough. We wish to know the number of men and the [respective] amounts of cotton, floss silk, and raw silk."

The first part gives the following equations, in which x=rolls of cotton and
y=number of men:

$$x = \frac{y}{6} \times 8 - 160 \longrightarrow 6x = 8y - 960$$

$$x = \frac{y}{7} \times 9 + 560 \longrightarrow 7x = 9y + 3,920,$$

or

$$6x - 8y = -960$$
$$7x - 9y = 3,920.$$

The coefficients are arranged as follows:

$$\begin{vmatrix} 8 & 9 \\ 6 & 7 \end{vmatrix} = 56 - 54 = 2.$$

This is the same as

[6] It is somewhat surprising that this problem is not treated in any earlier study of Chinese mathematics.

$$D = \begin{vmatrix} 6 & -8 \\ 7 & -9 \end{vmatrix} = -54 - (-56) = +2.$$

Indeed,

$$\begin{vmatrix} 6 & -8 \\ 7 & -9 \end{vmatrix} = \begin{vmatrix} 7 & 6 \\ 9 & 8 \end{vmatrix} = \begin{vmatrix} 8 & 9 \\ 6 & 7 \end{vmatrix}.$$

$$D_x = \begin{vmatrix} -960 & -8 \\ 3{,}920 & -9 \end{vmatrix} = 8{,}640 + 31{,}360 = 40{,}000.$$

The number 8,640 is computed by multiplying 160×54 or $(160 \times 6) \times 9$; and 31,360 is computed by multiplying 560×56 or $(560 \times 7) \times 8$; finally,

$$x = \frac{D_x}{D} = \frac{40{,}000}{2} = 20{,}000.$$

The preceding shows clearly that a new method has been developed.

The value of y is not found by determinants, but by means of the equalization method:

$$-x = -\frac{8y}{6} + 160$$

$$x = \frac{9y}{7} + 560$$

$$\overline{}$$

$$0 = \frac{9y}{7} - \frac{8y}{6} + 720$$

$$0 = 54y - 56y + 30{,}240$$

$$2y = 30{,}240$$

$$y = 15{,}120.$$

The arrangement for computing D and D_x is shown in the diagrams below and on the next page.

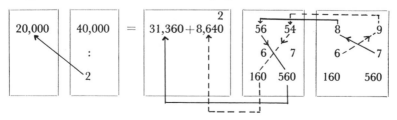

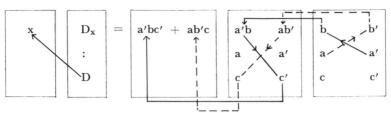

In the second part, $z=liang$ (ounces) of floss silk (there are 16 *liang* to the *chin*).

$$\begin{cases} z=\dfrac{y}{8}\times 150+16,500 \\ z=\dfrac{y}{9}\times 170+14,400 \end{cases} \longrightarrow \begin{cases} 8z-150y=132,000 \\ 9z-170y=129,600. \end{cases}$$

$$D=\begin{vmatrix} 150 & 170 \\ 8 & 9 \end{vmatrix}=1,350-1,360=-10.$$

This is the same as

$$\begin{vmatrix} 8 & -150 \\ 9 & -170 \end{vmatrix}=-1,360+1,350=-10.$$

$$D_z=\begin{vmatrix} 132,000 & -150 \\ 129,600 & -170 \end{vmatrix}=-22,440,000+19,440,000=-3,000,000.$$

$$z=\frac{D_z}{D}=\frac{-3,000,000}{-10}=300,000.$$

Ch'in uses the absolute differences 3,000,000 and 10. In the third part, $w=chin$ (pounds) of raw silk.

$$\begin{cases} w=\dfrac{y}{4}\times 13-6,804 \\ w=\dfrac{y}{5}\times 14 \end{cases} \longrightarrow \begin{cases} 4w-13y=-27,216 \\ 5w-14y=0. \end{cases}$$

$$D=\begin{vmatrix} 4 & -13 \\ 5 & -14 \end{vmatrix}=-56+65=9;$$

Ch'in gives:

$$\begin{vmatrix} 13 & 14 \\ 4 & 5 \end{vmatrix}=65-56=9.$$

$$D_w = \begin{vmatrix} -27{,}216 & -13 \\ 0 & -14 \end{vmatrix} = 381{,}024.$$

$$w = \frac{D_w}{D} = \frac{381{,}024}{9} = 42{,}336.$$

All this gives the impression that a kind of determinant was known in Sung times, and that we have in Ch'in a forerunner of Cramer. It is important that the same pattern is applied three times in this problem, showing that it was a kind of general method. This is the only place where this method is applied.

Ch'in calls his method *ying-fei* 盈朒. At first sight this seems to be the "rule of false position" (*ying-pu-tsu* 盈不足),[7] but it is obvious that the method applied here has nothing to do with the solution of simple linear equations by guesses. However, the pattern of solution resembles the method used by Ch'in, as we can see if we solve the equation $ax + b = 0$ by means of "double false position."

Suppose that g_1 and g_2 are two guesses, f_1 and f_2 the discrepancies.[8]

We substitute:

$$\begin{cases} ag_1 + b = f_1 \\ ag_2 + b = f_2. \end{cases}$$

The solution is

$$x = \frac{f_1 g_2 - f_2 g_1}{f_1 - f_2} = \frac{\begin{vmatrix} f_1 & g_1 \\ f_2 & g_2 \end{vmatrix}}{f_1 - f_2}.$$

Ch'in's method is used in the case

$$\begin{cases} ax + by = c \\ a'x + b'y = c', \end{cases}$$

and the general solution for x is

<hr>

[7] *Ying* means surplus; *fei* means "the young crescent moon" and hence "shortage, deficiency."
[8] Needham (1), vol. 3, pp. 117 ff, and Mikami (1), p. 14.

$$x = \frac{\begin{vmatrix} c & b \\ c' & b' \end{vmatrix}}{\begin{vmatrix} a & b \\ a' & b' \end{vmatrix}}.$$

Now we shall try to demonstrate that determinants were perhaps derived from this rule of false position.

The *regula falsae positionis* occurs for the first time in the *Chiu-chang suan-shu*, of which the seventh book is entitled "*Ying-pu-tsu.*"[9] This is a reference to a type of problem of the form shown in the following paragraph. In this transcription, z_1 is *ying* 盈 (surplus), and z_2 is *pu-tsu* 不足 (deficiency).

(I) $\begin{cases} a_1 x = y + z_1 \\ a_2 x = y - z_1. \end{cases}$

The solution is

$$x = \frac{z_1 + z_2}{a_1 - a_2} \qquad y = \frac{\begin{vmatrix} a_1 & a_2 \\ z_1 & z_2 \end{vmatrix}}{a_1 - a_2} = \frac{a_1 z_2 + a_2 z_1}{a_1 - a_2}.$$

Other forms are[10]

(II) $\begin{cases} a_1 x = y + z_1 \\ a_2 x = y + z_2 \end{cases}$ $x = \dfrac{z_1 - z_2}{a_1 - a_2}; \quad y = \dfrac{a_2 z_1 - a_1 z_2}{a_1 - a_2}.$

(III) $\begin{cases} a_1 x = y - z_1 \\ a_2 x = y - z_2 \end{cases}$ $x = \dfrac{z_2 - z_1}{a_1 - a_2}; \quad y = \dfrac{a_1 z_2 - a_2 z_1}{a_1 - a_2}.$

(IV) $\begin{cases} a_1 x = y + z_1 \\ a_2 x = y \end{cases}$ $x = \dfrac{z_1}{a_1 - a_2}; \quad y = a_2 x.$

(V) $\begin{cases} a_1 x = y - z_1 \\ a_2 x = y \end{cases}$ $x = \dfrac{z_1}{a_2 - a_1}; \quad y = a_2 x.$

Ch'in Chiu-shao extended this rule to the general cases

$\begin{cases} ax - by = -c \\ a'x - b'y = c' \end{cases}$ $\begin{cases} ax - by = c \\ a'x - b'y = c', \end{cases}$

[9] *TSCC*, pp. 111 ff; for a translation, see Vogel (2), pp. 70 ff.
[10] For details, see Yushkevitch (4), pp. 26 ff; Vogel (2), pp. 128 ff.

and gave the solution

$$x = \frac{b'c + bc'}{a'b - ab'}.$$

The value for y is computed by equalization of the equations. Since both numerator and denominator are computed by tabulation and cross-multiplication of the coefficients, it seems that we have here the first step toward a method for solving simultaneous linear equations by means of determinants. In general, Ch'in's method can be stated as follows:

$$(I) \quad \begin{cases} x = \dfrac{y}{a} \times b - c \\ x = \dfrac{y}{a'} \times b' + c' \end{cases}$$

$$x = \frac{\begin{vmatrix} a'b & ab' \\ c & c' \\ b & b' \\ a & a' \end{vmatrix}}{} = \frac{a'bc' + ab'c}{a'b - ab'}.$$

Compare with Cramer's rule: for

$$\begin{cases} ax - by = -ac \\ a'x - b'y = a'c', \end{cases}$$

the determinants are

$$D = \begin{vmatrix} a & -b \\ a' & -b' \end{vmatrix} = -ab' + a'b, \quad D_x = \begin{vmatrix} -ac & -b \\ a'c' & -b' \end{vmatrix} = ab'c + a'bc',$$

and x is given by

$$x = \frac{a'bc' - ab'c}{a'b - ab'}.$$

$$(II) \quad \begin{cases} x = \dfrac{y}{a} b + c \\ x = \dfrac{y}{a} b' + c' \end{cases}$$

$$x = \frac{\begin{vmatrix} a'b & ab' \\ c & c' \\ b & b' \\ a & a' \end{vmatrix}} = \frac{a'bc' - ab'c}{a'b - ab'}.$$

Compare with Cramer's rule: for

$$\begin{cases} ax - by = ac \\ a'x - b'y = a'c', \end{cases}$$

the determinants are

$$D = \begin{vmatrix} a & -b \\ a' & -b' \end{vmatrix} = -ab' + a'b, \quad D_x = \begin{vmatrix} ac & -b \\ a'c' & -b' \end{vmatrix} = -ab'c + a'bc',$$

with x given by

$$x = \frac{a'bc' - ab'c}{a'b - ab'}.$$

(III) $\begin{cases} x = \dfrac{y}{a} b - c \\ x = \dfrac{y}{a'} b'. \end{cases}$

$$x = \frac{\begin{vmatrix} a'b & ab' \\ c & 0 \\ b & b' \\ a & a' \end{vmatrix}} = \frac{ab'c}{a'b - ab'}.$$

In this case the subtraction is inverted: $ab'c - 0$. Compare with Cramer's rule: for

$$\begin{cases} ax - by = -ac \\ a'x - b'y = 0, \end{cases}$$

the solution is

$$x = \frac{\begin{vmatrix} -ac & b \\ 0 & -b' \\ a & -b \\ a' & -b' \end{vmatrix}} = \frac{ab'c}{a'b - ab'}.$$

Although we cannot pretend that Ch'in Chiu-shao is the inventor of Cramer's rule, nevertheless his method is an important extension of the rule of false position, and it is not impossible that determinants, as developed in Japan, have their roots in Sung algebra.

12

Series and Progressions

There is an elaborate discussion of series in Li Yen,[1] of which the first part is devoted to a general historical outline, and the rest to China. In the oldest problems, those in the *Chou-pei suan-ching* and the *Chiu-chang suan-shu*, no general methods are employed; thus the mathematical value of the methods used is very small.[2] Except for some generalization in the *Chang Ch'iu-chien suan-ching*, the first scholar to give more serious attention to series was Shên Kua (c. 1078).[3] The next treatment appears in Ch'in Chiu-shao's work,[4] where there are several problems on arithmetical progressions.

For the arithmetical progression a, $a+v$, $a+2v$, . . . , $a+(n-1)v$, Ch'in makes use of the formula $L=a+(n-1)v$. In V, 9 we find the derived formula

$$v = \frac{L-a}{n-1}.$$

The summation of a series is dealt with in several problems. In VIII, 2 Ch'in uses the ancient formula $S=(a+L)n/2$. In VII, 3 the formula for S is given for the case where L is unknown:

$$S = \frac{(n-1)n}{2} v + na.$$

Indeed,[5] if

$$S = \frac{(a+L)n}{2} \text{ and } L = a+(n-1)v,$$

then

[1] Li Yen (6'), vol. 1, pp. 315 ff.
[2] See Needham (1), vol. 3, p. 138, note *b*; Vogel (2), p. 120; and the thorough discussion in Yushkevitch (4), pp. 78 ff.
[3] See Li Yen (6'), pp. 337 f; Yushkevitch (4), p. 80.
[4] For material on subsequent developments in this area, the reader is referred to the bibliographical references cited. See also Hsü Ch'un-fang (3'), pp. 22 ff.
[5] Li Yen considers this formula as an interpolation formula.

$$S = \frac{[a+a+(n-1)v]n}{2} = \frac{[2a+(n-1)v]n}{2}$$

$$= \left[a + \frac{(n-1)v}{2} \right] n = an + \frac{(n-1)v}{2} n.$$

Problem VII, 9 says: "We begin with a 'pointed pile' of pine trees [Figure 47]. We do not happen to know the number. [Starting] from the top we remove [the trees] as far as the middle. We see that there are 9 trees forming the width of the top layer. How many trees were there originally and how many are there laid by?"

The v of the series is 1. Ch'in gives the formula

$$S = \frac{2m(2m-1)}{2}$$

where

$$m = \frac{1+L}{2}.$$

The rationale is

$$m = \frac{1+L}{2} \longrightarrow 2m-1 = L$$

$$n = L \qquad \longrightarrow n = 2m-1$$

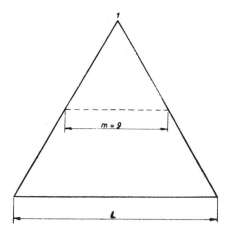

Figure 47

$$S=\frac{1+L}{2}\times n=m(2m-1)$$

An interesting combination of both formulae for S and L is shown in VIII, 3:

$$\begin{cases} S=\dfrac{(a+L)n}{2} \; ; \\ L=a+(n-1)v. \end{cases}$$

Find a formula for n.

$$S=\frac{[a+a+(n-1)v]n}{2} \; ;$$

$$2S=[2a+(n-1)v]n$$
$$2S=2an+n^2v-nv$$
$$2S=(2a-v)n+n^2v,$$

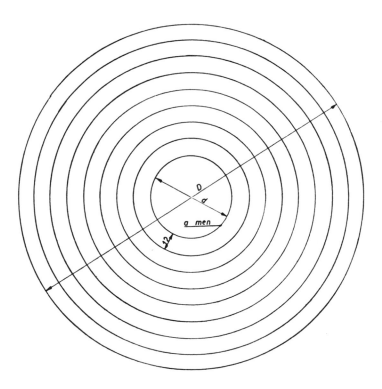

Figure 48

or,[6]

$vn^2 + (2a - v)n - 2S = 0.$

Problem VIII, 4 reads: "They draw up an army [of 12,500 men] in a circular camp of 9 rows [Figure 48]. Each man occupies a circle of 6 feet. The rows are at a distance from each other which amounts to twice the occupied circles. From these men they take away a fourth part; the camp cannot be made smaller. It is ordered to make use of the original camp and to draw up the remaining men. Find (1) the inner and outer circumference of the original camp; (2) the number of men drawn up; (3) the number of feet occupied by each man after sending away the others; (4) the number of men on the outer periphery."

There are n rows; S=the original number of men; p=space per man; d=inner diameter; D=outer diameter; a=number of men on the inner periphery.

$$a = \frac{S - 2\pi(n-1)n}{n}.$$

Indeed,

$$S = \frac{(a+L)n}{2} \text{ and } L = a + (n-1)v.$$

Substitute L in S:

$$S = \frac{[a + a + (n-1)v]n}{2}.$$

We calculate the value of v. The difference between the radii of any two successive circles is 12 feet. Thus, the difference between the circumferences is represented by $2\pi(r+12) - 2\pi r = 2\pi \times 12.$

The difference in the number of men $= (2\pi \times 12)/6 = 4\pi.$

$$S = \frac{[2a + (n-1)4\pi]n}{2} = an + (n-1)n \times 2\pi,$$

from which

[6] The problem gives the equation $6n^2 + 234n - 5,200 = 0$, with the the solution $n = 9$. See Chapter 13.

$$a = \frac{S - 2\pi(n-1)n}{n}.$$

The inner circumference is thus represented by

$$C_i = ap = \frac{S - 2\pi(n-1)n}{n} \, 6.$$

The outer circumference is $C_o = ap + 4\pi(n-1)p$. The number of men on the outer circumference is $a + 4\pi(n-1)$. When the number of men is reduced to $3/4\,S$, the number of men on the outer periphery is

$$P = \frac{\frac{3}{4}S + 2\pi n(n-1)}{n}$$

and the place occupied by each man $= C_o/P$.

We may conclude that Ch'in Chiu-shao was well aware of all the possibilities for use of the formulae for arithmetical progressions in practical calculations. There is no information about his forerunners.[7] Certainly none of the older mathematical books was as advanced in this area, but on the other hand Yang Hui's and Chu Shih-chieh's methods were much more highly developed.

[7] We have only Shên Kua's formula.

13

Numerical Equations of Higher Degree

One of the most important contributions of the Chinese to the development of algebra was undoubtedly their method for solving numerical equations of higher degree.

The first Western description of this method and a comparison between it and the Horner-Ruffini procedure appeared in A. Wylie's "Jottings on the Science of the Chinese; Arithmetic"(1852).[1] Horner's method was then of recent date (published in 1819), so that Biernatzki, in his translation of Wylie's article (1856), did not mention Horner's name.[2] As Wylie's "Jottings" were known in Europe only through Biernatzki's translation, we are not surprised to find that Cantor was not convinced; as he says: "An approximation method for equations of higher degree seems to have existed in which a similarity with the so-called Horner method is thought to be discernible,[3] but which in the text before us has been treated too sketchily to permit us to venture either supporting or denying this opinion."[4] Indeed, Wylie gives only one example: $-x^4 + 1{,}534{,}464x^2 - 526{,}727{,}577{,}600$, and in Biernatzki's translation the minus sign before x^4 is omitted and the procedure is changed.[5]

It was Mikami who in 1912 provided the first full explana-

[1] Wylie (1), pp. 186 ff. Wylie remarks: "It appears some have thought proper to dispute the right of Horner to the invention, and it will perhaps be an unexpected occurrence to our European friends to find a third competitor coming forward from the Celestial Empire, with a very fair chance of being able to establish his claim to priority" (p. 188).
[2] He speaks only of "the solution by European methods" (p. 87).
[3] With reference to Matthiessen, *Grundzüge der antiken und modernen Algebra der litteralen Gleichungen,* Leipzig, 1878, pp. 964 f.
[4] "Es scheint dabei eine Annäherungsmethode für Gleichungen höherer Grade bestanden zu haben, in welcher man eine Ähnlichkeit mit der sogenannten Horner'schen Näherungsmethode entdecken will, die aber wenigstens in unserer Vorlage zu dürftig behandelt ist, als dass wir es wagten diese Meinung zu stützen oder zu wiederlegen." Cantor (2), p. 586.
[5] We shall see other examples of Biernatzki's carelessness in the chapter on indeterminate analysis.

tion of the procedure followed by Ch'in Chiu-shao for solving
the equation $-x^4 + 763,200x^2 - 40,642,560,000 = 0$.[6] He says:
" . . . who can deny the fact of Horner's illustrious process
being used in China at least nearly six long centuries earlier
than in Europe . . . ?"[7] In Gauchet, on the other hand, we
read:[8] "The method of root extraction generally applied by
the Chinese has sometimes been compared to that used by
Horner for the solution of numerical equations; but any analogy
between the two procedures should take into account the
difference between these ideas from a mathematical standpoint." This gave rise to serious doubt about the value of the
Chinese algorithm, expressed by Loria in two of his works.[9]
He criticizes the equation of the tenth degree as it is stated in
Ch'in's work and says: "But we still have too meager information as to the details of the work for us to be able to affirm
confidently that Horner's method was known to the Chinese in
the 13th century. We can only say that this method, or one practically identical with it, was known at that time, and we must
await further evidence before affirming or denying the priority
of the Chinese in its discovery."[10] D. E. Smith, under the
influence of Loria's judgment, speaks of "a method which, in
its basic principles, seems to be that which Horner first used in
England in 1819, but about which (in the Chinese form) there
is considerable doubt." This was written in 1931; but in his
article on "Chinese Mathematics," published in 1912, Smith
had expressed his conviction that it was possible to "find the
detailed solution of a numerical higher equation by the method

---

[6] Mikami (1), pp. 74–77.

[7] Ibid., p. 77.

[8] "On a comparé parfois la méthode chinoise de l'extraction généralisée
des racines à celle de Horner pour la résolution des équations numériques,
mais l'analogie des procédés ne doit pas faire oublier la différence des
idées au point de vue mathématique." Gauchet (2), p. 549.

[9] Whose articles on Chinese mathematics are "so misleading as to be almost useless" Needham (1), vol. 3, p. 1.

[10] Loria (2), p. 521. His statement that "the result stated by Ch'in, $x = 9$,
does not satisfy his equation at all" (ibid., p. 520) is based on a mistake made
by Mikami. Other of Loria's utterances prove how dangerous it is to give
one's opinion when sinologically at sea, without access to original sources.

rediscovered by Horner in 1819, the only essential difference being in the numerals employed."[11] This doubt about the value of the Chinese method remained unresolved for some years.[12] In the meantime, Chinese scholars such as Li Yen and Ch'ien Pao-tsung investigated the original mathematical texts; and many important studies, which will be referred to in this chapter, were published in Chinese. One of the most recent surveys of Chinese mathematics is the third volume of Needham's *Science and Civilisation in China,* a work of great importance; however, the section on numerical higher equations[13] does not give a full historical description of the Chinese method, as it is largely restricted to the origin of the method in the *Chiu-chang suan-shu,* to which the author, in collaboration with Wang Ling, earlier devoted a monograph [Wang and Needham (2)]. Thus historians of mathematics, having no access to Chinese studies, cannot yet make a general and accurate judgment as to Horner's method in China. The explanation in Yushkevitch is much more useful, although it too stops short of an extensive description of the algorithm as it was worked out in Chinese mathematical textbooks.[14]

"The solution of numerical higher equations for approximate values begins, as far as we know, in China. It has been called the most characteristic Chinese mathematical contribution."[15] Ch'in Chiu-shao was certainly not the originator of this method, and he does not seem to have been the first to apply it to equations of all degrees. Yang Hui 楊輝, who employed this procedure,[16] refers to Liu I 劉益 (c. 1080) and Chia Hsien 賈憲 (c. 1200) as his predecessors; these mathematicians were able to

[11] Smith (3), p. 249; (2), p. 599.
[12] The refutation of Loria's judgment by Mikami (1') had no influence in Europe, because it was published only in Japanese.
[13] Vol. 3, pp. 126 ff.
[14] Yushkevitch (4), pp. 66 ff. This work will attempt to give a full account of the problems in the *Shu-shu chiu-chang* and to describe the Chinese method as it is given there.
[15] Needham (1), vol. 3, p. 126.
[16] Lam Lay Yong's *A Critical Study of the Yang Hui Suan Fa* is in press. See also Lam Lay Yong (2) and (3).

solve numerical equations higher than the second degree.[17] Although they demonstrated their ability to solve only the special cases $x^3=a$ and $x^4=a$, nevertheless, if Yang Hui's information is reliable, the procedure was the same as that used by Ch'in Chiu-shao for more complex equations.[18] As all the great mathematicans of the Sung and Yüan were well acquainted with this algorithm, and as we cannot ascertain any mutual influence among them, the method must have originated some time before, and it was perhaps an extension of the procedures employed by such men as Liu I and Chia Hsien. Although Ch'in Chiu-shao was not the originator, he deserves credit for explaining the method in a clear and systematic way.[19] Almost the same methods are applied to full equations of higher degree by Li Yeh 李冶[20] and Chu Shih-chieh 朱世傑.[21]

In order to convey to the reader a true idea of the contents of the Chinese text, we shall first set down a literal representation and translation. This is by no means an explanation of the mathematical approach or the logical background of the procedure, but merely the successive dispositions on the counting board.

The equation solved here is $-x^4 + 763,200x^2 - 40,642,560,000 = 0$.[22] The disposition on the counting board is called "*Chêng-fu k'ai san-ch'êng-fang t'u*" 正負開三乘方圖 (diagram of the solution of a "positive-negative" equation of the fourth degree)[23] (III,1).

---

[17] See Li Yen (6'), vol. 1, pp. 250 ff; Ch'ien Pao-tsung (2'), pp. 44 ff. The dates of Chia Hsien are very uncertain; Needham gives 1100, Li Yen 1200.
[18] For the development of this method from the root-extraction procedures in the *Chiu-chang suan-shu*, see Wang Ling and Needham (2).
[19] See Yushkevitch (4), p. 67.
[20] See Li Yen (6'), vol. 1, pp. 254 ff; Ch'ien Pao-tsung (2'), pp. 121 ff.
[21] See Li Yen (6'), vol. 1, pp. 255 ff.
[22] This equation is very well suited for a first explanation, because the root is a whole number. Examples with approximate roots will follow. The same example is dealt with in Li Yen (8'), pp. 190 ff and (17',) pp. 49 ff; in Mikami (1), pp. 74 ff; and in Ch'ien Pao-tsung (7'), p. 132, where the equation is reduced to $- x^4 + 7,632x^2 - 4,064,256 = 0$. Ch'ien Pao-tsung gives the unreduced equation in (2'), pp. 48 f.
[23] For the explanation of this term, see the section on "Classification of Equations."

The method says: "The sign of the quotient [*shang* 商] is positive; the sign of the constant term [*shih* 實] is negative; the sign of the following term is positive [*ts'ung* 從]; the sign of the first term [*yü*] is negative [*i* 益]."[24] See Diagrams 24–44, in which the symbols *a, b, c,* and so on represent the Chinese terms.

0	Quotient (=root) (*Q*)	
−40642560000	Constant term (*C*)	*e*
0	*fang* 方 (coefficient of *x*)	*d*
+763200	*ts'ung-shang-lien* 從上廉 (coefficient of x^2)	*c*
0	*ts'ung-hsia-lien* 從下廉 (coefficient of x^3)	*b*
−1	*i-yü* 益隅 (coefficient of x^4)	*a*

c is moved   2 columns.
a is moved   4 columns.
Q is moved 1 column.

Diagram 24

00	*Q*
−40642560000	*e*
0	*d*
763200	*c*
0	*b*
−1	*a*

c is moved again 2 columns.
a is moved again 4 columns.
Q is moved again 1 column.

Diagram 25

800	*Q*
−40642560000	*e*
0	*d*
763200	*c*
0	*b*
−1	*a*

The first number of the root is found to be 800.
Multiply this number by *a*.
Add to *b*.

Diagram 26

[24] The *yü* is in fact the coefficient of the term of the highest degree.

800	Q
−40642560000	e
0	d
763200	c
−640000	c'
−800	b
−1	a

Multiply *b* by *Q*.
Subtract from *c*.

Diagram 27

800	Q
−40642560000	e
0	d
123200	c
−800	b
−1	a

Multiply *c* by *Q*.
Add to *d*.

Diagram 28

800	Q
−40642560000	e
98560000	d
123200	c
−800	b
−1	a

Multiply *Q* by *d*.
The result is a positive
number.
Add to *c*.[25]

Diagram 29

800	Q
−40642560000	e
78848000000	e'
98560000	d
123200	c
−800	b
−1	a

To the negative *e* add the
positive *e'*.
The remainder is a positive
constant term, called *huan-ku*
換骨.[26]

Diagram 30

800	Q
+38205440000	e
98560000	d
123200	c
−800	b
−1	a

Multiply *Q* by *a*.
Add to *b*.

First Transformation

Diagram 31

800	Q
+38205440000	e
98560000	d
123200	c
−1600	b
−1	a

Multiply *Q* by *b*.
Transfer to *c*.
Add up.

Diagram 32

[25] The meaning of *hsiang-hsiao* 相消, used at this point in the Chinese text,
is "ausgleichen." It is translated by "to add (algebraically)."
[26] These technical terms are difficult to translate. *Huan* means "to sub-

800	Q
38205440000	e
98560000	d
123200	c
−1280000	c'
−1600	b
−1	a

Add up the positive and negative c and c'.

Diagram 33

800	Q
38205440000	e
9856000	d
−1156800	c
−1600	b
−1	a

Multiply Q by c.
Add to d.

Diagram 34

800	Q
+38205440000	e
98560000	d
−925440000	d'
−1156800	c
−1600	b
−1	a

Add d to d'.

Diagram 35

800	Q
+38205440000	e
−826880000	d
−1156800	c
−1600	b
−1	a

Multiply Q by a.
Add to b

Second Transformation

Diagram 36

800	Q
+38205440000	e
−826880000	d
−1156800	c
−2400	b
−1	a

Multiply Q by b.
Add to c.

Diagram 37

800	Q
38205440000	e
−826880000	d
−3076800	c
−2400	b
−1	a

Multiply Q by a.
Add to b.

Third Transformation

Diagram 38

stitute, to change"; it refers to the number that takes the place of the constant term.

800	Q
38205440000	e
−826880000	d
−3076800	c
−3200	b
−1	a

Move d back 1 column; c, 2 columns; b, 3 columns; a, 4 columns. In Q, set up the next figure.

Fourth Transformation

Diagram 39

800	Q
38205440000	e
−826880000	d
−3076800	c
−3200	b
−1	a

Divide e by d; as the next figure in Q take 40. Multiply 40 by a (40=Q′). Add to b.

Diagram 40

840	Q
38205440000	e
−826880000	d
−3076800	c
−3240	b
−1	a

Multiply Q′ by b. Add to c.

Diagram 41

840	Q
38205440000	e
−826880000	d
−3206400	c
−3240	b
−1	a

Multiply Q′ by c. Add to d.

Diagram 42

840	Q
38205440000	e
−955136000	d
−3206400	c
−3240	b
−1	a

Multiply Q′ by d. Add to e. No remainder.

Diagram 43

840	Q
00000000000	e
−955136000	d
−3206400	c
−3240	b
−1	a

Q is the solution = 840.

Diagram 44

The impression of extreme complexity is misleading. The operations can be performed on a counting board very quickly; only when written down do all the different transformations appear cumbersome.[27] These diagrams must not be taken as the written method, but simply as pictorial representations of successive configurations on a counting board.

Let us now try to explain the procedure in our own algebraical language.

$$F \equiv -x^4 + 763,200x^2 - 40,642,560,000 = 0. \tag{1}$$

First of all we determine the number of figures in the root.[28] The method is the same as that used in our own mathematical textbooks for determining the number of figures in a square or cube root.[29] To locate this number on the counting board we move the coefficient of x^4 4 columns to the left at once. We find 3 partitions of 4 or fewer digits, which gives 3 digits in the root. The first number of the root, determined by trial and error, is 8 hundreds,[30] or 800. Thus $x = 800 + y$ and $y = x - 800$.

Suppose that $100p = x$ (here, $100 \times 8 = 800$). Then (1) can be changed into[31]

$$-(100p)^4 + 763,200 \times (100p)^2 - 40,642,560,000 = 0, \tag{2}$$

or

$$-1 \times (100^4) \times p^4 + 763,200(100^2) \times p^2 - 40,642,560,000 = 0.$$

This is the reason why we find the disposition on the counting board shown in Diagram 45.

The zeros are shown by the place-value system of the counting board. The solution of the equation goes as follows:

[27] Compare with the explanation of the multiplication $2,519 \times 43$ in Yoshino (1), which covers about six pages, although it can be performed in a couple of minutes.
[28] Without decimals, of course.
[29] That is, the number of figures of a cube root is equal to the number of partitions of 3 digits. See Mikami (1), p. 75.
[30] Ch'in does not give his method for finding this number; see, however, the Section on "General Characteristics of Ch'in Chiu-shao's method."
[31] This explanation follows Li Yen (8'), pp. 198 f and (17'), pp. 49 ff.

```
            800
   −40642560000
           000   (100¹)
   7632000000    (100²)
     0000000     (100³)
   −100000000    (100⁴)
```

Diagram 45

First Transformation

Q	a	b	c	d	e
		−800	−640,000 *(27)*	98,560,000	78,848,000,000 *(30)*
800	−1	0	763,200	0	−40,642,560,000
	−1	−800	123,200	98,560,000	38,205,440,000
		(27)	*(28)*	*(29)*	*(31)*
	x^3	x^2	x	C (constant term)	R

The italic numbers in parentheses indicate the corresponding diagrams. The instructions at the bottom of each diagram indicate precisely these operations.

$$F \equiv (x-800)\,(-x^3-800x^2+123,200x+98,560,000) +$$
$$38,205,440,000 = 0.$$

The root $x = 800+y$, from which $y = x-800$, and
$$F \equiv y(-x^3-800x^2+123,200x+98,560,000)+38,205,440,000 = 0.$$

Again we divide the first quotient by $y = x-800$.

Second Transformation

		−800	−1,280,000 *(33)*	−925,440,000 *(35)*
800	−1	−800	+123,200	+98,560,000
	−1	−1,600	−1,156,800	−826,880,000
		(32)	*(34)*	*(36)*
	x^2	x	C	R

$$F \equiv y[y(-x^2-1,600x-1,156,800)-826,880,000]$$
$$+38,205,440,000 = 0.$$

Again we divide the second quotient by $y = x - 800$.

Third Transformation

		−800	−1,920,000
800	−1	−1,600	−1,156,800
	−1	−2,400	−3,076,800
		(*37*)	(*38*)
	x	*C*	*R*

$F \equiv y\{y[y(-x-2,400) - 3,076,800] - 826,880,000\}$
$+ 38,205,440,000 = 0.$

Again we divide the third quotient by $y = x - 800$.

Fourth Transformation

		−800
800	−1	−2,400
	−1	−3,200
		(*39*)
	C	*R*

$F \equiv y[y\{y[y(-1) - 3,200] - 3,076,800\} - 826,880,000] +$
$38,205,440,000 = 0.$

If we work this out, we get:

$F \equiv -y^4 - 3,200y^3 - 3,076,800y^2 - 826,888,000$
$+ 38,205,440,000 = 0.$

The accompanying table summarizes and indicates the meaning of the place-values on the counting board.[32]

Thus

$F \equiv -1 \times (100)^4 q^4 - 3,200 \times (100)^3 q^3 - 3,076,800 \times$
$(100)^2 q^2 - 826,880,000 \times (100)q + 38,205,440,000 = 0.$

If $10q = r$, then

---

[32] After Li Yen (8′), p. 198.

$$\begin{array}{r} \underline{8} \end{array}$$

$$-1 \times (100)^4 \qquad\qquad + 763,200 \times (100)^2 \qquad\qquad\qquad\qquad - 40{,}642{,}560{,}000$$
$$- 800 \times (100)^3 - 640{,}000 \times (100)^2 + 98{,}560{,}000 \times (100) + 78{,}848{,}000{,}000$$

$$-1 \times (100)^4 + 123{,}200 \times (100)^2 + 98{,}560{,}000 \times (100) + 38{,}205{,}440{,}000$$
$$- 800 \times (100)^3 - 1{,}280{,}000 \times (100)^2 - 925{,}440{,}000 \times (100)$$

$$-1 \times (100)^4 - 1{,}600 \times (100)^3 - 1{,}156{,}800 \times (100)^2 - 826{,}880{,}000 \times (100) + 38{,}205{,}440{,}000$$
$$- 800 \times (100)^3 - 1{,}920{,}000 \times (100)^2$$

$$-1 \times (100)^4 - 2{,}400 \times (100)^3 - 3{,}076{,}800 \times (100)^2 - 826{,}880{,}000 \times (100) + 38{,}205{,}440{,}000$$
$$- 800 \times (100)^3$$

$$-1 \times (100)^4 - 3{,}200 \times (100)^3 - 3{,}076{,}800 \times (100)^2 - 826{,}880{,}000 \times (100) + 38{,}205{,}440{,}000$$

$$F \equiv -1 \times (10)^4 r^4 - 3{,}200 \times (10)^3 r^3 - 3{,}076{,}800 \times (10)^2 r^2$$
$$-826{,}880{,}000 \times (10)r + 38{,}205{,}440{,}000 = 0.$$

On the counting board we get the configuration shown in Diagram 46. We find that $y = 40 + z$ or $z = y - 40$.

800	
38205440000	
-8268800000	(10^1)
-307680000	(10^2)
-3200000	(10^3)
-10000	(10^4)

Diagram 46

40		-40	$-129{,}600$	$-128{,}256{,}000$	$-38{,}205{,}440{,}000$
	-1	$-3{,}200$	$-3{,}076{,}800$	$-826{,}880{,}000$	$+38{,}205{,}440{,}000$
	-1	$-3{,}240$	$-3{,}206{,}400$	$-955{,}136{,}000$	0
		(41)	(42)	(43)	(44)
	y^3	y^2	y	C	

$F \equiv (y - 40)(-y^3 - 3{,}240y^2 - 3{,}206{,}400y - 955{,}136{,}000) = 0.$ F is divisible by $y - 40$ and $z = 0$; hence $z = y - 40 = 0$ and $y = 40$. As $x = 800 + y$, we get $x = 800 + 40 = 840$.

In the *Shu-shu chiu-chang* we find twenty-six equations of the second or higher degree, all solved by the method that is used here. There are twenty equations of the second degree, one of the third degree, four of the fourth degree, and one of the tenth degree (See Table 2, following this chapter). This large number of equations allows us to generalize Ch'in's method.

Suppose we have the general equation of the nth degree:

$$F(x) \equiv a_0 x^n + a_1 x^{n-1} + a_2 x^{n-2} + \ldots + a_{n-1} x + a_n = 0.$$

Suppose we divide $F(x)$ by $(x - c)$, the quotient again by $(x - c)$, the new quotient again by $(x - c)$ . . . until the last quotient is a constant. This constant is a_0, because the first coefficient of all quotients is a_0.

Or:

$$F(x) \equiv (x-c) \times q_1(x) + r; \quad r = f(c)$$
$$q_1(x) \equiv (x-c) \times q_2(x) + r_1$$

$$\vdots$$

$$q_{n-2}(x) \equiv (x-c) \times q_{n-1}(x) + r_{n-2}$$
$$q_{n-1}(x) \equiv (x-c) \times a_0 + r_{n-1}.$$

We multiply these identities respectively by 1, $(x-c)$, $(x-c)^2$, $\ldots$, $(x-c)^{n-1}$.

$$F(x) \qquad\qquad \equiv (x-c) \times q_1(x) \qquad + r \qquad\qquad [r = f(c)]$$
$$q_1(x)(x-c) \qquad \equiv (x-c)^2 \times q_2(x) \qquad + r_1 x(-c)$$
$$q_2(x)(x-c)^2 \quad\ \equiv (x-c)^3 \times q_3(x) \qquad + r_2(x-c)^2$$

$$\vdots$$

$$q_{n-2}(x)(x-c)^{n-2} \equiv (x-c)^{n-1} \times q_{n-1}(x) + r_{n-2}(x-c)^{n-2}$$
$$q_{n-1}(x)(x-c)^{n-1} \equiv (x-c)^n \times a_0 \qquad\quad + r_{n-1}(x-c)^{n-1}$$

$$F(x) \equiv f(c) + r_1(x-c) + r_2(x-c)^2 + \ldots r_{n-2}(x-c)^{n-2}$$
$$+ r_{n-1}(x-c)^{n-1} + a_0(x-c)^n$$

If $x = c + h$ or $x - c = h$, we get

$$F(c+h) \equiv f(c) + r_1 h + r_2 h^2 + \ldots + r_{n-1} \cdot h^{n-1} + a_0 h^n.$$

Going on to the next digit, we suppose that

$$x = c + h + k.$$

We get

$$F(c+h+k) \equiv f(c+h) + r'_1 k + r'_2 k^2 + \ldots + r'_{n-1} k^{n-1} + a_0 k^n.$$

For example, given

$$f(x) \equiv x^5 - 5x^4 + 11x^3 - 16x^2 + 19x - 20 = 0,$$

find $f(2+h)$.

Successive divisions by $(x-2)$ give:

1	-5	$+11$	-16	$+19$	-20
1	-3	$+5$	-6	$+7$	-6
1	-1	$+3$	0	$+7$	
1	$+1$	$+5$	$+10$		

```
1    +3    +11
1    +5
1
```

Thus

$f(2+h) \equiv 1h^5 + 5h^4 + 11h^3 + 10h^2 + 7h - 6 = 0.$

This method is exactly the same as that described by Horner. Of course, Horner provides a full analysis of his method, perhaps overly intricate and sophisticated;[33] moreover, his field of application is much wider than that of the Chinese mathematicians. But in fact the principle is the same. To use Gauchet's words again, "the difference between these ideas from a mathematical standpoint" is apparent, but perhaps it is simply the difference between medieval and modern mathematics. The identity of these procedures is now generally accepted, even in the West.[34]

It is almost certain that Horner was "blissfully unaware that he had rediscovered an ancient Chinese computation scheme."[35] Mikami leaves open the possibility of Chinese influence in the West: "We of course don't intend in any way to ascribe Horner's invention to a Chinese origin, but the lapse of time sufficiently makes it not altogether impossible that the Europeans could have known of the Chinese method in a direct or indirect way . . . ."[36] There are at present no serious historical arguments for this allegation, because we have no hint of evidence either in favor of the Chinese or in favor of Ruffini's and Horner's indepence of their influence.

Representation of an Equation—General Terminology

The coefficients of the different powers of x are arranged on

---

[33] For that reason we do not follow Horner but give preference to the modern representation.

[34] See Cajori (1), pp. 74 f; Yushkevitch (4), p. 69; Hofmann (1), p. 77; Struik (1), p. 73. For a comparison of Horner's method and the Chinese procedure, see Wylie (1), p. 189; Yabuuchi (3'), p. 65; Li Yen (8'), p. 198.

[35] Struik (1), p. 73.

[36] Mikami (1), p. 77.

the counting board, with the highest power below and the
constant term at the top, as in Diagram 47. The coefficients
are indicated by special terms given in Diagrams 48–51.

C	(x^0)
.	
.	
.	
a_3	(x^{n-3})
a_2	(x^{n-2})
a_1	(x^{n-1})
a_0	(x^n)

Diagram 47

Diagrams 48–51. Coefficients of x in equations of the second, third,
fourth, and tenth degrees

Second Degree

Shang 商	Root	
shih 實	a_2	
fang 方	a_1	x
yü 隅	a_0	x^2

Diagram 48

Third Degree

Shang	Root	
shih	a_3	
fang	a_2	x
lien 廉	a_1	x^2
yü	a_0	x^3

Diagram 49

Fourth Degree

Shang	Root	
shih	a_4	
fang	a_3	x
shang-lien 上廉	a_2	x^2
hsia-lien 下廉	a_1	x^3
yü	a_0	x^4

Diagram 50

Tenth Degree

Shang	Root	
shih	a_{10}	
fang	a_9	x
shang-lien	a_8	x^2
tz'ŭ-lien 次廉	a_7	x^3
ts'ai-lien 才廉	a_6	x^4
wei-lien 維廉	a_5	x^5
hsing-lien 行廉	a_4	x^6
hsiao-lien 爻廉	a_3	x^7
hsing-lien 星廉	a_2	x^8
hsia-lien	a_1	x^9
yü	a_0	x^{10}

Diagram 51

The positive and negative signs are placed next to the coefficient by means of the characters *chêng* 正 (positive) and *fu* 負 (negative); if the coefficient is zero, this is indicated by *hsü* 虛 (void). Negative terms are also called *i* 益 and positive terms *ts'ung* 從. In all probability positive terms were indicated by means of red rods on the calculating board, negative terms by means of black ones.

According to Li Jui 李銳 (1768–1817), Ch'in Chiu-shao made use of red and black ink for writing these coefficients in the counting-board diagrams of his manuscript,[37] but this practice was not followed in the printed editions.

Classification of Equations

Equations are classified according to their degree. The following terminology is used:

Solving an equation of the second degree is indicated by the phrase *k'ai-p'ing-fang* 開平方 (extract the square root). No distinction is made in the terminology between $ax^2 + c = 0$ and $ax^2 + bx + c = 0$. Solving an equation of the third degree is indicated by *k'ai-li-fang* 開立方 (extract the cube root); an equation of the fourth degree by *k'ai-san-ch'êng-fang* 開三乘方;[38] and one of the tenth degree by *k'ai-chiu-ch'êng-fang* 開九乘方.

Equations with the first coefficient equal to 1 have no special name. When the first coefficient $a \neq 1$, the term *lien-chih* 連枝 ("joined branches") is used. For example,

(III, 9) $1,404x^2 - 57,931,200 = 0$
(II, 9) $400x^4 - 2,930,000 = 0$
(VI, 4) $16x^2 + 192x - 1863.2 = 0$.

Equations having only even powers of x are called *ling-lung* 玲瓏 ("harmonious alternating").[39] For example,

(III, 1) $-x^4 + 763,200x^2 - 40,642,560,000 = 0$
(IV, 5) $x^{10} + 15x^8 + 72x^6 - 864x^4 - 11,664x^2 - 34,992 = 0$.

[37] See Li Yen (8'), p. 189.
[38] See Chapter 6.
[39] Wylie, in (1), p. 187, makes a mistake in considering *ling-lung k'ai-fang* 玲瓏開方 as the general term for "Horner's process" in China.

When in the equation $ax^2-b=0$ $(b>0)$, a and b are squares, Ch'in indicates the solution by the phrase *"k'ai t'ung-t'i lien-chih p'ing-fang"* 開同體連枝平方 ("extract the homogeneous joined branch square root"). He makes use of the following method: suppose $a=p^2$ and $b=q^2$; then the equation becomes $p^2x^2-q^2=0$, and the positive root $x=q/p$. This method is applied in IV, 2:

$$24,649x^2-41,912,676=0$$

$$x=\frac{\sqrt{41,912,676}}{\sqrt{24,649}}$$

$$p^2-24,649=0 \longrightarrow p=157$$

$$q^2-41,912,676=0 \longrightarrow q=6,474$$

$$x=\frac{q}{p}=\frac{6,474}{157}=41\frac{37}{157}$$

$$24,649x^2-6,200,100=0$$

$$x=\frac{\sqrt{6,200,100}}{\sqrt{24,649}}=\frac{2,490}{157}=15\frac{137}{157}.$$

This method is also applied by multiplying $ax^2-b=0$ by a, giving $a^2x^2-ab=0$ and hence $x=\sqrt{(ab)}/\sqrt{a^2}=\sqrt{(ab)}/a$. In III, 7 we find the equation $121x^2-43,264=0$; this is multiplied by 121, giving $121^2x^2-5,234,944=0$, and $121x=2,288$, from which $x=18\frac{10}{11}$.[40] Ch'in Chiu-shao says (p. 167): "The t'ung-t'i rule [is as follows]: First extract the square root of the *yü* 隅 [a]; we get a number called *t'ung-yü* 同隅 [common a]; with this 'common a' multiply the constant term [b] and extract the square root; this is the dividend; take the 'common a' as divisor and divide."

The term "common a" indicates clearly that this method was applied only when two equations with the same coefficient of x^2 are given. In IV,2 we have $24,649x^2-41,912,676=0$ and $24,649x^2-6,200,100=0$; in III,7: $121x^2-6,230,016=0$ and $121x^2-43,264=0$. There is indeed some progress when the methods just described are applied in this case.[41]

[40] The operation is quite useless here, because $43,264 = 208^2$ and it is possible to move immediately to $x = \sqrt{43,264/121} = 208/11 = 18\frac{10}{11}$.

[41] The explanation in Ch'ien Pao-tsung (2'), pp. 52 f is somewhat in-

Transformation rules are called *fan-fa* 翻法, *huan-ku* 換骨, and *t'ou-t'ai* 投胎. When we transform the given equation into an equation of lower degree, it happens that the constant term changes from negative to positive;[42] this term is called *huan-ku*[43] (the changed term, literally, "exchanging the bones") and the whole process is called *fan-fa* (inversion method).[44] (The last term is used only once in Ch'in's work, but it occurs also in that of Li Yeh and Yang Hui.)[45] Ch'in Chiu-shao says (p.126): "Take the [algebraic] sum of the negative constant and the positive number. This number has a remainder; it is a positive constant. This is called *huan-ku*."

For example, in Problem III,1, the equation

$$-x^4+763,200x^2-40,642,560,000=0; \text{ as } x=800+y$$

is changed into

$$-y^4-3,200y^3-3,076,800y^2-826,880,000y+38,205,440,000=0.$$

If after transformation the value of the constant term remains negative but decreases, this term is called *t'ou-t'ai* (literally, "reborn").[46] For example, in Problem IV,8 the equation

$$0.5x^2-152x-11,552=0; \text{ as } x=300+y$$

is changed into

$$5,000y^2+14,800y-12,152=0$$
$$-11,552>-12,152.$$

Ch'in says (p. 204): "add 600 as a *t'ou-t'ai* to the constant term" $(=-600-11,552=-12,152)$.

---

complete, because it does not lay stress on this fact. The same is true for Li Yen (8'), p. 201.

[42] Note that in a given equation the constant term is always negative.

[43] This is a Taoist term meaning "a thorough change of character and disposition."

[44] This explanation follows Ch'ien Pao-tsung (2'), pp. 53 f; see also Li Yen (8'), p. 199.

[45] Ch'ien Pao-tsung (2'), p. 53.

[46] See Ch'ien Pao-tsung (2'), p. 54; Li Yen (8'), p. 199. *T'ou-t'ai* literally means "to be reborn into another state of existence."

Approximate Roots

Three kinds of approximate roots are used in Ch'in Chiu-shao's work.[47] The first consists of *rounded whole numbers*. For example, in Problem VIII,1, $\sqrt{8,000}=89+ \approx 89+1=90$.

The second kind is derived by *adding a fraction to the whole part of the root*. In this case, approximate values of the square root are calculated by means of the inequalities

$$a+ \frac{r}{2a+1} < \sqrt{a^2+r} < a+ \frac{r}{2a}.$$

The second inequality is used as an approximation formula in the *Sun Tzŭ suan-ching*:

$$\sqrt{a^2+r} \approx a+ \frac{r}{2a}.$$

The first is used in the *Chang Ch'iu-chien suan-ching* (c. 475) and the *Wu-ching suan-shu* (sixth century).[48] Ch'in Chiu-shao makes use of the first approximation formula,

$$\sqrt{a^2+r} \approx a+ \frac{r}{2a+1}.$$

In Problem III, 8,[49]

$$\sqrt{640} \approx 25 \frac{15}{2 \times 25+1} = 25 \frac{5}{17} ;$$

$$\sqrt{1000} \approx 31 \frac{39}{2 \times 31+1} = 31 \frac{13}{21} .$$

Both formulae are also used in the work of the Arabian mathematician Al-Kâšî (?–c. 1436).[50]

There is no example in Ch'in's mathematics of the formula

$$\sqrt[3]{a^3+r} \approx a+ \frac{r}{3a^2+3a+1},$$

[47] The matter is discussed in Li Yen (6'), vol. 1, pp. 252 ff, and (8'), pp. 199 ff; in Ch'ien Pao-tsung (8'), pp. 163 f and (2'), pp. 51 ff; in Hsü Ch'un-fang (3'), pp. 81 ff.
[48] See Li Yen (6'), vol. 1, pp. 247 f.
[49] The number 25 5/17 = 25.2941; the exact value is 25.2982. The number 31 13/91 = 31.6190; the exact value is 31.623.
[50] See Yushkevitch (4), p. 244.

or

$$\sqrt[3]{a^3+r} \approx a + \frac{r}{(a+1)^3-a^3},$$

as it appears in Leonardo Fibonacci's work (1202).[51] Ch'in, however, does employ an approximate formula for the root of an equation of the fourth degree:

$$\sqrt[4]{a^4+r} \approx a + \frac{r}{4a^3+6a^2+4a+1}$$

or

$$\sqrt[4]{a^4+r} \approx a + \frac{r}{(a+1)^4-a^4}.$$

Generally expressed, the formula is

$$\sqrt[n]{a^n+r} \approx a + \frac{r}{(a+1)^n-a^n}.$$

The approximation $\sqrt{(a^2+r)} \approx a+r/(2a)$ is also given in Āryabhaṭa I,[52] and the formula was known to the Babylonians.[53] For the cube root, Āryabhaṭa I and Brahmagupta[54] employed $\sqrt[3]{(a^3+r)} \approx a+[r/(3a^2)]$; Hsia-hou Yang 夏候陽[55] (c. 500) used

$$\sqrt[3]{a^3+r} \approx a + \frac{r}{3a^2+1},$$

and Fibonacci[56] used the formula

$$\sqrt[3]{a^3+r} \approx a + \frac{r}{3a^2+3a+1}.$$

An-Nasawî[57] (c. 1025) made use of both the formulae

[51] We find this formula in the work of Chu Shih-chieh; see Li Yen (6'), vol. 1, p. 256.
[52] See Yushkevitch (4), p. 116.
[53] Ibid.
[54] Ibid.
[55] Li Yen (6'), vol. 1, p. 249.
[56] Hofmann (1), vol. 1, p. 95; Zeuthen (1), p. 318.
[57] Hofmann (1), vol. 1, p. 70; Yushkevitch (4), p. 246; see also Suter (2), pp. 96 f.

$$\sqrt{a^2+r} \approx a + \frac{r}{2a+1} \quad \text{and} \quad \sqrt[3]{a^3+r} \approx a + \frac{r}{3a^2+3a+1}.$$

Al-Kâšî[58] (?– c. 1436) knew the general formula

$$\sqrt[n]{a^n+r} \approx a + \frac{r}{(a+1)^n - a^n}.$$

The transmission of these formulae among China, Islam, and Europe needs more thorough investigation before we can draw conclusions as to their origin; but that they are indications of historical contact among the three cultures cannot be denied.

Ch'in Chiu-shao applies the approximation formula in the following way: suppose we have the equation

$$F(x) \equiv a_0 x^n + a_1 x^{n-1} + a_2 x^{n-2} + \ldots + a_{n-1} x + a_n = 0,$$

and the whole part of the positive root is a. After applying Horner's method to $F(x)$ for $y = x - a$, we get

$$F(x) \equiv b_0 y^n + b_1 y^{n-1} + b_2 y^{n-2} + \ldots + b_{n-1} y + b_n = 0.$$

If the root is $a + (p/q)$, Ch'in Chiu-shao takes

$$\frac{p}{q} = \frac{-b_n}{b_0 + b_1 + b_2 + \ldots b_{n-1}}.$$

EXAMPLES

In Problem II,9, $x^4 - 7,325 = 0$, or in general $x^4 - a = 0$. The value of $a = 9$, and the root is $x = a + (p/q)$.

$$\left(a + \frac{p}{q}\right)^4 = a^4 + 4a^3 \frac{p}{q} + 6a^2 \left(\frac{p}{q}\right)^2 + 4a \left(\frac{p}{q}\right)^3 + \left(\frac{p}{q}\right)^4 = 0;$$

$$x^4 - a = a^4 + 4a^3 \frac{p}{q} + 6a^2 \left(\frac{p}{q}\right)^2 + 4a \left(\frac{p}{q}\right)^3 + \left(\frac{p}{q}\right)^4 - a = 0;$$

$$4a^3 \frac{p}{q} + 6a^2 \left(\frac{p}{q}\right)^2 + 4a \left(\frac{p}{q}\right)^3 + \left(\frac{p}{q}\right)^4 = a - a^4.$$

In the approximation formula, it is supposed that

[58] Hofmann (1), vol. 1, pp. 70; Yushkevitch (4), p. 244.

$$\frac{p}{q} = \left(\frac{p}{q}\right)^2 = \left(\frac{p}{q}\right)^3 = \left(\frac{p}{q}\right)^4.$$

Hence

$$(4a^3+6a^2+4a+1)\,\frac{p}{q} = a-a^4;$$

$$\frac{p}{q} = \frac{a-a^4}{4a^3+6a^2+4a+1}.$$

For $x^4-7,325=0$, we get $a=9$, and thus

$$\frac{p}{q} = \frac{7,325-6,561}{4\times9^3+6\times9^2+4\times9+1} = \frac{764}{3,439}.$$

However, no separate calculation is necessary. For example, $f(x) \equiv x^4-7,325=0$. Applying Horner's method, we get

$$f(9+h) \equiv h^4+36h^3+486\,h^2+2,916h-764.$$

Suppose that $h=a/b$. Then

$$f\left(9+\frac{a}{b}\right) \equiv \left(\frac{a}{b}\right)^4 + 36\left(\frac{a}{b}\right)^3 + 486\left(\frac{a}{b}\right)^2 + 2,916\left(\frac{a}{b}\right) - 764;$$

$$\frac{a}{b} = \frac{764}{1+36+486+2,916} = \frac{764}{3,439}.$$

In Problem III,8, $-x^4+15,245x^2-6,262,506.25=0$. We find that $x=20$, and

$$f(20+h) \equiv -h^4-80h^3+12,845h^2+577,800h-324,506.25=0.$$

Here, $h=p/q$, and[59]

$$\frac{p}{q} = \frac{324,506.25}{-1-80+12,845+577,800} = \frac{1,298,025}{2,362,256}.$$

The third method of approximation is *that which uses decimal numbers*. For example, in Problem VI, 4,

$$16x^2+192x-1,863.2=0; \qquad x=6.35$$
$$36x^2+360x-13,068.8=0; \qquad x=14.7.$$
$16x^2+192x-1,863.2=0$ is solved as follows:

[59] In Li Yen (6'), vol. 1, p. 252, the coefficient of h^2 is given as 14,045 instead of 12,845 and the indices of x are omitted.

	16	192	−1,863.2
6		96	1,728
	16	288	− 135.2
		96	
6			
	16	384	
6			
	16		

$$f(x) \equiv y(16y + 384) - 135.2$$
$$\equiv 16y^2 + 384y - 135.2$$

Ch'in says: "Move back the *fang* [coefficient of x] once and the *yü* [coefficient of x^2] twice. The next figure of the root is 3 *fên* [0.3]." Hence $0.16y^2 + 38.4y - 135.2 = 0$.

	0.16	+38.4	−135.2	
3		+ 0.48	+116.64	
	0.16	38.88	− 18.56	
3		0.48		$0.16z^2 + 39.36z - 18.56 = 0$
	0.16	+39.36		
3				
	0.16			

Ch'in says: "Move back the *fang* once, the *yü* twice. The next figure is 5 *li* [0.05]." Hence $0.0016v^2 + 3.936v - 18.56 = 0$.

	0.0016	+3.936	−18.56
5		+0.008	+19.72
	0.0016	+3.944	+1.16

Thus we have $16x^2 + 192x - 1,863.2 = 0$; $x_1 = 6$, and we derive $16y^2 + 384y - 135.2 = 0$.
Put $y = y'/10$; then

$$16 \times \frac{y'^2}{100} + 384 \times \frac{y'}{10} - 135.2 = 0,$$

or $0.16y'^2+38.4y'-135.2=0$, where $y'=3\rightarrow y=y'/10=0.3$. We derive $0.16z^2+39.36z-18.56=0$. Put $z=z'/10$; then

$$0.16\times\frac{z'^2}{100}+39.36\times\frac{z'}{10}-18.56=0,$$

where $z'=5\rightarrow z=z'/10=0.5\rightarrow y=0.05$, and the solution is

$$y=0.3+0.05=0.35;$$
$$x=6+0.35=6.35.[60]$$

It is not impossible that this method was known to the authors of the earlier *Chiu-chang suan-shu*,[61] but it is not explicit; there is only the obscure phrase *"i mien ming chih"* 以面命之, which usually means "leave the remainder as such."[62] Liu Hui (third century) says in his commentary on the *Chiu-chang* that we can find the decimals of the root by application of this procedure, and Li Shun-fêng clearly states the same in his commentary. But there is no elaborated method before Ch'in Chiu-shao.[63]

Solution of the Equation $a_0x^n+a_1x^{n-1}+\cdots+a_{n-1}x+a_n=0$ with $a_0\neq1$

Sometimes all the coefficients of the equation are divided by a_0 before applying Horner's method. For example, in Problem III, 7,

$$121x^2-43,264=0. \tag{1}$$

[60] See Li Yen (8'), pp. 199 f; and Hsü Ch'un-fang (3'), pp. 82 ff.
[61] See Needham (1), vol. 3, p. 85, note *b*, and p. 127 and note *b*; Vogel (2), p. 116.
[62] The idea that this approximation method was known to the authors of the *Chiu-chang suan-shu* is defended by Wang Ling (5), for whose opinion we must rely on Needham (1), vol. 3, p. 85, note *b*; as Wang's doctoral dissertation is unavailable, his arguments are not known.
[63] Mikami (1) says; "The progress just described equally applies to the evaluation of the decimal part of the root, as later Chinese as well as Japanese mathematicians have done, but Ch'in did not try anything of the sort" (p. 77). This statement is incorrect.

Ch'in says: "If the root extraction leaves a remainder, apply the joined branch (*lien-chih* 連枝) method.[64] Multiply the *shih* 實 [43,264] by the *yü* 隅 [121] and you get the definite *shih* [*ting-shih* 定實]; take 1 as *yü*." Thus when

$$y^2 - (43{,}264 \times 121) = 0 \tag{2}$$

we find

$$y = 2{,}288$$
$$\text{and } x = \frac{2{,}288}{121} = 18\frac{110}{121} = 18\frac{10}{11}.$$

The same method is applied in II, 9; III, 9; VI, 4; etc.[65]

General Characteristics of Ch'in Chiu-shao's Method

(1) The general representation of an equation on the counting board puts the term of the highest degree below and the constant term on the top. For example, $x^4 - 3x^3 + 2x^2 - 7x - 20 = 0$

-20
$- 7$
$+ 2$
$- 3$
$+ 1$

is represented as in the diagram. Ch'in Chiu-shao always makes the constant term negative (in former times it was positive) and adds it to the other terms, thus equalizing the whole equation to zero. This is the method used by Thomas Harriot and René Descartes in the seventeenth century. The great importance of this notation was that it permitted the easy application of Horner's method. "In China the method of writing equations of the type $f(x) = 0$ constituted the basis for a unified application of the Horner method."[66] The equality

64 See the section on "Classification of Equations."
65 See Li Yen (8′), pp. 200 f; Hsü Ch'un-fang (3′), pp. 84 f; Ch'ien Pao-tsung (2′), p. 53.
66 "In China bildete die Schreibweise der Gleichungen vom Typ $f(x) = 0$ die Grundlage für die einheitliche Anwendung des Hornerschen Schemas." Yushkevitch (4), p. 69.

sign was not used; for, although it was very important in the development of algebra in Europe, it was of no use when the operations were performed on a counting board.[67] "The equal sign did not exist, but this lack was compensated for by the method of notation itself."[68] The missing degrees in an equation were indicated by zeros, a method still used in our application of Horner's procedure.

(2) Although the Chinese mathematicians were very well acquainted with negative numbers, they never gave negative solutions to their equations. The reason is that all their equations were derived from practical problems, and therefore, as D. Struik rightly observes, "the equation always expressed something concrete, such as a measure of land."[69]

(3) The value of the digits of a root was found "by easy trial and error."[70] In all probability the mathematicians relied on simple tables of powers, probably memorized. Loria criticizes this method,[71] but there was nothing difficult about it.[72]

(4) There is no evidence of any attempt to construct a general theory of equations, or perhaps it would be better to say that there is no theoretical description of the method. Indeed, what we have before us are only the successive situations on the counting board; it is difficult, however, to agree with Gauchet, who calls this "a more prolific procedure, let us not say the

---

[67] For that reason, the author does not agree with the statement in Gauchet (2): "Ignorance of the equal sign and the lack of any equivalent form of notation seem to have been the handicap of the ancient writers. (L'ignorance du signe = et l'absence de toute notation équivalente semble avoir été le déficit des anciens auteurs)" (p. 549, note 4).

[68] "Ein Gleichheitszeichen existierte nicht, doch wurde dieser Mangel durch die Schreibweise selbst ausgeglichen." Yushkevitch (4), p. 68.

[69] Struik (1), p. 74.

[70] Struik (1), p. 72; see also Yushkevitch (4), p. 46.

[71] Loria (3), vol. 1, p. 160: "...And he begins by stating: 'It can be seen that the first digit of the root is 8'; how 'it can be seen' he does not say... (...e comincia affermando: 'si vede che la prima cifra della radice è 8'; come 'si veda' egli non dice...)."

[72] Moreover, we read in Horner (1): "The root is manifestly a little greater than 2. Make it $x = 2 + z$..." (p. 326); "...we immediately perceive that the first figure of the root is 3..." (p. 324).

knowledge of a method for solving the numerical equations; that idea is really too alien to the ancient Chinese." It is obvious that Ch'in Chiu-shao was acquainted with the general algorithm, as he is able to apply it to all kinds of equations. The counting board is simultaneously an operational system and a notational system. Although the background of Chinese mathematics seems to be a real "Chinese puzzle," we can ask ourselves how it is possible to use a mechanical aid for solving problems without any idea of the theoretical background of the method adopted. "Trial and error" is helpful only for simple calculations, not for procedures that are rather intricate (at least for medieval times), such as Horner's method. We can solve a problem with the help of a counting board, but we cannot invent a method with it.

(5) As stated earlier, the counting board diagrams given in Ch'in's work are a kind of pictorial representation of the successive dispositions of the counting rods. There seems to be only one description of the solving of numerical equations on the counting board, namely, A. Westphal's "Über die Chinesische Swan-pan" (1876).[73] This account follows his explanation.

Suppose that there is a general equation of the fourth degree: $Ax^4 + Bx^3 + Cx^2 + Dx - E = 0$. Suppose that $x = a + b$. On the counting board we have the disposition shown in Diagram 52.

	a
$-(4A^3b + 6Aa^2b^2 + 4Aa^3b + Ab^4$ $+ 3Ba^2b + 3Bab^2 + Bb^3$ $+ 2Cab + Cb^2 + Db)$ or $Aa^4 + Ba^3 + Ca^2 + Da - E$	
	$Aa^3 + Ba^2 + Ca + D$
	$Aa^2 + Ba + C$
	$Aa + B$
	A

Diagram 52

[73] Mikami gives another explanation in (2), pp. 387 f, but not in the same clear and general way as does Westphal.

The first expression could also take the form $Aa^4+Ba^3+Ca^2+Da-E=0$, or $Ax^4+Bx^3+Cx^2+Dx=E$. As $x=a+b$, we get:

$A(a+b)^4+B(a+b)^3+C(a+b)^2+D(a+b)=E$,

from which

$Aa^4+Ba^3+Ca^2+Da-E=Aa^4+Ba^3+Ca^2+Da-A(a+b)^4-B(a+b)^3-C(a+b)^2-D(a+b)=-(4A^3b+6Aa^2b^2+4Aa^3b+Ab^4+3Ba^2b+3Bab^2+Bb^3+2Cab+Cb^2+Db)$.

Westphal takes E as positive. As this is not Ch'in's method, here it is negative.

Compare our own practical system:

	A	B	C	D	$-E$
a		aA	Aa^2+Ba	Aa^3+Ba^2+Ca	$Aa^4+Ba^3+Ca^2+D$
	A	$Aa+B$	Aa^2+Ba+C	Aa^3+Ba^2+Ca	$Aa^4+Ba^3+Ca^2+Da$
				$+D$	$-E$

The next disposition on the counting board is shown in Diagram 53.

a	
$-(4A^3b+6Aa^2b^2+4Aa^3b+Ab^4+3Ba^2b+3Bab^2+Bb^3+2Cab$ $+Cb^2+Db$	(5)
$4Aa^3+3Ba^2+2Ca+D$	(4)
$6Aa^2+3Ba+C$	(3)
$4Aa+B$	(2)
A	(1)

Diagram 53

In the Horner procedure the representation would be that shown in Diagram 54.

All these operations in Diagram 54 *a*an easily be set up on the counting board, as in Diagram 53. We start our multi-

	A	Aa+B	Aa^2+Ba+C	Aa^3+Ba^2+Ca +D	Aa^4+Ba^3+ Ca^2+Da−E	(5)
		Aa	2Aa^2+Ba	3Aa^3+2Ba^2 +Ca		
a						
	A	2Aa+B	3Aa^2+2Ba +C	4Aa^3+3Ba^2 +2Ca+D		(4)
		Aa	3Aa^2+Ba			
a						
	A	3Aa+B	6Aa^2+3Ba+C			(3)
		Aa				
a						
	A	4Aa+B				(2)
a						
	A					(1)

Diagram 54

plications and additions from rank 1 and go up to 4, then from
1 to 3, next from 1 to 2.

The third disposition on the counting board is shown in
Diagram 55. The Horner scheme would appear as in Diagram
56.

a+b	
0	(5)
4Aa^3+3Ba^2+2Ca+D+6Aa^{2}b+3Bab+Cb+4ab^2+Bb^3+Ab^3	(4)
6Aa^2+3Ba+C+4aAb+Bb+Ab^2	(3)
4Aa+B+Ab	(2)
A	(1)

Diagram 55

	(1)	(2)	(3)	(4)	(5)
	A	$4Aa+B$	$6Aa^2+3Ba$ $+C$	$4Aa^3+3Ba^2+2Ca$ $+D$	$-4A^3b-6Aa^2b^2$ $-4Aa^3b-Ab^4$ $-3Ba^2b-3Bab^2$ $-Bb^3-2Cab$ $-Cb^2-Db$
		Ab	$4Aab+Bb$ $+Ab^2$	$6Aa^2b+3Bab+Cb$ $+4Aab^2+Bb^2$ $+Ab^3$	$4Aa^3b+3Ba^2b$ $+2Cab+Db+$ $6Aa^2b^2+3Bab^2$ $+Cb^2+4Aab^3$ $+Bb^3+Ab^4$
b	A	$4Aa+B$ $+Ab$	$6Aa^2+3Ba$ $+C+4Aab$ $+Bb+Ab^2$	$4Aa^3+3Ba^2+2Ca$ $+D+6Aa^2b+3Bab$ $+Cb+4aAb^2+Bb^2$ $+Ab^3$	0

Diagram 56

From the preceding comparisons between Horner's method and the procedure used on the counting board, it may be obvious that there is not the slightest difference between them. If we consider the Chinese system, we cannot detect the influence of mechanical computation on the algebraic procedure; the device influenced the algebraic notation, but there is no reason to believe that it determined the method as such. However, Chinese algebra was limited by the possibilities of the counting board, and from the thirteenth century on that system was partly responsible for the stagnation of mathematics. The influence of the counting board was restrictive at a certain time; it was never creative.

Origin of the Method
Mikami[74] drew attention to the fact that the method for square and cube root extraction used in the *Chiu-chang suan-shu* is the

[74] Mikami (1), p. 25, and (7), p. 80.

same as that used by the Sung mathematicians for solving numerical equations; in other words, that the so-called Horner method was applied as early as Han times.[75] Since there is a very convincing monograph on this subject by Wang Ling and Needham,[76] there is no need to repeat their arguments here. It is obvious that this procedure is a Chinese invention and that "the *t'ien-yüan* method was one of the greatest achievements of the ancient Chinese mathematicians."[77] The method was not known in India;[78] in Islamic countries we find it used for cube root extraction in the work of An-Nasawî (c. 1025);[79] and extended by Al-Kâšî (?–c. 1436) to all kinds of roots.[80] If Fibonacci knew Horner's procedure,[81] it is very probable that he learned of it from Arabian mathematicians, who perhaps borrowed it from their Chinese predecessors.

[75] Sarton (1) says that Ch'in invented this method (p. 626). This is incorrect.
[76] Wang Ling and Needham (2). Other interesting data can be found in Needham (1), vol. 3, pp. 126 ff; Yushkevitch (4), pp. 41 ff and pp. 67 ff; Vogel (2), pp. 113 ff; Lam Lay Yong (1).
[77] Yushkevitch (4), p. 69.
[78] Yushkevitch (4), p. 116.
[79] Hofmann (1), vol. 1, p. 70; Yushkevitch (4), p. 242; on An-Nasawi, see Suter (2), pp. 96 f.
[80] Hofmann (1), vol. 1, p. 70; Yushkevitch (4), pp. 242 ff.
[81] It is not clear whether this is true or not. Only a thorough study can elucidate the historical relationship among Chinese, Arabian, and European mathematics. See Hofmann (1), vol. 1, p. 95.

Table 2.   Equations Solved by Ch'in Chiu-shao

	Problem	Approximation	Bibliography
1. *K'ai-p'ing-fang* 開平方:			
a = 1			
Second degree:			
$ax^2 + c = 0$			
$x^2 - 7,056 = 0$	III, 2		
$x^2 - 100 = 0$	III, 3		
$x^2 - 2,039,184,000,000 = 0$	III, 4		
$x^2 - 1,000 = 0$	III, 8	$31\frac{13}{21}$	Mikami (1), p. 78
$x^2 - 6,250 = 0$	III, 8		
$x^2 - 90 = 0$	III, 8		
$x^2 - 5,062.5 = 0$	III, 8		
$x^2 - 640 = 0$	III, 8	$25\frac{5}{17}$	Li Yen (8'), p. 199; Li Yen (17'), p. 51; Li Yen (6'), vol. 1, p. 252.
$x^2 - 8,000 = 0$	VIII, 1	$89 + \approx 90$	Li Yen (8'), p. 199; Li Yen (17'), p. 51; Li Yen (6'), vol. 1, p. 252; Ch'ien Pao-tsung (2'), p. 52.
Second degree:			
$ax^2 + bx + c = 0$			
$x^2 + 82,655x - 22,698,100,000 = 0$	III, 5		
$x^2 + 2x - 399 = 0$	VIII, 1		
2. *Lien-chih* 連枝:			
$a \neq 1$			
Second degree:			
$ax^2 + c = 0$			
$24,649\,x^2 - 41,912,676 = 0$	IV, 2	$41\frac{37}{157}$	Ch'ien Pao-tsung (2'), p. 52; Ch'ien Pao-tsung (8'), p. 164.
$24,649\,x^2 - 6,200,100 = 0$	IV, 2	$15\frac{135}{157}$	Ch'ien Pao-tsung (2'), p. 52 f.
$121x^2 - 6,230,016 = 0$	III, 7		
$121x^2 - 43,264 = 0$	III, 7	$18\frac{10}{11}$	Li Yen (8'), p. 200; Li Yen (6'), vol. 1, p. 253 f; Ch'ien Pao-tsung (2'), p. 53; Ch'ien Pao-tsung (8'), p. 164.
$1,404x^2 - 57,931,200 = 0$	III, 9	$203 + \approx 204$	Ch'ien Pao-tsung (2'), p. 52.

Table 2 (continued)

	Problem	Approximation	Bibliography
Second degree: $ax^2 + bx + c = 0$ $9x^2 + 5,100x -$ $322,500 = 0$	III, 6	$57\dfrac{853}{2045}$	Mikami (1), p. 71; Ch'ien Pao-tsung (2'), p. 51.
$528,381x^2 +$ $360,099,600x -$ $18,933,652,500 = 0$	III, 6	$49\dfrac{20,276,319}{412,406,319}$	Ch'ien Pao-tsung (2'), p. 51 f.
$16x^2 + 192x -$ $1,863.2 = 0$	VI, 4	6.35	Li Yen (6'), vol. 1, p. 253; Li Yen (8'), p. 199 f; Li Yen (17'), p. 52; Ch'ien Pao-tsung (8'), p. 163; Ch'ien Pao-tsung (2'), p. 52; Ch'ien Pao-tsung (7'), p. 134.
$36x^2 + 360x -$ $13,068.8 = 0$	VI, 4	14.7	Li Yen (8'), p. 199; Li Yen (6'), vol. 1, p. 253; Li Yen (17'), p. 52; Ch'ien Pao-tsung (2'), p. 52; Ch'ien Pao-tsung (8'), p. 163.
$0.5x^2 - 152x -$ $11,552 = 0$	IV, 8	$366\dfrac{412}{429}$	Li Yen (8'), p. 199 Ch'ien Pao-tsung (2'), pp. 51 and 54; Ch'ien Pao-tsung (8'), p. 163; Yabuuchi (3'), p. 67.
$6x^2 + 234x -$ $2,600 = 0$	VIII, 3		
Third degree: $ax^3 + c = 0$ $4,608x^3 -$ $72,000,000,000,000$ $= 0$	VI, 4		
Fourth degree: $ax^4 + c = 0$ $400x^4 - 2,930,000$ $= 0$	II, 9		Ch'ien Pao-tsung (8'), p. 159.

	Problem	Approximation	Bibliography
3. *Ling-lung* 玲瓏: $ax^{2n} + bx^{2n-2}$ $\ldots + hx^2 + k = 0$ Fourth degree: $-x^4 + 763,200x^2 -$ $40,642,560,000 = 0$	III, 1		Li Yen (8'), pp. 190 ff; Ch'ien Pao-tsung (2'), pp. 49 ff; Ch'ien Pao-tsung (8'), pp. 132 ff;

Table 2 (continued)

	Problem	Approximation	Bibliography
			Mikami (1′), pp. 82 f;
			Mikami (1), pp. 74 ff;
			Mikami (2), p. 387;
			Chang P'êng-fei (1′), pp. 9 f;
			Yushkevitch (4), p. 67;
			Loria (1), vol. 3, pp. 160 f;
			Struik (1), p. 73;
			Cajori (1), p. 74.
$-x^4 +$ $15{,}245x^2 -$ $6{,}262{,}506.25 = 0$	III, 8	$20\dfrac{1{,}298{,}025}{2{,}362{,}256}$	Li Yen (6′), vol. 1, p. 252; Li Yen (8′), p. 199; Li Yen (17′), p. 51; Ch'ien Pao-tsung (2′), p. 51 and p. 54; Ch'ien Pao-tsung (8′), p. 163; Yabuuchi (3′), p. 67.
$-x^4 +$ $1{,}534{,}464x^2 -$ $526{,}727{,}577{,}600 = 0$	IV, 6		Wylie (1), pp. 187 f.
Tenth degree: $x^{10} + 15x^8 + 72x^6 -$ $864x^4 - 11{,}664x^2 -$ $34{,}992 = 0$	IV, 5		Ch'ien Pao-tsung (8′), p. 159; Mikami (1′), p. 82; Mikami (1), pp. 71 ff; Loria (3), vol. 1, p. 159.

V

**The Chinese Remainder Theorem:
A Monograph**

14

Indeterminate Analysis of the First Degree outside China: General Historical Survey

A general algebraic rule states that in order to solve a system of equations one needs as many equations as there are unknowns. If there are n equations with more than n variables, we can in general find an infinite number of solutions. These solutions form a set which can be expressed by a simple formula.

Indeterminate equations of the first degree, which are the simplest form, have attracted much attention from the Chinese scholars. These equations of the form $ax+by=c$ are not dealt with in the works of Diophantos of Alexandria (active about the middle of the third century), who treats only equations of higher degree. Linear indeterminate equations are dealt with in the works of Âryabhaṭa, Brahmagupta, Mâhavîra, and Bhâskara in India. The *remainder problem* is a special form of simultaneous indeterminate linear equations with the general structure $N \equiv r_1 \pmod{a} \equiv r_2 \pmod{b} \equiv r_3 \pmod{c} \equiv \ldots \equiv r_n \pmod{n}$. It was developed in China, as well as in India, the Arabic world, and Europe. Since indeterminate equations of higher degree are not dealt with in China, we will restrict our historical survey to linear indeterminate equations and the remainder problem.[1]

Summary

SOLUTION OF $AX + BY = C$

The earliest attempts to solve this linear equation by a general procedure are to be found in India from about the fifth century on in the works mentioned in the preceding paragraph. In China there were attempts in the *Chang Ch'iu-chien suan-ching*

[1] For a modern algebraical representation of the remainder theorem, see Chapter 17, section on "Theoretical Representation of Ch'in Ch'in-shao's Method."

(the "hundred fowls" problem), which gives exact results but a very incomplete expression of the method. In Europe it was only from the seventeenth century on that this kind of problem was studied.

THE REMAINDER PROBLEM

This problem, also known as "the Chinese remainder theorem," first appears in the *Sun Tzŭ suan-ching* in the fourth century and finds its culminating point in the work of Ch'in Chiu-shao (1247). In India there were Brahmagupta (c. 625) and Bhâs-kara (twelfth century), who developed the *kuṭṭaka* method. In the Islamic world, Ibn al-Haitham treats this kind of problem, and he may have influenced Leonardo Pisano (Fibonacci) in Italy. After the thirteenth century we do not find much further investigation in China, India, or the Islamic world. But from the fifteenth century on there is a marked increase in European research, which reached its apogee in the studies of Lagrange, Euler, and Gauss. This problem still appears in all our modern books on theory of numbers and is called the Chinese remainder problem.[2]

The reader may wish in the course of this section to refer to Table 3, which summarizes the chronology of indeterminate analysis.

History of the Development of Indeterminate Analysis of the First Degree outside China

INTRODUCTORY NOTE

With this outline of the evolution of the remainder problem, we

[2] A list of the most recent works that mention the remainder problem would include Le Veque (1), pp. 35 f; Rademacher (1), pp. 22 f; J. Hunter (1), p. 55; Grosswald (1), pp. 49 ff. In the last work Dickson's mistake about the Sun Tzŭ problem in Nichomachus (see note 13) is repeated. In China, studies of ancient mathematics, begun at the end of the eighteenth century, have resulted in new investigations of indeterminate problems. But their interest is merely historical, since they give us access to the older works but no new insights into indeterminate analysis that are on the same level as European studies at that time.

Table 3.   Chronology of Indeterminate Analysis

Year	Europe	China	India	Islam
100				
200	Diophantos of Alexandria			
300		*Sun Tzŭ suan-ching (?)*		
400		Ho Ch'êng-t'ien Tsu Ch'ung-chih Hsia-hou Yang		
500		Chang Ch'iu-chien	Âryabhaṭa Bhâskara I	
600		Chên Luan Liu Hsiao-sun	Brahmagupta	
700		Li Shun-fung		
800		I-hsing Lung Shou-i	Mahâvîra	
900				Abû-Kâmil
1000		Hsieh Ch'a-wei Shên Kua		Ibn al-Haitham Abû Bakr al-Farajî
1100			Bhâskara II	
1200	Leonardo Pisano (Fibonacci)	Ch'in Chiu-shao Yang Hui Chou Mi		
1300	Isaac Argyros	Ku Hêng Yen Kung		
1400	Regiomontanus Elia Misrachi Munich MS			
1500	Michael Stifel Göttingen MS	Chou Shu-hsüeh Ch'êng Ta-wei		
1600	Van Schooten Beveridge			

are going beyond the scope of this work. But there are several reasons for doing so.

First, we can evaluate Ch'in Chiu-shao's method only if we compare it with other works treating the same problem. Many of these works are not easily accessible, and for the greater part there is no analysis of the contents. A scientific comparison is possible only if we are able to compare the original texts, without any translation into modern algebraical (and thus general) language. For such translation always gives an inaccurate idea of the original text.[3]

A comparison need not be historical; it can also be a comparison of patterns. In setting up a pattern, it is not legitimate to make use of interpretations of the original texts, particularly if they are intended to give evidence that a certain people have developed an idea and cannot be considered inferior to another people, and so on—all relics of an unscientific nationalism.[4]

From the logical point of view a pattern must be built up in a certain structure. A comparison can be made only if we have knowledge of a general set of possibilities for solving the

---

[3] A distinction must be made between rhetorical algebra and algorithmic algebra. Rhetorical algebra can be general, even without a general notation. From the methodological point of view we are reminded of what Nesselmann (1) said more than a century ago: "The rule that a historian, with respect to his documentation and authorities, will cite them correctly and accurately in the first place, and secondly quote only *what he has actually seen,* seems so obvious that it can be easily considered superfluous to waste any words over such trivial matters. And yet, how often and how thoroughly are both rules violated!" (Die Regel, dass ein historischer Schriftsteller, was seine Belege und Autoritäten anlangt, erstens dieselben richtig und genau, zweitens aber *nur das citirt, was er gesehen hat,* scheint so natürlich zu sein, dass man es leicht für überflüssig halten kann, über einen so trivialen Gegenstand noch Worte zu machen. Und doch, wie oft und wie vielfach wird gegen beide Regeln gefehlt!) (p. 35); and "Nothing is more common and more natural, when reading ancient books, than that we substitute our own point of view for that of the old author whom we are reading." (Nichts is also bei der Lektüre alter Werke gewöhnlicher und natürlicher, als dass wir unsern eigenen Standpunkt dem alten Schriftsteller, den wir lesen, substituiren) (p. 37).

[4] Many studies of the last century in particular display this characteristic, having as their aim "national glory" or "glory of the true religion" and an extreme historicism as their foundation.

problem.[5] Each of the solutions given for the problem in the
course of ages can be considered as a subset of this general set.
There is evolution only if subset A (the first in time) is a subset
of a subset B. Such comparison is impossible without real
insight into the specific structure of each subset. For this reason
we need access to the original texts.[6]

GREECE[7]

The oldest indeterminate problems known in Europe are all
of higher degree and are mostly of the form $Ay^2+1=x^2$, known
as the Pell equation.[8] Of these problems the so-called Cattle
Problem, doubtfully attributed to Archimedes, is most famous,[9]
and "the solution is more complicated than that of any in the
extant works of Diophantos."[10] It is supposed that Diophantos
of Alexandria (active A.D. 275) may have treated of the problem
in one of his lost books.[11] But as D. E. Smith pointed out, the

[5] At least as far as this specific mathematical knowledge has developed.
[6] This logical comparison is worked out in Chapter 21.
[7] In this section we have to restrict ourselves to the remainder theorem,
although mention will be made of the greatest mathematicians who have
done general studies of indefinite analysis.
[8] For further information, see Smith (1), vol. 2, pp. 452 ff and bibliographical notes.
[9] See also Archibald (1), pp. 411 ff with bibliographical notes; Smith (1),
vol. 2, p. 453 and p. 584; and Heath (1), pp. 142 ff.
[10] Archibald (1), p. 414.
[11] Smith (1), vol. 2, p. 453, and Tannery (1), p. 370. Diophantos treated
only indeterminate equations of higher degree. Needham (1), vol. 3, p.
122 writes: "Curiously, the algebra of Diophantos, in so far as it touches
this subject, deals with indeterminate quadratic equations almost solely."
There is however a very simple explanation for this fact. Diophantos did
not require whole numbers as solutions for his indeterminate problems,
and with this condition the solving of indeterminate equations of the first
degree has no sense: " . . . After all, they present no difficulties so long as no
integral solutions are required." (Sie machen ja, wenn die Ganzzahligkeit
der Lösungen nicht verlangt wird, gar keine Schwierigkeit). Tropfke
(1), vol. 3, p. 101. Those who disdain oriental mathematics should read
Hankel (1), pp. 164 f, where there is a realistic evaluation of Diophantos:
"With our author not the slightest trace of a general, comprehensive method
is discernible; each problem calls for some special method which refuses
to work even for the most closely related problems. For this reason it is
difficult for the modern scholar to solve the 101st problem even after having

solution of indeterminate problems as a form of recreation must be very old in Greece.[12] The greatest of the mathematicians, however, who applied themselves to indeterminate analysis was undoubtedly Diophantos of Alexandria.[13] After Diophantos, however, the decline of mathematics in Europe had begun and there was nothing more until the time of Fibonacci (c. 1202). For a millennium the Dark Ages covered Europe, but evolution was going on in China and India.

INDIA

The study of India's contribution to the solution of indeterminate equations is very important, because many scholars assume an Indian origin for a considerable part of Chinese mathematical knowledge. A later chapter will attempt to compare the *ta-yen* rule and the Indian *kuṭṭaka*,[14] but for the moment we shall restrict ourselves to a description of the Indian methods.

The general purpose of this branch of algebra[15] seems to be astronomical.[16] Datta and Singh distinguish three varieties of indeterminate problems[17], namely,

1. $N = ax + R_1 = by + R_2$. Putting $|R_1 - R_2| = c$, we can reduce the problem to: $by - ax = \pm c$.

---

studied 100 of Diophantos's solutions." (Von allgemeineren umfassenden Methoden ist bei unserem Autor keine Spur zu entdecken; jede Aufgabe erfordert eine ganz besondere Methode, die oft selbst bei den nächstverwandten Aufgaben ihren Dienst versagt. Es ist deshalb für einen neueren Gelehrten schwierig, selbst nach dem Studium von 100 Diophantischen Lösungen, die 101. Aufgabe zu lösen).

[12] Smith (1), vol. 2, p. 584.

[13] Dickson, in (1), vol. 2, p. 58, mistakenly says that Nicomachus of Gerasa (active c. 90) dealt with the famous Sun Tzŭ problem. Needham (1), vol. 3, p. 34, note *a*, discusses this question and shows that the problem must be derived from Isaac Argyros (middle fourteenth century), a Byzantine monk. For further information, the reader is referred to Needham. Dickson's error is reproduced in Ganguli (1), p. 113; Sen (1), p. 495; and Grosswald (1), pp. 49 f.

[14] On the name *kuṭṭaka*, see Datta and Singh (1), pp. 89 ff.

[15] "The *kuṭṭaka* was considered so important by the ancient Hindu algebraists that the whole science of algebra was once named after it." Datta and Singh (1), p. 88.

[16] See the very important paper by Van Der Waerden (1).

[17] Datta and Singh (1), p. 89.

2. $(ax \pm \gamma)/\beta = y$ (x and y must be positive integers).[18]

3. $by + ax = \pm c$. The general problem is $by \pm ax = \pm c$, which can be reduced to four kinds of equations:

$by - ax = +c$	(a)
$by - ax = -c$	(b)
$by + ax = +c$	(c)
$by + ax = -c$	(d)

A general condition of solvability is that a, b, and c be prime to each other.[19]

Āryabhaṭa I[20] solves only equations (a) and (b). In his work, now commonly known as the *Āryabhaṭīya*, there is a section on mathematics,[21] containing two stanzas (32–33) on indeterminate analysis.[22] There has been much discussion about the real meaning of the text, and a correct interpretation is not easy.

The translation of the text is roughly as follows:[23]

1. Divide the divisor having the greater *agra*[24] by the divisor having the smaller *agra* (*Adhikāgrabhāgahāram chindyāt ūnāgrabhāgahāreṇa*).

---

[18] This problem could be reduced to (1).

[19] This rule was known to the greater part of the Indian mathematicians. See Datta and Singh (1), p. 92.

[20] For a general account of Āryabhaṭa's life and work, see Smith (1), vol. 1, pp. 153 ff.

[21] Called *Ganita*.

[22] The text was lost for a long time. Colebrooke (1) tells us that "a long and diligent research in various parts of India failed of recovering the algebraic and other works of Āryabhaṭa" (p. v). It was after its rediscovery published by Kern; there are translations by Rodet (1), Kaye (1), and Clark (1). Other studies (much more valuable than the European ones), most including a translation, are Ganguli (1) and (2); Datta and Singh (1), pp. 93 – 101; Sen Gupta (1); Mazumdar (1); Datta (1) and (2); Sen (1).

[23] As all the existing translations must include many interpolations in order to make sense, I have avoided translating anything more than the text itself. I am much indebted to Dr. J. Deleu (University of Ghent), who examined this translation and gave me much information about the grammatical structure of the text. The division of the text is mine, and is made only to facilitate references.

[24] I have not translated the technical term *agra*, because it gave rise to very different interpretations, which will be explained later. As in Chinese, many technical terms are not to be taken in their general meaning.

2. Mutually divide the remainders (*śeṣaparasparabhaktaṃ*).
3. Multiply by the *mati* and add to the difference between the *agras* (*mātiguṇam agrāntare kṣiptam*).
4. Multiply the one below by the one above and add the ultimate [or the lowest] one (*adha upariguṇitam antyayuk*).
5. Divide by the divisor having the smaller *agra* (*ūnāgracche-dabhājite*).
6. Multiply the remainder by the divisor having the greater *agra* (*śeṣam adhikāgracchedaguṇaṃ*).
7. Add the *agra* which divides both to the greater *agra* (*dvic-chedāgraṃ adhikāgrayutam*)

ANALYSIS

The translations differ from each other in many points. L. Rodet and W. Clark both rely on the explanation given by the commentator Parameśvara (sixteenth century);[25] Clark also relies on the parallel text in Brahmagupta, XVIII, 3–5.[26] Ganguli interprets the text starting from the mathematical formulae; Datta follows "the interpretation of the rule by Bhâskara I (525), a direct disciple of Âryabhaṭa I."[27]

There is disagreement about the meaning of the term *agra*. Ganguli and Datta translate it as "remainder," while for Clark it is a specific technical term. This enables us to distinguish the interpretations from each other. As the translations of Rodet and Kaye[28] do not seem to be based on an understanding of the real meaning of Âryabhaṭa's rule, they may be ignored. The translations of Ganguli and Datta differ from each other, as the former adds the term "quotient" in (3), whereas the latter adds "residue" (see below).

As a thorough analysis of the interpretations would be beyond the scope of this study, we give an example of each interpretation together with some notes.

[25] Kern's edition of 1875 contains Parameśvara's commentary; other editions are U. N. Singh, Muzaffapur, 1906, with commentary by Nilakaṇṭha (1500); and K. Sastri, Trivandrum, 1930/31.
[26] Colebrooke (1), p. 325.
[27] Datta and Singh (1), pp. 93 ff.
[28] Kaye's translation of *mātiguṇam agrāntare kṣiptam* as "an assumed number

Clark's translation is as follows: "Divide the divisor which gives the greater *agra* by the divisor which gives the smaller *agra*. The remainder is reciprocally divided (that is to say, the remainder becomes the divisor of the original divisor, and the remainder of this second division becomes the divisor of the second divisor, etc.). (The quotients are placed below each other in the so-called chain.) (The last remainder) is multiplied by an assumed number and added to the difference between the *agras*.[29] Multiply the penultimate number by the number above it and add the number which is below it. (Continue this process to the top of the chain.) Divide (the top number) by the divisor which gives the smaller *agra*. Multiply the remainder by the divisor which gives the greater *agra*. Add this product to the greater *agra*. The result is the number which will satisfy both divisors and both *agras*."[30]

Parameśvara's example, quoted by Clark, is the following:

$$\begin{cases} 8x \equiv 4 \ (\text{mod } 29) \\ 17x \equiv 7 \ (\text{mod } 45). \end{cases}$$

Find x.

$$8x - 29y = 4 \tag{a}$$

(māti) together with the original difference is thrown in" makes no sense, even from the grammatical point of view. See Clark (1), p. 44: "It omits altogether the important word *guṇam* (multiplied)." Moreover, as Kaye was dominated by the idea that all Indian mathematical knowledge must be of Greek origin, he tried to prove that the basis of Âryabhaṭa's method was to be found in Euclid's method for finding the greatest common divisor. It may show some relation with Âryahbaṭa's method, but the application of the results is quite different. Kaye's statement is indeed very shallow, and as he failed entirely to give the correct explanation of Âryabhaṭa's text, his importance is only historical. Kaye's interpretation is followed by Mazumdar (1). For a general criticism of Kaye's work, see Ganguli (2).

[29] The text given by Brahmagupta is very close to Âryabhaṭa's, but here Brahmagupta has: "The residue [of the reciprocal division] is multiplied by an assumed number such that the product, having added to it the difference of the remainders, may be exactly divisible [by the residue's divisor]. That multiplier is to be set down [underneath] the quotient last." Clark (1), p. 42.

[30] Ibid. p. 43.

Applying the Euclidean algorithm for finding the G.C.D. to 29 and 8 (until the remainder is 1), we have:

$$\begin{array}{ccccc} 3 & 1 & 1 & 1 \\ \hline 29 & 8 & 5 & 3 & 2 \\ 24 & 5 & 3 & 2 \\ \hline 5 & 3 & 2 & 1 \end{array}$$

This allows us to draw up the following "chain" of equations:

$$8x = 29y + 4$$
$$8x_1 = 5y + 4 \qquad \text{with } x_1 = x - 3y$$
$$3x_1 = 5y_1 + 4 \qquad \text{with } y_1 = y - 1x_1$$
$$3x_2 = 2y_1 + 4 \qquad \text{with } x_2 = x_1 - 1y_1$$
$$1x_2 = 2y_2 + 4 \qquad \text{with } y_2 = y_1 - 1x_2.$$

The last equation can easily be solved by inspection:[31]

$$x_2 = 2y_2 + 4; \qquad y_2 = \frac{x_2 - 4}{2}.$$

It is obvious that the smallest positive solution is $x_2 = 6$, giving $y_2 = 1$. According to Âryabhaṭa's rule the following chain is to be drawn up:

			in general:
3	$73 = 3 \times 20 + 13$	q_1	$x = q_1 y + x_1$
1	$20 = 1 \times 13 + 7$	q_2	$y = q_2 x_1 + y_1$
1	$13 = 1 \times 7 + 6$	q_3	$x_1 = q_3 y_1 + x_2$
1	$7 = 1 \times 6 + 1$	q_4	$y_1 = q_4 x_2 + y_2$
6		x_2	
1		y_2	

One of the solutions of the problem (but not the smallest one) is $x = 73$, $y = 20$. For finding the smallest solution, we divide 73 by 29 to find x_0; the remainder is the *agra* required:[32]

---

[31] A method still used in our mathematical textbooks.
[32] This agrees with the general solution of $Ax - By = C$, being $x = x_0 + Bt$ and $y = y_0 + At$ (x_0 and y_0 are the smallest solutions). From which: $x_0 = X - Bt$, if X is any solution of the equation.

$73 - 29n = 15$ ($agra_1$).

In the same way we can find $agra_2 = 11$ as the smallest solution of

$17x - 45y = 7$.   (b)

According to Clark's interpretation, $agra$ is the smallest solution for x in an equation with two unknowns.[33] In fact, "the rule [given by Âryabhaṭa] applies only to the third process. . . . The solution of the single indeterminate equation is taken for granted and is not given in full."[34]

Thus,

$p = 15$ is the $agra$ of $8x = 29y + 4$;
$q = 11$ is the $agra$ of $17x = 45z + 7$.

We must now find a value of x satisfying both equations. The general solution[35] of (a) is $x = 15 + 29t$; of (b), $x = 11 + 45t'$. Thus, $29t + 15 = 45t' + 11$ and $45t' - 29t = 4$.

We solve in the same manner as above, and find $t' = 22$, $t = 34$. The general solution is

$x = 29t + 15$,
$x = 29 \times 34 + 15 = 1{,}001$.

This equation that Clark gives, following Parameśvara's commentary, is very close to the original text. Ganguli, however, holds that Âryabhaṭa's method is not identical with Brahmagupta's and that "for the same reason the interpretation given by the Sanskrit commentator Parameśvara cannot be accepted as correct."[36] This matter is by no means decided, and I do not want to enter an arena that is not mine.

---

[33] Clark defines $agra$ as "the remainders which constitute the provisional values of x, that is to say, values one of which will satisfy one condition, one of which will satisfy the second condition of the problem." This statement is somewhat confusing, but one will find the explanation in the mathematical representation. Anyhow, the "values of x" are the smallest values, a fact that is not pointed out by Clark.

[34] Clark (1), p. 47.

[35] This explanation is neither in Âryabhaṭa, nor in Clark. Clark says only: "Then in accordance with the rule $34 \times 29 = 986$ and $986 + 15 = 1{,}001$."

[36] Ganguli (1), p. 115.

Ganguli gives a very different explanation. In his opinion the problem closely resembles the Sun Tzŭ problem[37] and should have the general form

$$N \equiv R_1 \ (\text{mod } A) \equiv R_2 \ (\text{mod } B).$$

The treatment, however, seems to be very different from that of the *ta-yen* rule,[38] as Ganguli states:

"Āryabhaṭa's problem in indeterminate analysis appears to be exactly similar to the one given by Sun Tzŭ. Āryabhaṭa considers only *two* divisors, while Sun Tzŭ contemplates any number of divisors.[39] This difference may, at first sight, appear to be of no importance. But it is fundamental. Accordingly Āryabhaṭa's solution cannot be extended so as to give a solution of Sun Tzŭ's problem."[40]

Ganguli's explanation is very extensive, and it is impossible to give more than a summary of it here. The general problem is[41] $N = Ax + R_1 = By + R_2$, or $Ax + R_1 = By + R_2$, from which $Ax = By + R_2 - R_1$. Putting $R_2 - R_1 = C$, we have $Ax = By + C$. There are two possibilities: $B > A$ or $A > B$. Let us take an example:[42]

$$N \equiv 4 \ (\text{mod } 29) \equiv 7 (\text{mod } 45).$$

We apply the Euclidian algorithm:[43]

---

[37] It is a pity that Ganguli is also a victim of the "disease of historicism"; his chief purpose seems to be to prove that Āryabhaṭa owed his methods neither to Greece nor to China.

[38] As Matthiessen (1) already explained. It is difficult to understand Needham (1), vol. 3, p. 122, note *e*: "The argument of Matthiessen that they were very different does not carry conviction." This matter will be discussed again in Chapter 18. See also Yushkevitch (1), p. 145: "At any rate, the Indian method is quite different from the Chinese one. (Allerdings ist die indische Methode von der chinesischen verschieden)."

[39] This is not entirely true, because there is only one problem having three divisors. But the method is applicable to any number of divisors.

[40] Ganguli (1), p. 114. See Chapter 15.

[41] Ganguli equates the term *agra* with the remainders of the problem.

[42] Ganguli gives only a theoretical explanation without any example.

[43] Clark stops this mutual division when the remainder becomes 1, Ganguli when it becomes 0.

1	1	1	4	3	
45	29	16	13	3	1
29	16	13	12	3	
16	13	3	1	0	

From the chain[44]

$$\begin{array}{cccccc} q_1 & q_2 & q_3 & q_4 & t & r_3 \\ | & | & | & | & | & | \end{array}$$

or 1   1   1   4   1   3

we compute[45]

$t = y_2 = 1$ $x_2 = 3 \times 1 + 3 = 6$

$y_1 = 4 \times 6 + 1 = 25$ $x_1 = 1 \times 25 + 6 = 31$

$y = 1 \times 31 + 25 = 56$ $x = 1 \times 56 + 31 = 87.$

Take α and β as the smallest solutions.

As $x = Bm + \alpha \longrightarrow \alpha = x - Bm = 87 - 45 \times 1 = 42$

$y = Am + \beta \longrightarrow \beta = y - Am = 56 - 29 \times 1 = 27$

$N = B\beta + R_2 = 45 \times 27 + 7 = 1,222.$

[44] This t or *mâti* is any assumed number (zero or any positive integer). The value of $y_2 = t$.
The general rule is: $B = AQ_1 + r_1 \to A(x - Q_1 y) = r_1 y + C \to Ax_1$
$= r_1 y + C$ with $x_1 = x - Q_1 y$. Indeed,
$Ax = By + C$
$Ax = (AQ_1 + r_1) y + C$
$Ax = AQ_1 y + r_1 y + C$
$A(x - Q_1 y) = r_1 y + C.$
$A = Q_2 r_1 + r_2 \to r_2 x_1 = r_1 (y - Q_2 x_1) \to$
$r_2 x_1 = r_1 y_1 + C$ with $y_1 = y - Q_2 x_1$
$r_1 = Q_3 r_2 + r_3 \to r_2 (x_1 - Q_3 y_1) = r_3 y_1 + C \to$
$r_2 x_2 = r_3 y_1 + C$ with $x_2 = x_1 - Q_3 y_1$
$r_2 = Q_4 r_3 + r_4 \to r_4 x_2 = r_3 (y_1 - Q_4 x_2) + C \to$
$r_4 x_2 = r_3 y_2 + C$ with $y_2 = y_1 - Q_4 x_2$.
Suppose that $r_4 = 1$; $r_4 x_2 = r_3 y_2 + C \to x_2 = r_3 y_2 + C$. Take $y_2 = t \to$
$x_2 = r_3 t + C$ giving the solution of the last equation.
[45] In general:
$x = Q_1 y + x_1$ from $x_1 = x - Q_1 y$
$y = Q_2 x_1 + y_1$ from $y_1 = y - Q_2 x_1$
$x_1 = Q_3 y_1 + x_2$ from $x_2 = x_1 - Q_3 y_1$
$y_1 = Q_4 x_2 + y_2$ from $y_2 = y_1 - Q_4 x_2$
$x_2 = r_3 t + C$ from $x_2 = r_3 y_2 + C$
$y_2 = t.$

The working out of the problem is not greatly different from Clark's. But this is true only of the method. The problem to which it is applied is quite different:[46]

Ganguli	Clark
$N = Ax + R_1$	$mx = Ay + R_1$
$N = By + R_2$	$nx = Bz + R_2.$

Clark criticizes Ganguli's explanation of the text and states: "I cannot help feeling that the Sanskrit is stretched in order to make it fit the formula."[47]

Datta and Singh[48] rely on the interpretation of Bhâskara I (525), who was a direct disciple of Âryabhaṭa I.[49] The translation is as follows:

"Divide the divisor corresponding to the greater remainder by the divisor corresponding to the smaller remainder. The residue (and the divisor corresponding to the smaller remainder) being mutually divided (*) the last *residue* should be

---

[46] Of course, Clark's representation can be reduced to Ganguli's, but A, B, R_1, R_2 cannot be positive integers. This method was worked out by Mahâvîra (850) and Śrîpati (1039). See Datta and Singh (1), p. 137.

[47] Clark (1), p. 50. Ganguli also gives an explanation of the case in which $A > B$.

[48] Strictly speaking, this is Datta's interpretation, first published in Datta (1), and takee np in Datta and Singh (1).

[49] This seems to be a reliable reference. I have not seen this commentary. The following translation is given by Datta and Singh (1), p. 99: "Set down the dividend above and the divisor below. Write down successively the quotients of their mutual division, one below the other, in the form of a chain. Now find by what number the last remainder should be multiplied, such that the product being subtracted by the (given) residue (of the revolution) will be exactly divisible (by the divisor corresponding to the remainder). Put down that optional number below the chain and then the (new) quotient underneath. Then multiply the optional number by that quantity which stands just above it and add to the product the (new) quotient (below). Proceed afterwards also in the same way. Divide the upper number (i.e., the multiplier) obtained by this process by the divisor and the lower one by the dividend; the remainders will respectively be the desired *ahargana* and the revolutions." And note 1: "The above rule has been formulated with a view to its application in astronomy." Srinivasiengar [(1), pp. 96 ff] accepts Datta's explanation, which is very clear.

multiplied by such an optional integer that the product being
added (in case the number of quotients of the mutual division
is even) or subtracted (in case the number of quotients is odd)
by the difference of the remainders (will be exactly divisible
by the last but one remainder. Place the quotients of the
mutual division successively one below the other in a column;
below them the optional multiplier and underneath it the
quotient just obtained) (**). Any number below (i.e., the
penultimate) is multiplied by the one just above it and then
added to that just below it. Divide the last number (obtained
by doing so repeatedly) by the divisor corresponding to the
smaller remainder; then multiply the residue by the divisor
corresponding to the greater remainder and add the greater
remainder. (The result will be) the number corresponding to
the two divisors."

According to Datta, the part from (*) to (**) can also be
rendered as follows: ". . . (until the remainder becomes zero),
the last quotient[50] should be multiplied by an optional integer
and then added (in case the number of quotients of the mutual
division is even) or subtracted (in case the number of quotients
is odd) by the difference of the remainders. (Place the other
quotients of the mutual division successively one below the other
in a column: below them the result just obtained and underneath it the optional integer.). . ."

The problem is $N=ax+R_1=by+R_2$. If $R_1-R_2=c$, and
$R_1>R_2$, we have $by=ax+c$. If $R_2>R_1$, we have $ax=by+c$.
Suppose that $R_1>R_2$. The equation is $ax+c=by$. There are
four subcases:
1. The last $r=0$
a) The number of quotients is even[51]
b) The number of quotients is odd
2. The mutual division is stopped at a remainder $r_p\neq0$
a) The number of quotients is even
b) The number of quotients is odd.

[50] As we have seen, Ganguli agrees with "quotient."
[51] The first quotient must be neglected, as is usual with Āryabhaṭa.

The result is exactly the same as in Ganguli's explanation in case (1), but the difference between the interpretations lies in the subdivision of the cases,[52] all corresponding to the general rule.

It seems very likely that Âryabhaṭa's problem is indeed $N \equiv R_1 \pmod{A} \equiv R_2 \pmod{B}$ and that it in some way resembles the Sun Tzŭ problem. But as Clark states, "The general method of solution by reciprocal division and formation of a chain is clear,[53] but some of the details are uncertain and we do not know to what sort of problems Âryabhaṭa applied it."[54] The last question is not difficult to answer; they were chronological problems, as is demonstrated in a very important paper by Van Der Waerden.[55]

Which were the problems Âryabhaṭa was able to solve? They were all of the type

$$by - ax = \pm c. \tag{1}$$

If $R_1 > R_2$, solve the equation $by = ax + c$, or $y = (ax + c)/a$ $(c > 0)$. If $R_1 < R_2$, solve the equation $ax = by + c$, or $x = (by + c)/a$ $(c < 0)$. Equation (1) is arranged so that c is always positive.

Bhâskara I extends this rule[56] to a direct solution of $y = (ax - c)/b$ $(c < 0)$. Brahmagupta provides some improvements on Âryabhaṭa's rule. The most important is: ". . . it is not necessary to continue the operations of mutual division until the remainder becomes zero. We may stop at any stage if we can solve the resulting reduced equation [by inspection]."[57] An-

---

[52] Clark's interpretation being a more general and more complicated case of the same problem.

[53] This is very important because it gives a general rule, indispensable for solving indeterminate equations, especially when the moduli are very large, as in chronological problems.

[54] Clark (1), p. 50.

[55] As early as 1874, Hankel (1) wrote: "[it] owes its origin probably to the very same chronological-astrological problems to which it is so frequently applied. (... verdankt ihren Ursprung vermuthlich denselben chronologisch-astrologischen Aufgaben, auf welche sie von ihnen vielfach angewandt wird...)" (p. 197).

[56] See Datta and Singh (1), pp. 99 f, where some secondary rules are given.

[57] Ganguli (1), p. 130.

other is that we can transform the equation $by=ax+c$ to $ax=by-c$ "so that we shall have to start with the division of b by a,"[58] which makes it unnecessary to have $c>0$.[59]

Bhâskara I was the first to state the rule for solving the special indeterminate equation $by=ax\pm1$, the so-called "constant pulverizer."[60] If the equation $(ax\pm1)/b=y$ is solved,[61] the solution of $(ax\pm c)/b=y$ can easily be derived. Suppose the solutions of $(ax\pm1)/b=y$ are $x=a$ and $y=\beta$. Then $b\beta=a\alpha\pm1$. If we multiply by c, we get $b(c\beta)=a(c\alpha)\pm c$, and $x=c\alpha$, $y=c\beta$ is a solution of $(ax\pm c)/b=y$.

Almost the same rule is given by Brahmagupta and Bhâskara II.[62] For example

$$\frac{17x+5}{15}=y.$$

Solving the equation $\frac{17x+1}{15}=y$,[63] we find that $x=7$, $y=8$.

Multiply by 5:

$$x=7\times5=35 \quad -n\times15=5$$
$$y=8\times5=40 \quad -n\times17=6.$$

This method is adopted by Âryabhaṭa II (950) as the general rule. As he always assumes that $t=0$, this is an important simplification. Moreover, he was the first to formulate clearly the general solutions $x=x_0+bt$ and $y=y_0+at$. Finally, he

[58] Datta and Singh (1), p. 103.

[59] There was little change in the general rule after Âryabhaṭa; Brahma-gupta and Mahâvira are only links in the historical chain. Brahmagupta's work was for a long time the oldest one known, because Âryabhaṭa's work was lost. Colebrooke (1) translated it in 1817; Mahâvira's work was edited and translated by M. Rangacarya (1); see also Aiyar (1). It is impossible to treat these works in detail, because our purpose is only to make an objective comparison between Indian and Chinese methods.

[60] Or *sthira-kuṭṭaka*. For an explanation of this term, see Datta and Singh (1), pp. 117 f. This explanation of the "constant pulverizer" follows Datta and Singh (1), pp. 118 ff.

[61] This equation is solved in the same way as the general equation.

[62] See also Ayyangar (1).

[63] Given by Bhaskara II. See Datta and Singh (1), p. 120; Taylor (1), p. 112; Gurjar (1), pp. 115 f.

provided an interesting series of reducing rules, also applied by Bhâskara II.[64]

Brahmagupta, Bhâskara II, and Nârâyana (1350) give rules for the solution of the equation $by + ax = \pm c$.[65] We shall pass over all the special cases the Indian mathematicians have solved and investigate the "general problem of remainders."[66]

$$N = a_1 x_1 + r_1 = a_2 x_2 + r_2 = a_3 x_3 + r_3 = \ldots = a_n x_n + r_n.$$

The general solution is given by Bhâskara I and Bhâskara II:

$$a_1 x_1 + r_1 = a_2 x_2 + r_2. \tag{1}$$

Suppose that a_1 is the smallest value for x_1; the general solution is $x_1 = a_1 + a_2 t$,

and

$$N = a_1(a_1 + a_2 t) + r_1 = (a_1 a_1 + r_1) + a_1 a_2 t.$$

We can equalize to the third equation:

$$N = (a_1 a_1 + r_1) + a_1 a_2 t = a_3 x_3 + r_3,$$

or

$$a_1 a_2 t + (a_1 a_1 + r_1) = a_3 x_3 + r_3.$$

Suppose that a_2 is the smallest solution for t; the general solution is $t = a_2 + a_3 u$, and

$$N = a_1 a_2 a_3 u + (a_1 a_2 a_2 + a_1 a_1 + r_1).$$

We can proceed in the same way:

$$N = a_1 a_2 a_3 a_4 v + (a_1 a_2 a_3 a_3 + a_1 a_2 a_2 + a_1 a_1 + r_1) \ldots ,$$

until all the equations are used.[67]

EXAMPLES

1. To solve the equations[68]

---

[64] All these rules are included in Datta and Singh (1), pp. 111 ff; Gurjar (1), p. 115 ff gives the same rules with examples.
[65] See Datta and Singh (1), pp. 120 ff.
[66] See Datta and Singh (1), 131 ff; Yushkevitch (1), pp. 145 f.
[67] A thorough analysis suffices to prove Yushkevitch correct in his statement that the Indian and Chinese methods are completely different.
[68] Given by Bhâskara I. See Datta and Singh (1), p. 133.

$N = 8x + 5 = 9y + 4 = 7z + 1.$

For solving $N = 8x + 5 = 9y + 4$, we use the pulverizer to find that $x = 1$, $y = 1$, and $N = 13$.[69] The general solutions are $x = 1 + 9t$ and $y = 1 + 8t$. Substituting in the original equations, we get:

$8(1 + 9t) + 5 = 7z + 1$
or $72t + 13 = 7z + 1.$

The solution for t is 36, or in general $t = 36 = a + 7u$, from which $t_0 = 1$ $(a + 7u = 36 \longrightarrow a = 1)$. $N = 72 \times 1 + 13 = 85$ is the general solution of the problem.

2. $N = 2x + 1 = 3y + 1 = 4z + 1 = 5u + 1 = 6v + 1 = 7w$.[70] Bhâskara I gives the answer 721.

3. $N \equiv 5 \pmod 6 \equiv 4 \pmod 5 \equiv 3 \pmod 4 \equiv 2 \pmod 3$.[71]

Were the Indian mathematicians aware of the conditions of solvability? In his commentary on Brahmagupta, Pṛthûdakaswâmî investigates the case where the moduli are not rela-

---

[69] Following Bhâskara II's method. Here the equation is reduced to $8x + 1 = 9y$.

$\dfrac{8}{9} = (0, 1, 8)$

0	1
1	1
1	
0	

In the same way, we treat $72t + 12 = 7z$

$\dfrac{72}{7} = (10, 3, 2)$

10
3
12
0

$3 \times 12 + 0 = 36 = t.$

[70] Also from Bhâskara I. This problem is also treated by Ibn al-Haitham (c. 1000). See Dickson (1), p. 59, and Wiedemann (1), p. 83; Fibonacci includes it in his *Liber Abacci*, p. 281.

[71] See Datta and Singh (1), pp. 134 f.

tively prime.[72] The text says: "Wherever the reduction of two divisors by a common measure is possible, there 'the product of the divisors' should be understood as equivalent to the product of the divisor corresponding to the greater remainder and quotient of the divisor corresponding to the smaller remainder as reduced (i.e. divided) by the common measure. . . ."

Suppose that $N = a_1x_1 + r_1 = a_2x_2 + r_2$, and that the greatest common denominator of a_1 and $a_2 = g$.[73] Then $N = (a_1a_1 + r_1) + a_1a_2t$ can be written as

$$N = (a_1a_1 + r_1) + \frac{a_1a_2}{g}t.$$

We shall conclude this treatment of Indian methods with an evaluation.

1. There is a striking continuity in mathematical works on indeterminate analysis, which puts any study of their contents on a sure basis. This is not the case in China.

2. The methods are on a very high level,[74] and a large number of them are still used in our mathematical textbooks.

3. Modern Indian scholars are of the opinion that all knowledge of indeterminate analysis must have been derived from Indian works. Historical influence seems to have been very great, but internal analysis gives a more reliable basis for comparison than vague historical statements. Chapter 18 of this work is devoted to a comparison of the Indian *kuṭṭaka* and the Chinese *ta-yen* rule. For this reason a thorough analysis of the Indian methods was a necessity.[75]

---

[72] See Datta and Singh (1), p. 132 (the text) and Yushkevitch (1), p. 146 (interpretation).

[73] Where $r_2 - r_1$ is divisible by G.C.D. (a_1, a_2). This general condition is not expressed.

[74] ". . . As for methods, the Indian works have an aspect of generality that brings them close to the works of modern authors, something that neither Greek nor Arab mathematics succeeded in attaining." (. . . Quant aux méthodes, les travaux des Indiens ont un caractère de généralité qui les rapproche de ceux des modernes, et auquel ni les mathématiques des Grecs, ni celles des Arabes n'ont réussi à s'élever.) Woepcke (1), p. 32.

[75] For all the sources relied on in this study, the reader is referred to the general bibliography.

ISLAM

Little information concerning the Chinese remainder problem is to be found in Islamic mathematical works. Abû Kâmil al-Miṣrî[76] (c. 850–930) treats of indeterminate problems, but they are all of the "hundred fowls" type. "Abû Kâmil's procedure in this work, however, is less systematic, and he finds his solutions by trial."[77] As we shall see in the next chapter, the "hundred fowls" problem is to be found for the first time in the work of Chang Ch'iu-chien (c. 475). Similar problems appear in the works of Mahâvîra (ninth century) and Bhâskara II (twelfth century).

In the work of Ibn al-Haitham there is a real remainder problem,[78] identical in every respect to a problem of Bhâskara I[79] and a problem in Fibonacci (1202).[80]

Ibn al-Haitham's problem reads as follows: "Find a number, that divided by 2, 3, 4, 5, 6 has the remainder 1, and divided by 7 has no remainder." The author gives two methods. The first is $N = 2 \times 3 \times 4 \times 5 \times 6 + 1 = 721$, the solution used by Bhâskara I. However, this is not the smallest solution, because Ibn al-Haitham does not take into account the fact that some of the divisors are not relatively prime.[81] This method is of course not a general one.

The second method is the equation $N = \frac{3}{4}(6 + 2n \times 7)20 + 1$, where n is an integer such that $6 + 2n \times 7$ is a multiple of 4. If $n=1$, then $N = \frac{3}{4} \times 20 \times 20 + 1 = 301$. If $n=2$, then $N=721$, and

---

[76] His full name is Sogâ ben Aslam ben Muhamet ben Sogâ, Abû Kâmil. For a general description of his work, see Suter (2), p. 43. His algebraic work is translated by J. Weinberg (1), with the exception of the indeterminate problems. These have been translated by Suter (1) *(Das Buch der Seltenheiten der Rechenkunst.)* According to Weinberg (1), the Staatsbibliothek München has another work of Abû Kâmil on indeterminate problems. See also A. Mieli (1), p. 108. There is a discussion of some of Abû Kâmil's problems in Loria (1), p. 147, and (3), p. 153; Yushkevitch (4), pp. 232 ff; Tropfke (1), vol. 3, pp. 103 f; other articles on the algebra of Abû Kâmil are Karpinski (2) and (3).

[77] Martin (1), p. 8.

[78] Translated by Wiedemann (1). See Dickson (1), vol. 2, p. 59. There is a general description of his life and work in Suter (2), pp. 91–95.

[79] Datta and Singh (1), p. 133.

[80] (1), vol. 1, p. 281.

[81] See Yushkevitch (4), p. 146.

so on. It is obvious that $\frac{3}{4}(6+2n\times7)20$ is divisible by 2, 3, 4, 5, and 6. That $\frac{3}{4}(6+2n\times7)20+1$ is divisible by 7 follows from $\frac{3}{4}(6+2n\times7)20+1=90+30n\times7+1=91+30n\times7=(13+30n)\times7$.[82] The solutions are 301, 721, ...

The work of Abû Bakr al-Farajî[83] is entirely within Diophantos's sphere of influence.[84] Abû Bakr includes some indeterminate problems of the first degree,[85] but as Woepcke states: "Actually, it is only the statement of these problems that is indeterminate; the author makes them determinate at the outset by arbitrarily choosing the values of one or more unknowns. In most cases, however, he points out what is arbitrary in the solution provided. Like Diophantos, from whom several of these problems are borrowed, the author does not exclude fractional values. Thus we cannot think of this as a method for the solution of indeterminate equations of the first degree similar to those of Indian or modern mathematicians."[86] Consequently this kind of problem is not of use in our inquiry.[87]

[82] The same result, 301, was found in India by Sûryadeva Yajvâ. See Datta and Singh (1), p. 133. Fibonacci gives the following solution:
$N-1 \equiv 0 \pmod 2 \equiv 0 \pmod 3 \equiv 0 \pmod 4 \equiv 0 \pmod 5 \equiv 0 \pmod 6$.
Take $N-1 = \text{L.C.M.} (2, 3, 4, 5, 6) = 60; \quad 60 \equiv 4 \pmod 7$.
We have to find $N-1 \equiv 6 \pmod 7$.
$N-1 = 120 \equiv 1 \pmod 7$
$N-1 = 180 \equiv 4 \pmod 7$
$N-1 = 240 \equiv 5 \pmod 7$
$N-1 = 300 \equiv 6 \pmod 7$
The solution is $N = 300+1 = 301$.
Fibonacci gives the general solution
$N = 301+n\times2\times3\times4\times5\times6\times7 = 301+n\times420$.
[83] Muh. b. el-Hasan, Abû Bekr, el-Karchî (died c. 1029).
[84] Diophantos's work was translated by Abu'l Wafâ (940–998) into Arabian.
[85] See Woepcke (1), p. 10.
[86] Ibid. ("En vérité, ce ne sont que les énoncés de ces problèmes qui soient indéterminés; l'auteur rend ces problèmes tout de suite déterminés, en choisissant arbitrairement la valeur d'une ou de plusieurs inconnues. Cependant, il fair ressortir, dans la plupart des cas, ce qu'il y a d'arbitraire dans la solution donnée. Comme Diophante, auquel plusieurs de ces problèmes sont empruntés, l'auteur n'exclut pas des valeurs fractionnaires. Il ne s'agit donc pas ici d'une méthode semblable à celle des Indiens ou des modernes, pour la résolution des équations indéterminées du Ier degré.")
[87] It is very likely that there still exist other Arabian works on indeterminate

EUROPE

INTRODUCTORY NOTE

If one were to write a detailed history of indeterminate analysis in Europe, it would be necessary to examine all the existing works on arithmetic before modern times.[88] Some of these writers treat the remainder problem, for example, Peurbach, Koebel, Jacob, Rudolff, Cardano, and Tartaglia. However, since a large number of these works say nothing new about the matter, we must, in order to keep this outline within reasonable limits, restrict ourselves to the works mentioned in Dickson (1), who omits only the important Göttingen manuscript. It is necessary to remember that the purpose of this survey is to gather material for a comparison with the Chinese *ta-yen* rule.

The oldest example of an indeterminate problem in Europe appears in the *Propositiones ad acuendos juvenes,* attributed to Alcuin (730?–804);[89] however, as this problem has nothing to do with the remainder problem, it is important only for the investigation of possible historical relationships.

LEONARDO PISANO (FIBONACCI)

In the *Liber Abbaci* (1202) of Leonardo Pisano (c. 1170–c. 1250), a contemporary of Ch'in Chiu-shao, there are two problems in which the *ta-yen* rule is used.[90] Because the solution of the problems is of help in making a truly scientific comparison,[91] the whole text is provided here:

"Dividat excogitatum numerum per 3, et per 5, et per 7; et

---

analysis. One need only examine Suter (2), where there is a list of 528 names of authors on Arabian mathematics and astronomy. Most of their works exist only in manuscript form, and the greater part have never been studied or translated.

[88] A list of some of them is given in Hofmann (1), pp. 142–145.

[89] The text is published in Migne (1). See Vogel (3), p. 223.

[90] Vol. 1, p. 304.

[91] And not the sort of comparison Van Hee makes, between Ch'in Chiu-shao (1247) and Gauss (1801): "The operations are the same. But what a difference in theory! In the mind of the great German mathematician a sure, methodical progression, stripped of useless details; in the case of the yellow algebraists, obscurity, repetition, fumbling, and, it seems, no idea of uni-

semper interroga, quot ex unaquaque divisione superfuerit. Tu vero ex unaquaque unitate, que ex divisione ternarii superfuerit, retine 70; et pro unaquaque unitate, que ex divisione quinarii superfuerit, retine 21; et pro unaquaque unitate, que ex divisione septenarii superfuerit, retine 15. Et quotiens numerus super excreverit tibi ultra 105, eicias inde 105; et quod tibi remanserit, erit excogitatus numerus. Verbi gratia: ponatur, quod ex divisione ternarii remaneant 2; pro quibus retineas bis septuaginta, id est 140; de quibus tolle 105, remanent tibi 35. Et ex divisione quinarii remanent 3; pro quibus retine ter 21, id est 63, que adde cum predictis 35, erunt 98. Et ex divisione septenarii remaneant 4; pro quibus quater 15 retinebis, id est 60; que adde cum 98 predictis, erunt 158; ex quibus eice 105, remanebunt tibi 53; que erant excogitatus numerus.

"Procedit enim ex hac regula pulchrior divinatio, videlicet ut si quis tecum noverit hanc regulam; et aliquis ei privatim dixerit aliquem numerum, tunc ille tuus consocius, non interrogatus, tacite dividat numerum sibi dictum per 3, et per 5, et per 7, predicta ratione; et quod ex qualibet divisione remanserit, per ordinem tibi dicat; et sic poteris scire numerum sibi privatim dictum."[92]

formity capable of tying this most interesting system under discussion to the general principles of numbers." (Ce sont les mêmes opérations. Mais quelle différence dans la théorie! Chez le puissant mathématicien allemand une marche méthodique, sûre, et sobre de détails inutiles; chez les algébristes jaunes, des obscurités, des redites, des tâtonnements et, semble-t-il, aucune idée d'ensemble capable de relier l'intéressant système en discussion aux principes généraux des nombres.) (12), p. 448. It is a pity that Van Hee never saw the "Chinese" solution Fibonacci gives for the remainder problem.

[92] "Let a contrived number be divided by 3, also by 5, also by 7; and ask each time what remains from each division. For each unity that remains from the division by 7, retain 70; for each unity that remains from the division by 5, retain 21; and for each unity that remains from the division by 7, retain 15. And as much as the number surpasses 105, subtract from it 105; and what remains to you is the contrived number. Example: suppose from the division by 3 the remainder is 2; for this you retain twice 70, or 140; from which you subtract 105, and 35 remains. From the division by 5, the remainder is 3; for which you retain three times 21, or 63, which

The second problem is as follows:[93]

"Precipe ut numerum, quem in corde suo posuerit, dividat per 5, et per 7, et per 9 ad modum antecedentis regule: et singulariter interroga, quid ex unaquaque divisione remaneat; et pro unaquaque unitate, que ex divisione quinarii remanserit, retine 126; et pro qualibet unitate, ex septenario remanente, 225; et pro qualibet ex novenario, 280; et semper cum summa excreverit, ita ut possit inde extrahere 315, eice ea inde quotienscumque poteris; et quod tibi in fine remanserit, erit quesitus numerus."[94]

The first problem is $N \equiv 2 \pmod 3 \equiv 3 \pmod 5 \equiv 4 \pmod 7$.[95]
The solution is:

	r=1	r	
mod 3	70	2	$2 \times 70 = 140 - 105 = \quad 35$
mod 5	21	3	$3 \times 21 = \quad 63 \qquad 63$
mod 7	15	4	$4 \times 15 = \quad 60 \qquad 60$
			$\overline{\qquad\qquad 158}$
			-105
			$\overline{\qquad\qquad 53}$

you add to the above 35; you get 98; and from the division by 7, the remainder is 4, for which you retain four times 15, or 60; which you add to the above 98, and you get 158; from which you subtract 105, and the remainder is 53, which is the contrived number.

"From this rule comes a more pleasant riddle [game], namely if someone has learned this rule with you; if somebody should say some number privately to him, then your companion, not interrogated, should silently divide the number for himself by 3, by 5, and by 7 according to the above-mentioned rule; the remainders from each of these divisions, he says to you in order; and in this way you can know the number said to him in private."

[93] Vol. I, p. 304.

[94] "Bid someone to divide a number of which he is thinking by 5, by 7, and by 9 according to the preceding rule; and separately ask for the remainder from each division; and for each unity that remains from the division by 5, retain 126; for each unity that remains from the division by 7, retain 225; and for each unity that remains from the division by 9, retain 280; and each time as the sum should become too large, so that you can subtract 315, subtract it as many times as you can; and what finally remains should be the number asked for."

[95] The Sun Tzŭ problem is $N \equiv 2 \pmod 3 \equiv 3 \pmod 5 \equiv 2 \pmod 7$. The two problems are not entirely identical, as we find in Tropfke (1), vol. 3, p. 103; Cantor (2), vol. 2, p. 26.

For the second problem, only the *yen-shu*[96] are given:

mod 5   126
mod 7   225
mod 9   280

The *yen-mu*=315.

On page 231 of the *Liber Abbaci* the following problems are given:

1.
$N \equiv 1 \ (\text{mod } 2) \equiv 1 \ (\text{mod } 3) \equiv 1 \ (\text{mod } 4) \equiv 1 \ (\text{mod } 5) \equiv$
    $1 \ (\text{mod } 6) \equiv 0 \ (\text{mod } 7)$

The solution can be obtained as follows:[97]

$N-1 \ \ = \text{L.C.M.} \ (2, 3, 4, 5, 6) = 60$
    $60 = 7 \times \ \ 8 + 4$
$2 \times 60 = 7 \times 17 + 1$
$3 \times 60 = 7 \times 25 + 5$
$4 \times 60 = 7 \times 34 + 2$
$5 \times 60 = 7 \times 42 + 6$
and $N = 5 \times 60 + 1 = 7 \times 42 + 7$ or $N = 301$.
The general solution is $N = 301 + 420n$.[98]

2.
$N \equiv 1 \ (\text{mod } n) \equiv 0 \ (\text{mod } 11)$, where $n = 2, 3, \ldots, 10$.

3.
$N \equiv 1 \ (\text{mod } n) \equiv 0 \ (\text{mod } 23)$, where $n \equiv 2, 3, \ldots, 22$.

[96] In the terminology of Ch'in Chiu-shao; for the explanation of this term, and of *yen-mu*, see Chapter 17.
[97] "Eritque numerus ille 60; quem divide per 7, superant 4, qui vellent esse 6. Ideo quia totus numerus per 7 dividatur; ergo numerus, qui fuerit unum minus eo, cum per 7 dividatur, 6 inde superare necesse est, hoc est 1, minus septenario numero: quare duplicetur 60, vel triplicetur, vel multiplicetur per alium quemlibet numerum, donec multiplicatio ascendat in talem numerum, qui cum dividatur per 7, remaneant inde 6, etc." (p. 282).
[98] The same problem appears in Bhâskara I (522) and Ibn al-Haitham (c. 1000). This last or another Arabian mathematical work may have been Fibonacci's source.

4.
$N \equiv 1 \pmod{2} \equiv 2 \pmod{3} \equiv 3 \pmod{4} \equiv 4 \pmod{5} \equiv$
$\quad 5 \pmod{6} \equiv 0 \pmod{7}$.[99]

5.
$N \equiv 1 \pmod{2} \equiv 2 \pmod{3} \equiv 3 \pmod{4} \equiv 4 \pmod{5} \equiv$
$\quad 5 \pmod{6} \equiv 6 \pmod{7} \equiv 7 \pmod{8} \equiv 8 \pmod{9} \equiv$
$\quad 9 \pmod{10} \equiv 0 \pmod{11}$.[100]

Fibonacci does not give the slightest theoretical or general explanation of his method for the solution of the remainder problem, and for this reason his whole treatment is on a level no higher than that of Sun Tzŭ.[101] He does indeed state the *ta-yen* rule, but there is no evidence that he was able to apply it generally. Perhaps he knew only the *yen-shu* of the moduli 3, 5, 7, and 9. In any case his work includes no problem with moduli that are not relatively prime. His algebraical language is entirely rhetorical,[102] according to the custom of his time; and for the purposes of comparison with Ch'in Chiu-shao this fact is very important.

There are several investigations of Fibonacci's sources, and it seems to be beyond all doubt that the greater part of his problems were derived from Arabian mathematicians, namely Al-Khwârizmî and Abû-Kâmil.[103] The latter's work also contains indeterminate problems,[104] and perhaps the *ta-yen*

[99] Comes also from Bhâskara I. See Datta and Singh (1), p. 134. Bhâskara's methods, however, are not the same.

[100] There are some other indeterminate problems in the *Liber Abacci*, on p. 281 and p. 282, but as the method has nothing to do with the *ta-yen* rule, they are not discussed here. These problems can be found also in Dickson (1), vol. 2, p. 59.

[101] See Chapter 15. See also Tropfke (1), vol. 3, p. 105, who says of the indeterminate problem on p. 281 of the first part: "The type of solution does not exceed guessing by very much. Leonardo in a different place...is also familiar with the *ta-yen* rule." (Die Art der Lösung geht nicht viel über Raten hinaus. Leonardo kennt aber auch...an anderer Stelle die Regel *Ta Yen*.)

[102] As is that of Ch'in Chiu-shao; but the latter gives clear diagrams.

[103] Smith (1), vol. 2, p. 382; Yushkevitch (1), pp. 371 ff; see also Bartolotti (1).

[104] However, the *ta-yen* rule is not given in the extant work of Abû-Kâmil.

rule may be derived from an Arabian source. Several scholars
have noticed the remarkable similarity between the *ta-yen* rule
and the problems dealt with in the work of European mathematicians.[105] Perhaps it is not possible to settle this question
until the relationships among the Chinese, Indian, and Islamic
worlds have been studied in more detail and the role of Central
Asia has been elucidated.[106]

ISAAC ARGYROS (1318–1372)

In some editions of the Εισαγωγη Αριθμητικη *(Eisagogè Arith-
mètikè)* of Nichomachus of Gerasa[107] there is an appendix entitled προβληματα Αριθμητικα *(Problèmata Arithmètika),* of which
the fifth problem is the same as the one in Sun Tzǔ's work.
Needham discusses the matter[108] and on the basis of convincing
arguments attributes the problem to Isaac Argyros.[109] Moreover, an analysis of the text seems to justify this interpretation,
since the context highly resembles that of the problems in

---

[105] See Cantor (1), vol. 2, p. 26; Tropfke (1), vol. 3, p. 103; Yushkevitch
(1), p. 380.

[106] Extreme historicism should be avoided. From a logical point of view,
one can try to find a chronological relationship, laying stress on the historical dimension of ideas; but this kind of relationship is very difficult to prove
and involves much supposition and conjecture. Moreover, cultural nationalism sometimes obstructs a scientific judgment, as has often happened in
the case of this particular problem. Another approach is to try to give as
objective a description as possible, in order to discern a pattern. The
validity of the pattern can be determined from comparison. Such an
approach lays stress on the mathematical dimension. For this reason, it
should be pointed out that Fibonacci's knowledge of indeterminate analysis
is far beneath that of Ch'in Chiu-shao, in spite of the fact that he might have
been a greater mathematician than Ch'in and the fact that his work was
an important link in the evolution of world mathematics, which was perhaps
not the case for Ch'in. Curtze in (3), p. 82, concludes from Fibonacci's
text: "Thereby the *ta-yen* rule is proven to have been known to Leonardo
to the extent that it was known to the Chinese, albeit without proof."
(Damit ist also die Regel *ta-yen* in dem Umfange, welchen die Chinesen
kannten, aber ohne Beweis, als Leonardo bekannt nachgewiesen.) This
is true if we accept only the Sun Tzǔ problem for comparison, but Ch'in
Chiu-shao was Fibonacci's contemporary.

[107] So in the *Codex Cizensis*, a Byzantine manuscript of the late fourteenth–
early fifteenth centuries.

[108] (1), vol. 3, p. 34, note *a*.

[109] Ibid.

Fibonacci's *Liber Abbaci*. The title of the problem is *"Μεθοδος, δι'ης αστειως ευρησεις, οιον αριθμον εχει τις επι νουν"* (Method for finding out successfully of which number somebody is thinking). This reminds one of the *"excogitatus numerus"* (contrived number) and the *"Precipe ut numerum, quem in corde suo posuerit . . ."* (bid someone to divide a number of which he is thinking . . .) of Fibonacci.

The translation of the text is as follows:[110]

"If you wish to know no matter what number between 7 and 105 which somebody has in mind, you will find it by the following method. Let the man who has the number in mind mentally subtract 3 as many times as possible, and let him say aloud the remainder under 3—if there is one of course. When he has said it, keep in mind the number 70 for each unity. Thus, if 1 remains, only 70; if 2 remains, two times 70 or 140; if zero remains, you keep nothing. You must pay attention, with the subsequent remainders of subtractions, to this fact whereby there remains no unity. After that let him in the same way subtract 5 as many times as possible and let him say to you this remainder under 5, and take 21 for each unity, and add this number by yourself to the first one, if there is any. After that let him subtract the number 7 in the same manner and let him say the remainder under 7; take 15 for each unity, add all the numbers you have and subtract 105 from the sum as many times as possible—if it is possible of course; and what remains from this is the number you are searching for."

After this, Isaac Argyros gives the example $N = 28$:

$$28 - 9 \times 3 = 1 \longrightarrow \qquad 70$$
$$28 - 5 \times 5 = 3 \longrightarrow 3 \times 21 = 63$$
$$28 - 4 \times 7 = 0 \longrightarrow \qquad \underline{0}$$
$$\overline{133} \quad -105 = 28.$$

THE MUNICH MANUSCRIPT OF THE FIFTEENTH CENTURY (C. 1450)

After Leonardo Fibonacci, no studies on the remainder problem are known to have been produced in Europe until, in the

[110] See Hoche (1), pp. 152 f. I am much indebted to Drs. van Omme-slaeghe for giving me this translation of the Greek text.

fifteenth and sixteenth centuries, mainly in Germany, interest was awakened again. In 1870 Gerhardt published a part of a manuscript kept in the Staatsbibliothek at Munich,[111] and Curtze published another part of it (problems 268–272) in 1895.[112] A recent publication of the whole text is Vogel (3).

The text seems to have been written by more than one author, but the major part was from the hand of a certain Frater Fredericus.[113]

One of the indeterminate problems says:[114]

"Quidam dominus dives habet 4 bursas denariorum, in unaquaque tantum quantum in alia de denariis, quos vult distribuere in viam eleosine quator ordinibus scilicet czeilen [sic] pauperum. In primo ordine pauperum sunt 43 pauperes, in secundo sunt 39 pauperes, in tercio sunt 35 pauperes, in quarto sunt 31 pauperes. Primam bursam distribuit equaliter primo ordini, in fine tamen remanent sibi 41, ita quod ad complendum ordinem deficiunt sibi 2 denarii. Secundam bursam distribuit equaliter secundo ordini, in fine tamen non potest complere, sed habet in residuo 33, sicque ad complendum ordinem deficiunt ei 6 denarii. Terciam bursam distribuit equaliter tercio ordini, in fine tamen remanent sibi 25, sicque ad complendum ordinem deficiunt ei 10 denarii. Quartam bursam distribuit quarto ordini equaliter, in fine tamen est residuum 17 denariorum, sicque ei 14 denarii deficiunt ad complendum ordinem. Queritur nunc, quod fuerunt denarii in una bursa?

"In summa questio habet hoc: invenias usum numerum, qui dum dividitur per 43, post integra quocientis manent in residuo 41; item dum dividitur per 39, manent in residuo 33; item dum dividitur per 35, manent in residuo 25; item dum dividitur per

[111] (1). pp. 141 ff.

[112] (1), pp. 31–74 ("Ein Beitrag zur Geschichte der Algebra in Deutschland im fünfzehnten Jahrhundert").

[113] For more details about this work, see Curtze (1), and Tropfke (1), vol. 3, p. 105.

[114] Curtze (1), p. 64.

31, in residuo manent 17; licet autem non sit solum unus numerus, qui talis est, verum infiniti sunt signabiles. 5458590."[115]

There is not the slightest indication of any method. Curtze[116] supposes, however, that the author was familiar with the *ta-yen* rule, but it is very peculiar that he did not find the smallest solution.[117] Indeed, the solution of $\mathcal{N} \equiv 41$ (mod 43) $\equiv 33$ (mod 39) $\equiv 25$ (mod 35) $\equiv 17$ (mod 31) is $\mathcal{N} = 1,819,500 + 1,819,545\ n$.

In the same manuscript there is also a German text, which is much more interesting.[118] "I wish also to know how many

[115] "A certain rich man has 4 purses of silver coins; in each of them there is the same amount, which he wishes to distribute on the way...[?]. In the first group of poor men there are 43 paupers, in the second there are 39 paupers, in the third there are 35 paupers, in the fourth there are 31 paupers. The first purse he distributes equally to the first group; in the end he has a remainder of 41 coins, or he lacks 2 coins to complete the first group. The second purse he distributes equally to the second group, in the end, he cannot complete the distibution, but he has a remainder of 33, or a shortage of 6 coins. The third purse he distibutes equally to the third group; in the end, however, he has a remainder of 25, or a shortage of 10 coins. The fourth purse he distributes equally to the fourth group; in the end, however, he has a remainder of 17 coins, or a shortage of 14 coins. Find the amount of coins in one purse. In résumé this is the problem: find a number that, divided by 43, after you get the whole quotient, has a remainder of 41; when divided by 39, has a remainder of 33; when divided by 35, has a remainder of 25; when divided by 31, has a remainder of 17. Although there should not be only one number that is like that—indeed the solutions are of an infinite number—[a solution is] 5458590."
[116] (1), p. 67.
[117] Curtze (1) gives the solution by the *ta-yen* rule, and says: "The fact that we are justified in assuming that the solution was arrived at in this manner derives from the evidence that later teachers of arithmetic, such as Rudolf (active 1525), Koebel (active 1520), and Simon Jacob (active 1550) taught the extension into four simultaneous equations, without however indicating how they have found the 'use numbers' [Hülfszahlen, or Chinese *yen-shu*] for certain numbers." (Dass wir berechtigt sind, die Auflösung als in dieser Weise erfolgt anzunehmen, ergiebt sich daraus, dass spätere Rechenlehrer wie Rudolf (n. fl. 1525), Koebel (n. fl. 1520) und Simon Jacob (n. fl. 1550) die Erweiterung auf 4 gleichzeitige Gleichungen lehren, ohne jedoch anzugeben, wie sie die für bestimmte Zahlen mitgetheilten Hülfszahlen gefunden haben.)" (p. 67). Anyhow, if it was a Chinese problem, nobody should agree with Curtze's conclusion.
[118] Curtze (1), pp. 65 f; Gerhardt (1), p. 141; Vogel (3), p. 138 (no. 311).

coins he has in his purse or in his mind.[119] Do it so. Let him count the coins he has by threes, then by fives, then by sevens, and as often as there remains 1 with 3, note 70; and as often as there remains 1 with 5, note 21, and with 7, note 15. After that, add up these numbers, and from this sum subtract the radix;[120] that is, multiply 3 by 5 and 7; this will be 105, as many times as you can [subtract], and what remains, so much he has in mind or in his purse.[121] This example does not go higher[122] than as far as the radix goes, that is, to 105, and one does not have to take more than that."

The last sentence seems to mean that the number asked for must be smaller than 105, and thus only the smallest solution is kept.

The following passage is the most interesting one on the remainder problem that has been preserved from this period, and the first European explanation of the rule.

"You ask, why does one take 70 for 3, and 21 for 5, etc. Do it in this way. If you want the number for 3, multiply 5 by 7, and what you find, divide it by 3; and if the remainder is 1, this number is right;[123] however, if the remainder is more than 1, double the same number, and after that divide by 3, and if the remainder is more than 1, add the same number. Do it as long as the remainder becomes 1.[124] In the same manner, if you want the number for 5, multiply 3 by 7; you get 21; divide it by 5; the remainder is 1; for this reason 21 is the right num-

---

As the text has been published only in medieval German, it is translated here into English.

[119] The text says: "Wie vil pfenning in dem peutel oder im 'synn' hast." "*Synn*" is in another place spelled as "*sinn*." I think it means "mind"; it recalls the "contrived number" (*excogitatus numerus*) of Fibonacci.

[120] A part of this text is in Latin; here the text says: "et ab ista summa subtrahe radicem."

[121] To this point, this is exactly the same problem as the one in the *Sun Tzŭ suan-ching* and in the *Liber Abacci*.

[122] "Item das exempel get nit höher."

[123] 5×7 is the *yen-shu* of 3.

[124] 2: $35 \equiv 2 \pmod 3$
$35 \equiv 2 \pmod 3$
$\overline{70 \equiv 4 \pmod 3} \equiv 1 \pmod 3.$

ber for 5.[125] If you want the number for 7, multiply 3 by 5; you get 15; divide this by 7; the remainder is 1; thus 15 is the right number for 7.[126] The same way for the others [he indicates the accompanying diagram]."

70	21	15	15	10	6	40	45	36	28	21	36
3	5	7	2	3	5	3	4	5	3	4	7
	105			30			60			84	
21	28	36	63	36	28	128	175	120	216	225	280
2	3	7	2	7	9	5	6	7	5	8	9
	42			126			210			360	
1144	936	1782									
9	11	13									
	1287										

This text seems to be of great importance for the history of indeterminate analysis in Europe. The author does indeed know a general procedure for solving the problem; and he is far beyond the level of Fibonacci, since he is surely making use of more than conjecture. Whether this is an example of *"die reine Regel"* (the pure rule), as Curtze says, is another question. The author indeed solves the congruences, but he does not solve them in a general way, as Ch'in Chiu-shao does. He has to build up his "congruence factors" with great patience, looking always for the remainder 1. That seems to be the only reason why his moduli go no higher than 13.[127] The second characteristic of his method is that the moduli must always be relatively prime in pairs. It is difficult to understand why Curtze wishes to prove that there is absolutely no Chinese influence in Europe.[128] In the first place, this is without real

[125] 3×7 is the *yen-shu* of 5.
$21 \equiv 1 \pmod 5$.
[126] 3×5 is the *yen-shu* of 7.
$15 \equiv 1 \pmod 7$.
[127] For that reason, one could consider this as an example of the rule ("die Regel"), but not of the pure rule ("die reine Regel"). For comparison: instead of multiplying 2,345 with 8,457, one can also add up 2,345 times 8,457; this is also a rule, but not a pure rule.
[128] Curtze (1), p. 66.

importance, because it is beyond all doubt that in 1247 Ch'in Chiu-shao solved indeterminate problems in which the moduli are not relatively prime in pairs. On the other hand, the remainder problem became known in Europe through Fibonacci's work (1202), no matter from where he learned it. The Italian works were very well known in Germany, as one can see from the use of several words belonging to Italian mathematical terminology.[129] Is it unreasonable to suppose that the manuscript of Munich derived (perhaps over many links) from Fibonacci? It is in any case to the credit of the author of the manuscript that he was the first to understand the problem in a more general way than his predecessors.

There is another problem in the same manuscript: $N \equiv 2$ (mod 3) $\equiv 2$ (mod 5) $\equiv 3$ (mod 7). Here again we have the problem of Fibonacci (and of Sun Tzŭ). The author calls his method the *"regula posicionis, et dicitur regula falsa"* (the rule of position, also called the false rule, i.e., the rule of false position). This is of course entirely wrong.[130]

REGIOMONTANUS[131]

From the correspondence between Regiomontanus and Bianchini we know something about the knowledge of indeterminate analysis in the fifteenth century. In 1463 Regiomontanus posed the following problem in one of his letters: "Quero numerum, qui si dividatur per 17 manent in residuo 15, eo autem diviso per 13, manent 11 residua, et ipso diviso per 10 manent tria residua: quero, quis sit numerus ille."[132] Bian-

---

[129] Ibid., p. 33.

[130] It is strange to find the same in China. In the *Ssŭ-k'u ch'üan-shu chien-ming mu-lu* of 1782, one can read that the *ta-yen* rule relies on the *t'ien-yüan* algebra, but that the rule of false position, a European method, is much better. Even the *Tzŭ Yüan* gives the same information about the origin of the rule of false position, while it is clear that it was transmitted from China to Europe. See Needham (1), vol. 3, p. 118, and Ch'ien Pao-tsung (11').

[131] Zinner (1) has an extensive study on the life and work of Regiomontanus.

[132] De Murr (1), p. 99; Curtze (2), p. 219. "I ask for a number that divided by 17 has the remainder 15; divided by 13 has the remainder 11; divided by 10 has the remainder 3. I ask you, which is this number."

chini's answer (1464) says: "...Huic quesito multe responsiones dari possent cum diversis numeris, qui propositionem concluderunt, ut 1103, 3313 et alii multi. Sed in hoc non curo laborem expendere, in aliis numeris invenire."[133] From the last sentence, it is obvious that Bianchini did not know the general rule.[134] Regiomontanus wrote in reply: "...bene reddidistis numerum quesitum minimum 1103, secundum autem 3313. Satis est, nam infiniti sunt tales, quorum minimus est 1103. Huic si addiderimus numerum numeratum ab ipsis tribus divisoribus, scilicet 17, 13 et 10, habebitur secundus, item eodem addito resultat tertius, etc."[135]

In the margin Regiomontanus drew the diagram

$$
\begin{array}{rr}
17 & 170 \\
13 & \underline{13} \\
10 & 510 \\
 & \underline{17} \\
 & \overline{2210}
\end{array}
$$

[133] "To this problem many solutions can be given with different numbers, which agree with the problem, such as 1,103, 3,313, and many others. However I do not want to go to the the trouble of finding out other numbers."

[134] Curtze (2), p. 237, n. 1; Tropfke (1), vol. 3, p. 105; Cantor (1), vol. 2, p. 287: "...and even though Bianchini makes it known, through his next remark to the effect that he does not want to go to the trouble of finding further solutions, that he was not aware of the general solution 2,210 $n +$ 1,103, it can under no circumstances be assumed that such questions can find their answers through the mere fumbling of researchers unfamiliar with such matters."(...Und wenn auch Bianchini durch die nachfolgenden Worte, er wolle sich die Mühe nicht geben, weitere Lösungen zu suchen, zu erkennen gibt, dass er die allgemeine Auflösung 2210 $n +$ 1103 nicht besass, so ist doch keineswegs anzunehmen, dass solche Fragen durch blosses Herumtasten ihre Beantwortung finden konnten, ohne dass den Bearbeitern jemals vorher ähnliche Gegenstände vorgelegen hätten.) In any case, it is fortunate for Bianchini that he was not Chinese.

[135] "You have rightly given the smallest number asked for as 1,103, and the second one as 3,313. This is enough, because such numbers, of which the smallest is 1,103, are infinite. If we should add a number computed [by multiplying] these three divisors, namely 17, 13, and 10, we should have the second; in the same manner, by adding the same, the third one results, and so on."

Curtze concludes: ". . .this, together with the marginal notation, makes it obvious that Regiomontanus possessed a complete solution of this problem, as during his lifetime a great many such problems were widely circulated."[136] This is possible, but Regiomontanus does not explain how he found the first solution. And this was indeed not so simple. If we solve the problem according to the *ta-yen* rule, we get

ting-mu	17	13	10
yen-shu	130	170	221
yen-mu		2210	
chi-shu	11	1	1
ch'êng-lü	14	1	1

$14 \times 130 = 1820 \qquad \times 15 = 27,300$
$ 1 \times 170 = 170 \qquad \times 11 = 1,870$
$ 1 \times 221 = 221 \qquad \times 3 = \underline{663}$
$$29,883 - 13 \times 2,210 = 1,103$$

According to Curtze, Regiomontanus performed all these computations by mental arithmetic; but in the margin he did a simple multiplication. And how did he solve the congruences? Curtze's allegation "that he also knew thoroughly the remainder problem, the *ta-yen* rule of the Chinese . . ."[137] is not at all convincing. If we reject all unprovable suppositions, Regiomontanus says only that the numbers Bianchini found[138] are right, but that all the other numbers can easily be found by adding the least common multiple of 17, 13, and 10 to the first number. To pretend that he knew a general method simply because it was known elsewhere in his time[139] is not at all scientific. It is of course impossible to prove the contrary; without texts to rely on, all is mere conjecture.

[136] "Hieraus in Verbindung mit der Randglosse ist klar, dass Regiomontan die vollständige Lösung dieses Problems besass, wie denn zu seiner Lebenszeit dergleichen Aufgaben vielfach umliefen." (2), p. 254, *n.* 1.
[137] ". . .dass er auch das Restproblem, die Regel *ta-yen* der Chinesen, vollständig beherrschte." (2), p. 189.
[138] And, as we read, by "going to trouble" (*laborem expendere*), that is, without a general method.
[139] That is, to the author of the Munich manuscript.

ELIA MISRACHI

Elia Misrachi (1455–1526), a Jewish mathematician, included some remainder problems[140] in his *Sefer-Hamispar*.[141]

1. "What number has the remainder 1 when divided by 2, the remainder 2 when divided by 3, the remainder 3 when divided by 4, the remainder 4 when divided by 5, the remainder 5 when divided by 6 and the remainder 0 when divided by 7?"[142] It is beyond all doubt that this problem is derived from Leonardo Pisano; the method is also entirely the same.

2.
$N \equiv 1 \pmod{2} \equiv 2 \pmod{3} \equiv 3 \pmod{4} \equiv 0 \pmod{5}$ (variant of 1).
3.
$N \equiv 1 \pmod{2} \equiv 1 \pmod{3} \equiv 1 \pmod{4} \equiv 1 \pmod{5} \equiv 1 \pmod{6} \equiv 0 \pmod{7}$.[143]
4.
$N \equiv 1 \pmod{2} \equiv 1 \pmod{3} \equiv 1 \pmod{4} \equiv 0 \pmod{5}$.
5.
$N \equiv 1 \pmod{2} \equiv 2 \pmod{3} \equiv 3 \pmod{4} \equiv 1 \pmod{5} \equiv 5 \pmod{6} \equiv 1 \pmod{7} \equiv 7 \pmod{8} \equiv 8 \pmod{9} \equiv 1 \pmod{10}$.
6.
$N \equiv 0 \pmod{2} \equiv 2 \pmod{3} \equiv 0 \pmod{4} \equiv 0 \pmod{5} \equiv 2 \pmod{6} \equiv 3 \pmod{7} \equiv 4 \pmod{8} \equiv 5 \pmod{9} \equiv 0 \pmod{10}$.

The methods Elia Misrachi used in solving these problems do not include anything new.

MICHAEL STIFEL[144]

Dickson says: "Michael Stifel gave the correct result that if x has the remainders r and s when divided by a and $a+1$, respectively, then x has a remainder $(a+1)r + a^2 s$ when divided by $a(a+1)$."[145]

---

[140] Wertheim (1), pp. 60 f; Steinschneider (1), p. 477.
[141] Not to be confused with the *Sefer-Hamispar* by Abraham ibn Esra (1093–1167).
[142] The same problem is given by Bhâskara I, with a slight variation. See Datta and Singh (1), pp. 134 f. It appears in Fibonacci (1), vol. 1, p. 282.
[143] The same in Bhâskara I, Ibn al Haitham, and Fibonacci.
[144] For the life and works of Stifel, see Müller (1) and Hoppe (1).
[145] Dickson (1), vol. 2, p. 60.

In the *Arithmeticae Liber I* (1544), fol. 38, the rule is of course stated in full. However, everything about this work gives the impression that even for Stifel, the problem was a kind of game.[146] Dickson's modern representation is not correct; the last clause should read: " . . .then $(a+1)r+a^2s$ has the remainder x when divided by $a(a+1)$."[147]

Euler discusses Stifel's problem and demonstrates that this particular case can be deduced from his own general rule.[148]

THE MANUSCRIPT OF GÖTTINGEN (c. 1550)[149]

This is the most interesting study on the remainder problem that we have dealt with so far.[150] The general rule is given in Latin, the rest of the text in medieval German. The text is much too long to translate here, so this account will include only its mathematical explanation.[151] It consists of an entire solution of the remainder problem when the moduli are relatively prime in pairs; even a system for solving the congruences is given. There is also an attempt to solve the problem when the moduli are not relatively prime, but this is not a general rule.

1. With moduli relatively prime in pairs:
$\mathcal{N} \equiv 5 (\text{mod } 7) \equiv 7 (\text{mod } 8) \equiv 6 (\text{mod } 9) \equiv 0 \ (\text{mod } 11)$.
Solution:

(1)	(2)
$8 \times 9 \times 11 = 792$	$7 \times 9 \times 11 = 693$
$792 - n \times 7 = 1$	$693 - n \times 8 = 5$

---

[146] He writes: "Jam si numerus a te electus, qui mihi sit occultus..." (fol. 38, b).

[147] "Et aggregatum illud divido, per numerum qui provenit ex multiplicatione duorum meorum numerorum primo receptorum (...) *tunc apparebit semper numerus a te electus, in residuo divisionis meae.*" (fol. 38, b). See also *Die Coss Christoffs Rudolfs* (1453), fol. 16, b.

[148] Euler (1), par. 18, pp. 27 f.

[149] The MS was published by Curtze (2), pp. 552–558. It is dated 1524. The German part was written by Andr. Alexander (1545). See Hofmann (1), vol. 1, p. 145.

[150] It is rather surprising that we do not find it in Dickson (1).

[151] The author calls himself "Initius Algebras," but this is of course only a pseudonym.

(3)   (4)
$7 \times 8 \times 11 = 616$ $7 \times 8 \times 9 = 504$
$616 - n \times 9 = 4$ $504 - n \times 11 = 9$.

For (1) the remainder is 1; the congruence is solved.[152] For (2) the remainder is 5; it is necessary to "reduce"[153] (*reducirn*). The method is as follows:[154]

$$5 + 5 = \quad 10$$
$$- \quad 8$$
$$2 + 5 = \quad 7$$
$$+ \quad 5$$
$$12 - 8 = \quad 4$$
$$+ \quad 5$$
$$9 - 8 = 1.$$

Count the number of "*loca*":[155] there are five fives. Five is the solution of the congruence.

For (3) we get:

---

[152] "So wir thailen 729 mit 7, restat 1, davon darf die nicht weither reducirt werden...sunder sie pleiben unvorwandelt" (p. 555).

[153] "Aber so wir thailen 693 mit 8, restat 5, die müssen wir reducirn in ein ander zal, so sie in die wirdt gethailt mit 8, das dann unitas pleibt."

[154] This means
$$693 \equiv 5 \pmod 8$$
$$693 \equiv 5 \pmod 8$$
$$2 \times \overline{693 \equiv 2} \pmod 8$$
$$693 \equiv 5 \pmod 8$$
$$3 \times \overline{693 \equiv 7} \pmod 8$$
$$693 \equiv 5 \pmod 8$$
$$4 \times \overline{693 \equiv 4} \pmod 8$$
$$693 \equiv 5 \pmod 8$$
$$5 \times \overline{693 \equiv 1} \pmod 8$$

[155] *Loca* means the number of fives added up.

$$4+4= \quad 8$$
$$+ \ \underline{4}$$
$$12-9= \quad 3$$
$$+ \ \underline{4}$$
$$7+4= \quad 11$$
$$- \ \underline{9}$$
$$2+4= \quad 6$$
$$+ \ \underline{4}$$
$$10-9=1.$$

The number of "*loca*" is seven.

For (4) we get:

$$9+9= \quad 18$$
$$-\underline{11}$$
$$7+9= \quad 16$$
$$-\underline{11}$$
$$5+9= \quad 14$$
$$-\underline{11}$$
$$3+9=12$$
$$-\underline{11}$$
$$1$$

The number of "*loca*" is five.

This method for solving the congruences is naturally suited only for small numbers. However, it must be said that it is very ingenious and simple. Moreover, the author grasped very well the reason why he could solve the congruences in this way: "This is the reason why we examine such 'reductions' by the '*loca*.' If we take the 'reduced number' 504, then the remainder is 9. If we double it (i.e. 504), the 9, as the remainder of it, also doubles. This is the reason why we said 9 and 9 is 18, we have 11 above the remainder, and there is left 7, while two times 504 is 1,008, which divided by 11, has the remainder 7, etc."[156]

[156] "Warumb wir aber solche Reductionen durch die loca examinirt haben, propter hoc fit, das ist die ursach. Wann, so wir nemen die reducirte zal eine, als 504, do pleiben 9, so wir zwispalten, duplirn sich auch 9, das residium in ir... etc." (p. 556).

The text gives the entire explanation of the method for finding unity.[157]

We multiply by the congruence factors:

$792 \times 1 = \quad 792$
$693 \times 5 = 3{,}465$
$616 \times 7 = 4{,}312$
$504 \times 5 = 2{,}520.$

We multiply with the remainders:

$\quad 792 \times 5 = \quad 3{,}960$
$3{,}465 \times 7 = 24{,}255$
$4{,}312 \times 6 = 25{,}872$
$2{,}520 \times 0 = \underline{\qquad 0}$
$\qquad\qquad 54{,}087 = N.$

The most interesting part of the text gives evidence that the author was far advanced beyond conjecture and understood the system thoroughly: "What is the reason why we add them up and multiply them by the remainders; each of the numbers is divisible by all the divisors except by its own corresponding divisor, [in which case] there remains 1. The same 1 of each number is, in the "reduced numbers," multiplied with its remainder, and becomes the remainder of it; and for this reason each 'reduced number' is divisible by all divisors, except its own, if the division leaves a remainder. Therefore the whole sum 54,087 also leaves a remainder when divided by those divisors."[158]

This means that the author had grasped the following system:

---

[157] The author says: "Desgleichen mögen wir thun mit 504, wiewol es nicht not, wann sie khein Residuanten hat, darin sie soll gemultiplicirt werden..." [We can do the same (solving the congruence) with 504 although there is no need, because it has no remainder which it must be multiplied with.] He is well aware of the fact that, if the remainder is 0, the congruence must not be solved.

[158] Pp. 556 f.

$$3960 \equiv 5 \ (\text{mod } 7) \equiv 0 \ (\text{mod } 8) \equiv 0 \ (\text{mod } 9) \equiv 0 \ (\text{mod } 11)$$
$$24255 \equiv 0 \ (\text{mod } 7) \equiv 7 \ (\text{mod } 8) \equiv 0 \ (\text{mod } 9) \equiv 0 \ (\text{mod } 11)$$
$$25872 \equiv 0 \ (\text{mod } 7) \equiv 0 \ (\text{mod } 8) \equiv 6 \ (\text{mod } 9) \equiv 0 \ (\text{mod } 11)$$
$$2520 \equiv 0 \ (\text{mod } 7) \equiv 0 \ (\text{mod } 8) \equiv 0 \ (\text{mod } 9) \equiv 0 \ (\text{mod } 11)$$

$$54087 \equiv 5 \ (\text{mod } 7) \equiv 7 \ (\text{mod } 8) \equiv 6 \ (\text{mod } 9) \equiv 0 \ (\text{mod } 11)$$
$$-n \times 5544 \equiv 0 \ (\text{mod } 7) \equiv 0 \ (\text{mod } 8) \equiv 0 \ (\text{mod } 9) \equiv 0 \ (\text{mod } 11)$$

$$4191 \equiv 5 \ (\text{mod } 7) \equiv 7 \ (\text{mod } 8) \equiv 6 \ (\text{mod } 9) \equiv 0 \ (\text{mod } 11).$$

This is the first text in Europe giving evidence that the whole system was known thoroughly, at least for the case in which the moduli are relatively prime in pairs. Only the method for solving the congruences is not ideal, because it is restricted to small numbers.

2. With moduli not relatively prime in pairs:[159]
$N \equiv 2 \ (\text{mod } 6) \equiv 6 \ (\text{mod } 8) \equiv 4 \ (\text{mod } 10) \equiv 8 \ (\text{mod } 14)$.
The solution is given as follows:

The L.C.M. (6, 8, 10, 14) = 840.
840 : 6 = 140 $140 \equiv 2 \ (\text{mod } 6)$[160]
840 : 8 = 105 $105 \equiv 1 \ (\text{mod } 8) \longrightarrow 630 \equiv 6 \ (\text{mod } 8)$
840 : 10 = 84 $84 \equiv 4 \ (\text{mod } 10)$
840 : 14 = 60 $60 \equiv 4 \ (\text{mod } 14) \longrightarrow 120 \equiv 8 \ (\text{mod } 8)$
140 + 630 + 84 + 120 = 974
974 − 840 = 134.[161]

[159] The author is well aware of the difference between the two methods, for he says: "So do wurden vorgeschlagen divisores, die do communicirn mit einander..." (p. 557).
[160] It is impossible to solve the congruence $140 \equiv 1 \ (\text{mod } 6)$ "und hierumb, das er nicht khan in die loca khumen, dar inn er aufging an der dritten stadt, darumb lassen wir in reducirt und gemultiplicirt sein mit sein restanten 2." Indeed, $140 = 2 \ (\text{mod } 6)$; $2 \times 140 \equiv 4 \ (\text{mod } 6)$; $3 \times 140 \equiv 0 \ (\text{mod } 6)$.
[161] Ch'in Chiu-shao's rule would give the following solution:

	6	8	10	14
	3	8	5	7
	$3 \times 8 \times 5 \times 7 = 840$			
yen-shu	280	105	168	120
chi-shu	1	1	3	1
ch'êng-lü	1	1	2	1
	280	105	336	120
	×2	×6	×4	×8
	560 +	630 +	1,344 +	960

$$= 3{,}494 - n \times 840 = 134.$$

This is not a general method. The reason why the right solution is found is obvious from the following scheme:

$140 \equiv 2 \ (\text{mod } 6) \equiv 4 \ (\text{mod } 8) \equiv 0 \ (\text{mod } 10) \equiv 0 \ (\text{mod } 14)$
$630 \equiv 0 \ (\text{mod } 6) \equiv 6 \ (\text{mod } 8) \equiv 0 \ (\text{mod } 10) \equiv 0 \ (\text{mod } 14)$
$\ \ 84 \equiv 0 \ (\text{mod } 6) \equiv 4 \ (\text{mod } 8) \equiv 4 \ (\text{mod } 10) \equiv 0 \ (\text{mod } 14)$
$120 \equiv 0 \ (\text{mod } 6) \equiv 0 \ (\text{mod } 8) \equiv 0 \ (\text{mod } 10) \equiv 8 \ (\text{mod } 14)$

$974 \equiv 2 \ (\text{mod } 6) \equiv 6 \ (\text{mod } 8) \equiv 4 \ (\text{mod } 10) \equiv 8 \ (\text{mod } 14).$

The second equation $974 \equiv 6 \ (\text{mod } 8)$ is right only because the sum of the two remainders 4 equals 8, being the modulus.[162]

In a last problem, the author gives the reason why this problem is unsolvable:

$N \equiv 4 \ (\text{mod } 5) \equiv 3 \ (\text{mod } 6) \equiv 2 \ (\text{mod } 8) \equiv 1 \ (\text{mod } 9).$

The lowest common multiple of the moduli is 360. The *yen-shu* are 72, 60, 45, and 40. For 60 we have $a \times 6 \equiv 1 \ (\text{mod } 60)$; this is unsolvable, because 60 is divisible by 6.

The reason for unsolvability can be derived from the problem at first glance. The equation $3 \ (\text{mod } 6) = 1 \ (\text{mod } 9)$ is impossible, the only possibilities being $3 \ (\text{mod } 6) = r \ (\text{mod } 9)$, where r is 0, 3, or 6 ; $3 \ (\text{mod } 6) = 2 \ (\text{mod } 8)$ is also unsolvable, because the first is odd and the second even.

The general condition for solvability when the moduli are not relatively prime was unknown to the author, namely, that $r_i - r_j$ must be divisible by the greatest common divisor of m_i and m_j.[163]

C. G. BACHET DE MEZIRIAC[164]

In his *Problèmes plaisans et delectables, qui se font par les nombres* (1612), Bachet de Méziriac gives a solution of the equation

---

[162] In the system of Ch'in Chiu-shao, this scheme becomes:
$\ \ 560 \equiv 2(\text{mod } 6) \equiv 0(\text{mod } 8) \equiv 0(\text{mod } 10) \equiv 0(\text{mod } 16)$
$\ \ 630 \equiv 0(\text{mod } 6) \equiv 6(\text{mod } 8) \equiv 0(\text{mod } 10) \equiv 0(\text{mod } 14)$
$1344 \equiv 0(\text{mod } 6) \equiv 0(\text{mod } 8) \equiv 4(\text{mod } 10) \equiv 0(\text{mod } 14)$
$\ \ 960 \equiv 0(\text{mod } 6) \equiv 6(\text{mod } 8) \equiv 0(\text{mod } 10) \equiv 8(\text{mod } 14)$

$3494 \equiv 2(\text{mod } 6) \equiv 6(\text{mod } 8) \equiv 4(\text{mod } 10) \equiv 8(\text{mod } 14)$

[163] Ch'in Chiu-shao did not know this condition either (see Chapter 17). However, it is fulfilled in his examples.

[164] For the life and works of Bachet de Méziriac, see Itard (1).

$Ax+By=C$. A full description can be found in Dickson.[165] The general treatment is as follows:

The solution of $Ax=By+C$ can be deduced from $Ax=By+1$. Making use of Euclid's algorithm for finding the G.C.D. of A and B, we get:

$$A=Bq_1+r_1 \tag{1a}$$
$$B=r_1q_2+r_2 \tag{1b}$$
$$r_1=r_2q_3+r_3 \tag{1c}$$
$$r_2=r_3q_4+r_4 \tag{1d}$$
$$\vdots \quad \vdots \quad \vdots$$

Suppose that $r_4=1$; then $r_2=r_3q_4+1$, and

$$r_2r_3+1-r_2\equiv0 \ (\text{mod } r_3)\equiv1 \ (\text{mod } r_2). \tag{1}$$

From the first equation,[166] we get $a=q_4r_3-q_4+1$, and

$$ar_3=\beta r_2+1. \tag{2}$$

Multiply (1c) by a:

$$r_1=r_2q_3+r_3 \tag{1c}$$
$$ar_1=ar_2q_3+ar_3.$$

Substitute (2):

$$ar_1=ar_2q_3+\beta r_2+1$$
$$ar_1=(aq_3+\beta)r_2+1.$$

Take $aq_3+\beta=\gamma$

$$ar_1=\gamma r_2+1. \tag{3}$$

Multiply (1b) by γ:

$$B=r_1q_2+r_2 \tag{1b}$$
$$\gamma B=\gamma r_1q_2+\gamma r_2.$$

---

[165] Dickson (1), vol. 2, pp. 44 f.
[166] We can prove the first equation by substituting r_2 by r_3q_4+1; then we have $r_2r_3+1-r_2=r_2 \ (r_3-1)+1=r_3q_4+1$, and $(r_3-1)+1=a \ r_3$, where $a=q_4r_3-q_4+1$.

Substitute (3):

$\gamma B = \gamma r_1 q_2 + a r_1 - 1$
$\gamma B = (\gamma q_2 + a) r_1 - 1.$

Take

$\gamma q_2 + a = \delta$
$\gamma B = \delta r_1 - 1$
$\delta r_1 = \gamma B + 1.$ (4)

Multiply (1a) by δ:

$A = B q_1 + r_1$ (1a)
$\delta A = \delta B q_1 + \delta r_1.$

Substitute (4):

$\delta A = \delta B q_1 + \gamma B + 1$
$\delta A = (\delta q_1 + \gamma) B + 1.$

Take $\delta q_1 + \gamma = \varepsilon$. From this we obtain the solutions

$x = \delta$
$y = \delta q_1 + \gamma = \varepsilon.$

Example: [167]

$67x = 60y + 1$
$67 = 60 \times 1 + 7$
$60 = 7 \times 8 + 4$
$7 = 4 \times 1 + 3$
$4 = 3 \times 1 + 1$
$a = q_4 r_3 - q_4 + 1 = 1 \times 3 - 1 + 1 = 3$
$\beta = \dfrac{a r_3 - 1}{r_2} = \dfrac{3 \times 3 - 1}{4} = 2$
$\gamma = a q_3 + \beta = 3 \times 1 + 2 = 5$
$\delta = \gamma q_2 + a = 5 \times 8 + 3 = 43 = x$
$\varepsilon = \delta q_1 + \gamma = 43 \times 1 + 5 = 48 = y.$

Matthiessen is quite right in stating: "It [Bachet's method] agrees completely with the *kuṭṭaka* method of the Indian math-

[167] See Dickson (1), vol. 2, p. 44.

ematicians."[168] The only difference is that Bachet does not use "the assumed number t,"[169] but the direct formula $a = q_4r_3 - q_4 + 1$.

In Bachet's work there are other remainder problems that Dickson does not mention. On page 127 we read: "I ask for a number that, when divided by 2, leaves a remainder of 1; when divided by 3, leaves a remainder of 1; and likewise when divided by 4, 5, or 6 every time leaves 1; but, when divided by 7, leaves nothing."[170]

Bachet states that the problem can be changed as follows: $N \equiv 1 \pmod{60} \equiv 0 \pmod 7$, where $60 =$ the least common multiple of $(2, 3, 4, 5, 6)$. But the condition of solvability is that 2, 3, 4, 5, 6 be relatively prime with 7. Bachet gives a proof of this condition.

Concerning his method, he says: "But inasmuch as the construction of this problem is rather difficult and the demonstration [of it] too long, I do not wish to include it here. Therefore until my book on elements [*Eléments arithmétiques*] is brought to light, you can by a little trial and error find the number sought for in this manner. You must double, triple, quadruple, and so continue to multiply the number 60, until you find a number that, increased by one, is a multiple of 7. Thus multiplying 60 by 5 will produce 300; adding 1 to this you will have 301, the number sought for."[171]

---

[168] "Sie [Bachet's method] stimmt mit der Cuttaca der Inder vollständig überein." Matthiessen (1), p. 79.

[169] See the section on India in this chapter.

[170] "Je demande un nombre qui estant divisé par 2, il reste 1; estant divisé par 3, il reste 1; et semblablement estant divisé par 4, ou par 5, ou par 6, il reste tousiours 1; mais estant divisé par 7, il ne reste rien." It is the same problem already met with in the works of Bhâskara I, Ibn al-Haitham, Fibonacci, and Elia Misrachi. This must have been a very popular problem, for it is given in the form of a story: "Une pauvre femme portant un panier d'oeufs pour vendre au marché...."

[171] "Mais d'autant que la construction de ce problème est assez difficile, et la démonstration trop longue je ne la veux apporter icy. Parquoy en attendant que mon livre des elemens soit mis en lumière on pourra tastonnant quelque peu trouver le nombre cherché en cette sorte. Il faut doubler,

On page 131 he gives the problem $\mathcal{N} \equiv 1 \ (\mathrm{mod}\ 2) \equiv 2 \ (\mathrm{mod}\ 3)$ $\equiv 3 \ (\mathrm{mod}\ 4) \equiv 4 \ (\mathrm{mod}\ 5) \equiv 5 \ (\mathrm{mod}\ 6) \equiv 0 \ (\mathrm{mod}\ 7),$[172] and the solution $\mathcal{N} = a \times 60 - 1 = \beta \times 7 + 1$ or $2 \times 60 = 17 \times 7 + 1$, and so $\mathcal{N} = 119$.

CASPAR ENS

In his *Thaumaturgus Mathematicus* (Munich, 1636), Caspar Ens gives the problem already met with in Fibonacci, vol.1, p. 281: "Numeros, ex quorum facta per 2.3.4.5.6 divisione unitam residua sit, per 7. vero nihil remaneat, in usum arithmeticum adinvenire."[173] He says that the same problem is treated by C. G. Bachet.[174] His own solution, however, is the same as Fibonacci's; on the other hand, his method does not agree with that of Fibonacci, but with that of Ibn al-Haitham: $\mathcal{N} = 2 \times 3 \times 4 \times 5 \times 6 + 1 = 721 - 420 = 301$.

FRANS VAN SCHOOTEN

In his *Exercitationum Mathematicarum liber primus,* published in 1657, there is, beginning on p. 407, a chapter entitled "De modo inveniendi numeros qui per datos divisi certos post divisionem relinquant."[175] The method Van Schooten[176] gives is on a very low mathematical level compared with the Göttingen manuscript of a century before. However, from the work of Van Schooten we can see that knowledge of the (simple) *ta-yen* rule was widespread at this time.

---

tripler, quadrupler, et ainsi continuellement multiplier le nombre 60, jusques à ce que l'on trouve un nombre qui accreu de l'unité soit mesuré par 7. Ainsi multipliant 60 par 5 viendra 300, auquel adjoustant 1 on aura 301 le nombre cherché."

"Le livre des elemens" was never published; it exists only in manuscript form. See Itard (1), p. 36 and p. 48.

172 The general solution of this kind of problem had not yet been discovered in Bachet's time. He says himself that Sfortunati (c. 1500) and Tartaglia (1500–1557) did not understand the method.

173 Ens (1), p. 70, Problema LIV.

174 "Quaestio a Gasparo Bacheto subtilissime pertractatur."

175 "Method for finding numbers that divided by given numbers leave certain numbers after division." A Dutch translation, *Eerste Bouck der Mathematische Oeffeningen,* was published by Van Schooten in 1660.

176 On the Van Schooten family, see Smith (1), vol. 1, p. 425; on the life and works of Van Schooten, see Hofman (2).

Van Schooten solves two remainder problems. The first is $N \equiv 1 \pmod 2 \equiv 1 \pmod 3 \equiv 1 \pmod 5 \equiv 0 \pmod 7$. He gives the general representation $dz = au + 1 = bx + 1 = cy + 1$, from which $dz - 1 = au + bx + cy$.

He gives the solution[177]

$2 \times 3 \times 5 = 30$	$30 + 1 = 31$	indivisible by 7
($\times 2$) $= 60$	$60 + 1 = 61$	indivisible by 7
($\times 3$) $= 90$	$90 + 1 = 91$	divisible by 7.

But, he says, as this method is too long, do it thus:

$30 = 4 \times 7 + 2$
$90 = 12 \times 7 + 6$ $91 = 12 \times 7 + 7$.

The other solutions can been found by adding 210, the least common multiple of 2, 3, 5, and 7. This problem is of course a very special one, and the method used cannot be general.

In the second problem,

$N \equiv 2 \pmod 7 \equiv 1 \pmod{11} \equiv 9 \pmod{13}$
or $7x + 2 = 11y + 1 = 13r + 9$. We find
$7x + 1 = 11y = 13z + 8$.

One has to take a multiple of 11 such that

this 11-fold $- 1 = $ 7-fold
this 11-fold $- 8 = $ 13-fold.

Then he takes 99, but he does not say why,[178]

$99 - 1 = 98 = 7 - $ fold
$99 - 8 = 91 = 13 - $ fold

From this, we derive:

$7x + 2 = 98 + 2 = 100$
$11y + 1 = 99 + 1 = 100$
$13r + 9 = 91 + 9 = 100$

[177] "E qua liquet, ad quaesitum numerum obtinendum, opus tantum esse quaerere numerum, qui dividi possit per 2, 3 & 5, et si unitate augeatur, dividi queat per 7" (p. 407).
[178] "Patet, si ad id sumatur 99, non cuplum ipsius 11. . ." (p. 408). "So is openbaer, indien men daer toe neemt 99, 't 9-vout van 11. . ." (p. 380).

and $N=100+n\times1{,}001$, 1,001 being the L.C.M.(7, 11, 13).

Van Schooten also states the *ta-yen* rule, but he attributes it to a certain Nicolaus Huberti a Persijn.[179] He repeats the two problems with the *ta-yen* method:

1. L.C.M. (2, 3, 5, 7) = 210

$$210 : 2 \quad = 105; \quad 105 : 2 = 52 \text{ with } r=1$$
$$210 : 3 \quad = 70; \quad 70 : 3 = 23 \text{ with } r=1$$
$$210 : 5 \quad = 42; \quad 42 : 5 = 8 \text{ with } r=2$$
$$210 : 7 \quad = 30; \quad 30 : 7 = 4 \text{ with } r=2.$$

If the remainder is not equal to 1, one has to find how many times the remainder must be taken so that, if divided by 5, the remainder is 1. He finds 3.

He has thus no special method for solving the congruence

$a \times 42 \equiv 1 \pmod 5$.

Solving the last congruence $a \times 30 \equiv 1 \pmod 7$, he finds 4.[180]

Divisors	Remainders	"Multiplicators"	Products
2	1	105	105
3	1	70	70
5	1	126	126
7	0	120	0
			301
			$-n \times 210 = 91$.

2. L.C.M. (7, 11, 13) = 1,001

$1{,}001 : 7 = 143; \qquad 143 : 7 = 20 \text{ with } r=3$
$a \times 143 \equiv 3 \pmod 7 \rightarrow a = 5; \quad 5 \times 143 = 715.$

$1{,}001 : 11 = 91; \qquad 91 : 11 = 8 \text{ with } r=3$
$\beta \times 91 \equiv 3 \pmod{11} \rightarrow \beta = 4; \quad 4 \times 91 = 364.$

$1{,}001 : 13 = 77; \qquad 77 : 13 = 5 \text{ with } r=12$
$\gamma \times 77 \equiv 12 \pmod{13} \rightarrow \gamma = 12; \quad 12 \times 77 = 924.$

---

[179] Niclaes Huberts van Persijn. Van Schooten calls Nicolaus Huberti the inventor of the method, and says that the latter communicated it to him.

[180] The solution of this congruence is not necessary, because the remainder is zero. The author of the Göttingen manuscript was well aware of this fact.

Divisors	Remainders	"Multiplicators"	Products
7	2	715	1,430 · '
11	1	364	364
13	9	924	8,316
			10,110
			$-n \times 1{,}001 = 100.$

This is indeed the *ta-yen* rule.[181] Van Schooten says: "The well-known arithmetician Symon Jacobs of Coburg[182] also writes on this in his great *Arithmetica*, but he does not show how to find the 'multiplicators.' "[183]

WILLIAM BEVERIDGE

In his *Institutionum chronologicarum libri II* (1669), Beveridge included a full explanation of the remainder problem for the case in which the moduli are relatively prime. It is an important fact that this was the first general proof of the *ta-yen* rule.

His problem reads thus: "Invenire numerum *P*, quo per datos quoscumque *A*, *B*, inter se primos diviso, residua sunt data *K*, *L*."[184] A more general problem is: "Invenire numerum *O* minimum, quo diviso per datos *M*, *B*, *A* inter se omnimodo primos, residua sint data *K*, *L*, *Z*."[185]

The rule that is given says: "Primo inveniatur *D* minimus multiplex *B* talis, quo per *A* diviso supersit unitas, et inveniatur etiam *C* minimum numeri *A* multiplex talis, quo per *B* diviso restat unitas."[186] Thus

$$D = a \times B \equiv 1 \pmod{A}$$
$$C = \beta \times A \equiv 1 \pmod{B}.$$

[181] Dickson (1), vol. 2, p. 60.
[182] Died in 1565.
[183] "De his tradidit quoque celebris Arithmeticus Simon Jacobi Coburgensis in Arithmetica sua majori, sed multiplicatores invenire non docet." (p. 410). Simon Jacob (1510?–1565) wrote *Rechenbuch auff den linien und mit Ziffern* (1557).
[184] Beveridge (1), p. 254: "Find a number *P*, of which the remainders are given *K*, *C*, when divided by some given *A*, *B*, which are relatively prime."
[185] Beveridge (1), p. 256.
[186] "First there is to be found a number *D* the smallest multiple of *B* such that it leaves unity when divided by *A*. . ."

Here α and β are to be as small as possible; no method is given for finding α and β.

Then find the products DK and CL and "denique summam horum productorum ex K in D et L in C, divides per F ex A in B factum, et residuus fiet P . . . ."[187]

$$P = DK + CL - n \times AB.[188]$$

The proof given by Beveridge is roughly as follows:[189]

$$D = \alpha \times B \equiv 1 \ (\text{mod } A)$$
$$D - 1 \equiv 0 \ (\text{mod } A)$$
$$DK - K \equiv 0 \ (\text{mod } AK) \equiv 0 \ (\text{mod } A)$$
$$DK \equiv K \ (\text{mod } A).$$

Also

$$C = \beta \times A \equiv 0 \ (\text{mod } A)$$
$$CL \equiv 0 \ (\text{mod } A);$$

thus

$$DK + CL \equiv K \ (\text{mod } A) + 0 (\text{mod } A) \equiv K \ (\text{mod } A).$$

In the same way $DK + CL \equiv L \ (\text{mod } B)$, from which: $DK + CL \equiv K \ (\text{mod } A) \equiv L \ (\text{mod } B)$. Since $AB \equiv 0 \ (\text{mod } A) \equiv 0 \ (\text{mod } B)$, we get the least number P while dividing $DK + CL$ by AB; the remainder is the smallest solution: $P = DK + CL - m \times AB$ (m being as great as possible). A similar proof is given for the second problem.[190]

In Beveridge's work only the method for solving the congruences is lacking, although the method was known in the sixteenth century, as is obvious from the Göttingen manuscript.

With the work of Beveridge, we come to the end of what might be called the prescientific phase of the remainder problem. That term is not meant to deny that there was real

[187] "After that divide the sum of these products $K \times D$ and $L \times C$ by F ($= A \times B$), and the remainder will be P."
[188] This subtraction is omitted in Dickson (1), vol. 2, p. 61. But without this, P is not the smallest number.
[189] In fact, the proof is given in words and is too extensive to reproduce here, where it is stated in a modern algebraical way.
[190] See also Dickson (1), vol. 2, p. 61.

insight on the part of some mathematicians. But all of them remained in the arithmetical phase of the problem,[191] and only a few tried to give an arithmetical proof of the particular problems they were solving.[192] But after Beveridge there was a gradual attempt to give a general algebraical proof.[193]

However, the old style arithmetic continued to be studied in Europe,[194] and secondhand proofs of the problem[195] contributed nothing new to a general solution. In Europe the Sun Tzŭ problem was considered as a game, even as late as the nineteenth century.[196]

One of the problems in Fibonacci's work (vol. 1, p. 282), derived in all probability from Ibn al-Haitham, who relied on Bhâskara I, is reproduced by Elia Misrachi;[197] we find the same problem in the works of Caspar Ens[198] and Daniel Schwenter.[199] Euler wrote in *Commentarii academiae scientiarum Petropolitanae 7* (1734/5), 1740:[200]

[191] Or algorithmic phase, for the method was used without giving proof, and in the first centuries after Fibonacci, even the *yen-shu* were transmitted without explanation.

[192] As in the case of the Göttingen manuscript.

[193] In Chapter 21 we shall examine Ch'in Chiu-shao's work in the light of the algebraical works of Euler, Lagrange, and Gauss in order to come to an evaluation of Ch'in's method. Of course, one has to bear in mind that Ch'in's work is algorithmic and that he gives no proof for his method.

[194] For instance, the article of De Rocquigny (1) in 1881.

[195] Dostor (1), Marchand (1), Domingues (1). Dostor's problem is a very special case, very simple to prove.

[196] For example, Schäfer (1), *Die Wunder der Rechenkunst. Eine Zusammenstellung der rätselhaftesten, unglaublichsten und belustigendsten arithmetischen Kunstaufgaben zur Beförderung der geselligen Unterhaltung und des jugendlichen Nachdenkens* (The Miracles of the Art of Arithmetic. A compendium of the most mysterious, unbelievable, and entertaining problems in the art of arithmetic, for the stimulation of social intercourse and youthful reflection), Weimar, 1842. His indeterminate problem is on p. 50 (no. 60).

[197] Lived in Constantinople (1455–1526). He also includes the other problems of Fibonacci (*Liber Abacci*, p. 281).

[198] *Thaumaturgus Mathematicus*, Munich, 1636, pp. 70–71.

[199] *Deliciae Physico-Math. oder Math. -u. Phil. Erquickstunden*, Nürnberg, 1636, p. 41.

[200] P. 46. The title of this chapter is: "Solutio problematis arithmetici de inveniendo numero qui per datos numeros divisus relinquat data residua." This chapter is also published in *Commentationes arithmeticae collectae I*, pp. 18–32.

"Reperiuntur in vulgaribus arithmeticorum libris passim huiusmodi problemata, ad quae perfecta resolvenda plus studii et sollertiae requiritur, quam quidem videatur. Quamvis enim plerumque regula sit adjecta, cujus ope solutio obtineri queat, tamen ea vel est insufficiens solique casui proposito convenit, ita ut circumstantiis quaestionis parum immutatis ea nullius amplius sit usus, vel subinde etiam solet esse falsa . . . Simili quoque modo ubique fere occurrit istud problema, ut invenia-tur numerus, qui per 2, 3, 4, 5 et 6 divisus relinquat unitatem, per 7 vero dividi queat sine residuo. Methodus vero idonea ad huiusmodi problemata solvenda nusquam exhibetur; solutio enim ibi adjecta in hunc tantum casum competit atque ten-tando potius absolvitur.

"Si quidem numeri, per quos quaesitus numerus dividi debet sunt parvi, prout in hoc exemplo, tentando non difficulter quaesitus numerus invenitur; difficillima autem foret istiusmodi solutio, si divisores propositi essent valde magni."[201]

This text from Euler may be considered to represent the little that European mathematics had produced in indetermi-nate analysis until the beginning of the eighteenth century. In India and China, however, this branch of mathematics had already developed to a high level. The idea of *ex oriente lux* is by no means established historically, but there is no doubt about *in oriente lux* where medieval indeterminate analysis is concerned.

[201] "In popular books on arithmetic, one can frequently find such problems which require more study and cleverness for solving them correctly than may seem. For although a rule is usually given, with the help of which the solution may be found, nevertheless, either this rule is insufficient and appropriate only for the special case represented, so that, if the circum-stances of the problem are slightly changed, it is no longer useful; or it re-peatedly shows itself to be false [...]. This problem is found in the same form almost everywhere, namely that a number should be found, that, if divided by 2, 3, 4, 5, and 6 has the remainder 1, but can be divided by 7 without remainder. Nevertheless, a method appropriate for solving such problems is nowhere explained: for the solution given satisfies this case, but it is rather solved by conjecture [by trying]. If the numbers which the number asked for has to be divided by are small, as in this example, the number asked for can be found without much difficulty by conjecture. Such a solution, however, will be very difficult if the divisors given are fairly large."

15

History of Indeterminate Analysis in China

1. The Chiu-chang suan-shu 九章算術

The date of the *Chiu-chang suan-shu* (Nine chapters on mathematical techniques) is unknown, but in all probability it was composed in the first century A.D., although some material must have come from the Former Han (206–23 B.C.) and even from the Ch'in (255–206 B.C.).[1]

Problem 13 of *chüan* 8 (pp. 135 f) gives a system of four equations with five variables and is the first indeterminate problem in Chinese mathematics.[2] This problem says: "There is a common well belonging to five families; [if we take] 2 lengths of [well] rope of family *A*, the lacking part equals 1 length of rope of family *B;* the lacking part remaining from 3 ropes of *B* equals 1 rope of *C;* the lacking part remaining from 4 ropes of *C* equals 1 rope of *D;* the lacking part remaining from 5 ropes of *D* equals 1 rope of *E;* the lacking part remaining from 6 ropes of *E* equals 1 rope of *A*. In all cases if one gets the missing length of rope, the combined lengths will reach [the bottom]. Find the depth of the well and the length of the ropes."

Expressed in modern algebraic terms, the problem is as follows:

$$2x + y = w$$
$$3y + z = w$$

[1] There exist three complete translations of the work: E. I. Berezkina, "Drevnekitajskij Traktat Matematika v devjati Knigax," *Journal of the History of Mathematics* (Moscow), 1957, *10*, 423–584; Wang Ling, *The Chiu-chang suan-shu and the History of Chinese Mathematics during the Han Dynasty*, Inaug. Diss., Cambridge, 1956 (Cambridge dissertations are under copyright, and I did not succeed in getting the consent of the author to consult or copy it). A recent German translation is K. Vogel, *Neun Bücher Arithmetischer Technik*, Braunschweig, 1968. For the problem of dating, see Needham (1), vol. 3, pp. 24 f, where all further information about the work can be found.
[2] For information about this problem, see Ch'ien Pao-tsung (4'), pp. 37 f; Hsü Ch'un-fang (2'), pp. 10 f; Yushkevitch (4), pp. 39 f; Vogel (2), p. 87 and p. 134; Ang Tian Se (1), p. 242.

$4z+u=w$
$5u+v=w$
$6v+x=w.$

This is indeed an indeterminate problem, but the method used has nothing to do with a general solution of indeterminate equations. By several multiplications, additions, and subtractions, the author draws up these fractions:

$$\frac{x}{w}=\frac{265}{721}; \ \frac{y}{w}=\frac{191}{721}; \ \frac{z}{w}=\frac{148}{721}; \ \frac{u}{w}=\frac{129}{721}; \ \frac{v}{w}=\frac{76}{721}.$$

From these it can easily be seen that $w=721$ is a solution, with $x=265$, $y=191$, $z=148$, $u=129$, $v=76$.[3] The general solution of this problem would be $w=721t$, $x=265t$, and so on, with $t>0$.[4]

2. The Sun Tzŭ suan-ching 孫子算經

The date of the *Sun Tzŭ suan-ching* (Mathematical classic of Sun Tzŭ) is very uncertain, and it cannot be dated more exactly than within a span including the San Kuo, Chin, and (Liu) Sung periods (between A.D. 280 and A.D. 473).[5]

The original work has long been lost, but the extant version is composed of extracts included in the *Yung-lo ta-tien* 永樂大典.[6] A single problem treated in this work, and not given in the *Chiu-chang suan-shu*, is the starting point of the famous remainder theorem[7] (Figure 49).

[3] For a full explanation of the method, see Hsü Ch'un-fang (2′), pp. 10 ff.
[4] For other problems of the same kind, see Yushkevitch (4), pp. 39 ff.
[5] See the special study by Wang Ling (4); Ch'ien Pao-tsung (4′), p. 45. There is a recent translation into Russian by E. I. Berezkina. Sun Tzŭ is not to be confused with the general Sun Tzŭ. See Wylie (1), p. 180n and Needham (1), vol. 3, p. 33. This error can be found in Hayashi (1), vol. 1, p. 310. Dickson, in (1), vol. 2, p. 57, places the *Sun Tzŭ suan-ching* in the first century, although Wylie states in (1), p. 181 that the third century is the most probable. Nothing is known about the life of Sun Tzŭ, or even whether he really existed.
[6] See Yüan Tung-li (1′). On the mathematical books in the *Yung-lo ta-tien*, see Li Yen (6′), vol. 2, pp. 47 ff. See also Wylie (2), pp. 91 f.
[7] No special bibliography is necessary, because there is no general work on Chinese mathematics that does not include the now well-known Sun Tzŭ problem. For a general description, see Mikami (1), pp. 32 ff; Li Yen (13′), pp. 114 ff (the remainder problem is on p. 125).

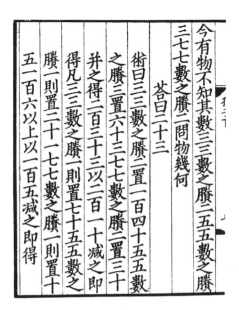

Figure 49. The famous Sun Tzŭ problem, the oldest instance of the remainder theorem. From the *Sun Tzŭ suan-ching* 孫子算經 (*T'ien-lu Lin-lang ts'ung-shu* 天祿琳琅叢書 ed.), *ts'e* 15, C, p. 10b.

The literal translation is as follows:

"We have things of which we do not know the number; if we count them by threes, the remainder is 2; if we count them by fives the remainder is 3; if we count them by sevens the remainder is 2. How many things are there? Answer: 23.[8] Method: if you count by threes and have the remainder 2, put 140.

If you count by fives and have the remainder 3, put 63.

If you count by sevens and have the remainder 2, put 30.

Add these [numbers] and you get 233.

From this subtract 210 and you have the result.

For each unity as remainder when counting by threes, put 70.

For each unity as remainder when counting by fives, put 21.

For each unity as remainder when counting by sevens, put 15.

If [the sum] is 106 or more, subtract 105 from this and you get the result."[9]

[8] Ch. 3, p. 10b.

[9] See Van Hee (12), p. 436; Wylie (1), p. 181; Mikami (1), p. 32.

In modern algebraic symbols:

$N \equiv 2 \pmod 3 \equiv 3 \pmod 5 \equiv 2 \pmod 7$.[10]

$70 \equiv 1 \pmod 3 \rightarrow 140 \equiv 2 \pmod 3$

$21 \equiv 1 \pmod 5 \rightarrow 63 \equiv 3 \pmod 5$

$15 \equiv 1 \pmod 7 \rightarrow 30 \equiv 2 \pmod 7$

$N \equiv 140 + 63 + 30 - n \times 105 = 23$.[11]

Hsü Ch'un-fang[12] gives a simple explanation of the Sun Tzŭ problem, and this may have been the way in which Sun Tzŭ found the somewhat mysterious numbers 70, 21, and 15:[13]

$$
\begin{aligned}
70 &= 3p+1 = 5p &&= 7p \\
21 &= 3p &= 5p+1 &= 7p \\
15 &= 3p &= 5p &= 7p+1 \\
105 &= 3p &= 5p &= 7p.
\end{aligned}
$$

(The symbol p has no algebraic meaning; $3p$ means treble; $5p$, quintuple, and so on.)

$$
\begin{array}{llll}
2 \times 70 &= 3p+2 = &5p &= 7p \\
3 \times 21 &= 3p &= 5p+3 &= 7p \\
2 \times 15 &= 3p &= 5p &= 7p+2 \quad (+)\\
\hline
233 &= 3p+2 = &5p+3 &= 7p+2 \\
2 \times 105 &= 3p &= 5p &= 7p \quad (-)\\
\hline
23 &= 3p+2 = 5p+3 &&= 7p+2.^{14}
\end{array}
$$

[10] Notice that the moduli are relatively prime in pairs.

[11] Another method of notation is followed in Li Yen (6'), vol. 1, pp. 122 ff. This notation comes from Hsü Chên-ch'ih (1):

$$\left| \frac{N}{3} \right. = 2 \qquad \left| \frac{N}{5} \right. = 3 \qquad \left| \frac{N}{7} \right. = 2 \qquad \text{or} \qquad \left| \frac{N}{(3,5,7)} \right. = (2,3,2).$$

[12] (3'), p. 13.

[13] As these numbers can easily be found by conjecture, and as Sun Tzŭ gives only one problem, it is not at all certain that he knew a general method for solving the congruences, as Ch'in Chiu-shao did long after him.

[14] See also Loria (1), p. 152, who treats immediately after the indeterminate problem the ridiculous problem of the "pregnant woman" (see Mikami (1), 33: "A pregnant woman, who is 29 years of age, is expected to give birth to a child in the 9th month of the year. Which should be her child, a son or a daughter?"). This does not diminish the value of the ta-yen rule of Sun Tzŭ. An instructive example of such silliness can easily be found in the famous "beast problem" of Michael Stifel [see Smith (1), pp. 327 f]. And Stifel was the greatest mathematician of his time!

Sun Tzŭ seems not to have seen that there is an infinite set of solutions, but even Ch'in Chiu-shao satisfied himself with a single solution.[15]

The remainder problem remained incomprehensible till Ch'in Chiu-shao, who was the first to give a full explanation of the problem.[16]

3. Ho Ch'êng-t'ien 何承天

Wylie[17] says: "In tracing the course of this process, we find it gradually becoming clearer, till towards the end of the Sung dynasty . . . ." As will be seen from this historical outline, Wylie is not at all right, for there is hardly anything between Sun Tzŭ and Ch'in Chiu-shao concerning the *ta-yen* rule.[18]

Ho Ch'êng-t'ien was a calendar-maker who lived between approximately A.D. 370 and 447. In the year 443 he drew up a new calendrical system, for which he used a method called *t'iao-jih fa* 調日法. Ch'ien Pao-tsung[19] states his opinion that Ho Ch'êng-t'ien may have been familiar with indeterminate analysis. In his work (now lost)[20] Ho made use of a *ch'iang-lü* 強率 and a *jo-lü* 弱率, a "strong ratio" and a "weak ratio." These are the fractions between which the "little excess" (*hsiao-yü* 小餘), that is, the difference between the length of a lunar month and an integral number of days, must lie. The

---

[15] The problem became known in Europe through Wylie's "Jottings." Biernatzki translated Wylie's article into German, but not without mistakes (see Chapter 16, Appendix). Matthiessen demonstrated the identity of the problem with Gauss's rule. (See ibid.) In China, Chang Tun-jên (1803) was the first to set down a full explanation of the method of Sun Tzŭ.

[16] See Huang Tsung-hsien's preface to his *Ch'iu-i-shu t'ung-chieh*.

[17] (1), p. 181.

[18] It will be seen that through a mistake of Matthiessen many of our historians of mathematics ascribe much to the Buddhist monk I-hsing that really originated in the work of Ch'in Chiu-shao.

[19] (4'), p. 46.

[20] All the information here comes from the *Sung-shu*, i. e., the history of the (Liu) Sung Dynasty, ch. 13, pp. 1b and 3a; from Li Hsin-ch'uan's *Chien-yen i-lai ch'ao-yeh tsa-chi* and Ch'in Chiu-shao's *Shu-shu chiu-chang*, vol. 2, ch. 3, pp. 57 ff. The texts are reproduced by Yen Tun-chieh in his article on "Mathematical knowledge in the calendar-calculations of the Sung, Chin, and Yüan dynasties," incorporated in Ch'ien Pao-tsung (2'), pp. 210 ff.

strong fraction has the value 26/49 and the weak one equals 9/17, and indeed, Ho's "little excess" (0.530585)[21] holds:

$$\frac{9}{17}(=0.529412)<0.530585<\frac{26}{49}(=0.530612).$$

The relation between 26/49 and 9/17 is[22] $(26 \times 17)-(9 \times 49)=1$. Ch'ien Pao-tsung[23] is of the opinion that the fraction 26/49 could be derived from 9/17 by solving the indeterminate equation $17x-9y=1$, which indeed gives the solution[24] $x/y=26/49$. This is of course not the same as the *ta-yen* rule, but only a little part of it, connected with the *ch'iu-i shu* 求一術 or "solving the congruences." Another question is whether Ho Ch'êng-t'ien was able to develop a fraction into a continued fraction and to find the approximate fractions of it.[25]

Ch'in Chiu-shao's Problem II, 3 on "adjusting the inequalities of the annual, lunar, and diurnal revolutions" makes use of the so-called *t'iao-jih fa* and attributes this method to Ho Ch'êng-t'ien.[26] The commentary of Sung Ching-ch'ang 宋景昌[27] states: "Since the *Shou-shih* 授時 method,[28] the *jih-fa* method is no longer used. Few know this method, and Li Jui 李銳, in his *Jih-fa shuo-yü ch'iang jo k'ao* 日法朔餘強弱考, was the only one who succeeded in explaining the secret not handed down."[29]

[21] According to Ho Ch'êng-t'ien.

[22] If P/Q and R/S are two successive approximate fractions, the difference $PS - QR$ equals $+1$ or -1, according to whether the second approximative fraction stands on an odd or an even rank.

[23] (4′), p. 46, and (2′), p. 212.

[24] Ch'in Chiu-shao says in his preface: "Only the *ta-yen* rule is not contained in the Nine Chapters, for no one has yet been able to derive it from other procedures. Calendar makers, in working out their methods, have made considerable use of it" (p. 1). This is of course no evidence that Ho Ch'êng-t'ien was one of these calendar makers.

[25] As we shall see in the study of Ch'in Chiu-shao's method, he solved the congruences by a method that is the same as our method of continued fractions.

[26] The text says: "With the *t'iao-jih fa*, a method like that of Ho Ch'êng-t'ien" (p. 58).

[27] (1′), p. 50.

[28] The *Shou-shih* calendar was drawn up by Kuo Shou-ching 郭守敬 in 1281. See Needham (1), vol. 3, p. 125.

[29] Li Jui (1768–1817) wrote this work in 1799.

The text incorporated in the *Sung-shih*[30] gives both the fractions 26/49 and 9/17, and says: "The *jih-fa* of Ho Ch'êng-t'ien is 752, and we find 15 as strong value and 1 as weak value."[31] The relation between these numbers is:[32]

$$(49 \times 15) + (17 \times 1) = 752.$$

The *Chien-yen i-lai ch'ao-yeh tsa-chi* 建炎以來朝野雜記 of Li Hsin-ch'uan 李心傳 (Sung) gives the *jih-fa* 10,002. The strong ratio is 201 and the weak ratio 9, with the following relation:[33]

$$(49 \times 201) + (17 \times 9) = 10,002.$$

One may suppose that Ho Ch'êng-t'ien solved indeterminate equations:[34] If $49x + 17y = 752$, then $x = 15$ and $y = 1$ is a solution. From this solution follows[35] $(26 \times 15) + (9 \times 1) = 399$. In the example given by Li Hsin-ch'uan, we find $49x + 17y = 10,002$, from which $x = 201$, $y = 9$. In the example given by Ch'in Chiu-shao: $49x + 17y = 16,900$, from which $x = 339$, $y = 17$.

This last example is important, for the whole computation $(26 \times 339) + (9 \times 17) = 8,967$ is in the *Shu-shu chiu-chang*.

Although there seems to be some possibility that Ho Ch'êng-t'ien was able to solve indeterminate equations, it appears to us not at all certain that he could really apply the *ch'iu-i shu*.[36] On the other hand we can derive Ho's numbers by that method:[37]

[30] See note 20.

[31] That is, 15 as "strong factor" and 1 as "weak factor." A synodic month has 29 399/752 days; 752 is the *jih-fa*. See Yabuuchi (3), p. 459.

[32] First pointed out by Li Jui.

[33] See Ch'ien Pao-tsung (2'), p. 212.

[34] As Ch'ien Pao-tsung admits, but Li Yen says: "The *t'iao-jih fa* is now lost, and it is impossible to define it" [(9'), p. 171].

[35] According to Li Jui (1'),

$$\frac{26 \times 15 + 9 \times 1}{49 \times 15 + 17 \times 1} = \frac{399}{752} = 0.530585.$$

However, Li Jui does not give any additional information about the origin of the factors 15 and 1.

[36] See Chapter 17.

[37] See *Sung-shu*, ch. 13, pp. 2a–2b.

1 lunation $= 29\frac{399}{752}$.

Develop the fraction in a continued fraction:

$$\frac{399}{752} = (0, 1, 1, 7, 1, 2, 15).$$

The approximate fractions are

$$\frac{0}{1}; \quad 1 + \frac{0}{1} = \frac{1}{1}; \quad \frac{1 \times 1 + 0}{1 \times 1 + 1} = \frac{1}{2}; \quad \frac{1 \times 7 + 1}{2 \times 7 + 1} = \frac{8}{15};$$

$$\frac{8 \times 1 + 1}{15 \times 1 + 2} = \frac{9}{17}; \quad \frac{9 \times 2 + 8}{17 \times 2 + 15} = \frac{26}{49}; \quad \frac{26 \times 15 + 9}{49 \times 15 + 17} = \frac{399}{752}.$$

If we take

$$\frac{9}{17} = \frac{9y}{17y}$$

then we have

$$\frac{26 \times 15 + 9y}{49 \times 15 + 17y} = \frac{399}{752}$$

if $y = 1$.

We can derive a set of interpolated fractions, when replacing 15 by $n = 1, 2, 3, \ldots, 15$, or in general:

$$\frac{26 \times n + 9y}{49 \times n + 17y}. \tag{1}$$

If we wish to equalize (1) to a fraction $N/16,900$ we have to take $n = x$, where x is an arbitrary value; we get

$$\frac{N}{16,900} = \frac{26x + 9y}{49x + 17y}.$$

The equation $49x + 17y = 16,900$ has the solutions $x = 339$ and $y = 17$; thus $N = 8,967$. And $8,967/16,900 = 0.53059$ could be considered as a kind of approximate fraction of $339/752 = 0.53058$.

All these considerations can make Ho's fractions less mysterious, but they cannot convince us that they were deduced in this way. Perhaps further investigation can elucidate this problem.

4. Tsu Ch'ung-chih 祖冲之

Tsu Ch'ung-chih (active 430–501) was the first in the history of mathematics to compute very accurate values for π. His work was called *Chui-shu* 綴術, and some have thought that this indicates the method Tsu Ch'ung-chih applied for computing the value of π. However, *chui-shu* is mentioned in the work of Ch'in Chiu-shao[38] "as the motto to some problems of calendrical computations."[39] The work of Tsu Ch'ung-chih has long been lost,[40] and all our information about it comes from the *Sui-shu* (History of the Sui dynasty).[41]

The text of the *Sui-shu* informs us that Tsu took a circle of diameter 100,000,000, which he considered as 10 feet. For the upper limit he found 31.415927 feet, for the lower limit 31.415926 feet,[42] and he concluded that the true value of π must lie between these two values. As *mi-lü* 密率 (precise value) he found 355/113 and as *yüeh-lü* 約率 (approximate value) 22/7.[43]

What should be especially interesting for our study is the method Tsu Ch'ung-chih used to derive the fractions 22/7 and 355/113 from the decimal numbers. We have not the slightest information about this method. Ch'ien Pao-tsung[44] thinks that he could have made use of the *ch'iu-i shu*. We see that (22×113)

---

[38] Problem II, 4 (p. 77) has the title *Chui-shu t'ui hsing* 綴術推星 (Investigation of the planets by the *chui* method); the preface says: "They calculated the laws of the heavenly bodies and called this *chui-shu*."

[39] Mikami (1), p. 50; For further information about this work, see Needham (1), vol. 3, p. 101.

[40] As early as the Sung. See Li Yen (14), p. 1.

[41] *Lü li chih* 律曆志, ch. 16, p. 3b. Other sources: Shên Kua, *Mêng-ch'i pi-t'an*, ch. 8, par. 6; the *Nan-shih* (History of the Southern Dynasties) and the *Nan-Ch'i shu* (History of the Southern Ch'i). There are many references to the π-value of Tsu; some of these are given in Needham (1), vol. 3, p. 102. There are biographies of Tsu Ch'ung-ch'ih in Li Yen (2), and Chou Ch'ing-chu (1').

[42] From which $\pi = 31.415927/10 = 3.1415927$, and so on.

[43] *Mi-lü* and *yüeh-lü* are the fractions between which the true value of π lies. There is a special study on the evolution of the computing of π: Mao I-shêng (1'), where Tsu Ch'ung-chih is treated on p. 417. See also Li Yen (2); Mikami (4); Chang Yung-li (1'); Hua Lo-kung (1').

[44] (4), p. 47.

$-(355 \times 7) = 1$; this means that 113 and 355 could have been derived from the indeterminate equation $22x - 7y = 1$. Mikami says: "He may have applied a process something like the indeterminate multipliers to determine three terms in an expansion."[45] According to Needham, "Ch'ien Pao-tsung brings forward some rather impressive arguments for thinking that Tsu expounded the approximation of fractions . . . ." "The 'precise value' fraction ($= 355/113$) was an extraordinary achievement for it is one of the continued-fraction convergents."[46] However, Yushkevitch does not agree with this interpretation: "It is well known that this value is one of the approximate fractions which result from developing π into a continued fraction. But this fact does not prove that the Chinese actually used continued fractions. This fraction may very well be obtained by other means."[47] Yushkevitch's conclusion seems to be justified because there are not enough data to conclude that continued fractions were used.

5. Chang Ch'iu-chien 張邱建

Chang Ch'iu-chien wrote his *Chang Ch'iu-chien suan-ching* 張邱建 算經 between 468 and 486, and was thus a contemporary of the Indian mathematician Âryabhaṭa (475–550).[48] In his work we find the famous problem of the "hundred fowls,"[49] which

[45] Mikami (1), p. 51.

[46] Needham (1), vol. 3, pp. 35 f and p. 101, note *h*.

[47] "Bekanntlich ist dieser Wert einer der Näherungsbrüche, die sich bei der Entwicklung von π in einen Kettenbruch ergeben, doch kann aus dieser Tatsache nicht geschlossen werden, dass die Chinesen Kettenbrüche verwendet hätten. Dieser Bruch kann durchaus auch auf andere Weise gewonnen worden sein." Yushkevitch (4), p. 59.

[48] For the dates of Chang Ch'iu-chien, see Needham [(1), vol. 3, p. 33], who relies mainly on Wang Ling (5), vol. 2, p. 66; for general contents, see Needham (1), vol. 3, p. 35. The work was printed in the Sung (1085). The text has been transmitted in a perfect state; see Van Hee (12), p. 442. There is a recent specialized study by Ang Tian Se (1). See also Ho Pêng-yoke (6).

[49] Ch. 3, C, p. 37a and following pages. See Van Hee (12). This problem is of great interest for the study of a possible relationship among Chinese, Indian, and Arabian mathematics.

consists of true indeterminate equations.[50] The problem is:[51] "A cock is worth 5 *ch'ien*,[52] a hen 3 *ch'ien*, and 3 chicks 1 *ch'ien*. With 100 *ch'ien* we buy 100 of them. How many cocks, hens, and chicks are there?"

Chang gives as the answer:

4 cocks,	18 hens,	78 chicks
8 cocks,	11 hens,	81 chicks
12 cocks,	4 hens,	84 chicks.

As his method he says only: "Increase the cocks every time by 4, decrease the hens every time by 7, and increase the chicks every time by 3."

In modern algebraic language we get (if cocks=x, hens=y, chicks=z):

$$5x + 3y + \frac{1}{3}z = 100$$

$$x + y + z = 100.$$

The general solution of these equations is[53]

$$x = 4t; \; y = 25 - 7t; \; z = 75 + 3t.$$

For $t = 0, 1, 2, 3$, the solutions are[54]

x	0	4	8	12
y	25	18	11	4
z	75	78	81	84.

[50] The problems in the *Chiu-chang suan-shu* were indeed not true indeterminate equations.

[51] Ch. C, pp. 54 ff. There are many translations of this problem: Mikami (1), p. 43; Van Hee (12), p. 443; Needham (1), vol. 3, p. 122; for a discussion of the problem in Chinese, see Ch'ien Pao-tsung (4'), p. 38; Hsü Ch'un-fang (2'), p. 14. In European languages there are Loria (1), p. 146, and (3), p. 153; Yushkevitch (3), p. 75; Ang Tian Se (1), pp. 244 ff.

[52] *Ch'ien* 錢 is a copper coin.

[53] We can solve the problem as follows: $5(100 - y - z) + 3y + 1/3z = 100$, or $3y + 7z = 600$, from which $y = (600 - 7z)/3$. For $z = 0$ we get $y = 200$, or $z = 0 + 3t; y = 200 - 7t$. Thus $x = 100 - y - z = 100 - (200 - 7t) - (0 + 3t) = -100 + 4t$. In order to get the first positive solution, we take $t = 25$, and get $x = 0; y = 25; z = 75$. The general solution if $0 < x, y, z < 100$ is: $x = 4t, y = 25 - 7t, z = 75 + 3t$.

[54] See Yushkevitch (4), p. 75.

Chang Ch'iu-chien found the right solutions, except the first one, although one can say that this solution is not positive.[55] But the striking fact is that he was familiar with the relation among the several solutions. It is a pity that we do not know the manner in which he found the first solution; of course it is possible that he did it by inspection. None of the commentators on his work understood Chang's rule till the time of Shih Yüeh-shun 時日醇 (1861).[56]

6. Chên Luan 甄鸞

Chên Luan, who was active c. 570, tried to solve the problem of the "hundred fowls," but without any success, for his method is really ad hoc, and not at all general. In his commentary on the *Shu-shu chi-i* 數術記遺 of Hsü Yüeh 徐岳 (A.D. 190), he gives the following problem: "If a cock costs 5 pieces, a hen 4 pieces, and 4 chicks 1 piece, how many cocks, hens, and chicks can be bought with 100 pieces so as to make 100 in all?" Chên gives

[55] "It can be very easily proved that the solutions he gives represent the only integral solutions in positive numbers. In addition to these there exists one non-negative solution, with $x = 0, y = 25$, and $z = 75$—the only solution which remained unknown to the Chinese mathematician." (Es lässt sich leicht nachprüfen, dass die von ihm angegebenen Lösungen die einzigen ganzzahligen Lösungen in positiven Zahlen darstellen. Darüber hinaus existiert noch eine nichtnegative Lösung mit $x = 0, y = 25$ und $z = 75$ als einzige, die von dem chinesischen Mathematiker unbeachtet geblieben ist.) Yushkevitch (4), p. 75. But, if he tried only to find positive solutions, he indeed found all of them.

[56] Van Hee (12), p. 445 says: "The author of the 'Mathematical Gleanings' had good reason to think such artifices unjustified and unjustifiable." (L'auteur des 'Miettes de mathématiques' trouve à juste titre tous ces artifices injustifiés et injustifiables.) Although this is quite correct in the case of the commentaries on Chang's work, it is not at all so in the case of Chang himself.

The author of the "Gleanings" is Ting Ch'ü-chung, who wrote the *Shu-hsüeh shih-i*, published as no. 12 of the collection *Pai-fu t'ang suan-hsüeh ts'ung-shu* (Pai-fu Hall collection of mathematical works), 1875. Van Hee is very careless in giving names, titles, and texts. Ang Tian Se writes: "As no process for finding these numbers has been indicated, Chang Ch'iu-chien might have found them by inspection. This is possible because whenever the number of cocks and the number of chicks are increased by 4 and 3, respectively, the number of chicks is increased by 7, thus keeping the total number of fowls together with its costs constant" [(1), p. 246].

his answer as 15 cocks, 1 hen, and 84 chicks; if the cock costs 4 pieces, the hen 3 pieces, and 3 chicks 1 piece, the answer becomes 8 cocks, 14 hens, and 78 chicks, as Chên solves it.[57]

The method given by Chên Luan[58] is: "Draw up the amount 100 *ch'ien*. Take 9 as divisor and divide it. You get the number of hens. Subtract the remainder 1 from the divisor 9; this is the number of cocks." Or, $100:9=11$ (y) and $9-1=8$ (x). This method does not make sense at all. Chiao Hsün 焦循 (1763–1820), in his *Chia-chien ch'êng-ch'u shih* 加減乘除釋, ch. 6,[59] says that the results given by Chên Luan are correct only by accident. Chiao Hsün's method, however,[60] is no more general than Chên's and his solutions are accidental, too.

7. Liu Hsiao-sun 劉孝孫

At the end of the sixth century, Liu Hsiao-sun wrote an explanation of the *Chang Ch'iu-chien suan-ching*, entitled *Chang Ch'iu-chien suan-ching hsi-ts'ao* 張邱建算經細草 (Detailed solutions of [the problems] in the *Chang Ch'iu-chien suan-ching*), in which he tried to give a method for solving the "hundred fowls" problem.[61] The text says: "Draw up the amount 100 *wên* as dividend and cocks: 1 and hens: 1. Multiply by 3; you get: 3 cocks and 3 hens. Add to 3 chicks, and together you get 9. This is the divisor. The division of the dividend 100 gives 11.

[57] Hsü Yüeh (1), p. 30. Translation from Mikami (1), p. 44. Yushkevitch (4), p. 74 says: "The problem of the fowls was probably formulated no later than the beginning of the third century. According to Chên Luan, who in 570 wrote a commentary on the...work of Hsü Yüeh of about 190, this work contains the solution of the following problem...." (Nicht später als zu Beginn des 3. Jahrhunderts dürfte die 'Aufgabe über die Vögel' entstanden sein. Zhen Luan zufolge, der um 570 einen Kommentar zum...Werk des Xu Yue aus der Zeit um 190 geschrieben hatte, enthält dieses Werk die Lösung der folgenden Aufgabe....) According to the works of modern Chinese historians of mathematics, such as Li Yen and Ch'ien Pao-tsung, this seems not to be correct. On the authenticity of the work, see Wylie (2), p. 92, but also Needham (1), vol. 3, p. 30.

[58] See Ch'ien Pao-tsung (4'), p. 39; Van Hee (12), p. 443; Ang Tian Se (1), p. 248.

[59] Published in 1799 in the collection *Tiao-ku lou ts'ung-shu*.

[60] The so-called *San-szŭ ch'a-fên* method. See Ch'ien Pao-tsung (4'), p. 39.

[61] See Van Hee (12), p. 443; Ch'ien Pao-tsung (4'), p. 39; Hsü Ch'un-fang (2') p. 14.

This is the number of hens. Subtract again the remainder 1 from the divisor 9. The remainder is 8. This is the number of chicks." The method is the same as that of Chên Luan, and its value is likewise nil.

8. Li Shun-fêng 李淳風

Li Shun-fêng (seventh century), "probably the greatest commentator of mathematical books in all Chinese history,"[62] also wrote a commentary on the work of Chang Ch'iu-chien.[63] But what he has to say on the "hundred fowls" does not amount to much: "As 3 chicks cost 1 cash, 1 chick costs 1/3 cash. One has to add 3 cocks and 3 hens and has 9." This is unmitigated nonsense!

Li Shun-fêng devised the *Chia-tzŭ yüan* 甲子元 calendrical system.[64] Ch'ien Pao-tsung is of the opinion that he made use of the *Ch'iu-i shu*,[65] but it is impossible to prove it.

9. I-hsing 一行

I-hsing was a Buddhist priest who lived from 683 to 727.[66] In the history of Ghinese mathematics (at least in Europe), he has received too much attention, mainly because of Matthiessen's error in ascribing to him much of the work of Ch'in Chiu-shao.[67] I-hsing drew up a calendrical system known as *Ta-yen*

[62] Needham (1), vol. 3, p. 38. Li Shun-fêng wrote a commentary on the *Chou-pei* and the *Chiu-chang*.
[63] See Mikami (1), p. 39.
[64] It was put into practice after the *Lin-tê* period (664–665).
[65] Ch'ien Pao-tsung (4'), p. 47.
[66] These dates are given in Mikami (1), p. 60.
[67] This mistake is repeated by many historians of mathematics who did not have access to Chinese sources: Dickson (1), vol. 2, p. 57; Cantor (1), p. 586. Matthiessen's mistake must be ascribed to the fact that Wylie (1), p. 182 had just discussed "Yih Hing" (as he transcribes the name of I-hsing), in his treatment of the first problem of Ch'in Chiu-shao about a passage in the "*Yih King*" (the *I-ching*). But Biernatzki (1), p. 79 confuses the name of I-hsing (Wylie's Yih Hing) with the title of the canonical *I-ching* (Wylie's Yih King) and so came to think that he was dealing with a book. This shows indeed that each letter has its importance! Yushkevitch (4), p. 76, seems to undergo some sort of aftereffect of Matthiessen's mistake when he says that "a series of similar problems were dealt with by the above-mentioned astronomer I-hsing..." (einer Reihe derartiger Aufgaben be-

li-shu 大衍曆書, and this title might have been suggestlve, but "it did not have much to do with indeterminate analysis," as Needham[68] says. All the works of I-hsing are lost and there is only a little (mathematical) information in the History of the T'ang.[69] Chang Tun-jên seems to have been of the opinion that I-hsing used indeterminate analysis.[70]

The text of the *T'ang-shu* gives only the *shang-yüan* 上元, or calendrical epoch, as 96,961,740 years.[71] Wang Ling reconstructed the problem as follows: "Whether I-hsing had actually used the method as Ch'in Chiu-shao did, depends on the following question: how did I-hsing obtain the number 96,961,740, the number of years between the beginning of the Grand Cycle, *shang-yüan*, and the 12th year of the *K'ai-yüan* period?

"Based on I-hsing's own record we have formed two indeterminate equations:

$$\frac{1,110,343 \times 60x}{3,040} = 60y + \frac{44,820}{3,040} \tag{1}$$

$$\frac{1,110,343 \times 60x}{3,040} = \frac{89,773z}{3,040} + \frac{49,107}{3,040}, \tag{2}$$

where $60x$ representing the number of years between the beginning of the Grand Cycle and the twelfth year of the *k'ai-yüan* reign period is to be found."

We can write this problem as follows:

$$N \equiv 0 \pmod{1,110,343 \times 60} \equiv 44,820 \pmod{60 \times 3,040}$$
$$\equiv 49,107 \pmod{89,773}.$$

---

handelte der bereits genannte Astronom Yi Xing. . .). And on p. 77: "I-hsing already extended Sun Tzŭ's method to include modules that are not relatively prime. . . ." (Bereits Yi Xing dehnte die Methode des Sun-zi auf den Fall nicht paarweiser teilerfremden Moduln aus. . . .)
[68] (1), vol. 3, p. 37.
[69] *Chiu T'ang-shu* (Old history of the T'ang dynasty), ch. 34, and *Hsin T'ang-shu* (New history of the T'ang dynasty), ch. 28A and B.
[70] According to Needham (1), vol. 3, p. 120, note *a*.
[71] This is the number of years between the beginning of the Grand Cycle and the twelfth year of the *k'ai-yüan* period (724).

It is not impossible that I-hsing solved this problem by means of the method we find in Ch'in Chiu-shao's work, but it is impossible to prove it.

10. Lung Shou-i 龍受益

According to the Treatise on Bibliography of the *T'ang-shu* 唐書, Lung Shou-i (active 785–803) wrote a work called *Suan-fa* 算法 (Mathematical methods), now lost.[72] In his preface, Chang Tun-jên mentioned the work *Ch'iu-i suan-shu hua ling ko* 求一算術化零歌. Li Yen[73] is of the opinion that this could be a work that influenced Ch'in Chiu-shao. Although this is not impossible, it remains mere conjecture.

11. Hsieh Ch'a-wei 謝察微

"Ch'êng Ta-wei 程大位 quoted fragments from the now lost book of Hsieh Ch'a-wei, who was probably a contemporary of Shên Kua and would have flourished in the last quarter of the 11th century."[74] He treats the problem of the "hundred fowls," but makes the same mistakes as Chên Luan.[75]

12. Shên Kua 沈括

In the *Mêng-ch'i pi-t'an* 夢溪筆談 (Dream creek essays) (1086) of Shên Kua, mention is made of the *ch'iu-i* method,[76] but without any explanation.[77]

The work of Ch'in Chiu-shao should come next in this chronological summary,[78] but it will be examined in detail

---

[72] Also mentioned in the *Sung-shih, I-wên chih*, where the name is Lung Shou-i and the work *Ch'iu-i suan-shu hua ling ko*, while in the *T'ang-shu, I-wên chih* the name is Lung Shou and the work *Suan-fa*. Ch'ien Pao-tsung [(4'), p. 48] thinks that these two are the same.

[73] (9'), p. 148.

[74] Needham (1), vol. 3, p. 79.

[75] See Ch'ien Pao-tsung (4'), p. 39; Ang Tian Se (1), p. 246.

[76] Ch. 18, par. 9. See Li Yen (9'), p. 128; Ch'ien Pao-tsung (4'), p. 48.

[77] Needham, in (1), vol. 3, p. 112, speaks about *ta-yen ch'iu-i shu*, but in the text there is only: "Numerous mathematical methods, such as the *ch'iu-i*...."

[78] Needham (1), vol. 3, p. 40, mentions the *Ta-yen hsiang-shuo* 大衍詳說 of Ts'ai Yüan-ting 蔡元定 (1180) and translates the title as "Explanation of Indeterminate Analysis." In his bibliography (ibid., p. 716) the same work

later on. The rest of this chapter will be devoted to Chinese mathematical works written after the *Shu-shu chiu-chang*.

13. Yang Hui 楊輝

Yang Hui treats indeterminate problems in his *Hsü-ku chai-ch'i suan-fa* 續古摘奇算法(1275).[79] The first problem is the Sun Tzŭ problem.[80] Next we find some variations on it; the text

is listed, with the word "algebra" to indicate the nature of the work. This must be a mistake. Needham refers to Forke (1), vol. 3, p. 204, who writes: "the *Ta-yen hsiang-shuo* on higher mathematics..." (das *Ta-yen hsiang-schuo* über höhere Mathematik...) and is perhaps misled by this quotation. The *Ta-yen hsiang-shuo* no longer exists, and must have been lost a long time ago. The work is not mentioned in ch. 28 of the *Shuo-fu* collection, where a list of mathematical works is included, compiled by Yu Mou (1127–1194). It is mentioned in the *Ming-shih, I-wên chih (Pu-pien, fu-pien, shang-ts'ê,* p. 359). But Mr. van der Loon informs me that this is no evidence that this work really existed in the Ming period, for the *Ming-shih, I-wên chih* is not a catalogue but only a list of books, including even the titles of lost works. Mr. van der Loon, who has thoroughly investigated the matter for me, assures me that the work has long been lost. I thank him for his kind assistance.

Even if the work had not been lost, I think that it would have little application to this study. Another work of Ts'ai Yüan-ting that has been preserved, the *Lü-lü hsin-shu* 律呂新書, speaks only about the permutations of the *kua*. As the *ta-yen* concept is derived from the *I-ching* [see Needham (1), p. 119 and note *j*], it is very likely that the title *Ta-yen hsiang-shuo* should be understood in this way.

[79] This is one of the five works that Juan Yüan calls *Yang Hui suan-fa* 楊輝算法 (Mathematical methods of Yang Hui).

[80] The commentary says "popularly called: *Ch'in wang an tien ping* 秦王暗點兵, The Prince of Ch'in's method of secretly counting soldiers" [cf. Needham (1), vol. 3, p. 122]. Yang Hui calls his method *chien kuan* 翦管, cutting tubes. Yen Tun-chieh says in Ch'ien Pao-tsung (2'), p. 164, that the sense of *chien kuan* is not explained, but that it seems to mean that, if a tube is cut several times in equal parts (but varying each time in length), we can find the original length of the tube (from the length of the pieces and the length of the surplus). This explanation makes sense, but no confirmation is found in the texts.

The Sun Tzŭ text is not given in the *Hsü-ku chai-ch'i suan-fa,* as preserved in the *I-chia t'ang* collection, but I found it in the *Chih-pu-tsu* collection, *ts'ê* 103, p. 11a and following pages. A recent study on Yang Hui by Lam Lay Yong is in press now, but it was impossible to obtain it for solving the complicated problem of the reconstruction of Yang Hui's works. On the other hand, Ho Pêng-yoke's article on Yang Hui, prepared for the *Dictionary of Scientific Biography*, was very helpful. As for the text, I relied on the *Chih-pu-tsu* edition. The first of Yang Hui's problems is not included in Lo T'êng-fêng (1'), who does give, however, the three other problems (p. 23a and following pages). I made also use of Li Yen (6'), vol. 1, p. 124 and following pages.

says: "Now I add four problems:[81] (1) We use an unknown number of workmen; a messenger pays the gratuities: if he gives one *chin* 斤 of meat to each three men, the remainder is 5 *liang* 8 *chu*; if he gives one *kuan* of money to each 5 men the remainder is 400 *wên*; if he gives one *tuo* 撥 of wine to each 7 men there is no remainder. Find what is paid to the workmen.

Answer: 98 men; money: 19 *kuan* 600 *wên*; wine: 40 *tuo*; meat: 32 *chin* 10 *liang* 16 *chu*.

Method: 3, remainder 2 (put 140); 5, remainder 3 (put 63); 7, remainder 0 (do not put anything).

Add these numbers (you get 203); subtract 105; the remainder is 98 workmen. Multiply the number of workmen by 200 and you get the amount of money; divide the number of workmen by 7 and you get the wine; divide by 3 and you get the meat."

Each workman receives 1/3 *chin* of meat. Since the remainder is 1/3 *chin* or 5 *liang* 8 *chu*,[82] the number of workmen is 2 (mod 3).

Each workman receives 1/5 *kuan* or 200 *wên*.[83] Since the remainder is 400 *wên*, there are 3 (mod 5) workmen. In the same manner there are 0 (mod 7) workmen.

The problem is $N \equiv 2$ (mod 3) $\equiv 3$ (mod 5) $\equiv 0$ (mod 7), and is only a variant of the Sun Tzǔ problem.

2. $N \equiv 1$ (mod 7) $\equiv 2$ (mod 8) $\equiv 3$ (mod 9)
mod 7 : if R = 1→288
mod 8 : if R = 1→441
        (if R = 2→882)
mod 9 : if R = 1→280
        (if R = 3→840)
$N = 288 + 882 + 840 - n \times 504 = 498$.
3. $N \equiv 3$ (mod 11) $\equiv 2$ (mod 12) $\equiv 1$ (mod 13).
4. $N \equiv 1$ (mod 2) $\equiv 2$ (mod 5) $\equiv 3$ (mod 7) $\equiv 4$ (mod 9).[84]

[81] To the Sun Tzǔ problem.
[82] 1 *chin* = 16 *liang*; 1 *liang* = 24 *chu*. See Vogel (2), p. 140.
[83] 1 *kuan* = 1,000 *wên*.
[84] These three problems are treated by Lo T'êng-fêng (see note 80); see further Chapter 16.

Problems (3) and (4) are solved in the same way. This proves that Yang Hui was able to calculate the *fan-yung*,[85] but nothing is known about his method. His knowledge of indeterminate analysis is obviously on a level no higher than that of the *Sun Tzŭ suan-ching*. This seems to be a proof that there was not the slightest contact between Ch'in Chiu-shao (1247) and Yang Hui (1275). Precisely the same *fan-yung* (for the moduli 2, 3, 5, 7, 9, 11,13) are given in the Munich manuscript (c. 1450), but here a method (although very primitive) is given for their computing.

14. Chou Mi 周密

Chou Mi, in his *Chih-ya t'ang tsa-ch'ao* 志雅堂雜鈔 (Miscellaneous notes from the Chih-ya hall) (1290), last *chüan*, paragraph *Yin-yang suan-shu* 陰陽算術[86] gives a description of the *ta-yen* rule in the form of directions for solving the problem of Sun Tzŭ: "The *kuei-ku* 鬼谷 computation,[87] also named the *ko-ch'iang* 隔牆 computation.[88] According to this method, take first some money, no matter how much;[89] count it by threes; if you get

[85] See Chapter 17.
[86] Ch. 9, pp. 2b–3a.
[87] The "*Kuei-ku* computation" is another name for the *ta-yen* rule, "presumably because of some attribution to the legendary philosopher Kuei-ku tzŭ," [Needham (1), vol. 3, p. 122]. The *Kuei-ku* (literally, valley of the ghosts) was a place in Ho-nan, where Master Kuei-ku was said to have lived. See also Needham (1), vol. 2, p. 206 and Kimm Chung-se (1).
[88] Needham (1), vol. 3, p. 122 translates as "behind the arras" computation. *Ko-ch'iang* means "across or behind the partition wall." The origin of the name is not clear.
[89] This seems to show that it was also a popular game, something like the *Regula Caecis* in Europe. See Smith (1), vol. 2, pp. 586 f. Chiao Hsün (1) says: "At present women and children consider it as a game" (ch. 2, p. 4a). Compare with Fibonacci (1), vol. 1. p. 304: "From this rule proceeds a more pleasant riddle" (Procedit enim ex hac regula pulcrior divinatio) and the Munich manuscript, where the problem has the title "Divinare."
    In his biography of Ch'in Chiu-shao, included in his *Kuei-hsin tsa-chih, hsü-chi* (1308), Chou Mi says nothing about the *ta-yen* rule. It is very likely that he never studied Ch'in's work, but that a (simplified) *ta-yen* rule was known as a game. However, as he gives it in an article on the *Yin-yang* mathematical methods (the divination technique of the *I-ching*), and as we have a similar problem in Ch'in Chiu-shao's work (I, 1), it is possible that the method was applied by fortune tellers. But the numbers used as moduli seem to indicate that the problem was derived from the *Sun Tzŭ suan-ching*.

the remainder 1, write down 70; if 2, write down 140; next count it by fives; if the remainder is 1, write down 21, if 2 write down 42; next count it by sevens; if the remainder is 1, write down 15; if 2 write down 30. Add up all these numbers; after that subtract 105, or, if it is still more [than 105] subtract 210. The remainder is the amount."

This is indeed the *ta-yen* rule, but restricted to the numbers of the Sun Tzŭ problem, and not at all certain proof of knowledge of a general method, as is evident from the added mnemotechnical stanza:[90]

"A child of 3 years when 70 is rare

[1 (mod 3) → 70]

At 5 to leave behind the things of 21 is still more rare[91]

[1 (mod 5) → 21]

At 7 one celebrates the Lantern Festival [*shang-yüan*] . . .

[1 (mod 7)→15]

[The *shang-yüan* falls on the first full moon on the fifteenth of the first lunar month.]

. . . Again they meet together

[This means, Add up the numbers found above.]

Han-shih, ch'ing-ming 寒食,清明 . . .

[The *han-shih* (literally, "cold meal") holiday was the day before the *ch'ing-ming*[92] and was the 105th day of the winter. Here it means that 105 is to be subtracted from the sum obtained, if it is at least 106 (*ch'ing-ming*).]

. . . Then you will get it."

Chou Mi adds: "According to the following rule, one gets the mutually multiplied numbers. If 3, then multiply 5 and 7 by

[90] This kind of poetry is not easy to translate, as it is merely a medium for memorizing the rule; and because it seems to have no strict sense, the translation is uncertain. The words in italics imply the numbers to be kept in mind.
[91] One can also translate: "From 5 [things], to leave behind 21...."
[92] It was a day on which only cold meals were eaten.

each other and double; if 5, multiply 3 and 7 by each other; if 7, multiply 3 and 5 by each other."

$$2 \times (5 \times 7) = 70$$
$$3 \times 7 = 21$$
$$3 \times 5 = 15.^{93}$$

If you take together [multiply] these [numbers], you get 105. $(3 \times 5 \times 7 = 105.)$

From these texts, it is obvious that there had been no progress since the time of Sun Tzǔ; that the work of Ch'in Chiu-shao giving the general rule was not known by Chou Mi (although he wrote Ch'in's biography); that such artifices as the stanzas were used to solve or to teach the problem; and that perhaps Chou's writing has more to do with games than with mathematics, although it is possible that its aim was training people in simple everyday computation.

There is also another conclusion that forces itself upon us. Yang Hui (1275) and Chou's writing (1290) did not include the *ta-yen* method. Li Yeh says nothing about it, nor does Chu Shih-chieh. The Ming mathematician Ch'êng Ta-wei gives only the Sun Tzǔ problem. Until the time of Chang Tun-jên (1803) there was no one who explained the *ta-yen* rule. For that reason one wonders if Ch'in's text was ever studied.

15. Yen Kung 嚴恭

Yen Kung was the author of the *T'ung-yüan suan-fa* 通原算法 (Traditional mathematics; literally, "Mathematics continuous with the origins") (1372),[94] containing some problems on in-

93 There is no explanation of this duplication.
94 This work, although very inferior to the works of the great mathematicians of the thirteenth and fourteenth centuries, has the good luck to be partly preserved in the original edition of the *Yung-lo ta-tien*, ch. 16,343 and 16,344. Needham (1), vol. 3, p. 50, gives the impression that the indeterminate equations included in this work are in the *Yung-lo ta-tien*, but this is not true. They are preserved in the *Chu-chia suan-fa chi hsü chi* 諸家算法及序記, which exists only in manuscript form; the author is unknown. The text is published in Li Yen (3'), pp. 25–43. On this work see Ting Fu-pao (1'), p. 42b (no. 165) and Li Yen (10'), p. 433.

determinate equations. Problem 28 of his work is as follows:[95] "You have an unknown number of coins. If you make 77 strings of them, you are 50 coins short; if you make 78 strings, it is exact. The number is asked for. Answer: 2,106.

"Method: Multiply 78 by itself and you have 6,084; next subtract the shortage 50 from 77; the remainder is 27; multiply with the first number and you get 164,268. On the other hand multiply 78 and 77 and you get 6,006. Subtract from the former number [as many times as possible]; the remainder is the number asked for."

The problem is: $N \equiv 27 \pmod{77} \equiv 0 \pmod{78}$. Let us find the solution to the first problem by Ch'in Chiu-shao's method. Since there are 50 coins missing the remainder is $77-50=27$.

1. $\theta = 77 \times 78 = 6,006$
2. *Yen-shu*: $M_1 = \theta/77 = 78$; $M_2 = \theta/78 = 77$.
3. *Ch'i-shu*: $N_1 = 78 - 77 = 1$; $N_2 = 77$.
4. Congruences:

$a \times 1 \equiv 1 \pmod{77}$ $\beta \times 77 \equiv 1 \pmod{78}$
$\quad a = 78$ $\beta = 77$.

5. *Fan-yung*: $a \times M_1 = 78 \times 78$; $\beta \times M_2 = 77 \times 77$.

6. $\sum_1^2 (raM) = 27 \times 78 \times 78 + 0 \times 77 \times 77 = 164,268$

7. $N = \sum_1^2 (raM) - n\theta = 164,268 - 27 \times 6,006 = 2,106$.

Yen Kung solves the problem as follows:[96]

$78^2 = 6,084$
$77 - 50 = 27$
$6,084 \times 27 = 164,268$
$78 \times 77 = 6,006$
$164,268 - n \times 6,006 = 2,106$.

This problem seems to give evidence that Yen Kung knew a general solution for indeterminate equations. However, the

[95] Indeterminate analysis is called *kuan-shu* 管數 ("tube numbers"), a term derived from Yang Hui's *chien kuan* ("cutting tubes"). See note 80.
[96] See Li Yen (9'), p. 126, note 1.

problem treated is very special. If we compare it with the next problem, we have to conclude that he did not know a general solution and had only methods for special cases.

Problem 29 reads: "We have an unknown number of tiles; if 34 pieces form a heap, the remainder is 5; if 36 pieces form a heap, the remainder is 7; how many tiles are there? Answer: 1,195."[97] Let us first solve the problem by Ch'in Chiu-shao's method. Note that the moduli are not relatively prime:

$N \equiv 5 \pmod{34} \equiv 7 \pmod{36}$.

$M_1 = 34$	$M_2 = 36$
$\mu_1 = 17$	$\mu_2 = 36$

$$\theta = 612$$

$N_1 = 2$	$N_2 = 17$
$\alpha \times 2 \equiv 1 \pmod{17}$	$\beta \times 17 \equiv 1 \pmod{36}$
$a = 9$	$\beta = 17$
$F_1 = 9 \times 36 = 324$	$F_2 = 17 \times 17 = 289$

$$\sum_1^2 (rF) = 5 \times 324 + 7 \times 289 = 3,643$$

$$N = 3,643 - n \times 612 = 583.$$

The solution given by Yen Kung is as follows:

$34\,(34+1) = 1,190$	$36\,(36-1) = 1,260$
$1,190 : 2 = 595$	$1,260 : 2 = 630$
$595 \times 7 = 4,165$	$630 \times 5 = 3,150$

$$4,165 + 3,150 = 7,315$$
$$34 \times 36 = 1,224$$
$$7,315 - 5 \times 1,224 = 1,195.$$

But 1,195 is only one of the possible solutions, and not the smallest one. Indeed, $1,195 \equiv 5 \pmod{34} \equiv 7 \pmod{36}$. If we write the general problem as

$$N \equiv r_1 \pmod{M_1} \equiv r_2 \pmod{M_2},$$

then it might be inferred from the preceding demonstrations that the general solution is

[97] The text (without explanation) is included in Li Yen (3'), p. 32, with the solution 2,195, which is incorrect.

$$N = \frac{M_1(M_1+1)}{2}\, r_2 + \frac{M_2(M_2-1)}{2}\, r_1 - nM_1M_2. \qquad (1)$$

Although there are some special problems that can be solved in this way, the solution is not at all general. Formula (1) is valid only if $M_2 - M_1 = 2$. Indeed, if we divide N by M_1, the remainder must be r_1. Replacing the division by a subtraction of n times M_1, we get:

$$\frac{M_2(M_2-1)}{2}\, r_1 - nM_1.$$

Replacing M_2 by $2 + M_1$, we get:[98] $\frac{2}{2} r_1 = r_1$.

17. Chou Shu-hsüeh 周述學

Chou Shu-hsüeh was the author of the *Shên-tao ta-pien li-tsung suan-hui* 神道大編, 曆宗算會(1558), which includes in the tenth *chüan* the following text:[99] "If it is not the rule of false position, but only all remainder factors[100] [as in the case where the moduli are] 3, 5, 7 or 7, 8, 9, consider these 1's[101] and the remainders; arrange them in order below each other [on the counting board]: if you subtract from these [numbers] the *k'uai-shu* 會數,[102] as many times as possible, as remainder you get the *tsung* 總[103] you search for. If you consider numbers like 2, 3, 4, they have no remainder factors. It is necessary to have the greatest common denominator 1 in order to find them."

This very uncertain translation seems to discuss something

---

[98] For example, $N \equiv 2(\mathrm{mod}\ 5) \equiv 3(\mathrm{mod}\ 7)$; $7 - 5 = 2$

$$N = \frac{5(5+1)}{2} \times 3 + \frac{7(7-1)}{2} = 87.$$

[99] The text is in Li Yen (6'), p. 127. It is extremely difficult to provide a correct interpretation for these obscure statements. The translation is given with all due reserve. In order to make it understandable I have referred to the Sun Tzŭ problem.

[100] The text says *yü-lü* 餘率. A possible meaning could be "the factor by which the remainder is to be multiplied."

[101] That is, the remainder factors.

[102] The *k'uai-shu* must be the least common multiple of the moduli.

[103] The *tsung* is the general number answering the problem.

like the Sun Tzŭ problem, but in a very inaccurate way.[104]
The last sentence clearly proves that Ch'in's work was not
known to Chou Shu-hsüeh.

18. Ch'êng Ta-wei 程大位

The *Suan-fa t'ung-tsung* (Systematic treatise on arithmetic)
appeared in 1593 and was written by Ch'êng Ta-wei, whose
literary name was Pin-ch'ü 賓渠.[105] Biot (3) did a detailed description of the contents, in which he included the Sun Tzŭ
problem,[106] giving only the following information: "This is a
case of things of which we do not know the number. Three questions. (These questions are of this form: someone asks for a
number such that, when it is divided by 3, there is a remainder of 2; by 5, a remainder of 3; by 7, a remainder of 2.)"
Ch'êng's work contains nothing that goes beyond the mathematical works of the Sung and Yüan; at the end of it there
is a list of mathematical works, most of which are no longer
extant.[107] There is also a useful glossary of the terminology used
in the work.[108] The *ta-yen* method is called "Han Hsin's method."[109] The Sun Tzŭ problem is stated thus: "Things with
unknown number." The Sun Tzŭ stanza (also called *Han Hsin
tien ping* 韓信點兵)[110] says:

"Three septuagenarians in the same family is exceptional

[104] For further information on Chou Shu-hsüeh, see Needham (1), vol. 3,
p. 51, p. 105, and p. 143.
[105] I used the edition prepared by Mei Ku-ch'êng (1680–1763), ch. 4, p.
7a. Dickson (1), vol. 2, p. 60 mentions only the name Pin Kue, relying
on Biot (3), p. 193.
[106] This seems to be the oldest statement of the Sun Tzŭ problem in Europe
(1839), but it seems that the problem drew no attention before Wylie (1852).
[107] Ch'in Chiu-shao is not mentioned.
[108] Vol. 1, pp. 2–3.
[109] Ch. 5, p. 21b and following pages. For a biography of the author, see
Hummel (1), p. 117. For further information on the work as a whole, there
are Li Yen (17'), p. 61 and pp. 82 ff; Ch'ien Pao-tsung (4'), pp. 8 f; Juan
Yüan (1'), ch. 31, p. 19b; Smith (1), vol. 1, p. 352 and vol. 2, pp. 114 ff;
Mikami (1), pp. 110 f; Wylie (2), p. 118.
[110] Needham (1), vol. 3, note *d*. "Han Hsin's method" refers to Han Hsin,
who was a famous Han general.

[1 (mod 3)→70]

Twenty-one branches of plum-blossom from 5 trees

[1 (mod 5)→21]

Seven brides in ideal union (*t'uan-yüan* 團圓) precisely the middle of the month

[1 (mod 7)→15]

Subtract 105 and you get it."
(The third line is an allusion to the *t'uan-yüan* or happy marriage feast, which took place on the fifteenth day of the eighth lunar month. On this day, daughters who were married returned to their parental homes to pay their respects to their parents. The meaning is thus 15). Then he gives the Sun Tzŭ problem and explains the method:

"Draw up the factors 3, 5, and 7. Multiply $3 \times 5 \times 7$ and you get 105 as the *man ch'ien-shu* 滿錢數 [literally, total amount].[111]

"$3 \times 5 = 15$; this is the congruence number [*shêng-i chih shuai* 剩一之衰, literally, the corresponding number giving 1 as remainder] of 7.[112]

"$3 \times 7 = 21$: this is the congruence number of 5. $5 \times 7 = 35$; $35 \times 2 = 70$: this is the congruence number of 3."[113] Further on there is the same text as in the *Sun Tzŭ suan-ching*.[114]

The mnemotechnical stanza of Ch'êng Ta-wei was handed down among the common people, as Chiao Hsün (1763–1820)

[111] This is of course the least common multiple; *ch'ien* means copper coin. It occurs also in the work of Yen Kung (1372) and might be an allusion to some folk game.
[112] The congruence number is the smallest number C that divided by one of the moduli m gives 1 as remainder, or: $C \equiv 1 \pmod{m}$.
[113] As always there is not the slightest indication why one has to multiply by 2.
[114] According to Biot (3), p. 207 there should be three problems of this kind. But the last two problems are determinate, and have nothing to do with the *ta-yen* rule. It seems to be an indication of the very low level of mathematics at the time of Ch'êng Ta-wei that three different problems are gathered under one title only because they begin with "Now you have *x*, of which the number is unknown...."

pointed out in his *T'ien-yüan-i shih* (1800): "By this time, among women and children, some consider it a game. . . . "[115] Moreover, it found its way to Japan.[116]

For the later period (after Ch'êng Ta-wei), there are only a few notes, giving evidence that understanding of the *ta-yen* method had been lost. In the "critical notes" of the *Ssŭ-k'u ch'üan-shu tsung-mu t'i-yao* the *ta-yen* rule is paraphrased as follows:[117] "For the remainders[118] find the *tsung-shu* 總數." Although one must indeed find the *tsung-shu*, this is no explanation of the rule. As for Problem I, 1 of Ch'in Chiu-shao (on the *I-ching* divination method) the only commentary is: "By the new method, he wanted to explain the divination-technique of the *I-ching*, but it is entirely different from the ancient sense."[119]

[115] Ch. 2, p. 4a.
[116] Li Yen (9'), p. 127 f.
[117] Ting Fu-pao (1'), p. 574b.
[118] Indicated as *ch'i-ling* 奇零.
[119] There are some other notes on a few more problems but they are all very superficial and have nothing to do with the mathematical contents.

16

Chinese Studies on the Ta-yen Rule
in the Nineteenth Century

As we can deduce from the historical outline, the *ta-yen* rule was almost forgotten during the Ming and Ch'ing periods, owing perhaps to the general decline of mathematics in China. But it is somewhat surprising that there exists no mathematical work dealing with the *ta-yen* rule of Ch'in Chiu-shao, while mention is made several times of the Sun Tzǔ problem. The downward turn of mathematics followed closely upon its great development during the Sung and Yüan dynasties, and it is perhaps a matter of sheer good luck that even the *Shu-shu chiu-chang* and other works have come down to us.[1] There are several papers that deal with this decline and try to explain it.[2] With the arrival of the Jesuits in China and the importation of European mathematics,[3] there was at first great admiration for the foreign sciences, but after a century of contact with the West[4] Mei Ku-ch'êng 梅穀成[5] (1681–1763) called attention to the existence of an autochthonous mathematics no less estimable than the foreign. From this time on, interest in ancient Chinese

[1] Perhaps only because they were incorporated in the *Yung-lo ta-tien* in the fifteenth century and in the *Ssŭ-k'u* collection in the eighteenth century.
[2] Li Yen(10'), pp. 13 ff; Mikami (1), pp. 108 ff.
[3] See Needham (1), vol. 3, pp. 52 f.
[4] Matteo Ricci arrived in Peking in 1601, and the translation of the first six books of Euclid was completed in 1607.
[5] The grandson of the well-known Mei Wên-ting 梅文鼎 (1633–1721), author of the *Li-suan shu-mu* 曆算書目 (Catalogue of works on calendar and mathematics). There is a special paper on Mei Wên-ting by Li Yen (6'), vol. 3, pp. 544 ff; see also Hummel (1), p. 570. Mei Ku-ch'êng was convinced that the loss of mathematical knowledge was owing to the ignorance of the Ming scholars. In fact, the interest in the history of Chinese mathematics began with Mei Wên-ting; although the rediscovery of the old works took place in the second half of the eighteenth century, "his labors served to...revive an interest in older Chinese mathematical discoveries" [Hummel(1), p. 571]. See also Mikami (6), p. 125. The best-known old-style mathematical work in Mei's time was the *Suan-fa t'ung-tsung* of Ch'êng Ta-wei (see Chapter 15).

mathematics began to grow, and it is to the credit of Tai
Chên 戴震 (1724–1777) that he gave impulse to the recovery
and the study of mathematics.[6] A detailed study of early mathematical methods was done by Chiao Hsün 焦循 (1763–1820)
in his *T'ien-yüan-i shih;* he also procured the work of Li Yeh 李
治[7] for Li Jui 李銳 (1765–1814), "and thus inspired the latter's
studies in Chinese algebra."[8] On the other hand, Juan Yüan 阮
元 (1764–1849) wrote his very important *Ch'ou-jên chuan*
(Biographies of mathematicians and astronomers) (1799);[9] it
was written with assistance of Chiao Hsün, Ling T'ing-k'an 凌
廷堪 (1757–1809),[10] and Ch'ien Ta-hsin 錢大昕 (1728–1804).[11]
"Juan Yüan's interest in mathematics helped to revive the
study of ancient Chinese mathematics and led to the recovery
of works in that field which had been neglected for centuries."[12] Lo Shih-lin 羅士琳 (?–1853), who was associated with
Juan Yüan, did an important study of Chinese algebra as
represented in the work of Chu Shih-chieh 朱世傑.[13] Shên
Ch'in-p'ei 沈欽裴 collated the *Shu-shu chiu-chang* of Ch'in Chiu-
shao; after his death, his disciple Sung Ching-ch'ang 宋景昌
completed the work.[14] Chang Tun-jên 張敦仁 was the first who

[6] See Hummel (1), pp. 695 ff. Tai Chên edited the *Suan-ching shih-shu* 算經
十書 (Ten mathematical classics). There is a new edition (1963), prepared
by Ch'ien Pao-tsung (12').
[7] Or Li Chih 李治 (1192–1279), a well-known Yüan mathematician, author
of the *Ts'ê-yüan hai-ching* 測圓海鏡 (Sea mirror of circle measurements)
(1248). This work was also known to Mei Ku-ch'êng.
[8] Hummel (1), p. 144. On Li Jui, see Chapter 4.
[9] On the meaning of the word *ch'ou,* see Needham (1), vol. 3, p. 3. On
Juan Yüan's work, see Van Hee (3); for criticism of this article, there is
Mikami (6). See also W. Franke (1); Vissière (1).
[10] See Hummel (1), p. 514. Li T'ing-k'an was a good friend of Juan Yüan.
[11] Ch'ien Ta-hsin was acquainted with Tai Chên. Li Jui was one of his
students, like Ku Kuang-ch'i 顧廣圻, the author of a preface to Ch'in
Chiu-shao's work, entitled *Shu-shu chiu-chang hsü* 數書九章序. (On Ku
Kuang-ch'i, see Chapter 4.)
[12] Hummel (1), p. 402.
[13] The work of Chu Shih-chieh is entitled *Ssŭ-yüan yü-chien* 四元玉鑑 (Jade
mirror of the four unknowns) and dated 1303; the study of Lo Shih-lin is
entitled *Ssŭ-yüan yü-chien hsi-ts'ao* (Detailed investigation of the Jade Mirror
of the Four Unknowns) (1837).
[14] See Chapter 4.

understood the *ta-yen* rule. From this time on, the study of the history of Chinese mathematics began to flourish. The discussion below is restricted to those scholars who elucidated indeterminate analysis.

1. Juan Yüan 阮元

In the *Ch'ou-jên chuan*(1799),[15] ch. 22, Juan Yüan gives a short description of the nine chapters of the *Shu-shu chiu-chang*. The whole first chapter is devoted to the *ta-yen* rule and gives a clear explanation of it:[16] "In this method one brings together the *yüan wên-shu* 元問數 (by twos)[17] and finds their common factor. After reducing they become the *ting-mu* 定母. First, mutually multiply all the *ting-mu* and you get the *yen-mu* 衍母; the mutual product is the *yen-shu* 衍數. Then, from the *ting-mu* subtract the *yen-shu*; the remainder is the *ch'i-shu* 奇數. Apply the *ta-yen ch'iu-i* 大衍求一 method, and you get the *ch'êng-lü* 乘率. Multiply with the *yen-shu* and you have the *yung-shu* 用數. Multiply these by the remainders given in the problem.[18] Add these up and you get the *tsung-shu* 總數. Subtract the *yen-mu* therefrom as often as possible; the remainder is the number searched for. As for the *ta-yen ch'iu-i* method . . . ."[19]

This is, indeed, after a space of time of 550 years the first quotation indicating that the author understood Ch'in's text. Juan Yüan grasped the quintessence of the method, since he gave a very good résumé of the main points. He fails to mention only the difficult problem of the "reduction."[20] In all possibility he found his information in the study by Chang Tun-jên.[21] Juan Yüan's other texts on Ch'in Chiu-shao are treated elsewhere.[22]

[15] See ibid. On this work see W. Franke (1); Van Hee (3), and Mikami (6).
[16] The explanation of all the technical terms occurring in the text is to be found in the translation of Ch'in's general method (Chapter 17).
[17] As is obvious from Ch'in Chiu-shao's text.
[18] The *yüan wên yü-shu* 元問餘數.
[19] From here on the text is taken from the *Shu-shu chiu-chang*.
[20] *Yüeh* 約.
[21] Ch'ien Ta-hsin collaborated with Juan Yüan; Ku Kuang-ch'i was a disciple of Ch'ien Ta-hsin and at the same time the protégé of Chang Tun-jên, author of a detailed study on the *ta-yen* rule.
[22] Juan Yüan provides some biographical notes on Ch'in Chiu-shao,

2. Chiao Hsün

In his *T'ien-yüan-i shih* Chiao Hsün 焦循 (1763–1820) deals with the *ta-yen* rule.[23] However, since his chief aim is to explain the *t'ien-yüan-i* 天元一 (celestial element unity), he turns his attention to this method. He knows that this concept is not the same in the work of Ch'in Chiu-shao as in that of Li Yeh. In Ch'in's work there are two places where the Celestial Element is used:[24] in the computing of the *yen-shu* and in the solving of the congruences; and Chiao deals with them. After that, he gives a correct explanation of the Sun Tzŭ problem.[25] However, his work is very confused and obscure. He tries to explain the reducing of the moduli, but he does not succeed. He attempts comparisons with Chang Ch'iu-chien[26] and the rule of false position, and so on, but all without success. Chiao Hsün was the first who applied himself to the old indeterminate problems(his work was published in 1799) and the first who was able to give a correct explanation of the Sun Tzŭ problem, although he failed to understand Ch'in Chiu-shao's method.[27] His work has only historical value.

3. Chang Tun-jên 張敦仁

After the long period of darkness for Chinese mathematics, Chang Tun-jên (1754–1834) was the first scholar who succeeded in understanding the *ta-yen* rule, which he dealt with in his *Ch'iu-i suan-shu* (Mathematical method for seeking unity[28]), published in 1803.[29] His work follows a very clear arrangement. The contents can be outlined thus:[30]

---

mainly derived from Chou Mi's *Kuei-hsin tsa-chih*; there is also a short description of the other methods used in the various chapters of Ch'in's work.

[23] B, p. 2a and following pages.

[24] For an attempt at explanation of this term, see Chapter 17.

[25] There is a diagram of it in Li Yen (6'), vol. 1, p. 144.

[26] See Chapter 15.

[27] There is a biography of Chiao Hsün in Hummel (1), pp. 144 ff.

[28] That is, indeterminate analysis.

[29] Some scholars give 1801, but Chang's preface is dated 1803.

[30] See Wylie (2), p. 99.

I. Reducing the moduli
A. Finding of the common divisors (*Ch'iu têng* 求等), with fcur examples
B. Reducing of the moduli (*Yüeh fên* 約分), with five examples
C. Further reduction ("compensation") (*Tsai yüeh* 再約), with five examples
D. Total reduction (*Lien-huan hsiang-yüeh* 連環相約), with three examples

II. *Ta-yen* rule
A. Finding of *yen-mu* and *yen-shu* (*Lien-huan hsiang-ch'êng* 連環相乘)[31] with two examples
B. Solving of the congruences (*Ta-yen ch'iu-i* 大衍求一), with three examples
C. Application of the rule (*ch'iu-i* 求一),[32] with six examples

III. Chronological applications (*yen-chi* 演紀), with five examples.

Chang Tun-jên tried to explain the very intricate method of reducing the moduli as it appears in the work of Ch'in Chiu-shao.[33] Here is an example:[34]

A=24   A'=?
B=30   B'=?
C=54   C'=?

The symbols *A'*, *B'*, *C'* are the reduced moduli, relatively prime in pairs.

$$(1) \quad \frac{24(e)\ 30(e)}{6(e)} \longrightarrow \frac{24(e)\ 5(o)}{1}$$
$$\frac{24(e)\ 54(e)}{6(e)} \longrightarrow \frac{24(e)9(o)}{3}$$
$$\frac{5(o)\ 9(o)}{1}$$

[31] The title means "mutually multiplying," but the purpose is to find the *yen-mu* and the *yen-shu*.
[32] *Ch'iu-i* was the name sometimes given to the *ta-yen* rule in general.
[33] See Chapter 17. We shall deal with Chang Tun-jên's explanation in treating Ch'in's method.
[34] See Chang Tun-jên (3), A, p. 5b.

(The symbol e=even ; o=odd; the number below the line is the greatest common divisor of the numbers above). One has to take all possible pairs of numbers, A and B, A and C, B and C, and find their greatest common divisor. One of the numbers must be divided by this greatest common divisor.

(2) 24   5   9

$$\frac{24 \quad 5}{1} \longrightarrow 24 \quad 5$$

$$\frac{24 \quad 9}{3} \longrightarrow \begin{bmatrix} 24 \longrightarrow 8(:3) \\ 9 \longrightarrow 27(\times 3) \end{bmatrix}$$

8   5   27
A'   B'   C'

If after the first reduction there remain numbers not prime in pairs, find their greatest common divisor, and divide the one and multiply the other.[35] Chang's explanation of solving the congruences is very clear: For example,[36] in Diagram 57, $a \times 5 \equiv 1 \pmod{23}$. The operation may be stopped at this point only if the above remainder is 1.[37] Draw up Diagram 58 and compute the quotients; the solution is $a = 14$ or $14 \times 5 \equiv 1 \pmod{23}$.

There is, however, one example in Chang's work that shows definitely that he did not entirely understand the reducing of the moduli. If $m_1, m_2, m_3 \ldots$ are the given moduli, not relatively prime in pairs, and $\mu_1, \mu_2, \mu_3 \ldots$ are the reduced moduli,

---

[35] In modern notation: $24 = 2^3 \times 3$; $30 = 2 \times 3 \times 5$; $54 = 2 \times 3^3$. Compare with Huang Tsung-hsien's method, explained later in this chapter.

[36] Chang Tun-jên (1'), A, p. 10b.

[37] The operation may be stopped only if the above remainder is 1. The reason is the following: if only the above remainder is considered, the number of quotients is always even, and thus the congruence factor always positive. For example, $5a \equiv 1 \pmod{23}$. If we stop the operation at the R_3 (below), the solution is $a = -9$, as $(-9) \times 5 \equiv 1 \pmod{23}$. We can change the factor into positive as follows:

$$\frac{(-9) \times 5 \equiv 1 \pmod{23}}{23 \times 5 \equiv 0 \pmod{23}}$$
$$14 \times 5 \equiv 1 \pmod{23}$$

We have the same result when requiring that the remainder must stand above.

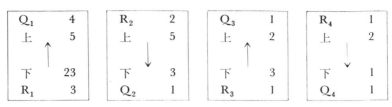

Diagram 57

L	R
4	1
1	4
1	5
1	9
.	14

L	R	
Q_1	天元	$1 = \alpha_1$
Q_2	$1 \cdot Q_1$	$= \alpha_2$
Q_3	$\alpha_2 \cdot Q_2 + 1 = \alpha_3$	
Q_4	$\alpha_3 \cdot Q_3 + \alpha_2 = \alpha_4$	
	$\alpha_4 \cdot Q_4 + \alpha_3 = \alpha_5$	

Diagram 58

then: $r_1 \pmod{m_1}$ may be substituted for by $r_1 \pmod{\mu_1}$ if and only if μ_1 is a divisor of m_1.[38] Chang Tun-jên was not aware of this condition, as is obvious from his example on page B.la:

$$N \equiv \ 9 \ (\mathrm{mod} \ 10) \equiv 19 \ (\mathrm{mod} \ 20) \equiv 10 \ (\mathrm{mod} \ 30) \equiv$$
$$19 \ (\mathrm{mod} \ 40) \equiv \ 9 \ (\mathrm{mod} \ 50) \equiv 19 \ (\mathrm{mod} \ 60) \equiv$$
$$69 \ (\mathrm{mod} \ 70) \equiv 59 \ (\mathrm{mod} \ 80).$$

He reduces the moduli in the following phases:[39]

A	B	C	D	E	F	G	H
10	20	30	40	50	60	70	80
1	20	30	40	50	60	70	80
1	1	3	40	50	60	70	80
1	1	1	40	50	60	70	80
1	1	1	1	50	15	35	80
1	1	1	1	5	3	7	80
1	1	1	1	1	3	7	400

[38] See Mahler (1), p. 120.
[39] The correct reduction would be: 25, 3, 7, 16. It is impossible to include all the details of Chang's example, but only the result is important.

yen-mu $= 3 \times 7 \times 400 = 8{,}400$

yen-shu $= F = 7 \times 400 = 2{,}800$

$\qquad\quad G = 3 \times 400 = 1{,}200$

$\qquad\quad H = 3 \times 7 = 21$

ch'i-shu $= F = 2{,}800 - n \times 3 = 1$

$\qquad\quad G = 1{,}200 - n \times 7 = 3$

$\qquad\quad H = 21 - n \times 400 = 21$

Congruences:

$a \times 1 \equiv 1 \pmod 3$	$a = 1$
$\beta \times 3 \equiv 1 \pmod 7$	$\beta = 5$
$\gamma \times 21 \equiv 1 \pmod{400}$	$\gamma = 381$
$1 \times 2{,}800 = 2{,}800$	$\times 19 = 53{,}200$
$5 \times 1{,}200 = 6{,}000$	$\times 69 = 414{,}000$
$381 \times \quad 21 = 8{,}001$	$\times 59 = \underline{472{,}059}$

$$939{,}259 - n \times 8{,}400 = 6{,}859. \quad [40]$$

Although Chang's solution is right, his method is wrong, The reason is that if $N \equiv 59 \pmod{400}$, or $N = 400z + 59$, then $N = 5 \times 8 \times 10 \times z + 50 + 9$, from which

$$N = 50x + 9$$
$$N = 80x' + 59.$$

[40] The correct solution is $E : 25$, $F : 3$, $G : 7$, $H : 16$.

Yen-mu $= 25 \times 3 \times 7 \times 16 = 8{,}400$.

Yen-shu		*Ch'i-shu*	
$E : 3 \times 7 \times 16 = 336$		$- 25n = 11$	
$F : 25 \times 7 \times 16 = 2{,}800$		$- 3n = 1$	
$G : 25 \times 3 \times 16 = 1{,}200$		$- 7n = 3$	
$H : 25 \times 3 \times 7 = 525$		$- 16n = 13$	

$11a \equiv 1 \pmod{25}$	$a = 16$
$1\beta \equiv 1 \pmod 3$	$\beta = 1$
$3\gamma \equiv 1 \pmod 7$	$\gamma = 5$
$13\delta \equiv 1 \pmod{16}$	$\delta = 5$

$16 \times 336 = 5{,}376$	$\times 9 = 48{,}384$
$1 \times 2{,}800 = 2{,}800$	$\times 19 = 53{,}200$
$5 \times 1{,}200 = 6{,}000$	$\times 69 = 414{,}000$
$5 \times 525 = 2{,}625$	$\times 59 = \underline{154{,}875}$

$$670{,}459 - 8{,}400n = 6{,}859.$$

Consequently, $N \equiv 59$ (mod 400) contains both $N \equiv 9$ (mod 50) and $N \equiv 59$ (mod 80). The choice of the remainders (9 and 59) results in a correct solution.

To end this chapter, let us examine one of the applications to chronology:[41] "Given in the 'Unicorn Virtue'[42] technique: the day divisor [*jih-fa* 日法] was 1,340, the year dividend [*sui-shih* 歲實] 489,428, and the lunation dividend [*shuo-shih* 朔實] 39,571.[43] It was determined on the basis of observation that on sexagenary day 1 [*jih-ch'ên chia-tzŭ* 日辰甲子], which was astronomical new year [*t'ien-chêng* 天正] and winter solstice [*tung-chih* 冬至] of sexagenary year 1 [*chia-tzŭ sui* 甲子歲], the epochal year [*yüan-nien* 元年] of the 'Unicorn Virtue' period, the minor remainder[44] amounted to 240, and the intercalary remainder[45] to 17,770. We wish the superior epoch [*shang-yüan* 上元][46] to be [simultaneously] a conjunction and winter solstice at midnight on sexagenary day 1 of the eleventh [civil calendar] month, astronomical new year of sexagenary year 1. Query: How many years will have elapsed between the superior epoch and the first year of the 'Unicorn Virtue' period?"

From the superior epoch (*shang-yüan*) to astronomical new year (*t'ien-chêng*) of the 'Unicorn Virtue' period, year 1 (5 December 663) there had passed:

(1) An integral number of sexagenary year cycles or $60x$ years. Since a year is $489,428/1,340$ days, the period in question is

[41] See Chang Tun-jên (1'), C, p. 2a. The same problem is dealt with in Li Yen (6'), vol. 1, pp. 161 ff, and in Ch'ien Pao-tsung (4'), pp. 59 ff. However, in Li Yen's text, a whole phrase has been omitted, a fact that makes the problem unintelligible. In Ch'ien Pao-tsung's version the character *jih* 日 is wrongly printed *yüeh* 曰. I am much indebted to Nathan Sivin for his kind assistance in elucidating this problem.

[42] *Lin-te* 麟德. The *Lin-te* calendar was drawn up by Li Shun-fêng 李淳風 (602–670). See Yabuuchi (3), p. 13.

[43] The day divisor is the common denominator for the year and month fraction of a day. A year is $489,428/1,340 = 365 \ 328/1,340 = 365.2448$ days. A month equals $39,571/1,340 = 29 \ 711/1,340 = 29.53060$ days.

[44] *Hsiao-yü* 小餘: an integral number of days plus $240/1,340$ day has passed since the *shang-yüan* 上元.

[45] *Jun-yü* 閏餘: an integral number of months plus $17,770/1,340$ days has passed since the *shang-yüan*.

[46] The moment in the past at which all cycles began simultaneously.

$$P = \frac{489,428}{1,340} \times 60x \text{ days.}$$

(2) An integral number of months (each being $39,711/1,340$ days) *plus* a remainder of $17,770/1,340$ days, or

$$P = \frac{39,571}{1,340} z + \frac{17,770}{1,340} \text{ days.}$$

(3) An integral number of sexagenary cycles of days *plus* a remainder of $240/1,340$ day, or

$$P = 60y + \frac{240}{1,340} \text{ days.}$$

Thus

$$P = \frac{489,428}{1,340} \times 60x = 60y + \frac{240}{1,340} = \frac{39,571}{1,340} z + \frac{17,770}{1,340}.$$

Solution:[47]

$$\frac{489,428}{1,340} \times 60x = 60y + \frac{240}{1,340}. \tag{1}$$

After reduction we get $122,357x = 335y + 1$. As $122,357 \equiv 82$ (mod 335), we have $[82 \text{ (mod 335)}]x \equiv 1$ (mod 335). We find that $x = 143$ and in general

$$x \equiv 143 \text{ (mod 335)} \longrightarrow 60x \equiv 8580 \text{ (mod 20,100).}$$
$$\frac{489,428}{1,340} 60x = \frac{39,571}{1,340} z + \frac{17,770}{1,340}. \tag{2}$$

We substitute for $60x$:

$$\frac{489,428}{1,340} [8,580 \text{ (mod 20,100)}] = \frac{39,571}{1,340} z + \frac{17,770}{1,340}.$$

After reducing

$$(489,428 \times 8,580) \text{ (mod } 489,428 \times 20,100) = 39,571z + 17,770. \text{(a)}$$

---

[47] This version follows the modern representation used by Ch'ien pao-tsung (4'), pp. 59 ff; the explanation in Li Yen (6'), pp. 160 ff, is closer to the texts (there are some misprints in it).

Now $489,428 \equiv 14,576 \pmod{39,571}$. After substituting and reducing (a),

$14,576 \times 8,580 = 125,062,080 \equiv 17,720 \pmod{39,571}$

$14,576 \times 20,100 \equiv 33,487 \pmod{39,571}$[48]

$17,720 \pmod{33,487} \equiv 17,770 \pmod{39,571}$

or

$0 \pmod{33,487} \equiv 50 \pmod{39,571}$

$33,487p \equiv 50 \pmod{39,571}$.

Solve the congruence $33,487a \equiv 1 \pmod{39,571}$: $a = 37,197$.

$p = 50a = 50 \times 37,197 \equiv 13 \pmod{39,571}$.

The smallest value of p is 13. Since $60x = 20,100p + 8,580$,

$60x = 20,100 \times 13 + 8,580 = 269,880$.[49]

4. Lo T'êng-fêng 駱騰鳳, Ch'êng Hung-chao 程鴻詔; Shih Yüeh-shun 時曰醇

Lo T'êng-fêng published his *I-yu lu* (The pleasant game of mathematical art) about 1820. In this work he gives the general *ta-yen* rule; its explanation is correct, but the "reduction of the moduli that are not relatively prime" is entirely inaccurate.[50]

---

[48] Indeed $8,580 \pmod{20,100} \leftrightarrow 20,100p + 8,580$.

$489,428 \times 20,100p + 489,428 \times 8,580 = 39,571z + 17,770$.

$489,428 = 39,571q + 14,576$.

$(39,571q + 14,576) \times 20,100p + (39,571q + 14,576) \times 8,580 = 39,571z + 17,770$.

$39,571 \times 20,100\,pq + 14,576 \times 20,100p +$

$\quad \parallel\!\parallel \qquad\qquad\qquad\quad \parallel\!\parallel$

$0 \pmod{39,571} \qquad 33,487 \pmod{39,571}$

$39,571 \times 8,580\,p + 14,576 \times 8,580 \equiv 17,770 \pmod{39,571}$.

$\quad \parallel\!\parallel \qquad\qquad\quad \parallel\!\parallel$

$0 \pmod{39,571} \qquad 17,720 \pmod{39,571}$

$39,571ap + 33,487p + 39,571b + 17,720 \equiv 17,770 \pmod{39,571}$

$39,571ap + 33,487p \equiv 50 \pmod{39,571}$

$\qquad\qquad 33,487p \equiv 50 \pmod{39,571}$

$\qquad\qquad\qquad p = 13$.

[49] Another example is dealt with in Li Yen (6'), vol. 1, p. 162.

[50] See p. 22a and examples on p. 33a. Lo gives the example

$N \equiv r_1 \pmod 4 \equiv r_2 \pmod 5 \equiv r_3 \pmod 6$.

His solution is: The G.C.D.$(4,6) = 2$, and he divides 4 and 6 by 2.

yen-mu $= 2 \times 5 \times 3 = 30$

He treats the three problems from Yang Hui's work and then the Sun Tzŭ problem.[51] He also explains Ch'in Chiu-shao's *ta-yen* rule[52] (without mentioning Ch'in's name), but in a very unusual and complicated way, if we compare it to Chang Tun-jên's clear explanation. The reader will find a general representation in Li Yen.[53] Other short texts are given in Shih Yüeh-shun's *Ch'iu-i-shu chih*[54] and in Ch'êng Hung-chao;[55] they do not contain anything new.

5. Huang Tsung-hsien 黃宗憲

In cooperation with Tso Ch'ien 左潛, Huang Tsung-hsien wrote an interesting study on the remainder problem, entitled *Ch'iu-i-shu t'ung-chieh* (Complete explanation of the method for finding unity), first published in the *Pai-fu* collection (1875).[56]

According to Van Hee, "he simplifies the computation with

yen-shu		15	6	10	
congruences		1	1	1	
fan-yung		15	6	10	
This should be:					
moduli	:	4	5	6	
reduced mod.	:	4	5	3	
yen-mu	:		60		
yen-shu	:	15	12	20, and so on.	

[51] Pp. 23 ff.

[52] P. 26a and following pages.

[53] (6'), vol. 1, pp. 146 ff.

[54] See Li Yen (6'), vol. 1, p. 149; the original text could not be located.

[55] Published before 1865.

[56] Van Hee (5) devoted a paper to this famous collection. On p. 147 of his article, he deals with Huang Tsung-hsien, but in a very cursory way and not without error. It seems very easy to prove that Van Hee never saw Ch'in Chiu-shao's work and that he derived all his information from Huang Tsung-hsien. In his reproduction of Ch'in's text, he even copied all of Huang's typographical errors and published Huang's completely changed version of Ch'in's Problem I, 7 with complete confidence in its authenticity. It is a pity that all the publications of Van Hee, who was one of the first scholars having access to original Chinese mathematical texts, were lacking in a rigorous scientific background. For that reason his work has value only as a general introduction, and historians of mathematics who do not have access to Chinese texts should be warned against placing too much confidence in his works. The *Ch'iu-i-shu t'ung-chieh* is no. 10 of the *Pai-fu* collection.

European methods."[57] But this is true only for the first part, where he avoids Ch'in's complex way of reducing the moduli that are not relatively prime in pairs and replaces it by the method of decomposition into prime factors. Ch'ien Pao-tsung[58] says that he does not have to find the *ch'i-shu* before searching for the *ch'êng-lü* and that he solves the congruences directly from the *ting-mu* and the *yen-shu;* this is so only in appearance, as we shall see presently. Ch'ien also says that Huang does not make use of the *t'ien-yüan-i;* this statement too is inaccurate.

In the work of Ch'in Chiu-shao there is an intricate method for reducing moduli not relatively prime in pairs. Even Chang Tun-jên[59] was not always able to apply it correctly. Huang's rule is as follows:[60]

1. Decompose all the moduli into their prime factors.
2. Retain only the factor with the highest exponent and strike out the others.[61]
3. If two prime factors have the same exponent, keep either one.[62]

He gives four examples. The first is the Sun Tzŭ problem (p. A, 2b):[63]

fan-mu: 3   5   7 (relatively prime)
ting-mu: 3   5   7.

The others are from the *Shu-shu chiu-chang:* I, 3 (p. A, 6a);[64] I, 7 (p. A, 13a);[65] and I, 8 (p. A, 17a).[66] They are shown in Diagrams 59–61.

[57] (12), p. 442.
[58] (4′). pp. 62 f.
[59] See par. 3.
[60] This is not a translation, but an outline, because the method is too complex for so simple a problem. See Hsü Ch'un-fang (6′), pp. 41 ff.
[61] Huang indicates the factors to be kept with the sign △.
[62] Huang is well aware of the fact that there are several possibilities if there are prime factors with the same exponent. On p. A, 24a he gives all the possibilities for a special problem.
[63] The *fan-mu* 泛母 are the given moduli; the *ting-mu* 定母 the reduced moduli (relatively prime!).
[64] For a comparison with Ch'in's method, see Chapter 17.
[65] Thii is Huang's altered version of the problem.
[66] If there seem to be too many examples for so simple a problem, it is because these are problems from Ch'in Chiu-shao, and they are thus interesting for comparison.

fan-mu	54	57	75	72
	2	3	3	2△
	×	×	×	×
his-mu[67]	3△	19△	5△	2△
	×		×	×
	3△		5△	2△
	×			×
	3△			3
				×
				3
ting-mu	27	19	25	8

Diagram 59

fan-mu	300	250	200
	2	2	2△
	×	×	×
hsi-mu	2	5△	2△
	×	×	×
	3△	5△	2△
	×	×	×
	5	5△	5
	×	×	×
	5		5
ting-mu	3	125	8

Diagram 60

fan-mu	130	120	110	100	60	50	25	20
	2	2△	2	2	2	2	5	2
	×	×	×	×	×	×	×	×
hsi-mu	5	2△	5	2	2	5	5	2
	×	×	×	×	×	×		×
	13△	2△	11△	5△	3	5		5
		×		×	×			
		3△		5△	5			
		×						
		5						
ting-mu	13	24	11	25	*fei-wei* 廢位[68]	*fei-wei*	*fei-wei*	*fei-wei*

Diagram 61

[67] The *hsi-mu* 析母 are the numbers decomposed into prime factors. *Hsi* literally means "to split, to divide."
[68] "Numbers to do away"; these are the numbers that do not play any further part in the computation.

SOLUTION OF THE CONGRUENCES

In the works of Li Yen and Ch'ien Pao-tsung[69] there is a general description of the method. Let us take a single example:[70]

$$a \times 291 \equiv 1 \pmod{391}.$$

Huang solves the congruences in Diagram 62. In this diagram

yen-shu	ting-shu		
$\alpha_0 = 1$			
13,585	391	$13{,}585 - 34 \times 391 = 291$	(1)
$\alpha_0 = 1$			
291	391	$391 - 1 \times 291 = 100$	
	$q_1 = 1$		
$\alpha_0 = 1$	$1 = \alpha_1$		
291	100	$291 - 2 \times 100 = 91$	
	$q_2 = 2$		
$\alpha_2 = 3$	$1 = \alpha_1$		
91	100	$100 - 1 \times 91 = 9$	
	$q_3 = 1$		
$\alpha_2 = 3$	$q_4 = \alpha_3$		
91		$91 - 10 \times 9 = 1$	
	$q_4 = 10$		
$\alpha_4 = 43$			
1			

Diagram 62

the "top corner numbers" can be found as follows: the first and the second number are always 1 (α_0).

$$\alpha_0 = 1 \qquad\qquad \alpha_0 = 1$$
$$a_1 = q_1 a_0 + 0 \qquad\qquad a_1 = 1 \times 1 + 0 = 1$$
$$a_2 = q_2 a_1 + a_0 \qquad\qquad a_2 = 2 \times 1 + 1 = 3$$
$$a_3 = q_3 a_2 + a_1 \qquad\qquad a_3 = 1 \times 3 + 1 = 4$$
$$a_4 = q_4 a_3 + a_2 \qquad\qquad a_4 = 10 \times 4 + 3 = 43.$$

Is this really a new method, as Huang Tsung-hsien thought,[71]

[69] Li Yen (6′) p. 151; Ch'ien Pao-tsung (4′), p. 64.
[70] The example is on p. A, 23a.
[71] See his introduction, p. 1a.

and one that is shorter than Ch'in's, as Ch'ien Pao-tsung says? And can we agree that the *t'ien-yüan-i* was not used?

Ch'in Chiu-shao's method is as follows: $13,585 - 34 \times 391 = 292$ (*ch'i-shu*); see Huang's method (1). The equation $a \times 291 \equiv 1 \pmod{391}$ is solved as in the successive parts of Diagram 63.

$Q_1 =$	1		$R_2 =$	91		$Q_3 =$	1		$R_4 =$	1
	291			291			91			91
	391			100			100			9
$R_1 =$	100		$Q_2 =$	2		$R_3 =$	9		$Q_4 =$	10

Q		
1	1	(天元) $1 \times 1 = 1$
2	1	$2 \times 1 + 1 = 3$
1	3	$1 \times 3 + 1 = 4$
10	4	$10 \times 4 + 3 = 43$
.	43	

Diagram 63

From this comparison we may conclude:
1. Huang makes no use of the *ch'i-shu*, but they are incorporated in the solving of the congruences.
2. The second "top corner number," which is always 1 (a_0), is nothing other than the *t'ien-yüan-i*.
3. There is not the slightest difference between the methods of Ch'in and Huang.
Huang Tsung-hsien states also that the number of solutions is infinite.[72] In the second part of his work he develops another method for solving the remainder problem.[73] But at his time (1875), indeterminate analysis had already developed in Europe to such a high level that Huang's method has no real interest for the history of mathematics.

For information on indeterminate analysis in Japan, the reader is referred to Smith and Mikami (1), p. 123, and Li Yen (6'), vol. 1, pp. 164 ff.

[72] P. A, 1a.
[73] See Ch'ien Pao-tsung (4'), pp. 65 f.

Appendix to Chapter 16

Historical Outline of the Investigation of the Chinese Ta-yen Rule in Europe

For the historian of mathematics who does not have access to the original sources or to the studies of modern Chinese scholars, it may be of interest to have at his disposal some critical notes on the studies of Chinese indeterminate analysis published in Western languages. Much of this information has to be used with caution, and many of the mistakes in these studies are long-lived, as we find them in many modern histories of mathematics.

The investigation began with studies by Chang Tun-jên (1801), Lo T'êng-fêng (1815), and others.[1] But as all these studies are in the old style, it seemed better to discuss them in relation to the evolution of indeterminate analysis in China, of which they represent a late phase.

In Europe, nothing was known about Chinese mathematics[2] until E. Biot published in 1839 in *Journal Asiatique*[3] a description of the contents of the *Suan-fa t'ung-tsung*,[4] from which he gives the Sun Tzǔ problem (but without solving it).[5] This must have been the first communication on the Chinese remainder problem, but it seems not to have been noticed in Europe. The most important paper was from the hand of Alexander Wylie. In 1852 he published his "Jottings on the Science of the Chi-

---

[1] See Chapter 16.

[2] In the *Histoire des Mathématiques...* of Montucla (1758), there is a chapter on China, but it deals only with Chinese astronomy. At that time practically all information about Chinese science was derived from the Jesuits. For instance, in the work *Mémoires concernant l'histoire, les sciences...des Chinois* in 15 volumes (1776–91), there is almost nothing on nonastronomical mathematics.

[3] Biot (3); the remainder problem is on p. 207.

[4] The *Suan-fa t'ung-tsung* (1593) is a mathematical treatise written by Ch'êng Ta-wei. For further information on this work, see Needham (1), vol. 3, pp. 51 f, and Ting Fu-pao (1'), Suppl., p. 44b.

[5] For the method used in this work, see Chapter 15.

nese" in the *North China Herald*,[6] in which he also dealt with the *ta-yen* rule of the *Sun Tzŭ suan-ching*. The most important point is that for the first time Ch'in Chiu-shao's rule was explained. Wylie included a full explication of Ch'in's first problem and some notes on the other problems. In 1842 the *Shu-shu chiu-chang* was published in the *I-chia t'ang* collection,[7] and this may have enabled Wylie to study the *ta-yen* rule. But Wylie also knew the work of Chang Tun-jên, published in 1803,[8] as is evident from his citation of this writer in *Notes on Chinese Literature*, another important work published in 1867.[9] There is a mistake in Wylie's "Jottings," where he says: "The second division of Tsin's work is on the calculation of astronomical terms which are also worked out by the *ta-yen*" (p. 184). This is true only for one problem.[10] It is possible that Wylie never read the whole work, as this error indicates, but that does not diminish his great merit in having provided a starting-point in Europe for studies of the history of Chinese mathematics,[11] and particularly of indeterminate analysis.

In 1856 Wylie's article was translated into German by K. L. Biernatzki[12] and in this way became known in Europe. This translation[13] contains several inaccuracies and mistakes.[14] On p. 78 Biernatzki omitted the *ch'iu-i* (solving of the congru-

---

[6] There are several reprints of this article, all listed in Needham's bibliography [(1), vol. 3, p. 800]. We used the reprint in *Copernicus*, 1882, 2, pp. 169–195.

[7] On this collection, see Hummel (1), p. 545.

[8] Namely, the *Ch'iu-i suan-shu*. See Chapter 16.

[9] The note on Chang Tun-jên is on p. 99.

[10] Namely, II,3.

[11] See Vacca (1).

[12] His source was the version published in the *Shanghai Almanac* (1853). "Because of a fortunate coincidence one copy appears to have come into the hands of Mr. K. L. Biernatzki in Berlin, but more detailed information on this subject is not available." (Ein Exemplar scheint durch einen glücklichen Zufall in die Hände des Herrn K. L. Biernatzki in Berlin gelangt zu sein, jedoch fehlen leider hierüber genauere Nachrichten.) Matthiessen (3), p. 254. It is indeed fortunate that Biernatzki obtained this copy; if he had not, the *ta-yen* rule might not have become known in Europe for a long time.

[13] It is not a mere translation. Biernatzki made many changes in the text.

[14] I have checked only the part treating the *ta-yen* rule.

ences)[15] and thus caused Cantor to think that the congruences had been solved by conjecture.[16] On p. 80 (corresponding to p. 182 in Wylie) the same process is omitted.[17] Matthiessen pointed out another mistake:[18] on p. 78 Biernatzki wrote: "35/3 leaves a balance, or *ki* of 2, that is, the multiplicator; $35 \times 2 = 70$, the helping number." The first 2 is the so-called *ch'i-shu*: the second 2 is the solution of the congruence $a \times 35 \equiv 1$ (mod 3). Biernatzki confused the two digits.[19] On p. 80 he showed that he did not understand Problem I, 1 of Ch'in Chiu-shao; and where Wylie, on p. 183, included an application (very interesting for grasping the sense of the problem), Biernatzki broke off his translation with the nonsensical statement: "This calculation . . . served to predict the future by means of number symbols; it constituted an arithmetical basis for the art of divination, which was very popular among the Chinese, as among all pagan peoples . . . " (p. 80).[20] His

[15] See Wylie (1), p. 181, who transcribes *ch'iu-i* as *Kue Yih*.

[16] Cantor (2), p. 587: "The manner in which these numbers were arrived at is not even hinted. The most plausible assumption is therefore that this was done by trial and error."(Wie diese Zahlen gewonnen wurden, ist auch nicht andeutungsweise gesagt, die Vermuthung liegt daher am nächsten, man werde sich durch Probieren geholfen haben.) It is highly probable that Biernatzki did not understand the rule, as it is not very clearly explained in Wylie.

[17] See also Matthiessen (3), p. 256, notes, and p. 259, note.

[18] (2), p. 271. ("35/3 lässt den Rest oder *Ki* 2, d.i. der Multiplicator; $35 \times 2 = 70$, d.i. die Hülfszahl.")

[19] See Matthiessen (2), p. 271.

[20] "Diese Rechnung...diente dazu, durch Zahlensymbole die Zukunft zu deuten; sie bildete eine arithmetische Grundlage für die bei den Chinesen, wie überhaupt bei heidnischen Völkern, sehr beliebte Wahrsagerkunst...," which caused Matthiessen (4) to say: "The continuation of the extract [quoted above]... in its unbelievable ignorance has fully discredited this notably interesting and remarkable theoretical introduction to I-hsing's indeterminate analysis." (Vollends aber hat die Fortsetzung des Auszuges... in unbegreiflichen Unverstande diese überaus interessante, merkwürdige theoretische Einleitung der unbestimmten Analytik von Yih-hing in Miscredit gebracht.) (On I-hsing, see note 21); and : "These presumptions ascribed to I-hsing rank with the greatest foolishnesses ever perpetrated in the history of mathematics...." (Diese Vermuthungen, welche dem Yih-hing angedichtet sind, gehören unter die grössten Thorheiten, welche je in der Geschichtschreibung der Mathematik begangen worden sind....)

translation is not always correct: he translates "a rule for the resolution of indeterminate problems" (Wylie, p. 180) as "the method for finding unknown quantities." It is a pity that Matthiessen, who did so much to correct Biernatzki's mistakes and to make the *ta-yen* rule known in Europe, was himself a victim of Biernatzki's carelessness.[21]

[21] It seems to be important to point out the origin of a serious mistake in the history of mathematics that has caused much of Ch'in Chiu-shao's work to be attributed to I-hsing. Wylie transcribes the name of the famous Buddhist monk I-hsing 一行 as Yih Hing, and the *I-ching* 易經, the Chinese classic on divination, as *Yih King*. In Biernatzki's translation both are written *I King*, and this was the reason why Matthiessen attributed Ch'in Chiu-shao's work (of which the first problem deals with the mathematical background of the *I-ching* divination method) to the monk I-hsing, whose contribution to indeterminate analysis is not very well known. (See Chapter 15.) Biernatzki says: "This work of I-hsing has been furnished with an exhaustive commentary... by Ch'in Chiu-shao." (Auch dieses Werk hat... Tsin Kiu Tschaou ausführlich commentirt.) There is no word of this in Wylie; but replace the name *Yih King* by I-hsing in Wylie (1), p. 182, line 14, and the problem becomes clear. Needham (1), vol. 3, p. 121 says: "Matthiessen was very much confused, attributing Ch'in Chiu-shao's first problem to I Hsing [The reason is obvious; it was Biernatzki who was confused!] and supposing that it was concerned with the numbers of workmen building dykes, instead of with the *I-ching* divination technique. These misunderstandings were faithfully reproduced by Dickson (1), vol. 2, p. 57" (note *b*). As for the "number of workmen," I think Needham is not entirely correct. What happened is this: Biernatzki's translation stops after the finding of the "fixed use numbers" (as Wylie calls the *ting-yung-shu*), and omits the application to an example, which is necessary in order to have remainders. Not having these remainders, Matthiessen could not work out the problem completely (that is also the reason why he entitled the first chapter "Von der Berechnung der Hülfszahlen" (on the computation of the use numbers). Moreover, as I-hsing was confused with *I-ching*, he could not have been aware of the nature of the problem; for the same reason he could not understand the text about "divination." For these reasons, Matthiessen took the third problem from Wylie, and added certain numerical values to it, because Wylie did not give the numbers which appear in the problem. He says: "To begin with, the application of these use numbers determined by I-hsing is to be *illustrated by an example*." (Es möge zunächst die Anwendung dieser von Yih-hing bestimmten Hülfszahlen an einem Beispiele erläutert werden.) Matthiessen (3), p. 259. Matthiessen gives the problem without numbers as Wylie and Biernatzki do; but after that he says: "Die disponibeln Arbeitskräfte *seien* 2, 3, 6, 12..." What he wrote is not "they are" (sein), but "let them be" (seien). Dickson did not see this distinction. From all this it is clear that Matthiessen never saw the

In 1858 Cantor published his *Zur Geschichte der Zahlzeichen*. He also had to rely on Biernatzki.[22] Cantor (p. 336) transcribed Biernatzki's problem into modern algebraical notation, whereupon it became evident that the solution did not follow. Cantor concluded: "It thus appears that particularly in the investigation of the indeterminate analysis the Chinese were rather less advanced than other contemporary cultural groups" (p. 336).[23] This error in Cantor's formulae was also a result of the mistakes in Biernatzki's translation. In 1875 Matthiessen wrote a letter[24] to Cantor, correcting Biernatzki's mistake and pointing out the identity of the *ta-yen* rule with Gauss's method.

In 1863 O. Terquem (1) translated Biernatzki's article into French.[25] After an explanation of the Sun Tzŭ problem, he says, on the subject of Ch'in's first problem, that he "is unable to understand what is said of it"; this is of course due to Biernatzki's errors.

In 1869 J. Bertrand (1) did another French translation of the same article. This is not just a translation, but a free version of the original. Even Wylie's name does not occur in the text.[26]

As the translator did not understand the problem, there is

---

original article, although he tried to find it, as he wrote: "According to a communication from the author [Wylie] dated 1874, the articles referred to can no longer be found in Shanghai." (Nach einer brieflichen Mittheilung des Verfassers vom Jahre 1874 sind die gedachten Artikel in Shanghai nicht mehr aufzutreiben.) Ibid., p. 254. I cannot but hold Matthiessen in esteem; he was incontestably a very clever mathematician, as he was able to understand the *ta-yen* rule with such inaccurate data as Biernatzki's translation provided him.

[22] "I was unable to procure the original." (Ich konnte mir das Original nicht verschaffen). (1), p. 395, note 2.

[23] Cantor (1). ("So scheinen gerade in Untersuchungen der unbestimmten Analytik die Chinesen hinter anderen gleichzeitigen Culturvölkern eher zurück gewesen zu sein.")

[24] Published in Matthiessen (2).

[25] Biernatzki's influence in Europe is surprising. None of the translators found Wylie's original article, although there was a reprint in the *Chinese and Japanese Repository*, published in London 1863–66.

[26] "M. Biernatzki, analyzing according to the Almanac of Shanghai mathematical writings that are already old in China." (M. Biernatzki, en analysant d'après l'almanac de Shanghai les écrits mathématiques déjà anciens en Chine.) (p. 319).

even more confusion in it than in Biernatzki's original article.[27] At the point where Bertrand could no longer follow the text, he said: "I shall go no further with this reproduction of the German translation, which I have not succeeded in understanding" (p. 473).

In 1874 H. Hankel (1) published his *Zur Geschichte der Mathematik in Alterthum und Mittelalter,* where in an appendix he dealt with Chinese mathematics (p. 405 ff). His sources were as a matter of course very deficient: only Gaubil, Biot, and Biernatzki were at his disposal. Hankel considered the *ta-yen* rule identical with the Indian *kuṭṭaka* (but did not give the slightest evidence).[28] Matthiessen did not agree with Hankel's opinion.[29]

Matthiessen, the first who was able to understand the true meaning of the *ta-yen* rule, published his first article in 1876. That he corrected Biernatzki's mistakes but that he was himself a victim of one of them, we have already seen.[30] Matthiessen is not responsible for several inaccurate statements about the works of Sun Tzǔ, I-hsing, and Ch'in Chiu-shao, as he had no access to sources other than Biernatzki's article.[31] However, the mathematical conclusions he drew from it are very important:

(1) He gave a correct explanation of the remainder problem, where the moduli are relatively prime in pairs (the Sun Tzǔ problem), although he did not know the *ch'iu-i* method (for solving the congruences), owing to an omission in Biernatzki's

[27] Even the name of "Tsin Kiu tschaou" (Ch'in Chiu-shao) is identified with "la dynastie des Tsin" (p. 472).

[28] Moreover, I-hsing was made an *Indian* Buddhist priest!

[29] This question will be discussed in Chapter 18. L. Sédillot published several studies from 1845 to 1868 [Sédillot (1), (2), (3).] and expressed the opinion already quoted in Chapter 2, "that these people had never known what mathematics is"; but "without the authority which any acquaintance with texts might have given him...." Needham (1), vol. 3, p. 1; see also ibid., p. 174, note *e*.

[30] See note 21.

[31] Although he sometimes gives information that is pure fiction; for example (3), p. 257: the work of Ch'in Chiu-shao is a work "in two parts, each of nine chapters" (in zwei Teilen, jeder von neun Kapiteln).

translation.[32]

(2) He demonstrated identity with Gauss's rule (*Disquisitiones arithmeticae,* par. 36)[33] but he also asked himself what the source of Gauss's rule was (it is certainly not the Chinese rule, as it is obvious from the historical outline[34] that the *ta-yen* rule was known in Europe about 1550). But Matthiessen did not know much about developments in Europe, except for the problem of Isaac Argyros[35] and the booklet by J. C. Schäfer (1). One need only read Euler's introduction[36] to see where the problem and the rule of Gauss came from.

(3) He made a comparison between the Chinese *ta-yen* rule and the Indian *kuṭṭaka*.[37] (This will be discussed again in Chapter 18).

(4) He was the first to understand the solution of the case in which the moduli are not relatively prime,[38] and he gives a very clear explanation of it in (3), pp. 260 f. He notices that this method is not found in Gauss.[39]

(5) He also gave the conditions that must be realized for the solution of this problem.[40]

He states the general conditions:

$$r_p = [r_q \bmod \delta \, (m_p, m_q)]$$

where δ is the greatest common divisor of the moduli m_p and m_q.

---

[32] There is a clear explanation in (3), pp. 256 f.

[33] Matthiessen (1), (2), (3), (4).

[34] See Chapter 14.

[35] Published as appendix to Nichomachus's work. See Hoche (1), pp. 152 f; see also Chapter 14.

[36] See Chapter 14.

[37] Matthiessen (1), p. 78.

[38] Matthiessen (3), p. 260; (4), p. 34.

[39] Matthiessen (4), p. 34: "However, he [Gauss] does not indicate how to proceed analogously in cases where the moduli are not relatively prime." (Er zeigt aber nicht, wie man auf ähnliche Weise verfahren solle, wenn die Moduln nicht relativ prim sind.) He also says: "In the same manner, Dirichlet bypasses this generalization in his *Lectures on the Theory of Numbers*" (Ebenso übergeht Dirichlet diese Verallgemeinerung in seinem *Vorlesungen über Zahlentheorie*), but this is incorrect. See Dirichlet (1), pp. 54 ff.

[40] However, this condition is not given explicitly in the Chinese text.

(6) In Matthiesen (5) and (6), there is a full explanation of the subject in modern algebraical notation, and a proof of the *ta-yen* rule.[41]

In 1880 Cantor's *Vorlesungen über die Geschichte der Mathematik* was published, including a chapter on Chinese mathematics.[42] As we have seen, Cantor said in 1858 that the Chinese *ta-yen* rule was incorrect, but in 1880, after reading Matthiessen's articles,[43] he was convinced of the validity of the rule and praised it.[44]

In 1893 H.G. Zeuthen published his *Die Geschichte der Mathematik im Altertum und im Mittelalter;*[45] the information on China is very poor for a work that came after Cantor. In 1905 G. Vacca published an article called "Sulla Matematica degli antichi Cinese."[46]

In 1912, the first article in English (after Wylie), from the hand of D. E. Smith, appeared in *Scientific Monthly*.[47] Smith "himself spent some time in China and Japan, made collections of mathematical books there, and had the advantage of intimate collaboration with Asian mathematicians, notably Mikami Yoshio."[48] He was the first after Wylie (1852) who had some access to original sources.[49] His article, entitled "Chinese Mathematics," is very general, but he was convinced of the autochthonous character of Chinese mathematics.[50] He knew Wylie's "Jottings" but not the interesting articles by

[41] This will be dealt with again in Chapter 17, par. *e*.

[42] Chap. XXXI.

[43] Cantor (2), pp. 586 f, footnotes.

[44] Cantor speaks of the "most fortunate ingenuity (glücklichster Scharfsinn)," but at this time no one knew about the *ch'iu-i* method for solving the congruences.

[45] First published in Danish; German translation, 1896.

[46] In 1904 he translated the *Chou-pei suan-ching*, probably the oldest Chinese mathematical work. See Needham (1), vol. 3, pp. 19–24. I have not seen Vacca's article mentioned here.

[47] Smith (2).

[48] Needham (1), vol. 3, p. 2.

[49] For sixty years all information on Chinese mathematics was second-hand.

[50] He "accepted impossibly early dates for the oldest Chinese mathematical books." Needham (1), vol. 3, p. 2, note *a*.

Matthiessen. He included the Sun Tzǔ problem without solution. Much of his information must have been borrowed from Mikami, who was preparing his *The Development of Mathematics in China and Japan* at this time.[51] He defended the view that there was no Indian influence on Chinese mathematics (p. 597) and that indeterminate problems in Europe must have been derived from the East (p. 599).[52]

In 1913 the Japanese scholar Yoshio Mikami (1875–1950) published his very important book, *The Development of Mathematics in China and Japan.*[53] For a general commendation of his work, one can trust Needham's balanced judgment: "Whatever criticisms may have been levelled therefore against Mikami's judgment, the fact remains that he occupied a position or vantage-point quite unique in the field, the only possible comparison being Alexander Wylie in an earlier generation."[54] As for indeterminate analysis, the great merit of Mikami is: (1) he explained the method for solving the congruences, not known in Europe before; (2) he gave a mathematical explanation of the whole process as given by Ch'in Chiu-shao. The failures of his treatment of the *ta-yen* rule are: (1) he did not work out any of Ch'in's own problems but dealt with Sun Tzǔ's problem by using Ch'in's method; (2) he provided less information than Wylie did about the nature of the problems in Ch'in's work; and (3) his representation of Problem I, 2 was completely inaccurate.[55]

[51] Although this article is very interesting as a new starting point for investigation on Chinese mathematics, it will not be possible to analyze it further on, as this account is restricted to the *ta-yen* rule. On p. 598 he mentions the *t'ai-yen ch'iu-yi-shu*. This must be derived from Mikami, who also transcribes *t'ai-yen* instead of *ta-yen*.

[52] There is information on the further development of Smith's views later in this chapter.

[53] His publications concerning the history of mathematics in China are dated from 1905 on. For a general bibliography see Yajima (1).

[54] Needham (1), vol. 3, p. 2.

[55] This criticism will be supported by an analysis of Ch'in's work. Bosmans's comment is ridiculous; he says: "This is no masterpiece" (Ce n'est pas un chef-d'oeuvre), for the reason that Mikami made some mistakes and inaccuracies in transcribing the names of the Jesuit missionaries, as if Chinese culture were identical with the contributions of the Jesuits. About

In 1911 L. Van Hee (1873–1951)[56] published the first of a long series of articles on Chinese mathematics. In a special article he described indeterminate analysis in China.[57] The value of his contribution is debatable.[58] In his "Les Cent Volailles . . ."(12), Van Hee gives what is meant to be a

Chinese mathematics there is not a single word in Bosmans's criticism.

Other papers of Mikami treating the *ta-yen* rule are (1'), (2'), (3').

[56] A biography is given in *Jesuiten, 9*, 1951, p. 25.

[57] Namely (12).

[58] It is very difficult to give a fair evaluation of the work of Van Hee. Needham (1), vol. 3, p. 1, characterized him thus: "Van Hee, in whom sinological competence wrestled with missionary disapproval...." When reading through the work of Van Hee, one can only wish for some knowledge of psychology, for only on that level could one account for this mishmash of valuable research and shallow allegations, of respect for China and scorn for its mathematics. In (9), p. 291, he says: "In the mind of the yellow mathematician, there is love of detail without great concern for synthesis...." (Chez le mathématicien jaune, c'est l'amour du détail, sans grand souci de la synthèse....)One might reply that in the mind of Van Hee, there is love for synthesis without great concern for detail. For instance, he almost always gives only the literary name (*tzŭ*) of the Chinese author under discussion, and does not include exact titles, dates, or other references, which makes the reading of his work difficult. What he says about Juan Yüan is entirely applicable to his own work: "The sources...are indicated in this treatise in a vague and incomplete manner; and the authors complacently refer to such and such a book without the slightest indication of the edition, volume, or page" [(3), p. 103]. Van Hee's work shows a great lack of scientific background. His methods for approaching the history of mathematics are very inferior:

(1) His historical arguments are extremely poor. He proclaims *ex cathedra* that all Chinese mathematical knowledge must be derived from elsewhere, whether from India, Arabia, or Europe; it is impossible, he says, that anything in Chinese mathematics should be original.

(2) He never exhibits an acquaintance with medieval Indian, Islamic, or European mathematics. A statement such as "in the mind of our algebraists from the Greeks to modern times, there has been a need for scientific generalization (chez nos algébristes depuis les Grecs jusqu'aux modernes, c'est le besoin de la généralisation scientifique)" [(9), p. 291] does not hold, and clearly proves that Van Hee did not understand the development of algebra in Europe from A.D. 300 to A.D. 1600.

(3) Although Van Hee seems to have possessed a good knowledge of the Chinese language, the sources on which he relies were of secondary importance, the greater part going back to the *Pai-fu t'ang suan-hsüeh ts'ung-shu* of 1875 and the *Ch'ou-jên chuan* of Juan Yüan (1799). There are many indications that he never saw the original editions of the works he discusses so contemptuously.

historical account of the "hundred fowls" problem. He says
that the problem dates from the Han, whereas it appears for
the first time in the *Chang Ch'iu-chien suan-ching* (c. 475), where
the correct solution is given. Van Hee includes the entirely
erroneous method of Liu Hsiao-sun (late sixth century)[59]
(without indicating the author, as usual), and rightly calls it
a "rather childish mechanism," but he does not include the
solution developed by Chang Ch'iu-chien himself, which was
correct. In addition, he says: "In time, and perhaps because of
foreign influences that comparison with Indian texts, for example, will elucidate, methods were discovered and problems
became infinitely varied." This very vague statement must

---

(4) His chief aim seems to be the glory of the Jesuit missionaries in China—
and for contrast he makes the background of Chinese knowledge as dark as
possible. His aversion for Chinese mathematics is so great that he writes:
"I say in addition that if the works of the Chinese mathematicians of every
era were to disappear, science would lose nothing in the way of mathematics." (J'ajoute que si les livres des mathématiques chinois de toute
époque disparaissaient, la science n'y perdrait rien comme mathématiques)
[(6), p. 259].

For these reasons it is difficult to understand Bosmans's estimate (2);
but Petrucci (1), after Van Hee's first article, wrote a very interesting
criticism of which the following points may be quoted:

"It seems to me unfair to reproach Chinese mathematicians for their
lack of a generalizing faculty, because it is necessary to realize that we are
presented with a collection of problems, and we must not forget that the
knowledge of a general law is made evident by the collection alone." (Il
me paraît injuste de reprocher aux mathématiciens chinois un manque
d'esprit de généralisation, car il faut tenir compte de ce que nous nous
trouvons devant un 'receuil de problèmes' et il ne faut pas oublier que la
connaissance de la loi générale est rendue évidente par ce receuil même.")
(P. 562.)

"Above all it is necessary, in my opinion, to take a historical viewpoint
and not allow oneself to be blinded by a modern European education in
mathematics." (Il faudra surtout, à mon avis, se placer au point de vue
de l'histoire et ne pas se laisser aveugler par une éducation mathématique
européenne et moderne.") (P. 564.)

Because Van Hee was the first European sinologist (after Wylie) who had
immediate access to Chinese sources, his authority and influence on historians of mathematics who were not also sinologists was very great, and—it
must be said—not always constructive.

[59] The works of Liu Hsiao-sun and other Chinese mathematicians mentioned
in this paragraph are covered in Chapter 15.

mean that Lo T'êng-fêng solved the problem correctly in 1815. Then Van Hee gives twenty-four problems on the same subject, calling them the *"problèmes principaux,"* while they are in fact the examples given by Shih Yüeh-shun in his *Pai-chi shu-yen* 百雞 術衍, published as no. 8 of the *Pai-fu t'ang suan-hsüeh ts'ung-shu.* In the second part of his paper, Van Hee gives the Sun Tzŭ problem (p. 436); then he quotes Huang Tsung-hsien's preface[60] and comes to Ch'in Chiu-shao's work. First of all he identifies the *t'ien-yüan* with the unknown *x*, which is not true for Ch'in's work.[61] Then he gives two "arguments" to prove that Ch'in's work was derived from India.[62] After that two problems from Ch'in's work are included with translations.[63] The texts are not derived from the *Shu-shu chiu-chang* but from Huang's quotation; Huang had changed the text (in the usual Chinese manner, without any indication), and Van Hee faithfully reproduced the text of Problem I, 7, giving only the problem and its solution.[64]

For Problem I, 3 he gives text, translation, and solution.[65] As for this solution, only the first computations are given, and there is not a word about the application of the real *ta-yen* rule. As information for those who have no access to the original

[60] Huang Tsung-hsien, who wrote the *Ch'iu-i-shu t'ung-chieh*, no. 10 of the *Pai-fu* collection, was the main source for Van Hee's paper, as stated earlier.
[61] See Chapter 17.
[62] The first argument, based on the fact that "the numbers are written from left to right, horizontally (les chiffres sont écrits de gauche à droite, horizontalement)" is ridiculous, since this notation derives from the counting-rod system. See Needham (1), vol. 3, pp. 8 ff. These counting-rod numerals appeared on coins in the third century B.C. (ibid., p. 70). The second argument to the effect that "the *ta-yen* resembles the Indian *Kuttikara*" will be discussed in Chapter 18.
[63] Van Hee says: "A translation of the three most important problems will suffice to give an exact idea of the rest" (La traduction des trois problèmes plus importants suffice pour se former une idée exacte du reste.) (P. 438). In fact he gives only two problems (in Huang Tsung-hsien there are indeed three of Ch'in's problems). "The most important" is only a phrase, since he never saw the others.
[64] Even his footnotes are from Huang Tsung-hsien.
[65] In a footnote on p. 440 he says: "The author has no idea of the possible simplifications." (L'auteur n'a aucune idée touchant les simplifications possibles.) This is not a general rule, as becomes clear from other problems.

sources, it is of little value.[66] Without transition it moves to the "hundred fowls" and then back again to the *ta-yen* rule. This article is extremely unsystematic, and the explanation in it is very confused; Van Hee never saw the *Shu-shu chiu-chang*.

Dickson's *History of the Theory of Numbers* (1919–1920) is interesting, because it was the first general work on number theory that also dealt with Chinese mathematics. He relied only on the most trustworthy sources of his time, namely Wylie, Matthiessen, and Mikami; of course all of Matthiessen's mistakes are repeated.[67]

The articles of G. Loria,[68] published from 1921 to 1929, are, as Needham says, "so misleading as to be almost useless."[69] As far as the *ta-yen* rule was concerned, "he suffered from an invincible suspicion that the Chinese *must* have borrowed all their ancient mathematical techniques from the West . . . ."[70] In (2), p. 519, he includes the Sun Tzŭ problem with the "valuable" historical criterion: "But as Sun Tzŭ does not give the slightest indication that he regards this problem as more interesting or valuable than its trivial companions, we may well question his claim to the discovery of *ta-yen*." His "Documenti relativi . . . " (1) is a mere restatement of Van Hee's publications. On p. 147 he speaks of Ch'in Chiu-shao's work as: "a treatise in which one recognizes foreign influence."[71] And

[66] In consequence note *g* on p. 120 in Needham (1), vol. 3 is to be taken with a grain of salt.
[67] All the mistakes in Matthiessen, including the figure of I-hsing and his method and the invented example are reproduced in good faith in Dickson's book. In Mikami (1) there is no mention of the indeterminate problems in I-hsing. Mikami's criticism of Dickson [Mikami (2′)] is chiefly a Japanese translation of Dickson's texts dealing with Chinese mathematics. After the part treating I-hsing, he merely expresses some doubt about the real content of the monk's work, saying: "I intend to look into this in the *Li-i* (Discussion on calendars) of the *T'ang-shu*." (p. 197), but whether he really did so is unknown. On p. 198 he says again that an investigation of the connection between Ch'in Chiu-shao and I-hsing is required; a mistake can indeed be obstinate, if one does not go back to the sources!
[68] Loria (1), (2), (3).
[69] Needham (1), vol. 3, p. 1, note *e*.
[70] Ibid.
[71] "Un trattato nel quale venne riconosciuto l'influsso straniero." I do

about Matthiessen's statement that the *ta-yen* rule is equivalent to Gauss's rule, he scornfully adds: "It is truly a pity that in Chinese literature there is no application except the one referred to above [Sun Tzŭ's problem]."[72] This is extremely poor information.

In his *Storia delle mathematiche* . . . (3), Chapter 9, "L'Enigma Cinese," he gives an outline of the history of Chinese mathematics. In his treatment of the *Sun Tzŭ suan-ching*,[73] after dealing with the indeterminate problem and his solution, he quotes the problem of *"una donna gestante,"*[74] which he derives from Mikami.[75] The latter, however, discussed the *Sun Tzŭ suan-ching* in a chapter of nine pages, giving this problem at the end as a curiosity, adding that Juan Yüan thought it was an addition to the text. Loria has much to say about this ridiculous problem; his intention of discrediting the *Sun Tzŭ suan-ching* is obvious.[76] On p. 158 he deals with Ch'in Chiu-shao's work.[77] Is it on purpose that he says almost nothing about the *ta-yen* rule, whereas there is a full explanation of it in Mikami (1)? The general impression given by Loria's work is that it was very strongly influenced by Van Hee.[78]

The first *History of Mathematics* in which Chinese mathematics was thoroughly discussed was published in 1923–1925 by D. E. Smith. In writing of the *Sun Tzŭ suan-ching*, he confuses

---

not understand why Loria did not do a comparison with Mikami's work, published ten years before.

[72] "E un vero peccato che nella letteratura cinese non se trovi altra applicazione all'infuori di quelle superiormente referita!"

[73] Pp. 151 f.

[74] "A pregnant woman."

[75] (1), p. 33.

[76] It may give a pedantic impression, but I counted the lines Loria devoted to the problems. In his treatment of Sun Tzŭ there are 66 lines, of which 13 are devoted to the indeterminate problem and 15 to the "donna gestante"!

[77] "The astronomer" (*l'astronomo*) (*sic*) Ch'in Chiu-shao.

[78] In (1') Mikami also wrote a critique of Loria's work, in which he reproached him for attributing all Chinese mathematical knowledge to India and Europe by mere conjecture, his only argument being that Greek and Indian mathematics surpassed Chinese knowledge. There is a discussion of the *ta-yen* on pp. 80 f.

it with the *Wu-ts'ao suan-ching*.[79] He treats the Sun Tzŭ problem in vol. 1, p. 380. He does not repeat the errors about I-hsing, but it is somewhat surprising that almost nothing is said about Ch'in Chiu-shao's *ta-yen* rule.

In 1931 Smith published another article entitled "Unsettled Questions concerning the Mathematics of China," which conflicts on many questions with his first publication.[80] It is obvious that this article was written under the influence of such authors as Van Hee and Loria.[81] As for Sun Tzŭ, Smith states: "What needs further investigation, however, is Sun Tzŭ's method of solving simultaneous linear equations and his work in indeterminate equations . . . . Viewed from the standpoint of the history of mathematics, each of these is doubtful, although each is possible" (p. 248).[82] In writing of Ch'in Chiu-shao, Smith expresses his doubts about the so-called Horner method, but he seems to refer only to the notorious equation of the tenth degree, dealt with also by Loria.[83] Why does he say nothing about the indeterminate analysis as dealt with in the *Shu-shu chiu-chang?*

Cajori's article on Chinese mathematics in his *A History of Mathematics* (1919) is entirely based on Mikami[84] and not affected by Van Hee's and Loria's unscientific conclusions.

In Tropfke's *Geschichte der Elementarmathematik* (1922), vol. 3, p. 99, there is a chapter on indefinite analysis, where something (but nothing of great interest) is said about China.

---

[79] Smith (1), vol. 1, p. 141. This mistake is incomprehensible, because there was adequate information about Sun Tzŭ at that time; but it is not repeated in the second part (p. 499).
[80] See Chapter 14.
[81] ". . .It may be added that similar doubts have been expressed by such Western historians as Van Hee, Vacca, and Loria, and by Sédillot..." (p. 246).
[82] Smith adds however: "It is, of course, possible that the considerable number of problems in indeterminate analysis found in the early Chinese texts are authentic and represent a kind of mathematical trait of the people" (p. 248).
[83] See Chapter 13.
[84] Cajori (1), pp. 71–77.

In 1927 G. Sarton began publishing his very important *Introduction to the History of Science,* in which he dealt with Ch'in Chiu-shao.[85]

In 1930 there was a Russian article, "Istorija razvitija matematiki v Kitae, a takže v Japonii" (History of mathematics in China and Japan), written by A. V. Marakuev.[86]

That the influence of critical notes on the authors of general histories of mathematics is not very great we can learn from Becker and Hofmann's *Geschichte der Mathematik* (1951), where all Smith's mistakes are repeated.

The most important modern studies before Needham come from the Russian author A. P. Yushkevitch.[87]

In 1955 he published "O dostiženijax kitajskix učenyx v oblasti matematiki" (On the achievements of Chinese scholars in the field of mathematics). In cooperation with B. A. Rosenfeld he wrote in German *Die Mathematik der Länder des Ostens im Mittelalter* (1960) and another work in Russian, to appear soon in English as *A History of Mathematics in the Middle Ages.*[88] Even in his first publication he made use of new Chinese sources, namely, the works of Li Yen and Ch'ien Pao-tsung, giving his study a reliable base and beginning a new period in the study of Chinese mathematics in Europe.[89] The most interesting characteristics of his work are that he tries to give an objective description of his subject matter and that he has no preconceptions about the level of advancement of Chinese mathematics.

An article of great importance from the mathematical point of view was published by K. Mahler under the title "On the

[85] Vol. 3, p. 626. Sarton is mistaken when he attributes to Ch'in Chiu-shao another work called the *Shu-shu ta-lüeh.* This was the name given to the work by Ch'ên Chên-sun and Chou Mi. See Chapter 4.

[86] Review by Gaspardone (1); I have not seen the work.

[87] It must be stated that Russian historians of science are doing very interesting work, as they have access to original sources with the assistance of sinologists and Chinese scholars. E.I. Berezhkina, who has already done several translations of Chinese mathematical writings, is at present working on a study of Ch'in Chiu-shao's work.

[88] German edition in 1964: *Geschichte der Mathematik im Mittelalter.* English edition in press.

[89] In his most recent works, he also relies on Needham (1).

Chinese Remainder Theorem" (1957), stressing the fact that in China the general case when the moduli are not relatively prime in pairs was well known.[90]

In the *History of Mathematics* by J. F. Scott (1958), there is a short summary of Chinese mathematics[91] that relies entirely on Mikami; it does not always stress the most important features of Chinese mathematics. Another interesting summary, clear and reliable, is given in D. J. Struik (1) (1964).

The most important and most thoroughly documented work available is the section on mathematics in Needham's *Science and Civilisation in China,* vol. 3 (1959), pp. 1–168.[92] Needham has in every case checked the surveys from which he begins with the original sources and in many cases has done new research of high quality. It is a very important fact that he makes full use of the studies written by modern Chinese scholars. The bibliography alone is a great achievement, surpassing everything else on this subject in Western languages. It is to be hoped that Needham's gigantic task may encourage thorough investigation of the field of Chinese mathematics.[93]

It is only on such a trustworthy base as this that really scientific work can be done.

It must be emphasized, however, that the progress in the study of Chinese mathematics made in the last decades would have been totally impossible without the studies of Chinese scholars. One of the greatest of these scholars was Li Yen 李儼 (1892–1963),[94] who first of all made enormous bibliographical contributions.[95] As for the *ta-yen* method, there is a very in-

---

[90] This article will be dealt with again in Chapter 17.

[91] Chapter 5, pp. 80–82.

[92] For reviews of Needham (1), vol. 3, see *Endeavour*, 1960, *19*, p. 118 (H. H. Dubs); *Isis*, 1960, *51*, p. 598 (Huang Su-shu); *Bibl. Orient.*, 1961, *18*, nos. 1–2, p. 106 (R.J. Forbes).

[93] The inspiration for writing the present book is entirely due to Needham's work, as well as to his personal encouragement and kind assistance, for which I am extremely grateful.

[94] There are short biographies at the end of Li Yen (13') and in Wong (1), pp. 310 f.

[95] See Li Yen (1'), (2'), (3'), (4'), (5'), (11'), (18'), (19') (20').

teresting monograph in (6′), vol. 1, pp. 122–174; on Ch'in Chiu-shao one can also consult Li Yen (8′), pp. 204 ff and (13′), pp. 210 ff. Li Yen has the great merit of having introduced modern scientific methods into Chinese research in this field.

Ch'ien Pao-tsung 錢寶琮 is the equal of Li Yen.[96] He treats of the *ta-yen* rule in several works.[97] In 1966 he published a monograph on mathematics in the Sung and Yüan periods, in which there is a splendid study of Ch'in Chiu-shao.[98] The present work owes much to this last book and would have been impossible without its support.

There are some other valuable articles, on the same level as those mentioned earlier, although much more concise, written by Hsü Ch'un-fang 許純舫.

Special papers devoted to indeterminate analysis are: Fu Chung-sun 傅種孫(1′); Kao Chün 高均(1′) (1920); Hsü Chên-ch'ih徐震池(1′) (1925); Chao Jan-ning 趙然凝(1′) (1944), and YenTun-chieh 嚴敦傑(2′) (1947). As Japanese contributions one should mention Yabuuchi (1′) (1944) and Fujiwara (1′) (1939–1940).

The Chinese studies have been treated in a very concise way, in contrast with the discussion of European works, but this necessarily represents an inverse proportion: as the Chinese scholars have access to the original sources and understand them profoundly both philologically and in terms of modern mathematics, no criticism is needed as to lack of information. Subsequent chapters will make extensive use of these works, and criticism of details will be included where pertinent.[99]

---

[96] Ch'ien Pao-tsung is a graduate of Harvard, and was at last report editor-in-chief of the *K'o-hsüeh shih chi-k'an*, published by the Institute for the History of Sciences in Peking.

[97] (4′), pp. 37–66; (3′), pp. 125–134; (9′), pp. 138–143.

[98] (2′), pp. 60–103. In the last work, which is the result of the collaboration of several scholars, there is a paper on the "trigonometry" of Ch'in, written by Pai Shang-shu (pp. 290–303).

[99] A large part of the European works has been useless for the purposes of this study.

17

Ch'in Chiu-shao's General Method

Translation of the Text (pp. 1–3)

"The general *ta-yen* computation method is as follows:

A.

1. Set up all the *wên-shu* [problem numbers].[1]

2. First of all, one has to join the numbers. . .[2] by twos and find their common factors.[3]

3.

a. Reduce the odd numbers, do not reduce the even ones.[4]

b. (Sometimes one reduces and gets 5, and the other number is 10. In this case, one has to reduce the even number and not the odd one.)[5]

4. Sometimes all the numbers[6] are even. After all the numbers have been reduced, we may keep only one even number.

5.

a. Sometimes after reducing all the numbers there still remain numbers with common factors.[7] Provisionally set them up until you can reduce them with the others [of the same

---

[1] The *wên-shu* 問數 are of course not the numbers asked for but the numbers given in the problem.

[2] I have omitted some sentences, because they do not relate to the general rule discussed here. Ch'in gives at this point a classification of the numbers in four groups: whole numbers, decimal fractions, fractions, and numbers of the type $N \times 10^n$. He also includes a definition of each kind of number and the way in which one has to apply the *ta-yen* rule to these special kinds of numbers. I shall return to these special rules in another part of this section. I have already treated the classification of numbers in Chapter 6. The only numbers treated of here are the so-called *yüan-shu* 元數, literally, "original numbers"; these are the whole numbers or natural numbers.

[3] Li Yen (6'), vol. 1, p. 129, says: "This sentence means: one has to find the least common multiple." This interpretation is not correct. Ch'ien Pao-tsung (4'), p. 49, gives the correct explanation.

[4] The meaning of the terms *chi* 奇 (odd) and *ou* 偶 (even) presents a very difficult problem in this context. See pp. 333 f.

[5] See pp. 334 ff.

[6] In the text: *yüan-shu* (whole numbers).

[7] Literally, "numbers of the same class." See Ch'ien Pao-tsung (2'), p. 70, note 2.

class].[8]

b. Finally, find the common factors of the ones you have provisionally set up and reduce them [by those common factors]. . . .

6. If you reduce the odd number, do not reduce the even number, but multiply it.[9] Or if you reduce the even number, do not reduce the odd number but multiply it.

7. Sometimes one can reduce the one as well as the other. But of the ones that still have common factors, again find their mutual divisor by the method of successive divisions.[10] By this mutual divisor reduce the one and multiply the other.

8. Then you get the *ting-shu* 定數. . . .[11]

9. In finding the *ting-shu,* do not allow two numbers to be mutually divisible, or one to become too large.[12] If one [of the numbers] becomes too large, then you have to 'compensate.'[13]

10. If you do not want to borrow, it is allowed to have the number 1 [as *ting-shu*].[14]

B.

11. You multiply the *ting-shu* with each other, and you get the *yen-mu* 衍母.[15]

---

[8] The text says: "*yü ch'i t'o* 與其他" (with the others of them): *ch'i* 其 indicates the *lei-shu* 類數. See note 7.

[9] Meaning, "by the number with which you reduce the other number," as is obvious from the examples. See pp. 345 ff.

[10] The text says *hsiang-chien* 相減 (literally, mutual subtraction); see Ch'ien Pao-tsung (2'), p. 71.

[11] Literally, "definite numbers" or "fixed numbers."

[12] This means, larger than the "problem number," from which it is derived.

[13] Literally, "borrowing use the superfluous." From the examples given, one can easily deduce the meaning "to compensate."

[14] This sentence is not quite clear. No Chinese commentator gives an explanation of it. The sense seems to be: the *ting-shu* 1 is allowed too, but you can also "borrow" from another *ting-shu*.

[15] Literally *yen-mu* means "extension mother." Needham (1), vol. 3, p. 120, translates as "multiple denominator." *Mu* is indeed used for the denominator of a fraction, but in this case there is no fraction at all (for this reason Needham put the term between quotation marks). As we shall see later, the terms *tzŭ* (son) and *mu* (mother) are also used for the numbers drawn up in columns, where the numbers of the first column are multiplied by those of the second one. As in many cases, fractions are drawn up in the same way (on the counting board); the meanings "numerator" and "de-

12.

a. You divide the *yen-mu* by all the *ting-shu* and you obtain the *yen-shu* 衍數.[16]

b. Or you set up all the *ting-shu* as *mu* 母 [factors][17] in the right column, and before all these, you set up the *t'ien-yüan* 天元 1 as *tzŭ* 子 [factor][18] in the left column. By the *mu* you mutually multiply[19] the *tzŭ* and you get the *yen-shu* too.

13. From all the *yen-shu* you subtract all the [corresponding] *ting-mu* as many times as possible.[20] The part that does not suffice any more [literally, the incomplete part], is called the *chi* 奇 [remainder].[21]

C.

14. On the *chi* and the *ting-shu* one applies the *ta-yen ch'iu-i* 大衍求一.[22] With this method one will find the *ch'êng-lü* 乘率.[23] (These of which one gets the remainder 1 are the *ch'êng-lü*.)[24]

15. The *ta-yen ch'iu-i* method says: Set up the *chi* at the right hand above, the *ting-shu* at the right hand below. Set up the *t'ien-yüan* 1 at the left hand above.

16. First divide the 'right below' by the 'right above,' and the

---

nominator" are secondary, although in modern Chinese they have indeed these specific meanings. For this reason, I have not tried to translate the term. "Factor" would be the best translation, but it is too general in this case.

[16] Wylie: "extension numbers"; Mikami: "operation numbers"; Biernatzki: "Erweiterungszahlen" (translated from Wylie); Needham: "multiple numbers."

[17] See note 6; "denominator" makes no sense in this case. Together with the *tzŭ*, they are the two factors of the multiplication.

[18] For the meaning of the *t'ien-yüan-i* (Celestial element unity), see pp. 345 ff.

[19] *Hu-ch'êng* 互乘, although not clearly indicated by this term, means "multiply with all the *mu*, excepting the one opposite to the *tzŭ* made use of." It can be rendered as "cross-multiplication" (Mikami).

[20] Literally, "subtract the whole *ting-mu*'s", that is, subtract the *ting-mu* as many times as possible.

[21] *Chi* = *chi-shu* 奇數(remainder), a term still used in modern mathematics.

[22] Nothing other than a method for solving the congruences (literally, "searching for one").

[23] Literally, "multiplying factors."

[24] Or $\alpha N \equiv 1 \pmod{\mu}$.

quotient obtained, multiply[25] it by the 1 of 'left below.'

17. Set it up at the left hand below [in the second disposition]. After this, in the 'upper' and 'lower' of the right column, divide the larger number by the smaller one. Transmit [the numbers to the following diagram] and divide them by each other. Next bring over the quotient obtained and [cross-] multiply with each other.[26] Add[27] the 'upper' and the 'lower' of the left column.

18. One has to go on until the last remainder [*chi*] of the 'upper right' is 1 and then one can stop. Then you examine the result on the 'upper left'; take it as the *ch'êng-lü*.[28]

19. Sometimes the *chi* [remainder] is already 1; this is then the *ch'êng-lü*.[29]

D.

20. Draw up all the *ch'êng-lü* and multiply with the corresponding *yen-shu*, and you will obtain the *fan-yung* 泛用.

E.

21. Add up the *fan-yung* and examine the *yen-mu*.

22. If [the sum of the *fan-yung*] surpasses the *yen-mu* by 1, then [the *fan-yung*] are the *chêng-yung* 正用.[30]

23. If [the sum of] the *fan-yung* surpasses a multiple of the *yen-mu* [with 1], then examine the *yüan-shu*.[31]

24. If two numbers [*yüan-shu*] are divisible by the same factor,[32] one subtracts half [the *yen-mu*] [from the *fan-yung*.][33]

---

[25] The text says *hsiang-shêng* 相生. Chiao Hsün (1), B, p. 3a says "*hsiang-shêng* = *hsiang-ch'êng* 相乘."

[26] See note 19.

[27] *Kuei* 歸 has here the special meaning of "to add."

[28] Mikami (1), p. 68, also translated a piece of this text (nos. 15–18).

[29] For the sake of readability, the translation has been simplified. When translating Chinese one cannot satisfy both the grammarian and the historian of science, as the former is interested in the form, the latter in the content.

[30] *Chêng-yung* could be translated as "correct use numbers."

[31] The "origin numbers," that is, before reducing.

[32] Literally, "numbers of the same class, odd or even." It is clear that "odd and even" means only "having the same divisor." See note 4.

[33] Although not in the text, this is beyond doubt, because one cannot subtract from the *yüan-shu*; the examples make the case clear.

25. If there are three numbers of the same class, one divides the *yen-mu* by 3 and diminishes the three numbers with it.

26. All these numbers are the *chêng-yung-shu* 正用數.

27. Sometimes you get 1 as *ting-mu*,[34] or if the *yen-shu* and the *yen-mu* are the same, there is no *yung-shu*.

28. We have to examine, among the *yüan-shu,* the ones that are mutually divisible, and borrow the *chêng-yung* from the largest numbers.

29. Among the *yüan-shu* find by twos the common factors. With the common factors reduce the *yen-mu* and you get the *chieh-shu* 借數.[35]

30. With the *chieh-shu* diminish the existing [*yung-shu*] and increase with this [amount] the nonexisting ones. These are the *ch'êng-yung*.

31. Or otherwise, if you wish to place numbers in the vacant places, these will be the *yen-mu*. Reduce the *yen-mu*.[36] Multiply the number obtained with a factor of choice.[37]

32. You borrow proportionally and put them in the places [which are empty].

33. Or, if you wish, after reducing, do not borrow. You are allowed to leave these places empty.[38]

F.

34. After that, multiply the *chêng-yung* with [*shêng*]-*yü* (賸)餘 [remainders].[39] You obtain the *tsung* 總.

35. Add up the *tsung*.

36. Subtract from it the *yen-mu* as many times as possible.

37. What is no more sufficient [to subtract the *yen-mu* from] is the *lü-shu* 率數 looked for."

---

[34] *Ting-mu* has the same meaning as *ting-shu*.
[35] Literally, the borrowed numbers.
[36] By the greatest common divisor.
[37] On the condition that the sum of these factors equals the greatest common divisor.
[38] As we shall see in the next section, this whole "reduction" is not necessary.
[39] These are the remainders as propounded in the problem.

Interpretation of the Text

There are only two works that attempt to give a full explanation of this text, namely, Li Yen (6'), vol. 1[40] and Ch'ien Pao-tsung (2'),[41] and this account owes much to these articles. Mikami (1) explains only a few passages of the text.[42]

Ch'in's text is given as an explanation of the first problem (I,1) but according to the commentary of Sung Ching-ch'ang,[43] this text formerly stood in front of the first problem, a fact that is quite natural, because it gives the general method of *ta-yen*.

In this chapter we shall give a theoretical reproduction in modern algebraical terms as well as a few examples to illustrate the method. In the last part of this section one will find a translation of the indefinite problems of Ch'in Chiu-shao.

THE *TING-SHU*

A.1–A.3: Take the "problem numbers" by twos, and reduce the odd numbers; do not reduce the even ones.

Li Yen, in (6'), p. 129, explains thus: "one wishes to reduce so as no longer to have a common factor." This is right, but it does not explain the peculiar use of the terms "odd" and "even." Ch'ien Pao-tsung, in (2'), pp. 70 f, states that *chi* and *ou* indeed mean "odd" and "even," but that in this context, Ch'in denotes only two different numbers.

In my opinion, *chi-ou t'ung-lei* 奇偶同類 is to be considered as a technical expression, with the general meaning "belonging to the same class," that is, divisible by the same factor. It is possible that the first subcategories of the set of integers were odd and even numbers.[44] But here the technical term is extend-

---

[40] First published in *Hsüeh-i*, 1925, 7, 2, pp. 1–45; reprint in (6'), vol. 1, pp. 122–174.

[41] In (2'), pp. 60–103.

[42] Mikami is in a sense quite right to eliminate all the really superfluous passages of the text. But as this is a monograph on Ch'in Chiu-shao, it should present the whole text. The reader interested only in the mathematical description of the method is referred to the section on "Theoretical Representation of Ch'in Chiu-shao's Method."

[43] (1'), p. 3.

[44] See Needham (1), vol. 3, pp. 54 f. This is also the subdivision of the *lo-shu* and the *ho-t'u* diagrams (ibid., p. 57).

ed to all kinds of numbers divisible by the same factor, a fact that is proved in Problem I, 2. Ch'in speaks of the *chi-ou t'ung-lei* (odd-even-same class) and says: "The 'tail digit' of 114,445 is 5, the 'tail digit' of 225,600 is 600; both belong to the *wu t'ung-lei* 五同類 (the five-folds)." From this statement it is obvious that *chi-ou* can be equal to all kinds of prime numbers.[45]

This can be illustrated by examples:
(1) Given the pair of numbers 6 and 4; the greatest common divisor=2. While reducing 6 with 2, one obtains 3 and 4 (relatively prime). While reducing 4 with 2, one obtains 6 and 2 (not relatively prime). The general purpose is thus *to reduce one of the two numbers in such a way that they become relatively prime.*

(2) $\dfrac{25 \quad 10}{5} \longrightarrow 5$ 10 (not relatively prime);

$\dfrac{25 \quad 10}{5} \longrightarrow 25$ 2 (relatively prime).[46]

A. 4: All the numbers are "even." This means that all the numbers are divisible by the same factor.
Example:[47]

$$\frac{130 \quad 120 \quad 110 \quad 100 \quad 60 \quad 50 \quad 25 \quad 20}{5} \longrightarrow$$

26   24   22   20   12   10   5   *20*

Reduce all the numbers by 5, except one (20).

A. 5: If you can still reduce with other numbers, go on.
Example (the same as A. 4):[48]

26	24	22	20	12	10	5	20	: 5
26	24	22	4	12	2	*5*	*20*	: 2
13	12	11	2	6	*2*	*5*	*20*	: 6
13	2	11	2	*6*	*2*	*5*	*20*	: 2
13	1	11	*2*	*6*	*2*	*5*	*20*	

[45] Here "odd-even" means "five."
[46] To represent the operation, the greatest common divisor is placed below the line under the numbers, and the reduction is indicated with → .
[47] This is problem I,8. It shows clearly what is meant by "even" and "odd," for in this case all the numbers are multiples of five.
[48] These numbers are not yet relatively prime in pairs, and the reduction must go on.

A.6–A.10: Further reduction: Example (the same as A.5): When you reduce one number, multiply the other with the same factor.

13	8	11	1	3	1	5	20

Indeed, $6 = 3 \times 2$. Then we obtain $2 \times 2 \times 2 = 8$. We must transpose 8 in the column of 120, because the reduced number must always be a divisor of the "problem number." The number 120 is the one divisible by 8, and $20 = 5 \times 4$. We can compensate the 5 by multiplying 5 with it while dividing 20. As the 4 itself is kept in 8,[49] we may reduce it to 1. The *ting-shu* are

13	*8*	*11*	*1*	*3*	*1*	*25*	*1*

One will notice that the *ting-shu* are all divisors of their corresponding "problem numbers."

This method, although the first in the history of mathematics that succeeded in solving the problem where the moduli are not relatively prime in pairs,[50] is in some cases very long-winded and complicated.[51] Represented in modern algebraical symbolism:

L.C.M.$(130, 120, 110, 100, 60, 50, 25, 20) = 2^3 \times 3 \times 5^2 \times 11 \times 13$.

These factors, being relatively prime, are divisors of at least one of the given numbers. The last ones can be substituted for by these factors; the remaining ones are substituted for by 1.[52]

130	120	110	100	60	50	25	20
13	8	11	1	3	1	25	1

There are other examples in Ch'in's work:[53]
Problem 1:

---

[49] This follows from the confused method itself. With an entire decomposition into factors this would not be the case.

[50] Of course without regard for some conditions and restrictions found by modern mathematicians.

[51] This is a consequence of the lack of rigid algebraic language.

[52] There is of course more than one possibility.

[53] According to the problems given by Ch'in Chiu-shao. The greater part are taken from Li Yen (6'), pp. 129 ff, but they can easily be deduced from the problems. They are not given in the text in such a concise manner.

1	2	3	4	(: 2)
1	*2*	3	2	(compensation)
1	1	3	4	(*ting-shu*)

Or: L.C.M. (1, 2, 3, 4) $= 3 \times 2^2$

$\longrightarrow$ 1   1   3   4.

Problem 2:[54]

225,600	111,036	1,373,340	(: 225,600)
225,600	487×19	487	(: 487)
225,600	*487×19*	1	(compensation)
225,600	19	487	(*ting-shu*)

Problem 3:

54	57	75	72	(: 3)
54	19	25	24	(: 6)
9	19	25	*24*	(compensation)
27	19	25	8	(*ting-shu*)

Or: L.C.M. (54, 57, 75, 72) $= 2^3 \times 3^2 \times 5^2 \times 19$

27   19   25   8

Problem 4:

12	11	10	9	8	7	6	(: 6)
2	11	10	9	8	7	*6*	(: 2)
1	11	5	9	*8*	7	6	
1	11	5	9	8	7	2	

In the third row here, 9 and 6 are not relatively prime; thus reduce 6 to 2. In the fourth row, 8 and 2 are not relatively prime; thus reduce 2 to 1.

1	11	5	9	8	7	1	(*ting-shu*)

Or L.C.M. (12, 11, 10, 9, 8, 7, 6) $= 2^3 \times 3^2 \times 5 \times 7 \times 11$

1	11	5	9	8	7	1	

[54] This is in fact a problem with fractions; the numbers given here are already reduced.

Problem 5:

83	110	135	(: 5)
83	110	27	(*ting-shu*)

Problem 6:

300	240	180	(:60)
300	4	3	(compensation)
100	4	9	(compensation)
25	16	9	

Or: L.C.M. $(300, 240, 180) = 2^4 \times 3^3 \times 5^2$

25	16	9

Problem 7:

300	250	200	(: 50)
6	250	4	(compensation)
3	250	8	(compensation)
3	125	16	

Or: L.C.M. $(300, 250, 200) = 2^3 \times 3 \times 5^3$

3	125	16

Problem 8: See p. 334.
Problem 9:

19	17	12

are relatively prime in pairs.[55]

B.11: *Yen-mu:* The *yen-mu* is the product of all the *ting-shu.*
Let $m_1, m_2, m_3, \ldots, m_k$ be the given "problem numbers." Let $\mu_1, \mu_2, \mu_3, \ldots, \mu_k$ be the derived *ting-shu.* L.C.M. $(m_1, m_2, m_3 \ldots) = \theta = \mu_1 \times \mu_2 \times \mu_3 \times \ldots$.

The *yen-mu* is nothing other than the least common multiple of the "problem numbers." Example (Problem 3):

Moduli:	54	57	75	72
ting-shu:	27	19	25	8

yen-mu $(\theta) = 27 \times 19 \times 25 \times 8 = $ L.C.M. $(54, 57, 75, 72)$.

B.12a: *Yen-mu* $(M_1, M_2, \ldots)$:

[55] For the general description of the method, see p. 357.

$$M_1 = \frac{\mu_1 \times \mu_2 \times \mu_3 \times \ldots}{\mu_1} = \mu_2 \times \mu_3 \times \ldots ,$$

$$M_2 = \frac{\mu_1 \times \mu_2 \times \mu_3 \times \ldots}{\mu_2} = \mu_1 \times \mu_2 \times \ldots ,$$

and so on.

$$M_1 = \frac{27 \times 19 \times 25 \times 8}{27} = 19 \times 25 \times 8 = 3{,}800 ;$$

$$M_2 = \frac{27 \times 19 \times 25 \times 8}{19} = 27 \times 25 \times 8 = 5{,}400 .$$

B. 12b:

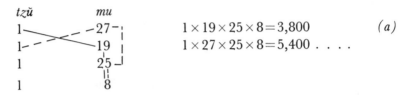

$$1 \times 19 \times 25 \times 8 = 3{,}800 \qquad (a)$$
$$1 \times 27 \times 25 \times 8 = 5{,}400 \ldots .$$

The ones placed in the left column are the famous *t'ien-yüan i* (the unity symbol). This *t'ien-yüan* 1 is not the same as that used in the solving of the congruences. It is suggested here that this 1 was nothing more than a place indicator on the counting board.

1	27		3,800	27		3,800	27
1	19		1	19		5,400	19
1	25		1	25		1	25
1	8		1	8		1	8

Diagram 64

When operation *a* in Diagram 64 is performed, the first 1 is the place where the product has to be set up; it cannot be set up in place of 27, as was usually done in operations (for instance, in the solving of simultaneous linear equations), because afterward the 27 is needed again. This is obvious from the diagrams on p. 5 (Problem I, 1). See Figure 50 and the accompanying diagram.

Figure 50. The "Celestial Element One" as place indicator. *SSCC*, Problem I, 1 (p. 5).

12	1	1	1
12	1	1	1
4	3	1	3
3	4	1	4

B. 13: *The chi*: From the *yen-shu* subtract the corresponding *ting-shu* as many times as possible:

$M_1 - p\mu_1 \ = N_1$

$M_2 - p'\mu_2 \ = N_2$

$M_3 - p''\mu_3 = N_3$.

This substitution is allowed because M_1/μ_1 and N_1/μ_1 have the same remainder (since $p\mu_1/\mu_1$ has no remainder). Example (Prob. 7):

300	250	200	
ting-shu : 3	125	16	
yen-mu : $3 \times 125 \times 16 = 6{,}000.$			
yen-shu : $M_1 = 125 \times 16 = 2{,}000.$			
$M_2 = 3 \times 16 = 48$			
$M_3 = 3 \times 125 = 375$			
chi-shu : $2{,}000 - p \times 3 = 2$			
$48 - p' \times 125 = 48$			
$375 - p'' \times 16 = 7.$			

C.14: The *ta-yen ch'iu-i*: The purpose of the *ta-yen ch'iu-i*[56] is to solve the following congruence:

$(ch'\hat{e}ng\text{-}l\ddot{u}) \times (chi\text{-}shu) \equiv 1 \ [\mathrm{mod} \ (ting\text{-}shu)]$

or:

$a N_1 \equiv 1 \ (\mathrm{mod} \ \mu_1).$

C.15: For the reader's information the method is given here as represented by Li Yen (6'), vol. 1, pp. 135 ff, and by Ch'ien Pao-tsung (2'), pp. 67 ff, and (4'), pp. 52 ff. One will find other information in Ch'ien Pao-tsung (7'), pp. 127 f, and in Mikami (1), pp. 65 ff.

Li Yen solves the following congruence by Ch'in's method:[57]

$a \times 65 \equiv 1 \ (\mathrm{mod} \ 83)$ or $a \times N_1 \equiv 1 \ (\mathrm{mod} \ \mu_1).$

[56] *Ch'iu-i* means "searching for unity."

[57] Li Yen uses here the special manner of representation derived from Hsü Chên-ch'ih (1'). $\alpha\, 65 \equiv 1 (\mathrm{mod} \ 83)$ is represented by

$$\left| \frac{65a.}{83} \right. = 1,$$

the vertical line indicating the remainder of the division. In (13') Li Yen follows the European manner of notation, which is derived from Gauss.

(I)   _t'ien-yüan_ 1 | 65 (_chi-shu_) $\alpha_0 = 1$ | $N_1 = 65$
                        | 83 (_ting-shu_) | $\mu' = 83$
                        | 1 | $q_1 = 1$

(II)   1 | 65 $\alpha_0 = 1$ | N_1
                       1 | 18 $\alpha_1 = q_1\alpha_0$ | r_1
                         | 1 | q_1

(III) | 3 | q_2
                       4 | 11 $\alpha_2 = q_2\alpha_1 + \alpha_0$ | r_2
                       1 | 18 α_1 | r_1

(IV)   4 | 11 α_2 | r_2
                       5 | 7 $\alpha_3 = q_3\alpha_2 + \alpha_1$ | r_3
                         | 1 | q_3

(V) | 1 | q_4
                       9 | 4 $\alpha_4 = q_4\alpha_3 + \alpha_2$ | r_4
                       5 | 7 α_3 | r_3

(VI)   9 | 4 α_4 | r_4
                      14 | 3 $\alpha_5 = q_5\alpha_4 + \alpha_3$ | r_5
                         | 1 | q_5

(VII) | 1 | q_6
                      23 | _1_ $\alpha_6 = q_6\alpha_5 + \alpha_4$ | r_6
                      14 | 3 α_5 | r_5

Diagram 65

The diagrams to draw up are shown in Diagram 65.[58]

The operation must go on until the remainder is 1. This explains the term "searching for unity." The general representation can be given as follows:[59]

Suppose we have to solve the congruence

$aN_1 \equiv 1 \pmod{\mu_1}$.

We develop μ_1 / N_1 in a continued fraction:

$$\frac{\mu_1}{N_1} = (q_1, q_2, q_3, \ldots, q_{n-1}, q_n).$$

The approximative fractions are

$c_1 = q_1$
$c_2 = q_2 c_1 + 1$

[58] The "turning over" of the representation from diagram to diagram can only be a consequence of the mechanical solving on the counting board.
[59] See Ch'ien Pao-tsung (2'), p. 68.

$$c_3 = q_3 c_2 + c_1$$
$$\vdots$$
$$c_n = q_n c_{n-1} + c_{n-2}.$$

One is surprised at the fact that none of the Chinese scholars has ever seen that this method is exactly the same as solving with the aid of continued fractions.

Indeed,

$$\frac{83}{65} = (1, 3, 1, 1, 1, 1, 3).$$

The approximative fractions are

$$\frac{1}{1}; \ 1 + \frac{1}{3} = \frac{4}{3}; \ \frac{4 \times 1 + 1}{3 \times 1 + 1} = \frac{5}{4};$$
$$\frac{5 \times 1 + 4}{4 \times 1 + 3} = \frac{9}{7}; \ \frac{9 \times 1 + 5}{7 \times 1 + 4} = \frac{14}{11};$$
$$\frac{14 \times 1 + 9}{11 \times 1 + 7} = \frac{23}{18} \ \left(\frac{p}{q}\right).$$

Ch'in gives:

$$a_0 = 1; \ a_1 = 1 \times 1 = 1; \ a_2 = 3 \times 1 + 1 = 4;$$
$$a_3 = 1 \times 4 + 1 = 5; \ a_4 = 1 \times 5 + 4 = 9; \ a_5 = 1 \times 9 + 5 = 14$$
$$a_6 = 1 \times 14 + 9 = 23.$$

This fully agrees with the numerators of the approximate fractions. As the number of approximate fractions is odd, the solution is

$$x = -qc \qquad x = -18 \times 1 = -18$$
$$y = pc \qquad y = 23 \times 1 = 23.$$

These two methods are entirely identical. Ch'in always considers the signs of p and q positive. As c is always 1, the solution for y is, if the number of approximate fractions is odd: $y = p$; if the number of approximate fractions is even:[60] $y = -p$.

Proof of this method is given by Ch'ien Pao-tsung:[61] Let

[60] In the work of Ch'in Chiu-shao, all the congruences are positive. How he solves the problem if the solution is negative, we shall see later.
[61] Ch'ien Pao-tsung (2′), p. 68.

$L_2=q_2$, $L_3=q_3L_2+1$, $L_4=q_4L_3+L_2$, ..., $L_n=q_nL_{n-1}+L_{n-2}$.

From $\mu_1/N_1=(q_1, q_2, q_3, \ldots, q_{n-1}, q_n)$ and the approximate fractions, we derive:

$r_1=\mu_1-N_1q_1=\mu_1-c_1N_1$ $\qquad\qquad\qquad$ $[c_1=q_1]$

$r_2=N_1-r_1q_2=N_1-(\mu_1-c_1N_1)q_2=c_2N_1-L_2\mu_1$

$r_3=r_1-r_2q_3=(\mu_1-c_1N_1)-(c_2N_1-L_2\mu_1)q_3=L_3\mu_1-c_3N_1.$

$\vdots$

$r_{n-1}=L_{n-1}\mu_1-c_{n-1}N_1$

$r_n=c_nN_1-L_n\mu_1$

from which: $c_nN_1\equiv r_n \pmod{\mu_1}$, with $r_n=1$:[62]

$c_nN_1\equiv 1 \pmod{\mu_1}$

$a=c_n$.

A very clear representation of Ch'in's method is given by Chang Tun-jên. For instance, $\alpha\times 65\equiv 1 \pmod{83}$ (see Diagram 66).

Q_1	1		R_2	11		Q_3	1
	65			65			11
	83			18			18
R_1	18		Q_2	3		R_3	7

R_4	4		Q_5	1		R_6	1
	11			4			4
	7			7			3
Q_4	1		R_5	3		Q_6	1

Diagram 66

We stop the division in that diagram if the upper remainder is 1.[63] Then we have the configuration shown in Diagram 67.

[62] Our mathematical textbooks explain it thus: let us take the case in which the number of the approximate fractions is odd. Then for the congruence $ax\equiv c\pmod{b}$, we develop in a continued fraction. Let p/q be the second-to-last approximate fraction.

$bp-aq=1$

$bpc-aqc=c$

$a(-qc)+b(pc)=c.$

Thus $x=-qc$; $y=pc$, and with $c=1$, $x=-q$ and $y=p$.

[63] This is the same as the Euclidian algorithm.

Q		
1	1	
3	1	$1 \times 1 = 1$
1	4	$1 \times 3 + 1 = 4$
1	5	$4 \times 1 + 1 = 5$
1	9	$5 \times 1 + 4 = 9$
1	14	$9 \times 1 + 5 = 14$
·	23	$14 \times 1 + 9 = 23$

Diagram 67

If the lower remainder is then 1, the operation must go on until the upper remainder is 1. Indeed, if the number of quotients is even, the congruence factor is positive. If the number of quotients is odd, the congruence factor is negative. In order to make it positive, we do one more division. For example: $a \times 5 \equiv 1$ (mod 23). If we stopped the mutual division at $R_3 = 1$, the congruence factor should be -9. If we go further, it becomes $+14$.

Indeed,

$$(-9) \times 5 \equiv 1 \ (\text{mod } 23)$$
$$\underline{23 \times 5 \equiv 0 \ (\text{mod } 23)}$$
$$14 \times 5 \equiv 1 \ (\text{mod } 23).$$

The same rule is applied in Ch'in Chiu-shao's work. Diagram 68 summarizes the method up to this point.

wên-shu	A	B	C
ting-shu	μ_1	μ_2	μ_3
yen-shu	M_1	M_2	M_3
ch'i-shu	N_1	N_2	N_3
ch'êng-lü	α	β	γ

Diagram 68

Suppose that $A = a^m b^n c^p d^q$ $m > m' > m''$

(I) $B = a^{m'} b^{n'} c^{p'}$ $n' > n > n''$

               $C = a^{m''} b^{n''} c^{p''} e^{r''}$ $p'' > p' > p$

(II) $\mu_1 = a^m d^q$
                $\mu_2 = b^{n\prime}$ $\mu_1{}', \mu_2{}', \mu_3{}'$ are relatively
                $\mu_3 = c^{p\prime\prime} e^{r\prime\prime}$ prime in pairs.

(III) $M_1 = \mu_2 \mu_3$
                $M_2 = \mu_1 \mu_3$
                $M_3 = \mu_1 \mu_2$

(IV) $N_1 = M_1 - g\mu_1$ $N_1 < \mu_1$
                $N_2 = M_2 - g'\mu_2$ $N_2 < \mu_2$
                $N_3 = M_3 - g''\mu_3$ $N_3 < \mu_3$

(V) $aN_1 \equiv 1 \pmod{\mu_1}$
                $\beta N_2 \equiv 1 \pmod{\mu_2}$
                $\gamma N_3 \equiv 1 \pmod{\mu_3}$

THE *T'IEN-YUAN-I* (CELESTIAL ELEMENT UNITY)

Needham, in (1), vol. 3, p. 42, says that it was "placed at the
left-hand top corner of the counting board before the beginning
of one of the most important parts of the operation [of indeterminate analysis.]" From this statement it is not at all clear what
t'ien-yüan-i really means. Ch'ien Pao-tsung says: "Ch'in Chiu-
shao does not explain what the *t'ien-yüan-i* in the left upper corner of the diagram (I) represents.[64] Chiao Hsün of the Ch'ing
in his *T'ien-yüan-i shih* states that this *t'ien-yüan-i* is 'one.' In my
own *Chung-kuo suan-hsüeh shih* (1932) I said that it represents the
remainder *R*. In fact, the 1 in the left upper corner is only the
unit 1; the two characters *t'ien-yüan* of the technical term are a
conventional expression. . . ."

Let us try to give another explanation, but—we must emphasize—with all due reserve.

When drawing up the approximative fractions, we proceed
as follows:

$$\frac{57}{25} = (2, 3, 1, 1, 3).$$

[64] (2'), p. 69. The diagram in Ch'in's text is

t'ien-yüan 1	20
	27

First approximate fraction $=\dfrac{2}{1}$,

Second approximate fraction $=2+\dfrac{1}{3}=\dfrac{6+1}{3}$.

In the explanation given earlier we had

$a_2=q_2a_1+1$; $a_2=3\times2+1=7$.

The 1 to be added to q_2a_1 is the *t'ien-yüan* 1. Without this 1 it is impossible to solve the congruences. But, without proof of the operation, the 1 seems in some way to be arbitrary, though general in its application. The finding of the Celestial Element was indeed the key for solving the congruences.

In our modern mathematical notation the Celestial Element seems not at all celestial, but very earthly. We denote the continued fraction given earlier as follows:

$$2+\cfrac{1}{3+\cfrac{1}{1+\cfrac{1}{1+\cfrac{1}{3}}}}$$

First approximate fraction $=\dfrac{2}{1}$; second $=2+\dfrac{1}{3}$.

At any rate it is quite clear that the congruences could not be solved without this 1.[65]

D.20: The *fan-yung*:

fan-yung=ch'êng-lü × *yen-shu*
$F_1=\alpha M_1$
$F_2=\beta M_2$
$F_3=\gamma M_3$.

---

[65] In my own experience with teaching mathematics for more than fifteen years, I was always surprised by the fact that some of my students were in many cases not able to develop approximate fractions, because they could not find the second one (for the others there exists a general rule). Now I understand that they could not find the "Celestial Element." I am grateful to those who elucidated this point by example. I am well aware of the fact that *t'ien-yüan* means something else in the algebra of Li Yeh. See Mikami (2), p. 386.

E. The *chêng-yung-shu*: The following part of the method is rather confusing and in some way useless, although it is mathematically correct. It is a method for reducing the *fan-yung*, which are in some problems very large. The only operation following the computation of the *fan-yung* is their multiplication with the given remainders. As the Chinese mathematicians like to show their ability in computing, they use very large numbers in many cases; on the other hand the numbers used in calendrical computations are large by nature. It is thus quite natural that they should try to reduce them. But the Chinese mathematicians did not always show great ability in reducing.

E.21–E.26: Add up the *fan-yung*: $F_1+F_2+F_3=\sum F$. Examine the *yen-mu*: $\mu_1\mu_2\mu_3=\theta$.

(a) If $\sum F-\theta=1\rightarrow F_1$, F_2, F_3 are the *chêng-yung*,[66]
(b) if $\sum F-q\theta=1$ (where $q>1$) "then examine the *yüan-shu*" $A, B, C, \ldots$ (the given numbers).

"If two numbers are divisible by the same factor, subtract half the *yen-mu* from each [*fan-yung*]."
    For example, $N\equiv 7$ (mod 15)$\equiv 2$ (mod 5)$\equiv 2$ (mod 7).

yüan-shu	15	5	7
ting-shu	3	5	7
yen-mu		105	
yen-shu	35	21	15
ch'i-shu	2	1	1
ch'êng-lü	2	6	8
fan-yung	70	126	120

---

[66] $\Sigma aM \equiv 1$ (mod $\mu_1\mu_2\mu_3$)
$aM_1 + \beta M_2 + \gamma M_3 = a\mu_2\mu_3 + \beta\mu_1\mu_3 + \gamma\mu_1\mu_2$
$\Sigma aM = \mu_3 (a\mu_2 + \beta\mu_1) + \mu_3 q + 1$
        $[\gamma\mu_1\mu_2 \equiv 1 (\mathrm{mod}\ \mu_3)]$
    $= \mu_3 (a\mu_2 + \beta\mu_1 + q) + 1$
    $\equiv 1$ (mod μ_3).
In the same way we can prove: $\Sigma aM = 1$ (mod μ_2) and $\Sigma aM = 1$ (mod μ_3). In order that the three conditions be filled at the same time it is necessary that $M = 1$(mod $\mu_1\mu_2\mu_3$).

$\sum F = 70 + 126 + 120 = 316$
$\sum F = 3 \times \theta + 1$
$\quad = 3 \times 105 + 1$.

Li Yen in (6'), p. 140, gives the following explanation: "If $\sum a M_1$ $= m\theta' + 1$ and $\theta = m\theta'$, where $m > 1$, then, if among the *yüan-shu* A, B, C, . . . corresponding to the *ting-shu* A', B', C', . . . , there are two or more numbers which have a common factor m, then the least common multiple θ of A', B', C', . . . or $m\theta'$, also contains this factor m.

  When we have found this m, then we make use of

$m\theta' = m_1\theta' + m_2\theta' + m_3\theta' + \ldots$

and

$m = m_1 + m_2 + m_3 + \ldots$

and distinguish, among the numbers in $\sum a M_1$ those that contain the common factor m, and according to the circumstances we subtract them. Then [we have]:

$\sum a M'_1 = m\theta' + 1$ and $m = 1$."
$\sum a M_1 \equiv 1 \pmod{\theta}$ or $\sum a M_1 = q\theta + 1$
$\sum a M'_1 + m\theta' \equiv 1 \pmod{\theta}$ or $\sum a M'_1 + m\theta' = q'\theta + 1$.

This is true, because $m\theta' \equiv 0 \pmod{\theta}$ and $\sum a M'_1 \equiv 1 \pmod{\theta}$. Applied to the original example, this means that $365 \equiv 1 \pmod{105}$, or $316 = 3 \times 105 + 1$. If $m = $ greatest common divisor $(15, 5) = 5$, $105 = M\theta' = 5 \times 21$. Among the terms of $\sum a M_1$, 70 and 120 contain the factor $m = 5$. From 70 and 120 we subtract (together) 5 times 21 (θ').

  For example,

$70 - 2 \times 21 = 28$	$\times 7 = 196$
$126 \qquad = 126$	$\times 2 = 252$
$120 - 3 \times 21 = 57$	$\times 2 = 114$
$\overline{211} \equiv 1 \pmod{105}$	$\overline{562} - n \times 105 = 37$.

$37 \equiv 7 \pmod{15} \equiv 2 \pmod{5} \equiv 2 \pmod{7}$.

This reduction makes sense only when the terms of $\sum a M_1$ are very large, as in I, 2. However, it is not applied very consistently in Ch'in's work.

E. 27–E. 33: If one of the *ting-shu* is 1, then the correspond-ing *yen-shu* $M_k = \theta$, because $M_k = \theta/1$; in this case there is no *yung-shu*. It can, however, be borrowed from another *yung-shu*. If among the *yüan-shu*, some have a common factor m, the *chieh-shu* will be $m_1\theta'$, $m_2\theta'$, ... Subtract them from the *chêng-yung* which are large enough, and put them in the empty places.

For example, I, 4 (see p. 336),

yüan-shu	A	B	C	D	E	F	G
	12	11	10	9	8	7	6
yung-shu	0	2,520	22,176	15,400	3,456	11,880	0

$m = $ G.C.D. $(12, 10, 8, 6) = 2$

$\dfrac{\theta}{m} = \dfrac{27,720}{2} = 13,860$

$22,176 - 13,860 = 8,316$ (C′)

$13,860 = 6,930$ (A′) $+ 6,930$ (G′).

The new *yung-shu* are

A′	B	C′	D	E	F	G′
6,930	2,520	8,316	15,400	3,465	11,880	6,930.

In fact, this change is a useless complication, and Ch'in says: "It is allowed to leave these places empty."[67]

F. 34: *Chêng-yung* × remainders = *tsung*.

$T_1 = F_1 r_1$
$T_2 = F_2 r_2$
$T_3 = F_3 r_3$

F. 35: Add up the *tsung*:

$T_1 + T_2 + T_3$

F. 36–F.37: Subtract the *yen-mu* as much as possible; the re-mainder is the number asked for. $N = T_1 + T_2 + T_3 - n\theta$, where n is as great as possible.

---

[67] See Problem I, 4, p. 382.

Résumé and Example[68]

(1)

$N \equiv r_1 \pmod{A}$	$N \equiv 10 \pmod{12}$
$\equiv r_2 \pmod{B}$	$\equiv 0 \pmod{11}$
$\equiv r_3 \pmod{C}$	$\equiv 0 \pmod{10}$
$\equiv \ldots$	$\equiv 4 \pmod{9}$
	$\equiv 6 \pmod{8}$
	$\equiv 0 \pmod{7}$
	$\equiv 4 \pmod{6}$

(2) Reducing of the moduli[69]

$\mu_1, \mu_2, \mu_3 \ldots$	$12 = 2^2 \times 3$	1	μ_1
ting-shu	$11 = 11$	11	μ_2
	$10 = 2 \times 5$	5	μ_3
	$9 = 3^2$	9	μ_4
	$8 = 2^3$	8	μ_5
	$7 = 7$	7	μ_6
	$6 = 2 \times 3$	1	μ_7

(3) *Yen-mu*

$\theta = \mu_1 \mu_2 \mu_3 \ldots$ $\qquad = 11 \times 5 \times 9 \times 8 \times 7 = 27{,}720$

(4) *Yen-shu*

$M_1 = \mu_2 \mu_3 \mu_4 \ldots$ $\qquad M_1 = 11 \times \ 5 \times 9 \times 8 \times 7 \times 1 = 27{,}720$
$M_2 = \mu_1 \mu_3 \mu_4 \ldots$ $\qquad M_2 = \ 1 \times \ 5 \times 9 \times 8 \times 7 \times 1 = \ 2{,}520$
$\qquad\qquad\qquad\qquad M_3 = \ 1 \times 11 \times 9 \times 8 \times 7 \times 1 = \ 5{,}544$
$\qquad\qquad\qquad\qquad M_4 = \ 1 \times 11 \times 5 \times 8 \times 7 \times 1 = \ 3{,}080$
$\qquad\qquad\qquad\qquad M_5 = \ 1 \times 11 \times 5 \times 9 \times 7 \times 1 = \ 3{,}465$
$\qquad\qquad\qquad\qquad M_6 = \ 1 \times 11 \times 5 \times 9 \times 8 \times 1 = \ 3{,}960$
$\qquad\qquad\qquad\qquad M_7 = \ 1 \times 11 \times 5 \times 9 \times 8 \times 7 = 27{,}720$

(5) *Chi-shu*

$N_1 = M_1 - p\mu_1$ $\qquad N_1 = 27{,}720 - 27{,}720 \times 1 = 0$

---

[68] The example is taken from Ch'in Chiu-shao, I, 4, but abridged for the sake of clarity.
[69] In this example, Ch'in does not give the method, only the results. The modern method is used here.

$$N_2 = M_2 - p'\mu_2$$
$$\vdots$$

$$
\begin{aligned}
N_2 &= 2{,}520 - & 229 \times 11 &= 1 \\
N_3 &= 5{,}544 - & 1{,}108 \times 5 &= 4 \\
N_4 &= 3{,}080 - & 342 \times 9 &= 2 \\
N_5 &= 3{,}465 - & 433 \times 8 &= 1 \\
N_6 &= 3{,}960 - & 565 \times 7 &= 5 \\
N_7 &= 27{,}720 - & 27{,}720 \times 1 &= 0
\end{aligned}
$$

(6) Solving of the congruences[70]

$$
\begin{aligned}
\alpha \times 0 &\equiv 1 \ (\mathrm{mod}\ \ 1) \\
\beta \times 1 &\equiv 1 \ (\mathrm{mod}\ 11) & \beta &= 12 \\
\gamma \times 4 &\equiv 1 \ (\mathrm{mod}\ \ 5) & \gamma &= 4 \\
\delta \times 2 &\equiv 1 \ (\mathrm{mod}\ \ 9) & \delta &= 5 \\
\varepsilon \times 1 &\equiv 1 \ (\mathrm{mod}\ \ 8) & \varepsilon &= 9 \\
\zeta \times 5 &\equiv 1 \ (\mathrm{mod}\ \ 7) & \zeta &= 3 \\
\eta \times 0 &\equiv 1 \ (\mathrm{mod}\ \ 1)
\end{aligned}
$$

(7) *Fan-yung*[71]

$$F_1 = \alpha M_1$$
$$F_2 = \beta M_2$$
$$\vdots$$

$$
\begin{aligned}
F_1 & \\
F_2 &= 12 \times 2{,}540 \\
F_3 &= 4 \times 5{,}544 \\
F_4 &= 5 \times 3{,}080 = 15{,}400 \\
F_5 &= 9 \times 3{,}465 = 31{,}185 \\
F_6 &= 3 \times 3{,}960 \\
F_7 &=
\end{aligned}
$$

(8) Multiply by the remainders.

$$
\begin{aligned}
T_1 & \\
T_2 & \quad \times 0 \\
T_3 & \quad \times 0 \\
T_4 &= 15{,}400 \times 4 = 61{,}600 \\
T_5 &= 31{,}185 \times 6 = 187{,}110 \\
T_6 & \quad \times 0 \\
T_7 &
\end{aligned}
$$

[70] The method for solving the congruences, which was explained earlier, is not given here.
[71] This operation is not necessary when the given remainder is zero.

A	12	11	10	9	8	7	6
μ	1	11	5	9	8	7	1
θ				27,720			
M	27,720	2,520	5,544	3,080	3,456	3,960	27,720
N	0	1	4	2	1	5	0
α		12	4	5	9	3	
F				15,400	31,185		
r		0	0	4	6	0	
T		0	0	61,600	187,110	0	

$$61,600 + 187,110 = 248,710$$
$$248,710 - 8 \times 27,720 = 26,950$$

Diagram 69

(9) Add up. $61,600 + 187,110 = 248,710$

(10) Diminish by the *yen-mu* as many times as possible.

$$248,710 - 8 \times 27,720 = 26,950.$$

In the other problems, we shall arrange the numbers as in Diagram 69.

Extension to All Kinds of Numbers

We have already seen that Ch'in Chiu-shao distinguishes four kinds of numbers: (1) the *yüan-shu* 元數 or whole numbers, (2) the *shou-shu* 收數 or decimal numbers,[72] (3) the *t'ung-shu* 通數 or fractions, (4) the *fu-shu* 復數 or multiples of 10^n. The *shou-shu*, *t'ung-shu*, and *fu-shu* can be reduced to *yüan-shu*.

SHOU-SHU

The rule says: "One has to see that the decimal fractions of the tail positions become numbers having only zeros [as tail position]," that is, that the decimal numbers be reduced to integral numbers. "When you got the *ting-shu*, apply the rule of the *yüan-shu*."

As there is no such example given, the rule is somewhat obscure. The most acceptable explanation seems to be this:

Suppose that $N \equiv 1 \pmod{1.25} \equiv 6 \pmod{1.7}$. Multiply in

[72] Needham (1), vol. 3, p. 86.

order to have whole numbers: $100N \equiv 100$ (mod 125)$\equiv 600$ (mod 170). We find that

$100N = 1,876$

$N = 18.76.$

The second method is: "Or, if you wish, set up the number that is the denominator. Reduce the decimal fractions of the *shou-shu*, and from this deduce the "problem numbers." Apply the rule of the *t'ung-shu* [fractions]."

Suppose that

$N \equiv 1$ (mod 1.25)$\equiv 6$ (mod 1.7).

This is equivalent to

$$N \equiv 1 \left(\text{mod } \frac{125}{100}\right) \equiv 6 \left(\text{mod } \frac{17}{10}\right).$$

T'UNG-SHU

"Set up the problem numbers. Find the common denominator, and bring in the numerators. Mutually multiply them. These are the *t'ung-shu*. Find the greatest common divisor; do not reduce one [of them], reduce all the other ones, and you get all the *yüan-fa-shu* 元法數.[73] Apply the rule of the *yüan-shu*." For example, in I, 2,

$365\frac{1}{4}$	$29\frac{499}{940}$	60
$\dfrac{1461}{4}$	$\dfrac{27,759}{940}$	$\dfrac{60}{1}$
$\dfrac{1,461 \times 940 \times 1}{940 \times 4 \times 1}$	$\dfrac{27,759 \times 4 \times 1}{940 \times 4 \times 1}$	$\dfrac{60 \times 940 \times 4}{940 \times 4 \times 1}$
$1,373,340$	$111,036$	$225,600$

G.C.D. $(1,373,340,\ 111,036,\ 225,600) = 12$

$114,445$	$9,253$	$225,600$

(These are the *yüan-fa-shu*).[74]

---

[73] The meaning of *yüan-fa-shu* 元法數 seems to be: "the numbers on which one can apply the rule of the *yüan-shu*."
[74] See Li Yen (6'), vol. 1, p. 131.

FU-SHU
"As for the numbers having a 'tail position' of 10 or more, find the greatest common divisor of all of them. Keep the one, and reduce the other."

For example, I, 6

300	240	180

G.C.D.=60

300	4	3

Note on Terminology

The correct explanation of the term *ta-yen* is given by Needham: "In the course of time indeterminate analysis acquired the name *ta-yen shu* 大衍術, derived from an obscure statement in the *I-ching*[75] that the 'Great Extension Number' is 50." And: "The reason of the adoption of this technical term is clear enough. In the classical method of consulting the oracle, which the *I-ching* (Book of Changes) describes . . . , one of the fifty stalks or rods is set aside before the forty-nine are divided into two random heaps symbolising the *Yin* and the *Yang*. It was very natural, therefore, that mathematicians, seeking for remainders of one by continued divisions, should have remembered this. . . ."[76]

The *wên-shu* 問數 are of course the numbers as given in the problem; the *ting-mu* 定母 (or *ting-shu*) are the "definite base-numbers" (Mikami),[77] that is, the numbers to which the *ta-yen* rule can appropriately be applied (the moduli reduced such

[75] Great Appendix, 1, ch. 9. See Wilhelm (1), vol. 2, p. 236 ("Die Zahl der Gesamtmenge ist 50"); Z.D. Sung (1), p. 291 ("Great Expansion").
[76] Needham (1), vol. 3, p. 119 and note *j*; Van Hee (12), p. 435, where he says: "One could translate as the 'Great Pulverizer.'" (On pourrait traduire 'Grand pulvérisateur'.) "Pulverizer" as a translation of the Indian term *kuṭṭaka* was proposed by Colebrooke (1). See Chapter 14. The Sanskrit word *kuṭṭaka*(agens) may be translated: "that which cuts,that which pulverizes; the cutter, the pulverizer," and is derived from the verbal root *kuṭṭ* (to cut, to pulverize). The Chinese word *yen*, however, means: "to spread out, to amplify, to extend," and cannot be a translation of the sanskrit *kuṭṭaka*.
[77] For Mikami's translation of the technical terminology, see (1), p. 65 f.

that they are relatively prime);[78] the *yen-shu* 衍數 are the numbers obtained by the first operation (Mikami therefore calls them "operation numbers"); the *ch'i-shu* 奇數 are the "residues" of the subtraction of n times the *ting-shu* from the *yen-shu*. The meaning *ta-yen ch'iu-i* 大衍求一 (*ch'iu-i* meaning "searching for unity") is obvious from the operation itself. *Ch'êng-lü* 乘率 means "multiplication factor." *Fan-yung* 泛用 or *fan-yung-shu* is hard to translate. *Fan* means "to float, to drift." It seems to have some such meaning as the "floating numbers" or "the numbers floating on the surface," the "numbers left," and may perhaps refer to the numbers left on the counting board. *Fan-yung-shu* could be translated as "the remaining use numbers."[79] Finally, the *tsung* 總 is the "general number" or the "chief number."

Theoretical Representation of Ch'in Chiu-shao's Method

WITH MODULI RELATIVELY PRIME

1. $N \equiv r_1 \pmod{m_1} \equiv r_2 \pmod{m_2} \equiv r_3 \pmod{m_3} \equiv \ldots \equiv r_n \pmod{m_n}$.

Find the value of N. A number k_z $(1 \leq z \leq n)$ is to be found such that

$$\frac{m_1 m_2 \ldots m_n}{m_z} k_z \equiv 0 \pmod{m_1} \equiv 0 \pmod{m_2} \equiv \ldots \equiv 1 \pmod{m_z} \equiv \ldots \equiv 0 \pmod{m_n}.$$

This is the same as solving the congruence

$$\frac{m_1 m_2 \ldots m_n}{m_z} k_z \equiv 1 \pmod{m_z},$$

from which we deduce

$$r_z \frac{m_1 m_2 \ldots m_n}{m_z} k_z \equiv r_z \pmod{m_z}.$$

---

[78] Wylie (1), p. 181, gives the literal, but somewhat strange translation "fixed parents." In the *Chiu-chang suan-shu* the term *mu* (mother) had a technical meaning in connection with fractions.

[79] I am not at all sure that this explanation makes sense.

2. Suppose that

$M = m_1 m_2 \ldots m_n.$

Then

$r_1 \dfrac{M}{m_1} k_1 \equiv r_1 \ (\mathrm{mod}\ m_1) \equiv 0 \ (\mathrm{mod}\ m_2) \equiv \ldots \qquad \equiv 0 \ (\mathrm{mod}\ m_n)$

$r_2 \dfrac{M}{m_2} k_2 \equiv 0 \ (\mathrm{mod}\ m_1) \equiv r_2 \ (\mathrm{mod}\ m_2) \equiv \ldots \qquad \equiv 0 \ (\mathrm{mod}\ m_n)$

$\vdots$

$r_z \dfrac{M}{m_z} k_z \equiv 0 \ (\mathrm{mod}\ m_1) \equiv 0 \ (\mathrm{mod}\ m_2) \equiv \ldots \equiv 0 \ (\mathrm{mod}\ m_z) \equiv \ldots$

$\ldots \equiv 0 \ (\mathrm{mod}\ m_n)$

$\vdots$

$r_n \dfrac{M}{m_n} k_n \equiv 0 \ (\mathrm{mod}\ m_1) \equiv 0 \ (\mathrm{mod}\ m_2) \equiv \ldots \equiv 0 \ (\mathrm{mod}\ m_z) \equiv \ldots$

$\ldots \equiv r_n \ (\mathrm{mod}\ m_n).$

The sum of these is

$\sum_1^n r_z \dfrac{M}{m_z} k_z \equiv r_1 \ (\mathrm{mod}\ m_1) \equiv r_2 \ (\mathrm{mod}\ m_2) \equiv \ldots \equiv r_z \ (\mathrm{mod}\ m_z) \equiv$

$\ldots \equiv r_n \ (\mathrm{mod}\ m_n);$

or

$\sum_1^n r_z \dfrac{M}{m_z} k_z \equiv r_z \ (\mathrm{mod}\ m_z).$

3. Suppose that a is the smallest solution of the problem; then the general solution is represented by the formula

$N = a + pM;\ p \geq 0.$

For, noting that

$a \equiv r_1 \ (\mathrm{mod}\ m_1) \equiv r_2 \ (\mathrm{mod}\ m_1) \equiv \ldots \equiv r_n \ (\mathrm{mod}\ m_n)$

$M \equiv 0 \ (\mathrm{mod}\ m_1) \equiv 0 \ (\mathrm{mod}\ m_2) \equiv \ldots \equiv 0 \ (\mathrm{mod}\ m_n),$

we have

$a + pM \equiv r_1 \ (\mathrm{mod}\ m_1) \equiv r_2 \ (\mathrm{mod}\ m_2) \equiv \ldots \equiv r_n \ (\mathrm{mod}\ m_n).$

If β is any solution of the problem, then $a = \beta - qM$ is the smallest solution (q as great as possible).

WITH MODULI NOT RELATIVELY PRIME[80]

The method given by Ch'in Chiu-shao is as follows:[81] If the moduli $m_1, m_2, m_3, \ldots, m_n$ are not relatively prime in pairs, expand each of them into its prime factors. Suppose that the least common multiple of $(m_1, m_2, \ldots, m_n) = 1^p \times 2^q \times 3^r \times 5^s \times \ldots$; find positive integers $\mu_1, \mu_2, \mu_3, \ldots, \mu_n$, being relatively prime in pairs, such that the least common multiple $(m_1, m_2, \ldots, m_n) = \mu_1 \mu_2 \ldots \mu_n$, and such that μ_i divides the corresponding m_1. Then: $N \equiv r_i \pmod{m_i}$ (with $i = 1, 2, 3, \ldots, n$) may be replaced by $N \equiv r_i \pmod{\mu_i}$, which can be solved by the first rule. The rule is valuable only on the general condition that $r_i - r_j$ is divisible by the greatest common divisor of (m_i, m_j). Several writers have given proofs of this case.[82]

SOLVING OF THE CONGRUENCES

Ch'in's method is the same as the modern method of continued fractions, which was explained earlier. However, Ch'in does not make use of negative solutions.

If $-a(a > 0)$ is the solution of $k_z a_z \equiv 1 \pmod{m_z}$, where

$$a_z = \frac{M}{m_z} - p m_z$$

giving $-a a_z \equiv 1 \pmod{m_z}$ Ch'in does the following addition:

---

[80] Theoretical explanations of the method are given by Matthiessen (1), (2), (3), (4), (5), (6); in Dickson (1); Van Hee (12); Needham (1), vol. 3; Mahler (1); Mikami (1); Ch'ien Pao-tsung (2′), (4′), (9′). (12′).

[81] We shall presently examine the conditions under which the solution is possible.

[82] See Lebèsgue (1), p. 56; Stieltjes (1), pp. 295 ff; and Mahler (1), who says: "This method is entirely different from that in Gauss's *Disquisitiones Arithmeticae*...." This statement seems not to be exact. Indeed, Gauss gives only the case where the moduli are relatively prime, and not the general case; however, this latter can be reduced to Gauss's rule (after reducing the moduli). Mahler adds: "I cannot remember finding it in Western books," but the first proof was given by Lebèsgue in 1859; another proof by Stieltjes in 1890.

$$-a\mathrm{a_z} \equiv 1 \ (\mathrm{mod\ m_z})$$
$$\mathrm{m_z a_z} \equiv 0 \ (\mathrm{mod\ m_z})$$
$$\overline{(\mathrm{m_z}-a)\mathrm{a_z} \equiv 1 \ (\mathrm{mod\ m_z})}$$

$m_z - a$ is positive, because $m_z > a$.

18

Ta-yen Rule and Kuṭṭaka

Many scholars are certain that the Chinese *ta-yen* rule is derived from the Indian *kuṭṭaka*. However, the arguments by which they support their statements are very unconvincing. In Europe Wylie was the first to state: "This appears to be the formula, or something very like it, which was known to the Hindoos under the name of *Cuttaca* [*kuṭṭaka*], or as it is translated 'Pulverizer,' implying unlimited multiplication, which is not so far from the meaning of the *ta-yen* or 'Great Extension.' "[1] Hankel says:[2] "If any proof is still necessary that a very close intercon-

[1] Wylie (1), p. 185 gives as his source the *Edinburgh Review*, November 1817. If the article in no. 57, November 1817–February 1818, vol. XXIX, pp. 141–164 was the only source for Wylie's statement, the comparison of the *ta-yen* rule and the *kuṭṭaka* allows one only to say: "something very like it," for there is no real proof of the similarity between the methods. It is likely that this is the reason Cantor says: "It is also a fact that this (*ta-yen*) method is absolutely different from the Indian pulverization (*kuṭṭaka*), with which [scholars] liked to compare before they understood it." (Es steht eben so fest, dass dieses Verfahren von der indischen Zerstäubung, mit welchem man es zu vergleichen liebte, bevor man es verstand, durchaus verschieden ist....) Cantor (2), vol. 1, p. 587. Biernatzki (1), p. 83 translates the statement of Wylie and adds: "From this it does not follow that the Chinese received their arithmetical researches ready-made from the Indians, or borrowed from them their elements of arithmetic, especially the Great Extension rule." (Daraus folgt aber nicht, dass die Chinesen ihre arithmetischen Forschungen von dem Hindus fertig übernommen, oder von ihnen Elemente der Arithmetik, insbesondere die grosse Erweiterungsregel, entlehnt haben.) This again is a statement without arguments.
[2] "Wenn es noch eines Beweises bedürfte, dass zwischen indischer und chinesischer Mathematik der engste Zusammenhang besteht, so ist die Regel *Ta jàn* (= Grosse Erweiterung), die schon im 3. Jahrhundert n. Chr. in dem *Suan-king* (= arithmetischer Klassiker) des Sun-tsè vorkommt, und im Anfange des 8. Jahrhunderts ausführlich behandelt wurde in dem *Ta jàn li shu* (= Sehr erweitertes Himmelzeichenbuch), dem Werke eines indischen [!] Buddhapriesters... Diese Regel *Ta jàn* ist aber nichts anderes, als die indische *kuṭṭaka*, von der hier ganz dieselben Anwendungen auf die Chronologie und die Berechnung gewisser Constellationen gemacht werden wie dort..." [(1), p. 407]. Hankel's opinion is a statement without arguments. It is so typical that I have taken the liberty of quoting him at length.

nection exists between Indian and Chinese mathematics, then
this can be found in the *ta-yen* rule (Great Extension), which
is mentioned in the third century A.D. in the *Suan-ching* (Arithe-
metical Classic) of Sun Tzŭ and which was discussed in detail
at the beginning of the eighth century in the *Ta-yen li-shu*
(Greatly Expanded Book of Heavenly Symbols), the work of an
Indian [sic] Buddhist monk . . . . This *ta-yen* is ultimately
nothing other than the Indian *kuṭṭaka,* which had precisely the
same applications to chronology and calculations of conjunctions." Matthiessen says that *ta-yen* and *kuṭṭaka* are two different
methods,[3] and he is the first to give a full explanation of both
methods. According to Matthiessen the Indian *kuṭṭaka* agrees
with the method of Bachet de Méziriac, whereas the *ta-yen* is
the same method as Gauss's congruences.[4] Matthiessen's comparison rests on solid grounds, because only an internal analysis
of both methods is able to yield a really scientific treatment of
the question. Matthiessen, however, did not know the Chinese
method of solving the congruences,[5] and this seems to be why
some scholars have considered the methods identical. The arguments in Van Hee (12) are extremely poor: "The foreign
influence manifests itself: (a) the numbers are written from
left to right, horizontally; (b) the *ta-yen*, or formula for solving
indeterminate problems, resembles the Indian *kuṭṭikara*."[6]
Tropfke (1) repeats Matthiessen's statement that *ta-yen* and
kuṭṭaka are distinct methods. Yushkevitch says: "The Indian
method is of course quite different from the Chinese one."[7]
According to Needham: "The Chinese *ta-yen* procedure was
similar to the *kuṭṭaka* . . . method in Indian mathematics

[3] Matthiessen (1).
[4] See Chapter 16, Appendix.
[5] Owing to the fact that Biernatzki did not translate this part of Wylie's
article; see chapter 16, Appendix.
[6] For the first argument, see Chapter 16, Appendix, note 62. Moreover,
in India there was also an inverse system of numerical notation in use; see
The Tjoe Tie (1) and Datta and Singh (1), pp. 21 ff. As for the second
argument, Van Hee gives only three types of equations from Mahâvirâ,
which prove nothing.
[7] "Allerdings ist die indische Methode von der chinesischen verschieden."
Yushkevitch (4), p. 145.

. . ." and "the argument of Matthiessen that they were very different does not carry conviction."[8] Hayashi must have been aware of this discrepancy: "Hankel . . . shows us that the *ta-yen* rule for the solution of indeterminate equations is the same as the *cuttaca d'hyana* of Indian mathematics, while Matthiessen . . . shows us that the *cuttaca* is the method of continued fractions and the *ta-yen* rule is the method of congruences of Gauss."[9]

This was all the information available in the works of European scholars. In India there are also some contributions to this problem, but it is a pity that their chief aim seems to be to prove the priority of Indian mathematics in this area. Ganguli (1) gives a description of the *ta-yen* rule and concludes: " . . . that this method is different from the Indian methods . . ."[10] Mazumdar treats the Chinese remainder problem, concludes that the congruences are solved only by inspection,[11] and states: "I absolutely fail to see how the Chinese method can stand in comparison with, or can be taken as the basis of, the elaborate process of the Indians."[12] S. N. Sen (1) gives the following arguments: (1) "Sun Tzŭ found the values of k_1, k_2, k_3[13] most probably by inspection as no process of finding them has been indicated. In simple cases inspection may be possible, but not when divisors are inconvenient numbers. Moreover, Sun Tzŭ gave an example with answer; Âryabhaṭa I, on the other hand, or more correctly, the school to which he belonged, gave correct and general rules for the solution of both linear and simultaneous indeterminate equations. The chronological argument of about a hundred years between the time of Sun Tzŭ and Âryabhaṭa I, to which some emphasis has been given, is hardly of any significance in view of the interest already referred to of the Vedic Hindus in indeter-

[8] Needham (1), vol. 3, p. 122 and note *c*.
[9] Hayashi (1), p. 310.
[10] Ganguli (1), p. 113.
[11] He is correct insofar as he limits himself to the Sun Tzŭ problem.
[12] Mazumdar (2), p. 11.
[13] That is, the solutions of the congruences.

minate problems." (2) The fact that I-hsing was a Buddhist astronomer is the strongest argument that the method must be derived from India.[14] His conclusion is: " . . . if there was any borrowing between China and India, it was not India but China at the receiving end."

For elucidating the relationship between *kuṭṭaka* and *ta-yen*, we must rely on the following principles:

(1) As the historical data are insufficient, it seems to be impossible to solve the puzzle by external evidence, and mere conjecture is unscientific.[15]

(2) It may be considered as a general rule that an algorithm borrowed from a foreign country is preserved more or less in its original state, and in many cases even the numbers are not changed.[16] This means that, if an algorithmic rule is transmitted, there may be changes in the problems themselves, but seldom in the method; and when two different cultures make use of different patterns to solve the same problem, it is entirely wrong to deduce the first method from the second, even if this is possible in our modern deductive systems.

(3) Before maintaining that one algorithm is deduced from another, it is absolutely necessary to make an internal analysis of both methods and to see if one of the methods implies the other one.

We have seen in Chapter 14 that the Indian method for solving the general problem

$$N \equiv r_1 (\mathrm{mod}\ m_1) \equiv r_2 (\mathrm{mod}\ m_2) \equiv r_3 (\mathrm{mod}\ m_3) \equiv \ \ldots$$
$$\mathrm{or}\ N = a_1 m_1 + r_1 = a_1 m_2 + r_2 = a_3 m_3 + r_3 = \ \ldots$$

was a method of successive substitution, giving the solution

$$N = a_1 a_2 a_3 a_4 v + a_1 a_2 a_3 \alpha_3 + a_1 a_2 \alpha_2 + a_1 \alpha_1 + r_1$$

[14] See, however, Mikami (1), p. 61: "The possibility of the Chinese mathematic having been influenced by the science of India may well be conjectured from the meager account here given [that is, about I-hsing]. As for exact information, we have none."

[15] Moreover, historical nationalism rises where historical science is unable to provide evidence, in default of historical data.

[16] The reason seems to be that an algorithm, as it is not part of a general deductive system, is not derived easily from another algorithm.

(restricting ourselves to four equations for the sake of convenience). The Chinese method is

$$N = r_1 m_2 m_3 m_4 k_1 + r_2 m_1 m_3 m_4 k_2 + r_3 m_1 m_2 m_4 k_3 + r_4 m_1 m_2 m_3 k_4.$$

The general method used by the Indians may be described as follows:[17]

We consider $ax + by = c$, in which we suppose a to be positive and in absolute value smaller than b.

If $a = 1$, then we have $x = c$; $y = 0$.

If $a > 1$, then we solve the equation in x. We find:

$$x = \frac{c - by}{a}.$$

We divide b by a and take the quotient such that $r > 0$. Then we have

$$x = \frac{c - (aq + r)y}{a} = -qy + \frac{c - ry}{a}.$$

In order that to an integer for y there belongs an integer for x, it is necessary and sufficient that $c - ry$ is a multiple of a. Suppose that $c - ry = at$. We have to find integers for y and t which satisfy

$$ry + at = c. \tag{1}$$

If $r = 1$, then $t = 0$; $y = c$; $x = -qc$. If $r > 1$, then we solve (1) in y, because $0 < r < a$, and so on.

The coefficients of each equation found in this way are two successive remainders obtained when searching for the greatest common divisor of a and b. These numbers are relatively prime; consequently we finally find an equation

$$u + kv = c, \tag{2}$$

which satisfies the relations $v = 0$ and $u = c$. Substitution for u

---

[17] The three usual methods for solving indeterminate equations of the first degree are: (1) the arithmetical method (a kind of solution by inspection); (2) the method by continued fractions; (3) the method by continued substitutions. This last is the Indian method.

and v by the values found in the equation preceding (2), and further substitution, give finally the values of x and y.

It is obvious that in principle the method of Āryabhaṭa is the same, although there are some differences in detail.[18]

The method of continued fractions relies on the following property:

If p/p' and q/q' are two successive approximate fractions, the difference $pq'-qp'$ equals $+1$ or -1 according to whether the second approximate fraction is on an odd or an even rank. If a/b is the given fraction (and thus the last approximate fraction) and p/q is the penultimate approximate fraction, the following rules hold:

1. If the coefficients a and b of the equation $ax+by=c$ have the same sign, we can expect them to be positive.

(a) If the number of approximate fractions is odd, then

$bp-aq=1$
$bpc-aqc=c$
$a(-qc)+b(pc)=c,$
from which $x=-qc$ and $y=pc$.

(b) If the number of approximate fractions is even, then

$bp-aq=-1$
$bpc-aqc=-c$
$a(qc)+b(-pc)=c,$
from which $x=qc$ and $y=-pc$.

2. If the coefficients a and b have different signs, we take a as positive and b as negative, and the equation is $ax-by=c$. In the same way we find

(a) If the number of approximate fractions is odd: $x=-qc$; $y=-pc$.

(b) If this number is even: $x=qc$; $y=pc$.

This method is applied in the Chinese *ta-yen* rule. As in $aN_1\equiv1 \pmod{\mu_1}$, if all numbers are positive, and the coefficients always have different signs, only method 2 is applied.

---

[18] The same interpretation is given by Yushkevitch (4), p. 145.

We have seen that 2(a) was reduced to 2(b) (see p. 357). The Chinese solved the equation:

$r_1 a N_1 \equiv r_1 \pmod{\mu_1}$.

By means of continued fractions they found $q=a$, and $x=qc=ar_1$ (the *tsung*). This is of course not the *ta-yen* rule, but only the procedure for solving the congruences of the several equations.

There does not seem to be any relationship between the general methods of the two systems. The Chinese method was used in Europe by Beveridge (1669), Euler (1740), and Gauss (1801). The Indian method was developed by Bachet de Méziriac (1612); we find it also in the work of Lagrange (1769). Another important problem is the application of the Euclidean algorithm. It is true that both the *ta-yen* and the *kuṭṭaka* make use of it. At the end of his discussion on Âryabhaṭa's rule, Rodet writes: "I do not know by what means there became widespread among historians of mathematics the belief that the Indians solved the problem before us by means of continued fractions . . . .' "[19] On the other hand, several others agree that what they used was continued fractions.[20]

The Indian method has been explained in Chapter 14. Let us now compare it with the Chinese method. For example, let us take the equation $100x+1=63y$, or $63y\equiv1 \pmod{100}$. In both procedures we start from the Euclidian algorithm

$$\frac{100}{63} = (1, 1, 1, 2, 2, 1, 3).$$

By the method of continued fractions we get the approximate fractions

$$\frac{1}{1}; \; 1+\frac{1}{1} = \frac{2}{1}; \frac{2\times1+1}{1\times1+1} = \frac{3}{2}; \frac{3\times2+2}{2\times2+1} = \frac{8}{5}$$

$$\frac{8\times2+3}{5\times2+2} = \frac{19}{12}; \frac{19\times1+8}{12\times1+5} = \frac{27}{17}$$

as $c=1$, $x=17$, and $y=27$.

[19] Rodet (1), p. 434.
[20] Smith (1), vol. 1, p. 156 (referring to Kaye); Hofmann (1), vol. 1, p. 60; Yushkevitch (1), pp. 141 f; Cantor (2), vol. 1, p. 590.

Ch'in's method

$$
\begin{array}{lll}
Q_1 & 1 & \\
Q_2 & 1 \;\big|\; 1 & =1\times1 \\
Q_3 & 1 \;\big|\; 2 & =1\times1+1 \\
Q_4 & 2 \;\big|\; 3 & =2\times1+1 \\
Q_5 & 2 \;\big|\; 8 & =3\times2+2 \\
Q_6 & 1 \;\big|\; 19 & =8\times2+3 \\
& \downarrow 27 & =19\times1+8
\end{array}
$$

Indian method

$$
\begin{array}{lll}
Q_1 & 1 \uparrow & 27=1\times17+10 \\
Q_2 & 1 & 17=1\times10+\ 7 \\
Q_3 & 1 & 10=1\times\ 7+\ 3 \\
Q_4 & 2 & 7=2\times\ 3+\ 1 \\
Q_5 & 2 & 3=2\times\ 1+\ 1 \\
Q_6 & 1 & 1=1\times\ 1+\ 0 \\
& 3 & 1=3\times\ 0+\ 1 \\
(r_5=Q_7) & & \\
t & 0 & (0) \\
C & 1 &
\end{array}
$$

In general:

$$
\begin{array}{ll}
Q_1 & a_0=1 \\
Q_2 & a_1=Q_1\times1 \\
Q_3 & a_2=Q_2 a_1+1 \\
Q_4 & a_3=Q_3 a_2+1 \\
Q_5 & a_4=Q_4 a_3+2 \\
Q_6 & a_5=Q_5 a_4+3 \\
\;\bullet & a_6=Q_6 a_5+4
\end{array}
$$

$$
\begin{array}{ll}
Q_1 & x =Q_1 y +x_1 \\
Q_2 & y =Q_2 x_1+y_1 \\
Q_3 & x_1=Q_3 y_1+x_2 \\
Q_4 & y_1=Q_4 x_2+y_2 \\
Q_5 & x_2=Q_5 y_2+x_3 \\
Q_6 & y_2=Q_6 x_3+y_3 \ (t) \\
\;r_5 & x_3=r_5 t +C \\
t & (y_3=t) \\
(C) &
\end{array}
$$

From this comparison it is obvious that the Indian method (1) is not the same as the Chinese method, and (2) has nothing to do with continued fractions. It is a method of continued substitutions, going over the values of x and y. In Ch'in's method we can find both values of x and y independent from each other. The only point of resemblance is the so-called Euclidean algorithm. By considering these methods identical one should make the same mistake as Kaye in his *Notes on Indian Mathematics* (1908), where he tries to derive the *kuṭṭaka* from Euclid, merely relying on this algorithm.

If our internal analysis holds, it makes no sense to accept the idea of a historical relationship between the Chinese *ta-yen* rule and the Indian *kuṭṭaka*.

19

Ch'in's Ta-yen Rule and Calendar Making

In his introduction, Ch'in Chiu-shao says: "Only the *ta-yen* method is not contained in the *Chiu-chang*, for no one has yet been able to derive it [from other procedures]. The calendar makers have made considerable use of it in working out their methods." From this it is clear that the source of indeterminate analysis is to be found in chronological problems. In the stanzas at the end of the preface he says again: "The calendar makers, although making use of it, use it without understanding 歷家雖用而不知." In the same preface Ch'in mentions the *Ta-yen* 大衍 calendar and the *Huang-chi* 皇極 calendar.[1] We may admit that in the *Ta-yen* calendar there was some use of the *ta-yen* rule; Ch'ien Pao-tsung[2] believes that the *Huang-chi* calendar also made use of this rule; although we cannot prove this, it is surprising that Ch'in mentions both of these calendars.[3] Chang Tun-jên 張敦仁 (1) says in his preface that the *ta-yen* method was used from the *Lin-te* calendar (665) on.[4]

The general problem in drawing up a calendar is the incommensurability of the sun's revolutions (day and year) and the lunar month. If the day has a duration D, the month a duration L, and the year a duration Y, the general problem is as follows: suppose that at a certain moment there have passed a total amount of days plus a remainder r_1, a total amount of months

[1] The *ta-yen* calendrical method was worked out by I-hsing in A.D. 727; "...though it did not have much to do with indeterminate analysis, it was known as the *ta-yen li* 大衍歷" [Needham (1), vol. 3, p. 37]. See, however, Needham, ibid., p. 120 and Wang Ling (4). There is more information in Yabuuchi (3), p. 456 and Hartner (1). For further information on I-hsing, see Chapter 15. The *Huang-chi* calendar was compiled by Liu Ch'o 劉焯 in about A.D. 600. It was modified in the T'ang dynasty by Li Shun-fêng 李淳風 (602–670) into the *Lin-te* 麟德 calendar.
[2] Ch'ien Pao-tsung (4'), p. 47.
[3] It is possible that Ch'in Chiu-shao connected this calendrical problem with the indeterminate problem of Sun Tzŭ, but Ch'in does not mention the Sun Tzŭ problem, even in his preface, whereas the greater part of the mathematicians of his time mention only this problem, for example, Yang Hui, Chou Mi and later Ch'êng Ta-wei.
[4] See also Wang Ling (4).

plus a remainder r_2, a total amount of years with a remainder r_3; there is to be found a certain moment when $r_1 = r_2 = r_3 = 0$, or:

$$N \equiv r_1 \pmod{D} \equiv r_2 \pmod{L} \equiv r_3 \pmod{Y}.$$

If an artificial period such as a week or a sexagenary period S is used, the number of days may be replaced by this period, or $N \equiv r_1 \pmod{S} \equiv r_2 \pmod{L} \equiv r_3 \pmod{Y}$. This was the usual Chinese representation of this kind of chronological problem. Needham[5] states: "The whole history of calendar-making . . . is that of successive attempts to reconcile the irreconcilable, and the numberless systems of intercalated months . . . and the like, are thus of minor scientific interest." This is true, but that does not alter the fact that it was a serious mathematical problem. The calendrical science consists in finding a longer period which is a multiple of both the month and the year.[6]

The computation of the length of this period is of course very simple, and depends on the accuracy of the measuring of the length of both the solar and lunar revolutions.[7] The problem is much more difficult if we wish to compute back from a given moment x to the point of conjunction of the various periods. This was the problem Chinese mathematicians tried to solve.

Ch'in Chiu-shao does not seem to have been very interested in calendrical problems as such, because in his examples he makes use of old calendrical systems.

In I, 2 he computes the conjunctions according to the Ku-li 古歷, the Old Calendar,[8] with these very inaccurate values: a tropical year $= 365 \frac{1}{4}$ days and a synodic month $= 29 \frac{499}{940}$ days. In II, 1 he uses the same year length. In II, 2 and 3 he ap-

[5] (1), vol. 3, p. 390.
[6] And if there is an artificial period such as the sexagenary cycle, of this also. See Pannekoek (1), p. 17. There is a very clear explanation in Sivin (3), pp. 12 ff.
[7] The modern method for finding a longer period is that of continued fractions. See Pannekoek (1), p. 17.
[8] This Old Calendar is the Ssŭ-fên calendar used during the later Han period. In fact this calendar originated before the Han dynasty. See Yabuuchi (3), p. 452; Maspero (1), p. 236.

plies the data of the *K'ai-hsi* 開禧 calendrical system in use in his own time,[9] where the year and month values are derived from the system of Ho Ch'êng-t'ien.[10] For details about these problems the reader is referred to Chapter 21.

It seems to be justified to conclude from these very poor data that Ch'in has derived the inspiration for elaborating his *ta-yen* rule from chronological problems.

However, there is another argument one can derive from his own preface, where he says: "In my youth I was living in the capital, so that I was enabled to study in the Board of Astronomy."[11] One of the tasks of the Board of Astronomy was to draw up the calendar. It is very likely that he learned there the application of the *ta-yen* rule.

It is a fact that the only access to the calendrical *ta-yen* method is Ch'in's work. Neither before him nor after him is there any other information, except some poor data about I-hsing.[12]

[9] With the length of the year = 365 79/325 and the length of the month = 28 8,967/16,900 days. See Ch'ien Pao-tsung (7'), p. 130.

[10] See Chapter 15. See also Ch'ien Pao-tsung (7'), p. 130, and (2'), p. 211; Li Jui (1'), pp. 2a and following pages.

[11] The *t'ai-shih* 太史. See Chapters 3 and 4. See also Nakayama (1), pp. 14 f.

[12] There is a short paper by Mikami (3'), pp. 69–71 on the relation between *ta-yen* and calendar, but it does not give much elucidation. There is more information in Li Yen (6'), pp. 154 ff, where some of Ch'in Chiu-shao's and Chang Tun-jên's chronological problems are included.

20

Ch'in's Ta-yen
Rule and Modern Mathematics

In Europe, the first proof of the *ta-yen* rule was given by Beveridge in 1669 for moduli that are relatively prime;[1] his rule is the same as the one that Euler and Gauss used later.[2] The influence of Beveridge's work does not seem to have been very great.[3] J. Wallis (1616–1703) of Oxford University set down a solution for calendrical problems (*De juliana periodo, disquisitio chronologica*),[4] but his method was empirical and he obtained his solutions "in a laborious manner."[5] Euler (1707–1783) solved the remainder problem in his "Solutio problematis arithmetici de inveniendo numero qui per datos numeros divisus relinquat data residua."[6] In the last paragraph[7] Euler says that the following method is much shorter and easier:

"A number is to be found that, when divided by *a, b, c, d, e,* which numbers I suppose to be relatively prime, leaves respectively the remainders *p, q, r, s, t.* For this problem the following numbers satisfy:

$$Ap + Bq + Cr + Ds + Et + m \times abcde$$

in which A is a number that divided by *bcde* has no remainder, by *a,* however, has the remainder 1; B is a number that divided by *acde* has no remainder, by *b,* however, has the remainder 1 . . . which numbers can consequently be found by the rule

[1] See Chapter 14.
[2] See Dickson (1), vol. 2, p. 61.
[3] He is not mentioned in Smith (1), Hofmann (1) or Tropfke (1).
[4] Chapter 54, pp. 450–455.
[5] Tropfke (1), p. 110. One might compare Wallis's method with that of Ch'in Chiu-shao, who solved such problems in a simple way.
[6] "Solution of the arithmetical problem about the finding of a number which leaves given remainders when divided by given numbers." First published in *Commentarii academiae scientiarum Petropolitanae*, 7 (1734/5), 1740, pp. 46–66; see his *Algebra*, vol. 2, Chapter 1, pp. 213 ff; *Commentationes arithmeticae collectae I*, pp. 11–20. The edition of 1915 is cited here.
[7] Euler (1), par. 29, p. 32.

given for two divisors."[8]

This last phrase refers to the solving of the congruences

$A \equiv 1 \pmod{a} \equiv 0 \pmod{bcde}$.

To find the congruence factor, Euler used the process of the greatest common divisor (the Euclidean algorithm) (par. 7).[9]

The methods used by N. Saunderson[10] and J. L. Lagrange[11] have nothing to do with the *ta-yen* rule and are of the same nature as the Indian *kuṭṭaka*.[12] Lagrange and Saunderson also treat the general problem in which the moduli are not relatively prime in pairs, and both give the condition: if $N \equiv r_1 \pmod{a} \equiv r_2 \pmod{b}$, then $r_1 - r_2$ must be divisible by the greatest common divisor of a and b.

The case in which the moduli are relatively prime is treated in full by Gauss (1777–1855) in *Disquisitiones arithmeticae* (1801).[13] The *ta-yen* rule was applied for the first time to the case in which the moduli are not relatively prime in pairs by Lesbègue (1859).[14] T. J. Stieltjes published a profound treatment in 1890.[15] K. Mahler (1) published a very interesting

[8] "Inveniendus sit numerus, qui per divisores *a, b, c, d, e,* quos numeros inter se primos esse pono, divisus relinquat respective haec residua *p, q, r, s, t.* Huic quaestioni satisfacit iste numerus
Ap + Bq + Cr + Ds + Et + m abcde,
in qua expressione *A* est numerus, qui per factum *bcde* divisus nihil relinquat, per *a* vero divisus relinquat unitatem; *B* est numerus, qui per *acde* divisus relinquat nihil, per *b* vero unitatem...; qui ergo numeri per regulam pro duobus divisoribus datam inveniri possunt."
[9] Euler's general method is in essence the same as the one used by M. Rolle (1652–1719), in his *Traité d'algèbre,* vol. 1, chap. 7, pp. 69–77.
[10] *The Elements of Algebra* (Cambridge, 1740). I used the French translation *Elémens d'algèbre* (1756), where his method is explained in part 1, pp. 314–327. There is a short description of it in Dickson (1), vol. 2, pp. 62 f.
[11] In *Oeuvres,* vol. 2, pp. 388 ff. Lagrange's method is similar to Bachet's (see Chapter 14), except for the fact that the latter uses continued fractions.
[12] However, for solving the congruences, the Indian mathematicians make use of a substitution method and not of continued fractions (see Chapter 18).
[13] I used *Carl Friedrich Gauss, Werke,* vol. 1 (Göttingen, 1863); there is an English translation of the chapter on indeterminate analysis of the first degree, in Smith (5), pp. 107 ff.
[14] Lebèsgue (1), pp. 54–58.
[15] First published in *Annales de la Faculté des Sciences de Toulouse,* 4, 1890, pp. 1–103, and reprinted in his *Oeuvres complètes,* vol. 2 (1918), pp. 265–378.

article on the Chinese remainder problem, where he stated: "This method is entirely different from that in Gauss's *Disquisitiones Arithmeticae,* and I cannot remember finding it in Western books."[16] The best article in Chinese is by Hsü Chênch'ih [(1) (1925)].

We shall presently compare Ch'in Chiu-shao's method to modern mathematical methods. In order to do so, we have to keep certain principles in mind. First of all, let us not forget that Ch'in's work appeared in 1247 and that the other material to be considered is all of fairly recent date. From a logical point of view such a comparison is impossible unless it is clearly defined. The only valuable comparison would seem to be an operational one, that is, a study of the methods whereby the mathematician goes from the data to the correct solution. It is true that Ch'in does not give proofs, that he does not relate the *ta-yen* rule to a general theory of numbers, and that his method is somewhat long-winded, but he proceeds from the data to the solution by a sure and general method. If we take this operational point of view, his method is more general than those used by Euler and Gauss. This comparison will be based on Stieltjes's text.

(1) "One can reduce the general case[17] to the case where A, $B, \ldots, L$ (the moduli) are relatively prime. For that purpose, change the least common multiple M of the moduli into

$$M = A' \times B' \times C' \times \ldots \times L',$$

where A', B', C', $\ldots$, L' are relatively prime and divide respectively $A, B, \ldots, L$.[18] It is clear that the solutions of the problem in question satisfy also the congruences

$$x \equiv a \pmod{A'}, \ x \equiv \beta \pmod{B'}, \ldots, x \equiv \lambda \pmod{L'}."[19]$$

Stieltjes gives the following example:

[16] In my opinion Ch'in Chiu-shao's method is more general than that of Gauss; but I do not consider them different in principle.

[17] That is, the case where the moduli are not relatively prime in pairs.

[18] Of course, Stieltjes gives full evidence for his method; the explanation of the reduction is on pp. 280 ff.

[19] Stieltjes (1), pp. 299 f.

$x \equiv 31 (\text{mod } 72 = 2^3 \times 3^2)$ $x \equiv 31 \ (\text{mod } 72)$
$x \equiv 22 \ (\text{mod } 105 = 3 \times 5 \times 7)$ $x \equiv 22 \ (\text{mod } 35)$
$x \equiv 50 \ (\text{mod } 77 = 7 \times 11)$ $x \equiv 50 \equiv 6 \ (\text{mod } 11)$
$x \equiv 337 \ (\text{mod } 399 = 3 \times 7 \times 19)$ $x \equiv 337 \equiv 4 \ (\text{mod } 19)$

Ch'in Chiu-shao's method is the same, but his way of reducing the moduli is overly intricate. However, he reaches his goal in all the problems he attempts.[20] On the other hand, he does not reduce the remainders.

(2) "Theorem 3. In order that the system of congruences

$x \equiv a \ (\text{mod } A), x \equiv \beta \ (\text{mod } B), \ldots, x \equiv \lambda \ (\text{mod } L)$

have solutions, it is necessary and sufficient that the differences

$a - \beta, a - \gamma, \beta - \gamma, \ldots, \varkappa - \lambda$

be divisible respectively by

$(A, B), (A, C), (B, C), \ldots, (K, L)$."[21]

"Here we have $M = 2^3 \times 3^2 \times 5 \times 7 \times 11 \times 19 = 525,680$ and $ABCD : M = 441$. Thus, if the remainders 31, 22, 50, 337 had been taken at random, there would be only one chance in 441 that the problem was possible. Thus it is advisable first to make sure whether the problem is possible or not. The numbers 9, 19, 306, 28, 315, 287 being divisible respectively by 3, 1, 3, 7, 21, 7, the problem is possible."[22]

In Ch'in's work, "no such necessary condition is mentioned in the Chinese text, but it is satisfied in the example which Ch'in Chiu-shao gives."[23] This is quite true, but can we answer the question of whether Ch'in knew this condition?

In I, 8 the product of the moduli is 25,740,000,000,000 and the least common multiple is 85,800; if Ch'in did not know the condition, he had one chance in 300,000,000 of finding a

[20] Chang Tun-jên, however, did not fully understand the method (see Chapter 16); Huang Tsung-hsien used the correct technique.
[21] Stieltjes (1), p. 298. A full proof of this theorem is given; another proof is in Mahler (1).
[22] Stieltjes (1), p. 300.
[23] Needham (1), vol. 3, p. 121.

solution; in I, 3 the probability is 1/162; in I, 4, 1/144; in I, 6, 1/3,600; in I, 7, 1/2,500. From this it is obvious that Ch'in was aware of this condition. On the other hand there is one problem (I, 2) in which he violates this condition (see p. 394).

(3) "Applying now Gauss's method, the auxiliary numbers $a', \beta', \gamma', \delta'$ are determined by the congruences

$$35 \times 11 \times 19a' = 43a' \equiv 1 \pmod{72}$$
$$72 \times 11 \times 19\beta' = -2\beta' \equiv 1 \pmod{35}$$
$$72 \times 35 \times 19\gamma' = 8\gamma' \equiv 1 \pmod{11}$$
$$72 \times 35 \times 11\delta' = -\delta' \equiv 1 \pmod{19}$$

from which $a' = -5, \beta' = 17, \gamma' = 7, \delta' = -1$." (p. 103)

Gauss gives his method in *Disquisitiones*, paragraph 36. In paragraph 28, the methods for solving the congruences from Euler and Lagrange are mentioned. Ch'in Chiu-shao's method agrees with that in Lagrange (1), vol. 2, p. 386, paragraph 7: "To find the numbers p_1 and q_1 which can satisfy the equation

$$pq_1 - qp_1 = \pm 1$$

develop the fraction p/q into a continued fraction, from which you deduce a series of fractions converging toward p/q and alternately greater or smaller than this fraction. . . . Take for p_1 the numerator of the fraction immediately preceding the fraction p/q, and for q_1 the denominator of the same fraction; if the fraction p_1/q_1 is smaller than the fraction p/q, you will have $pq - qp = 1$, and if $p_1/q_1 > p/q$, you will have $pq_1 - qp_1 = -1$."

Ch'in Chiu-shao's procedure agrees with this rule. There are a few points of difference; (1) The *ch'i-shu*[24] in Ch'in's work are always positive. For example, instead of $72 \times 11 \times 19\beta' = -2\beta'$, Ch'in would write: $= 33\beta'$. (2) The solutions of the congruences are always positive.[25]

The formula for getting these solutions is the same one, but Ch'in takes into account only the smallest solution. From the pragmatic point of view common to Chinese mathematicians, this is quite normal; in their eyes the other solutions are only unrealistic and useless.

21

General Evaluation of Ch'in Chiu-shao's Ta-yen Rule

The purpose of including a general history of indeterminate analysis of the first degree was to come to a logical evaluation of Ch'in's method. A historical comparison always depends on the historical sources at our disposal, and it is easy to forget that, although European mathematicians include such great men as Euler, Lagrange, and Gauss, the only contemporary of Ch'in Chiu-shao was Leonardo Pisano, who did not know the general rule for solving the remainder problem. In order not to fall a victim to a kind of historical journalism, it is necessary to set up a comparative method based on a clear and strict logical model. Petrucci's basic principle for evaluating Chinese mathematics has already been quoted: "Above all it is necessary to take *a historical viewpoint* and not allow oneself to be blinded by a modern European education in mathematics."

That Chinese mathematics is not general is a well-known objection. But we must not forget that there was no general algebra anywhere in the world before modern times.[1] To quote Petrucci again, "It seems to me unfair to reproach Chinese mathematicians for their lack of a generalizing faculty, because it is necessary to realize that we are presented with a collection of problems, and we must not forget that the knowledge of a general law is implied by the collection alone." In my opinion the ten indeterminate problems that Ch'in solves in the same algebraic pattern suffice to prove that he knows the general rule. That Chinese algebra was not deductive is entirely another matter, for an algebraic rule can be general without being deductive. Moreover, even the algebra of great mathematicians as Euler, Lagrange, Gauss, and others was only "logically deductive" and not at all "axiomatically deductive":

[1] The only general branch of mathematics was Greek geometry. In Diophantine algebra there is not the slightest general algebraic rule; see Tropfke (1), p. 100.

they gave logical proofs of their theorems, but they did not build up algebra on a limited set of axioms.[2]

In the history of mathematics, however, the role of algorithms cannot be overestimated. "The specificity of mathematics is that the great role in it belongs to the calculi, which according to certain rules allow to solve various problems arising in the practice of man."[3] And "*The quality of any calculus is determined, first of all, by its effectiveness, to be more exact, by its algorithmicity.*"[4]

Rybnikov gives these requirements for a valuable algorithm: "(1) The algorithm should represent, as Marx, the 'stratagem of action': it should contain an aggregate of *clear, exact and simple directions,* showing, step by step, what are the operations to begin with and what is to be done at every step after the foregoing step has already been done.

"(2) This stratagem of action should have a *general character,* that is, concern not any one problem, taken separately, but the whole class of homogeneous problems.

"(3) The algorithm should *solve the problems of this class,* that is to give answer to the question after the finite number of operations beginning from the initial data are fulfilled."[5]

It is obvious that Ch'in's *ta-yen* rule satisfies these three conditions.

Another point raised against Chinese mathematicians is that they do not prove their theorems. This is quite true if proof is taken to mean only deductive proofs. However, from the modern logical point of view there are two kinds of demonstra-

---

[2] It was only after the development of formal logic that mathematicians such as Hilbert and Russell were able to investigate the foundations of mathematics.

[3] Rybnikov (1), p. 162, who rightly adds: "However, the problem of clearing up the importance of these calculi and in general of the algorithmic methods is still far from being solved. Neither is this problem solved in the historical-mathematical aspect."

[4] It is interesting to read in the same article that even in the time of Descartes, "'method' was interpreted in a very simplified manner—as a universal solving algorithm, which has only to be strictly followed" (p. 143).

[5] Ibid., pp. 142 f.

tions: deduction and reduction. The general patterns are as follows:

Deduction	Reduction
If A is true, then B is true	If A is true, then B is true
A is true	B is true
Thus B is true	Thus A is true

The method of reduction is widely used in all sciences. On the lower levels of mathematics, when deduction was totally impossible, there was only the method of reduction. Bochenski states: "It is for example very well known that great mathematical discoveries were very often brought about in this way."[6] In the *ta-yen* algorithm of Ch'in Chiu-shao, several problems are given which are solved in the same pattern, and always with the correct solution, which can easily be controlled. If A represents the infinite set of problems which can be solved by the *ta-yen* rule, and B is the set of problems given in Ch'in's work, we can draw up the pattern of reduction: if A is right, then B is right; we state that B is right, thus A is right.

Of course this kind of logical proof does not suffice in our modern mathematical thought; but we should remember that the greater part of medieval mathematical methods are not algorithms but only ad hoc procedures, without general applicability (this is true for European as well as for Chinese mathematics).

Euclidean proofs are general, although Bochenski says that they were in fact also regressive. However, the most striking difference between Greek and Chinese mathematics is that the Chinese algebraists give no logical proof *for each step* in the solving of the problem; in other words, there is no logical chain. Their proofs require only (1) that the algorithm be generally applicable; (2) that if A is the set of data, and B is the set of solutions, one can by means of testing conclude that B satisfies A; or, logically speaking, that there is a bijection between A and B. But they do not analyze step by step the relation as given by the algorithm.

[6] Bochenski (1), p. 103.

Moreover, as it was not usual to include the logical background of a method, we are not well informed about the way these methods were built up; however, for difficult problems such as the remainder problem, it makes little sense to speak of "trial and error."

Modern mathematical proofs of theorems consist of the *analysis of the relation in a general way*. In the realm of *axiomatic deductive systems* modern mathematics goes far beyond the level even of eighteenth-century mathematics. Let us now try to build up a model that allows us to make a logical comparison. General principle: a solution A of a problem P ranks above a solution B of the same problem (1) if A and B are operationally comparable, and (2) if B is a subset of A. If we analyze the *ta-yen* rule, we find the following partial methods:

1. One problem with a solution ad hoc, without indication of the method
2. A separate problem, where the algorithm is restricted to special numbers
3. An algorithm restricted to a set of numbers
4. Proof of a special case
5. The general algorithm with moduli that are relatively prime, without solving of the congruences
6. Solving of the congruences
7. The general algorithm with moduli that are not relatively prime, without proof
8. The characteristics in item 7 with knowledge of the condition of solvability
9. The characteristics in item 5 with proof
10. The characteristics in item 9 with proof

The list is arranged so that each item ranks above and includes the one preceding it. Table 4 shows which items were known to the several mathematicians.

Before coming to a conclusion, we should once more remind ourselves that this comparison is made from an operational point of view and is restricted to the *ta-yen* rule. It has nothing to do with the general mathematical work of the authors.

Table 4. Methods of Solution for the Remainder Problem in Mathematical Works Surveyed

Mathematicians		Partial Methods									
		1	2	3	4	5	6	7	8	9	10
Sun Tzŭ	c. 400?	X	X								
Fibonacci	1202	X	X	X							
Ch'in Chiu-shao	1247	X	X	X	0	X	X	X	X*		
Yang Hui	1275	X	X	X							
Isaac Argyros	c. 1350	X	X								
Yen Kung	1372	X									
Munich MS	c. 1450	X	X	X	X	?					
Regiomontanus	c. 1460	X	?								
Göttingen MS	c. 1550	X	X	X	X	X	X†	X‡			
Ch'êng Ta-wei	1593	X	X								
Van Schooten	1657	X	X	X	0	X					
Beveridge	1669	X	X	X	X	X	0	0	0	X	
Euler	1743	X	X	X	X	X	X	0	0	X	
Gauss	1801	X	X	X	X	X	X	0	0	X	
Stieltjes	1890	X	X	X	X	X	X	X	X	X	X

Note: X: known; 0: not known.
*Not explicitly stated.
†Restricted to small numbers.
‡Unsuccessful attempt.

From Table 4 we derive this ranking:[7]

1. Stieltjes (1890)
2. Euler/Gauss (1743/1801)
3. Ch'in Chiu-shao (1247)
4. Beveridge (1669)
5. Göttingen MS (c. 1550)
6. Van Schooten (1657)
7. Munich MS (c. 1450)

[7] This method may seem to be pedantic, but it is founded on a sure and logical base. The aim is not to extol Ch'in Chiu-shao, and we consider a demonstration of the rule as more valuable than the algorithm. The purpose of such a comparison is rather to advance beyond the unscientific allegations that seek to defend the glory of country and religion, without the slightest argument. One can attack the method used here, but it is a very simple logical truth that a comparison is a matter of relations, and thus a matter of structure. Explicitness about the bases of comparisons and evaluations is a matter of scientific honesty.

8. Fibonacci (1202)
9. Yang Hui (1275)
10. Sun Tzŭ (c. 400?)
11. Isaac Argyros (c. 1350)
12. Ch'êng Ta-wei (1593)
13. Yen Kung (1372)
(?) Regiomontanus (c. 1460)

If we take into account the early date of Ch'in's work within the field of indeterminate analysis, we can see that Sarton did not exaggerate when he called Ch'in Chiu-shao "one of the greatest mathematicians of his race, of his time, and indeed of all times."[8] We must also remember that Ch'in goes far beyond the other Chinese mathematicians and that he had to develop the *ta-yen* method in a very unsuitable mathematical language. Indeed, "while we recognize the great achievements of modern mathematics, we must not underestimate the great work accomplished by the pioneers."[9]

[8] (1), vol. 3, p. 626.
[9] Konantz (1), p. 310.

22

Indeterminate Problems in the Shu-shu chiu-chang

This chapter will consist of translations of the ten remainder problems that are solved in Ch'in's work. As these problems are worked out at great length, it is impossible to give a literal translation of all of them. However, for the sake of objectivity, an integral translation of Problem I, 4 is included.[1] For the others the reader will find only the translation of the problem as such, with the solution transcribed into modern mathematical symbolism.

Problem I, 4. To Compute the Amounts of the Treasuries

(Note: In this problem several technical terms are used; they are explained in Part VI, "Socioeconomic Information.") Question: "There are seven treasuries of district cities. The daily revenues[2] in full strings [of *kuan*] are the same. The full amount of the annual tax is now being collected.[3] Because recently ready money has been scarce,[4] each treasury is allowed to adjust the *chiu-hui* [old rate][5] to the local standard [literally, "the market-hundred"] of the place in question. Treasury *A* has a remainder of 10 coins; *D* and *G* have a remainder of 4 coins; *E* has a remainder of 6 coins. In the other treasuries there is no remainder. The local standard of treasury *A* is 12 *wên* [per 100]. [These standards] proportionally decrease by 1 coin up to *G*. Find the daily draft of all the treasuries in 'full strings' as originally collected, with the nominal value and [the daily draft] according to the 'old standard' as actually

---

[1] This problem has been chosen because it is very clear.
[2] The amounts paid in taxes.
[3] A *kuan* is a string of 1,000 coins.
[4] That is, there are not enough coins in circulation; see Chapter 23, section on "Money and Currency in the Southern Sung."
[5] Ibid.

collected; and also the part collected in a great and in a small month."[6]

Note: When the coins in circulation became scarce, it was permissible to diminish the number of cash in a string according to the local standard. For example, in treasury *A*, 12 coins represent a nominal value of 100 coins. However, at this moment of inflation the official value of the money in circulation was 77 pro 100.

We have to find (1) the daily draft at the new rate; (2) the daily draft before the inflation; (3) the nominal value, that is, the value before the diminishing of the strings to 77/100.

"Answer: the daily draft of all the treasuries in 'full strings' as originally collected amounts to 26 *kuan* 950 *wên*.[7]

"The nominal value is 35 *kuan wên*."[8]

Treasury *A*:

(1) 224 *kuan* 583 *wên* (old standard)
(2) 6,737 *kuan* 500 *wên* (old standard)
(3) 6,512 *kuan* 902 *wên* (old standard)
(1) = the daily draft according to the "old standard"
(2) = great month
(3) = small month

(For the other treasuries, only the first amount is given.)

B : 245 *kuan*	*C* : 269,500 *kuan*
D : 299,444 *kuan*	*E* : 336,806 *kuan*
F : 385 *kuan*	*G* : 449,166 *kuan*

Note: If we know that the daily draft in "full strings" is 26,950 *kuan*, we find that nominal value = $26,950 \times 100/77 = 35,000$ *kuan*.

Local value *A*: at a rate of 12/100,

$$\frac{26,950 \times 100}{12} = 224,583 \ kuan.$$

---

[6] Respectively a month of 30 days and a month of 29 days.
[7] See Chapter 23, section on "Money and Currency in the Southern Sung".
[8] The number of coins *(wên)* on 35 strings *(kuan)*.

"Method: (Same as the preceding.) We solve by the *ta-yen* rule. Draw up the local standard of treasury A. Subtract from it the proportional decreasing number. We get all the original standards of all the treasuries. Find their mutual common divisors. Reduce the odd numbers, do not reduce the even numbers,[9] and you get the *ting-mu*. Multiply by each other all the *ting-mu* and this is the *yen-mu*. Reduce the *yen-mu* by the *ting-mu,* and you get the *yen-shu*. Those of which the *yen-shu* is the same as the *yen-mu*, are to be omitted; they have no [*yen-shu*]. (These which have no *yen-shu* we take as belonging to the same class). From [the *yen-shu*] subtract as many times as possible the *ting-mu ;* the remainders are the *chi-shu*. Apply the *ta-yen ch'iu-i* to the *chi-shu* and the *ting-shu* to find the *ch'êng-lü*. Multiply the *yen-shu* by these [*ch'êng-lü*] and you get the *yung-shu* [or *fan-yung-shu*]. Of those that have no [*yen-shu*], find the greatest common denominator of the *yüan-shu* belonging to the same class [i.e., mutually divisible]. Reduce the *yen-mu* and the numbers you get are the *chieh-shu* [borrowed numbers]. Next draw up the remainders of the treasuries that have remaining cash. Multiply by the original *yung-shu*. Add up and you get the *tsung-shu*. Subtract therefrom the *yen-mu* as many times as possible. The remainder is the number of 'full strings' of the daily draft of all treasuries. Multiply them by the number of days of a great and of a small month.[10] These are the dividends. Reduce them by the original standard and you get the 'old standard' value.

"Solution: Draw up the local standard of A: 12. Continually decrease by 1. You get 11, being the standard of treasury B; 10, being the standard of treasury C; 9, being the standard of treasury D; 8, being the standard of treasury E; 7, being the standard of treasury F; 6, being the standard of treasury G. You get the original standards of all the treasuries.

"Find the mutually common divisors. After finishing the reduction you get $A=1$, $B=11$, $C=5$, $D=9$, $E=8$, $F=7$, $G=1$.

[9] That is, reduce so that the numbers (moduli) are relatively prime.
[10] See note 6.

These are the *ting-mu*. Put 1 before them as 'factor'[11] [see Diagram 70].

12	A		1	1
11	B		1	11
10	C		1	5
9	D		1	9
8	E		1	8
7	F		1	7
6	G		1	1
			L	R
Original Standards				

Diagram 70

"First multiply all the *ting-shu* by each other; you get 27,720: this is the *yen-mu*. Next cross-multiply all the *ting-mu* by the factor 1. You get $A=27,720$, $B=2,520$, $C=5,544$, $D=3,080$, $E=3,465$, $F=3,960$, $G=27,720$. These are the *yen-shu*.[12]

27,720	1	A		0	A	1
2,520	11	B		1	B	11
5,544	5	C		4	C	5
3,080	9	D		2	D	9
3,465	8	E		1	E	8
3,960	7	F		5	F	7
27,720	1	G		0	G	1
yen-shu	*ting-mu*			*chi-*	*ting-*	
L	R			*shu*	*mu*	
27,720						

Diagram 71   Diagram 72

"Next investigate the *yen-shu*. Those which have the same

11 *Tzŭ* 子 can mean the numerator of a fraction. But in general *tzŭ* and *mu* 母 are two factors of an operation such as multiplication or division. See Chapter 17, note 15. For an explanation of the reduction applied here, see Chapter 17.

12 For example,
$$\frac{1 \times 11 \times 5 \times 9 \times 8 \times 7 \times 1}{11} = \textit{yen-shu B.}$$

yen-mu,[13] omit all of them. They are those which have no *yen-shu*. Next subtract as many times as possible the *ting-mu* from the original *yen-shu*, and you get the *chi-shu* [Diagram 71]. *A* has no [*chi-shu*], *B*=1, *C*=4, *D*=2, *E*=1, *F*=5, *G* has none [Diagram 72]. These are the *chi-shu*.

"Next investigate these which have the *chi-shu* 1, and consider 1 as *ch'êng-lü*. Some have [as *chi-shu*] the number 2 or more; use them as *chi-shu* in the right column. Put the *ting-mu* in the upper right space and the *t'ien-yüan* 1 in the upper left space.[14] Apply the rule *ta-yen ch'iu-i*. Examine by multiplication and division until the remainder is 1 and then stop [the operation]. Consider the result in the left upper space as the *ch'êng-lü*. *A* has none, *B*=1, *C*=4, *D*=5, *E*=1, *F*=3, *G* has none [Diagram 73]. These are the *ch'êng-lü*. Draw them up in the right column, opposite to the *chi-shu* in the left column.[15]

0	*A*	0
2,520	*B*	1
5,544	*C*	4
3,080	*D*	5
3,465	*E*	1
3,960	*F*	3
0	*G*	0
yen-shu	*ch'êng-lü*	

Diagram 73

0	*A*
2,520	*B*
22,176	*C*
15,400	*D*
3,465	*E*
11,880	*F*
0	*G*
yung-shu	

Diagram 74

"Multiply the opposite numbers of both columns. These are the *yung-shu*. The number in *A* has no *yung-shu*, *B*=2,520, *C*=22,176, *D*=15,400, *E*=3,465, *F*=11,880, *G* has none [Diagram 74].

"Next, in order to examine those that have no *yung-shu*—only *A* and *G* coincide in the same class—we arrange to borrow

[13] That is, where *yen-mu* and *yen-shu* are the same, or when the *ting-mu* = 1.
[14] The representation is

[15] The text says *tso-yen* 左衍; this should be *tso-hang* 左行.

them. We draw up the so-called *yüan-shu* and examine them [Diagram 75].

12	A	8	E
11	B	7	F
10	C	6	G
9	D	*yüan-shu*	

Diagram 75

"Now we examine $A=12$, $G=6$, $C=10$, and $E=8$; all are even; they are of the same class. The *yung-shu* of E is 3,465. This number is too small. We cannot borrow from it. Only the *yung-shu* of $C=10$, being 22,176, is the largest. Therefore we have to borrow from it. Then find the greatest common divisor of 12, 10, and 6; you get 2. By the greatest common divisor 2 reduce the *yen-mu* 27,720; you get 13,860; this is the *chieh-shu* [borrowed number]. Then subtract it from the *yung-shu* $C=$ 22,176; the remainder is 8,316; it is the *yung-shu* C. Then the number 13,860, which is lent, is the dividend. We consider the original greatest common divisor 2 as divisor, and divide by it. We get 6,930. This is the *yung-shu* A. We subtract the *yung-shu* A from the lent number [13,860]; the remainder is also 6,930; this is the *yung-shu* G. If you do not wish to make the *chieh-shu* of A and G, it is the same. We can make use of some [other] reductions if we wish to. Take 1/3 (from 13,860); you get 4,620, being the *yung-shu* A; take 2/3, you get 9,240, being the *yung-shu* G. Draw them up in the right column.

10	A	4,620		A	46,200
0	B	2,520		B	0
0	C	8,316		C	0
4	D	15,400		D	61,600
6	E	3,465		E	20,790
0	F	11,880		F	0
4	G	9,240		G	36,960
ting-shu		*yung-shu*			*tsung-shu*

Diagram 76   Diagram 77

"Then examine all the treasuries that have no remainder.

We get the three treasuries *B*, *C*, and *F* without remainder. First we omit their *yung-shu*. Then draw up the remainders of the four treasuries *A*, *D*, *E*, and *G* in the left column. Multiply them respectively by their own *yung-shu*. Then $A=46,200$; $D=61,600$; $E=20,790$; $G=36,960$. These are the *tsung-shu*, [Diagrams 76, 77].

"Add up these four *tsung-shu;* you get 165,550. As many times as possible subtract the *yen-mu* 27,720. The remainder is 26,950. It is the rate searched for. Reduced to *kuan* [1,000] it is 26 *kuan* 950 *wên*. It is the equal number of the daily drafts of all the treasuries . . . ."[16]

Problem I, 1

Question: The *I-ching* says:[17] "The number of the *ta-yen* is 50;[18] of that we use 49."[19] It also says: "Divide them and make two heaps to represent the two;[20] take one to represent the three.[21] Count them off by fours to represent the four seasons. Change three times and you complete a symbol;[22] change eighteen times and you complete a hexagram."[23] We wish to know the expansion rule and the value of the numbers.

Note: In the *I-ching* divination technique used here, the forty-nine stalks of the milfoil are divided into two arbitrary heaps, of which the right-hand heap is counted off by fours, threes, twos, and ones. In the original technique of the *I-ching* there is only a counting off by fours.[24]

[16] The last part, which has nothing to do with the *ta-yen* rule, is omitted.
[17] The Great Appendix, ch. 9.
[18] In Legge's translation: "The numbers of the Great Expansion (multiplied together) make 50." In Wilhelm: "Die Zahl der Gesamtmenge. . ."; in Wylie: "The Great Extension number. . . ."
[19] Wylie: "The Use number is 49."
[20] These 'two' are the emblematic lines, or "heaven and earth" (Legge), "Beide Grundkräfte" (Wilhelm), "The 2 spheres" (Wylie).
[21] Legge: "One is then taken (from the heap on the right) and placed (between the little finger of the left hand and the next), that there may thus be symbolized the three (powers of heaven, earth and man)."
[22] A *hsiao* 爻 or "symbol" is one of the lines of a hexagram.
[23] A *kua* 卦 is a hexagram.
[24] There is a full explanation in Wilhelm (1), vol. 1, pp. 236 f. See also Chao Jan-ning (1').

The general problem is $N \equiv r_1 \pmod 1 \equiv r_2 \pmod 2 \equiv r_3 \pmod 3 \equiv r_4 \pmod 4$.

Diagram 78 shows the computing of the *yung-shu*.

	A	B	C	D
yüan-shu	1	2	3	4
ting-shu	1	1	3	4
yen-mu			12	
yen-shu	12	12	4	3
chi-shu	1	1	1	3
ch'êng-lü	1	1	1	3
yung-shu	12	12	4	9

Diagram 78

Ch'in gives the *yung-shu* as 12, 24, 4, 9. But as $24 = 2 \times 12$, and the *yen-mu* is 12, this does not change the result. All this gives the impression that Ch'in tried to explain the magic numbers 49 and 50 of the *I-ching* in a mathematical way. From the other problems it is obvious that Ch'in had entirely mastered the *ta-yen* rule, but here, taking the second *yung-shu* as 24, he gets the sum $12 + 24 + 4 + 9 = 49$; whereas when following the general rule, we get $12 + 12 + 4 + 9 = 37$. On the other hand, when searching the *yen-shu*, he starts from moduli that are not relatively prime in pairs, instead of reducing the moduli as he does in the other problems. He gets

1	$2 \times 3 \times 4 = 24$	
2	$1 \times 3 \times 4 = 12$	$24 + 12 + 8 + 6 = 50$
3	$1 \times 2 \times 4 = 8$	
4	$1 \times 2 \times 3 = 6$	

He adds: "For this reason the *I-ching* says: 'The Great Expansion number is 50.'" It is not my belief that the *I-ching* has anything to do with mathematics. However, the system of counting off the milfoil stalks is analogous to the remainder problem. All this makes it likely that Ch'in distorted the *ta-yen* rule in order to explain the magic numbers 50 and 49. After getting the number 50, he says: "According to the mathematical rule, we cannot consider 50 as the *yung-shu*,[25] because the

[25] The *yung-shu* in this case is the number of stalks used.

yen-shu are divisible by 2." Indeed, the *yen-shu* may not have a common divisor. But this is so only if the moduli are relatively prime in pairs. Moreover, the counting off of the milfoil stalks was in fact done only by fours.[26] Wylie[27] translates: "Fifty being an even number, and consequently unsuitable for the Use number . . . ," but the text says: "We cannot consider 50 as the *yung-shu*, because we can divide them [the *yen-shu*] by 2 . . . ." Matthiessen gives the right explanation: "These expansion numbers [*yen-shu*], however, are not suitable for use as auxiliary numbers [*yung-shu*] because they have the common factor 2."[28]

Concerning the *yung-shu* 24, Wylie[29] says: "The second Origin Number 2 having been reduced to 1 at the beginning, once the Extension Parent 12 is consequently added to the second Expansion Number, and the Expansion Numbers then become the Fixed Use Numbers. . . ."[30] As already seen, this is not necessary; however, it is important to note that Ch'in was well aware of this fact. He says: "If we wish to make the 'milfoil number' approximate the Great Expansion number, unless it is 49 or 51, it does not go," stating that 49 is taken only because it agrees with the number given in the *I-ching*. And on p. 9 he clearly says that we can take 12 as *yung-shu* instead of 24, and in that case the *yung-shu* is 37, but that this number makes no sense because it does not agree with the number of milfoil stalks used. On pp. 9 and 10 it is stated that we get the same solution whether we use 37 or 49. From these statements it seems to be clear why Ch'in Chiu-shao deviates from his general rule.

As the divination technique is invariable, the moduli are always the same, and the *yung-shu* are preserved for all applications.

Ch'in Chiu-shao gives the following example. After dividing

[26] See Wilhelm (1), vol. 1, p. 236.
[27] (1), p. 182.
[28] "Diese Erweiterungszahlen eignen sich jedoch nicht zu einer Bestimmung der Hülfszahlen, weil sie den gemeinschaftlichen Factor 2 besitzen." Matthiessen (3), p. 258.
[29] (1), p. 183.
[30] This explanation is repeated in Matthiessen (3), p. 259.

the stalks into two heaps, suppose the left hand holds 33 stalks (of course unknown). If we count by ones, the remainder is always 1.[31] Then we find that $N \equiv 1 \pmod 2 \equiv 3 \pmod 3 \equiv 1 \pmod 4$.[32]

Solution: $N = 12 \times 1 + 12 \times 1 + 4 \times 3 + 9 \times 1 - n \times 12 = 45 - 12 = 33$

It is obvious that the general solution is $9 + k \times 12$, giving as values between 0 and 49: 9, 21, 33, 45.

For additional applications of the divination technique, the reader is referred to Wylie (1) and Chao Jan-ning (1').

Problem I, 2

Question: "In old calendars[33] they consider the solar year[34] of $365\frac{1}{4}$ days; the lunar month of $29\frac{499}{940}$ days, and the sexagenary cycle (*chia-tzŭ* 甲子)[35] of 60 days as one revolution. Suppose in the year [A.D.] 1246, in the eleventh month, the fifty-third day of the sexagenary cycle is the first day of the month; the winter solstice [or first day of the solar year] is the fifth day of the month and the fifty-seventh day of the sexagenary cycle; the ninth day of the month is the first day of the sexagenary cycle.[36] Required: the number of years, months,

---

[31] Here also Ch'in adapts his rule to the divination technique. Suppose we take $r = 0$; as the *yung-shu* is 12 ($= yen\text{-}mu$), the solution is the same: For example,

12	12	4	9
$\times$ 1	$\times$ 1	$\times$ 3	$\times$ 1
12 +	12 +	12 +	9 $\to$ 9
12	12	4	9
$\times$ 0	$\times$ 1	$\times$ 0	$\times$ 1
0 +	12 +	0 +	9 $\to$ 9

[32] For 3 (mod 3), see the preceding note.

[33] The given values for the length of year and month are these of the *Ssŭ-fên* 四分 calendar. (*Ssŭ-fên* means 1/4 and refers to 365 1/4.) This calendrical method has been used since the fourth century B.C. For details, see Yabuuchi (3) and Maspero (1). The information on the *Ssŭ-fên* calendar is preserved in the *Hou-Han-shu* (Records of the Later Han), ch. 12, pp. 3a–4b.

[34] The text says "winter solstice."

[35] The *chia-tzŭ* is an artificial sexagenary cycle. See de Saussure (1), p. 341.

[36] This text is corrupt, but as it is repeated on p. 17 of Ch'in's text before

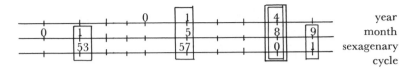

and days of the conjunction of the solar year, the lunar month, and the sexagenary cycle according to the Old Calendar; and the number of years already passed since the last conjunction of the three cycles, and of years not yet passed."[37]

The remainders used by Ch'in for calculating the *tsung-shu* are indeed $R_1=4$, $R_2=8$, and $R_3=0$. From this fact and the diagram it is obvious that the interpretation given here is correct.

yüan-shu[38]	$365\frac{1}{4}$	$29\frac{499}{940}$	60
	$\frac{1,461}{4}$	$\frac{27,759}{940}$	60
	$\frac{1,373,340}{3,760}$	$\frac{111,036}{3,760}$	$\frac{225,600}{3,760}$
	1,373,340	111,036	225,600
ting-mu	487	19	225,600
yen-mu		2,087,476,800	
yen-shu	4,286,400	109,867,200	9,253
chi-shu	313	4	9,253
ch'êng-lü	473	5	172,717
yung-shu	2,027,467,200	549,336,000	1,598,150,401
remainders	4	8	0
tsung-shu	8,109,868,800	4,394,688,000	—
sum		12,504,556,800	
N		$12,504,556,800-5\times2,087,476,800=2,067,172,800$	

Diagram 79

the application to the given remainders, we can reconstruct the original text: p. 10: 11th month/53rd sexagenary day/lunar cycle/5th day/57th sexagenary day/winter solstice/9th day/*chia-tzŭ*. p. 17: 11th month/1st day of lunar cycle/53rd sexagenary day/winter solstice/5th day/57th sexagenary /9th day/*chia-tzŭ*.
From the diagram it is obvious that the text on p. 17 is the correct one.
[37] The translation in Mikami (1), pp. 68 f makes no sense, because in this representation only the least common multiple of the cycles is to be found. Wylie's translation [(1), p. 183] is not correct either.
[38] The common denominator 3,760 is too large; it should be 940.

The problem is

$N \equiv 4 \ (\text{mod } 365\frac{1}{4}) \equiv 8 \ (\text{mod } 29\frac{499}{940}) \equiv 0 \ (\text{mod } 60).$

Ch'in's solution, which he obtains by the procedure shown in Diagram 79, is

$\dfrac{2,067,172,800}{225,600} = 9,163 \text{ years.}$[39]

This solution is incorrect. Shên Ch'in-p'ei said that the remainders as Ch'in gave them were wrong, but Sung Ching-ch'ang refuted this opinion.[40] Li Jui and Mao Yüeh-shêng stated that Ch'in's solution does not give the remainders of the problem, but they were not able to determine why this was so. Mao Yüeh-shêng calculated all the possible remainders of the year period with regard to the lunar period. As the "concordance cycle" (the Metonic cycle, *chang* 章) is 19 years,[41] there are only 19 possible remainders. Among these there is no remainder of 4 days such as the one in Ch'in's problem.[42] Moreover, as the year remainder with regard to the sexagenary cycle is $5\frac{1}{4}$ days, the remainder given in the problem [8] should be of the general form

$R = 5\frac{1}{4}x - t \times 60,$

or

$5\frac{1}{4}x - t \times 60 = 8$

$21x - 240t = 32.$

The equation is insolvable in integers, because 21 and 240 are not relatively prime[43] and thus $R = 8$ is impossible.

[39] This division does not give the number of years but the number of sexagenary cycles. See Sung Ching-ch'ang (1'), p. 4.
[40] Sung Ching-ch'ang (1'), p. 3 (according to our reconstruction, Sung is quite right). Shên Ch'in-p'ei was mistaken as to the year, because he supposed that it was the winter solstice between 1245 and 1246. In fact it was the winter solstice between 1246 and 1247.
[41] Or 365 1/4 × 19 = 24 499/940 × 235. See Yabuuchi (3), pp. 445 f.
[42] The possible remainders are given in Sung Ching-ch'ang (1'), p. 5.
[43] If the coefficients *a* and *b* are not relatively prime, the equation $ax + by = c$ does not have integer solutions.

Shên Ch'in-p'ei corrected the problem as follows:[44]

$$N \equiv 33\tfrac{3}{4} (\text{mod } 365\tfrac{1}{4}) \equiv 10\tfrac{499}{940} (\text{mod } 29\tfrac{499}{940}) \equiv 0 \text{ (mod 60)}.$$

We do not know how Shên found these remainders, but perhaps he did it by an arbitrary choice from a list drawn up beforehand.

Shên Ch'in-p'ei's solution of the problem is dealt with by Li Yen.[45]

INVESTIGATION

It is not easy to find the reason why Ch'in Chiu-shao did not see that his solution was totally wrong; it would have been very easy for him to prove it. The general reason why the solution is incorrect is that the general condition that $r_i - r_j$ must be divisible by the greatest common divisor of (m_i, m_j) is not fulfilled.

The equations

$$N \equiv R_1 \left(\text{mod } \frac{343,335}{940} \right) \equiv R_2 \left(\text{mod } \frac{27,759}{940} \right) \equiv 0 \left(\text{mod } \frac{56,400}{940} \right)$$

can be written as follows:

$940N \equiv 940R_1$ (mod 343,335) $\equiv 940R_2$ (mod 27,759) $\equiv 0$ (mod 56,400).

The conditions are[46]

$$\left| \frac{R_1 - R_2}{\text{G.C.D. } (343,335, \, 27,759)} \right| = 0 \quad \text{or} \quad \left| \frac{R_1 - R_2}{1,460} \right| = 0$$

$$\left| \frac{R_1 - R_3}{\text{G.C.D. } (343,335, \, 56,400)} \right| = 0 \quad \text{or} \quad \left| \frac{R_1 - R_3}{705} \right| = 0$$

$$\left| \frac{R_2 - R_3}{\text{G.C.D. } (27,759, \, 56,400)} \right| = 0 \quad \text{or} \quad \left| \frac{R_2 - R_3}{3} \right| = 0$$

from which we derive

---

[44] In Sung Ching-ch'ang (1′), pp. 6 ff.
[45] (9′), pp. 158 ff.
[46] The notation

$$\left| \frac{A}{B} \right.$$

means the remainder of A/B.

$R_1 - R_2 = 1,461x$
$R_1 - 0 = 705y$
$R_2 - 0 = 3z.$

From this

$705y - 3z = 1,461x$ (1)
$235y - z = 487x.$

Equation (1) has a possible solution:

$y = 3;\ z = 218;\ x = 1.$

From this we get

$R_1 = 705 \times 3 = 2,115$
$R_2 = 3 \times 218 = 654.$

Divide these numbers by 940 and we get:

$$R_1 = \frac{2,115}{940} = 2\frac{1}{4}$$

$$R_2 = \frac{654}{940}.$$

All the possible remainders are given by:

$$R_1 = 2\frac{1}{4} \times n$$

$$R_2 = \frac{654}{940} \times n,$$

where $n > 0$ and is an integer.
Taking $n = 15$, we get Shên Ch'in-p'ei's remainders:

$$R_1 = 2\frac{1}{4} \times 15 = 33\frac{1}{4}$$

$$R_2 = \frac{654}{940} \times 15 = 10\frac{410}{940}.$$

Let us solve Ch'in's problem (see Diagram 80), taking as remainders

$$R_1 = 2\frac{1}{4} \text{ and } R_2 = \frac{654}{940}$$

$$N \equiv 2\frac{1}{4}\left(\bmod\ 365\frac{1}{4}\right) \equiv \frac{654}{940}\left(\bmod\ 29\frac{499}{940}\right) \equiv 0\ (\bmod\ 60)$$

$$N \equiv \frac{2,115}{940}\left(\bmod\ \frac{343,335}{940}\right) \equiv \frac{654}{940}\left(\bmod\ \frac{27,759}{940}\right) \equiv 0\left(\bmod\ \frac{56,400}{940}\right).$$

$940\ N \equiv 2,115\ (\bmod\ 343,335) \equiv 654\ (\bmod\ 27,759) \equiv 0\ (\bmod\ 56,400)$.

yüan-shu	343,335	27,759	56,400
ting-mu	487	19	56,400
yen-mu		521,869,200	
yen-shu	1,071,600	27,466,800	9,253
chi-shu	200	1	9,253
ch'êng-lü	431	20	3,517
yung-shu	461,859,600	549,336,000	
remainders	2,115	654	0
tsung-shu	976,833,054,000	359,265,744,000	0
sum		1,336,098,798,000	
940 N	1,336,098,798,000 − n × 521,869,200 = 113,646,000		
	N = 120,900		

Diagram 80

$$N = 120,900\ \text{days} = 331\ \text{years} + \frac{2,115}{940}\ \text{days}$$

$$\text{or } 331y + 2\frac{1}{4}d\ = 4,049\ \text{months} + \frac{654}{940}\ \text{days}$$

$$= 2,015\ \text{sexagenary cycles}.$$

Problem I, 3

"For erecting a dyke they levy men from four prefectures, to which they apportion equal parts. The width is 2 *chang*.[47] The *li* value is 360 *pu* and the *pu* value is 5 feet 8 inches.[48] The number of men is fixed according to the 'properties' [*wu-li*][49] of the four prefectures. The 'property' of *A* is 138,600 strings [of cash]; of *B*, 146,300 strings; of *C*, 192,500 strings; and of *D*, 184,800 strings. For each 770 strings 'property' they levy one

[47] One *chang* is equal to 10 feet.
[48] The *li-fa* 里法 = the value of 1 *li* = about 1,890 feet English measure, or about 0.57 km. One *pu* 步 = one pace.
[49] For *wu-li*, see Chapter 23, section on "Taxes and Levies of Service."

man. Each man's task according to the 'spring standard' is 60 square feet. The prefecture which arrives first is first allocated work. When A and B together finish [their part], prefecture C has a remainder of 51 *chang*[50], and prefecture D has a remainder of 18 *chang*. In less than one day [more], their work is finished. We wish to know the length of the dyke and the [equal] parts built by the four prefectures."[51]

SOLUTION:

(1) Multiply the *wu-li* of each prefecture by the standard 60:

138,600 × 60 = 8,316,000 square feet/day
146,300 × 60 = 8,778,000 square feet/day
192,500 × 60 = 11,550,000 square feet/day
184,800 × 60 = 11,088,000 square feet/day.

(2) Multiply the number of workmen in a group by the breadth of the dyke:

770 × 20 = 15,400.

(3) Divide the numbers of (1) by the numbers of (2):

 8,316,000 : 15,400 = 540 feet = 54 *chang*
 8,778,000 : 15,400 = 570 feet = 57 *chang*
11,550,000 : 15,400 = 750 feet = 75 *chang*
11,088,000 : 15,400 = 720 feet = 72 *chang*.

Explanation: For example, A has a *wu-li* of 138,600 strings of cash. For each 770 strings they levy one man. Number of men:

$$\frac{138,600}{770}.$$

Each workman makes 60 square *pu*/day:

$$\frac{138,600 \times 60}{770}.$$

[50] Van Hee, in (12), p. 440, mistakenly gives this figure as 53 *chang*.
[51] The problem is also given in Huang Tsung-hsien (1'), p. 5b and following pages; there is a partial translation in Van Hee (12), p. 440 f.

Breadth of the dyke $= 20$ pu.

$$\text{Length} = \frac{138{,}600 \times 60}{770 \times 20} = 540 \text{ feet.}$$

Van Hee is correct in saying: "One can see that the operations are performed at great length. The author has no idea
whatever of the simplifications that are possible." [(12), p.
440]. This, however, is not a general rule. All computations
are worked out on the counting board, and this must be the
reason why such large numbers did not cause difficulties.
Chinese mathematicians always computed the dividend
($shih$ 實) first and then the divisor (fa 法) ; this too must have
some connection with the counting board.[52]

In Diagram 81, $N \equiv 0 \pmod{45} \equiv 0 \pmod{57} \equiv 51 \pmod{75} \equiv 18 \pmod{72}$.

yüan-shu	54	57	75	72
ting-mu	27	19	25	8
yen-mu			102,600	
yen-shu	3,800	5,400	4,104	12,825
chi-shu	20	4	4	1
ch'êng-lü	23	5	19	1
yung-shu	87,400	27,000	77,976	12,825
remainders	0	0	51	18
tsung-shu	0	0	3,976,776	230,850
sum			4,207,626	
N	4,207,626 − n × 102,600 = 1,026 chang			

Diagram 81

[52] Van Hee (12), p. 442, note 1 says: "The solution proceeds easily enough
from the method which will be given later, in full, with modern notation."
(La solution poursuit alors aisément d'après la méthode qui sera donnée
plus bas, tout au long, avec la notation moderne.) On pp. 449 ff he gives
Gauss's method. But he does not take notice of the fact that the moduli are
not relatively prime in pairs, and that consequently this problem cannot be
solved by Gauss's rule. The impression his article gives of Ch'in Chiu-shao's
rule is entirely wrong, because the merit of Ch'in's work—giving a solution
for the general remainder problem—is not conveyed by this misleading
representation.

The length of the dyke is $1,026 \times 4 = 4,104$ *chang*. As 1 *chang* = 10 feet and 1 *pu* = 5.8 feet, we get:

$$\frac{4,104 \times 10}{5.8} = 7,075 \ pu \ 5 \ ch'ih.$$

As 1 *li* = 360 *pu*, we get

$$\frac{7,075 \ pu \ 5 \ ch'ih}{360} = 19 \ li \ 235 \ pu \ 5 \ ch'ih.$$

The fourth part is 1,026 *chang* = 1,768 *pu* 5 *ch'ih* 6 *ts'un* = 4 *li* 328 *pu* 5 *ch'ih* 6 *ts'un*.[53]

Problem I, 5

"There are three farmers of the highest class. As for the rice they got by cultivating their fields, when making use of full *tou* 斗,[54] [the amounts] are the same. All of them go to different places to sell it. *A* sells to the official authorities of his own prefecture; *B* to the people of the villages of An-chi,[55] and *C* to a commercial agent[56] of P'ing-chiang . . . ."[57]

In all these places different standards for the corn measure are used. They are not given in the problem, but in the "solution" (*ts'ao*) and the "method" (*shu*):

"The official *hu* 斛[58] of the *Wên-hsi* hall[59] is 83 *shêng* 升;[60]

[53] It is a good illustration of Van Hee's style that he gives the following solutions:
"Part de A 1,026 perches
Part de B 1,768 pas 5 pieds 6 pouces
Part de C 4 lis 328 pas 5 pieds 6 pouces."
In fact, these three parts are the same; they are only given in different measures. The problem requires the part of each, but this is only 1/4 of the total distance. The numbers given by Van Hee are indeed in the text, but this is only a whim of Ch'in, who in a footnote adds that all parts are the same (p. 19).
[54] A *tou* is a dry measure for grain.
[55] In Chekiang.
[56] See Chapter 23, section on "Commercial Life in the Southern Sung."
[57] In Hunan.
[58] *Hu* is a dry measure.
[59] Name of a government office in the Sung.
[60] Ten *shêng* are equal to one *tou*.

the local *hu* of the district An-chi is 110 *shêng*; the market *hu* of the prefecture P'ing-chiang is 135 *shêng*."[61]

The remainders are:

$R_A = 3$ *tou* 2 *shêng*
$R_B = 7$ *tou*
$R_C = 3$ *tou*.[62]

"We wish to know the total amount of rice and the number of *shih* 石[63] sold by each of the three men."

The problem (see Diagram 82) is

$$N \equiv 0.32 \ (\text{mod } 0.83) \equiv 0.70 \ (\text{mod } 1.10) \equiv 0.30 \ (\text{mod } 1.35).$$

The relation of the measures is

1 *shih* = 10 *tou* = 100 *shêng*.

We reduce the *shih* to *shêng*:

$$N \equiv 32 \ (\text{mod } 93) \equiv 70 \ (\text{mod } 110) \equiv 30 \ (\text{mod } 135).$$

yüan-shu	83	110	135
ting-mu	83	110	27
yen-mu		246,510	
yen-shu	2,970	2,241	9,130
chi-shu	65	41	4
ch'êng-lü	23	51	7
yung-shu	68,310	114,291	63,910
remainders	32	70	30
tsung-shu	2,185,920	8,000,370	1,917,300
sum		12,103,590	
N	12,103,590 $- n \times 246{,}510 = 24{,}600$		

Diagram 82

24,600 *shêng* = 246 *shih*
A = 246 : 0.83 = 296 *shih* $R_A = 0.32$ *shih*
B = 246 : 1.10 = 223 *shih* $R_B = 0.70$ *shih*
C = 246 : 1.35 = 182 *shih* $R_C = 0.30$ *shih*.

[61] The same measure is given in VI, 3; see Chapter 23, section on "Harmonious Purchase."
[62] See Mikami (1), p. 69; Wylie (1), p. 184.
[63] *Shih* is a dry measure for grain.

Problem I, 6

"A military unit wins a victory. At 5:00 A.M.[64] they send three express messengers to the capital [where they] arrive at different times to announce the news. *A* arrives several days earlier at 5:00 P.M., *B* arrives several days later at 2:00 P.M., and *C* arrives today at 7:00 A.M.[65] According to [their] statements, *A* covered 300 *li* a day, *B* 240 *li* a day, *C* 180 *li* a day. Find the number of *li* from the army to the capital and the number of days spent by each messenger."

Messenger *A* starts at 5:00 A.M. and arrives at 5:00 P.M.: there is a whole number of working days and thus no remainder. Messenger *B* starts at 5:00 A.M. and arrives at 2:00 P.M. There is a remainder of 4.5 *shih* 時. [One *shih*=2 hours.] Here only the working day (excluding the night) is taken into account; therefore 1 *shih*=1/6 day, and as *B* covers 240 *li* a day, in 4.5 *shih* he covers $(4.5 \times 240)/6 = 180$ *li*.

Messenger *C* starts at 5:00 A.M. and arrives at 9:00 A.M. There is a remainder of 2 *shih*, and $(2 \times 180)/6 = 60$ *li*. Or, $R_A = 0; R_B = 180; R_C = 60$, and the problem (Diagram 83) is

$N \equiv 0 \pmod{300} \equiv 180 \pmod{240} \equiv 60 \pmod{180}$.

yüan-shu	300	240	180
ting-mu	25	16	9
yen-mu		3,600	
yen-shu	144	225	400
chi-shu	19	1	4
ch'êng-lü	4	1	7
yung-shu	576	225	2,800
remainders	0	180	60
tsung-shu	0	40,500	168,000
sum		208,500	
N		$208,500 - n \times 3,600 = 3,300$	

Diagram 83

---

[64] *Tsao-tien* 早點.
[65] The hours are given in horary characters; the Chinese day is divided into twelve parts of two hours each and indicated by the "twelve branches." For example: *shên* 申 = 3:00–5:00 P.M.; *shên-chêng* 申正 = 4:00 P.M.; *shên-mo* 申末 = 5:00 P.M.

Number of days:

$$A = \frac{3,300}{300} = 11 \text{ days}$$

$$B = \frac{3,300}{240} = 13\frac{3}{4} \text{ days}$$

$$C = \frac{3,300}{180} = 18\frac{1}{3} \text{ days.}^{66}$$

Problem I, 7

"There are three couriers: *A* covers 300 miles a day, *B*, 250 miles and *C*, 200 miles. First they send *C* to another place to leave a letter. Two days later there is another letter and they send *B* to pursue [*C*] and to hand it over. After half a day again there is a letter and thereupon they order *A* to hasten and to hand it over to *B*. It happens that the three men do not meet each other, but at the same time they reach that place together. We wish to know the distance between both places and the number of days in which *B* in fact should catch up with *C*, and *A* in fact should catch up with *B*."

Answer: The distance is 3,000 *li*; *B* should catch up with *C* in 8 days, and *A* with *B* in 2.5 days.

This text is very corrupt. In fact the problem as set forth makes no sense, because it contains a contradiction: indeed, it is impossible that the three couriers should arrive together and that the one should join the other on his way.

If we take into account only the fact that they join each other on the way, the solution is:

(1) *A* joins *B*: $\dfrac{250 \times 1/2}{300-250} = 2.5$ days

$2.5 \times 300 = 750$ *li*.

(2) *B* joins *C*: $\dfrac{200 \times 2}{250-200} = 8$ days

$8 \times 250 = 2,000$ *li*.

This solution is given by Ch'in. However, there can be no

66 There is an explanation of this problem in Ch'ien Pao-tsung (4'), pp. 50 ff.

doubt that Ch'in himself must have seen that they could not arrive together.

In the second part, he solves the equation (Diagram 84) $N \equiv R_1 \pmod{300} \equiv R_2 \pmod{250} \equiv R_3 \pmod{200}$.[67]

yüan-shu	300	250	200
ting-mu[68]	3	125	16
yen-mu		6,000	
yen-shu	2,000	48	375
chi-shu	2	48	7
ch'êng-lü	2	112	7
yung-shu	4,000	5,376	2,625

Diagram 84

The literal translation of the rest of the text is as follows:

1. Method

(*a*) Consider the distance from which *A* overtakes *B*. This is the *B*-factor. Consider the distance from which *B* overtakes *C*. This is the *C*-factor.

(*b*) Subtract as many times as possible the daily distance of *B* from the *B*-factor. The remainder is the *B*-remainder. Subtract as many times as possible the daily distance of *C* from the *C*-factor. The remainder is the *C*-remainder.

(*c*) With both the remainders multiply the corresponding *yung-shu*. Add up. This is the *tsung-shu*. As many times as pos-

2. Solution[69]

(*a*) Consider the 750 *li* from which *A* overtakes *B* in the foregoing solution. This is the *B*-factor. The 2,000 *li* at which *B* overtakes *C* is the *C*-factor.

(*b*) Subtract from these as many times as possible the daily distance covered by *B* and *C*. Now the distances covered by *B* and *C* both fit precisely. If we subtract them, there is no remainder.

(*c*) Although it is asserted that all arrive at the same time, nevertheless they refer to the distance covered in a whole

---

[67] The remainders are not clear in Ch'in's text.

[68] The figure 16 should be 8.

[69] In each problem Ch'in gives the method and the solution of the problem. Here they are placed side by side in order to show that the text is corrupt.

sible subtract the *yen-shu*. The remainder is the distance between the two places.

day; then reduce the 6,000 *li* by the two men *B* and *C,* and you get, 3,000 *li*. This is the number of *li* between the two places. It answers the question.

It is clear that parts 1(*c*) and 2(*c*) do not correspond to each other. It seems likely that some parts of the text are lacking, and that 2(*c*) is a later addition; at any rate, it makes no sense.

Ch'in Chiu-shao explains the method thus: "Find the solution with the *chün-shu* method; introduce the *ta-yen* method."[70] The *chün-shu* method is also mentioned in the *Chiu-chang suan-shu;*[71] its literal meaning is "impartial taxation,"[72] and it was applied to problems of pursuit.[73] Here the method is used for solving the first part of the problem. The second part is solved by means of the *ta-yen* rule. Huang Tsung-hsien[74] says that the point is that the problem can be solved in two ways.[75] However, the form "*I . . . ch'iu chih, . . . ju chih*" is found many times in Ch'in's work, and it is fairly certain that he was of the opinion that both methods were necessary. The question is: "*First* we wish to know the time and distance of the real meeting of *B* and *C* and of *A* and *B*." Ch'in does not speak of the meeting of *A, B,* and *C*; moreover, the word "real"[76] is very important in this context.[77] This part of the problem is solved by the *chün-shu* method. The second part of the question is: "*Next* we wish to know the distance in miles between the two places"; this must be related to the phrase: "It happens that the three men do not meet each other, but at the same time they reach that place together." This problem can be solved by the *ta-yen*

---

[70] *I chün-shu ch'iu chih; ta-yen ju chih* 以均輸求之. 大衍入之.

[71] Ch. 6.

[72] See Needham (1), vol. 3, p. 26.

[73] Other applications of this "rule" in Ch'in's work appear in V,8; VI,1; VII,6 and 7.

[74] (1), p. 13b.

[75] *Shih wei chün-shu k'o ch'iu. Erh ta-yen i k'o ch'iu* 是謂均輸可求. 而大衍亦可求.

[76] *Kuo* 果.

[77] It is also repeated in the "answer" *(ta)*.

rule, although its application is far from necessary. Van Hee[78] says: "*Kuo chi* 果及 is opposite to *Ou pu hsiang chi* 偶不相及; the latter means 'it happens that they do not meet,' the first 'if in reality (*kuo*) they should meet.'"

This seems to be the correct interpretation, although Van Hee does not draw the obvious conclusion from it.

The *ta-yen* problem can be formulated as follows:

$$N \equiv 0 \ (\text{mod } 300) \equiv 0 \ (\text{mod } 250) \equiv 0 \ (\text{mod } 200).$$

It is evident that $N = \text{L.C.M.} \ (300, 250, 200) = 3,000$.

Ch'in was aware of the fact that the remainders were all zero, although he derived them in the wrong way, as is obvious from 2 (a,b); but his *yen-mu* would have been the solution, if he had not mistakenly used the *ting-mu* 16 instead of 8. The deduction of 3,000 from 6,000 in 2(c) is nothing but an artifice.

As the problem is known in Europe as Van Hee's "Problème des Courriers,"[79] which is actually the reconstruction by Huang Tsung-hsien, Huang's interpretation is included here.[80]

According to Huang, B starts 3 days after C, and A starts 2 days after B.[81] That this reconstruction gives a possible solution is obvious from the fact that the following condition must be realized:

$$N = 300t_A = 250t_B = 200t_C,$$

or

$$6t_A = 5t_B = 4t_C,$$

from which $t_A = 10$, $t_B = 12$, $t_C = 15$ is one of the possible solutions. The problem is still

---

[78] (12), p. 439, note 2.

[79] (12), pp. 438 f.

[80] Van Hee, who used only Huang Tsung-hsien's text, was not aware that this was not Ch'in's original text. Moreover, as he did not give the solution of the problem, he has done an ill service to the history of science, because the text of this problem is the most corrupt one in Ch'in's work. And Van Hee classed it among the three most important problems; the reason is that Huang gives only these three problems.

[81] See Huang Tsung-hsien (1′), p. 13a and following pages.

$N \equiv 0 \pmod{300} \equiv 0 \pmod{250} \equiv 0 \pmod{200}$, and
$N = 3,000$

Huang has only adjusted the pursuit problem to the solution 3,000.[82] After that he gives himself an equivalent problem:

$N \equiv 100 \pmod{300} \equiv 200 \pmod{250} \equiv 100 \pmod{200}$.

The solution is $N = 750 \ li$.

Problem I, 8[83]
"We wish to lay a foundation for a building. There are provided four kinds of bricks: (1) large square bricks, (2) small square bricks, (3) *liu-mên* 六門,[84] and (4) city wall bricks. We instruct the workmen to select the proper ones, either horizontal or on their sides, but to make use only of one kind of brick for laying [the foundation], and to take care that they fit well. The workmen measure the surface with the bricks and the results are calculated. They say: If we measure with the large squares, the width is too great by 6 inches and the length is too small by 6 inches. If we use the small squares, the width is too great by 2 inches, and the length is too small by 3 inches. If we use the wall bricks lengthwise, the width is too great by 3 inches and the length is too small by 1 inch; if widthwise, the length is too small by 1 inch, and the width is too great by 3 inches; if we use the thickness, the width is too great by 5 *fên* 分 [0.5 inch] and the length is too great by 1 inch. If we use the *liu-mên* lengthwise, the width is too great by 3 inches and the length is too great by 1 inch; if lengthwise, the width is too great by 3 inches, the length is too great by 1 inch; if we use the thickness, the width is too great by 1 inch, the length is too great by 1 inch. Nothing fits; we cannot avoid using broken bricks to make up the deficiency.

"These [are the] four kinds of stones: the large squares have

[82] (1′), p. 15a.
[83] The problem is in Huang Tsung-hsien (1′), p. 16a, and in Ch'ien Pao-tsung (7′), p. 130.
[84] Literally "six gates."

a side of 1 foot 3 inches; the small squares have a side of 1 foot 1 inch; the wall bricks have a length of 1.2 feet, a width of 6 inches, and a thickness of 2.5 inches; the *liu-mên* have a length of 1 foot, a width of 5 inches, and a thickness of 2 inches. Find the length and width of the foundation."

		Size	Remainders	
			W	L
Large squares		1.3	+0.6	−0.6→+0.7
Small squares		1.1	+1.2	−0.3→+0.8
Wall bricks	L	1.2	+0.3	−0.1→+1.1
	W	0.6	+0.3	−0.1→+0.5
	T	0.25	+0.05	+0.1
Liu-mên	L	1	+0.3	+0.1
	W	0.5	+0.3	+0.1
	T	0.2	+0.1	+0.1

	Large squares	Small squares	Wall bricks L	Wall bricks W	Wall bricks T	Liu mên L	Liu mên W	Liu mên T
yüan-shu	130	110	120	60	25	100	30	20
ting-mu	13	11	9	3	25	1	1	1
yen-mu				85,800				
yen-shu	6,600	7,800	10,725	28,600	3,432			
chi-shu	9	1	5	1	7			
ch'êng-lü	3	1	5	1	18			
yung-shu	19,800	7,800	53,625	28,600	61,776			
re-mainders	60	20	30	30	5	30	30	10
tsung-shu	1,188,000	156,000	1,608,750	858,000	308,880			
sum				4,119,630				
N(width)			$4,119,630 - n \times 85,800 = 1,230 = 12.30$ feet					
re-mainders L	70	80	110	50	10	10	10	10
tsung-shu	1,386,000	624,000	5,898,750	1,430,000	617,760			
sum				9,956,510				
N(length)			$9,956,510 - n \times 85,800 = 3,710 = 37.10$ feet					

Diagram 85

Note: The words "foot" and "inch" as used here do not denote the English measures but are rather translations of the Chinese *ch'ih* and *ts'un*. Since 1 *ch'ih*=10 *ts'un*, in this problem 1 foot 3 inches=1.3 inches. The problem is summarized in the accompanying table and in Diagram 85.

Problem I, 9

"There is a rice shop which reports thieves have stolen 3 bins of a certain kind of rice. These bins were first full to the top, but their exact capacity is not known. The three vessels were located. It is found that in the left-hand one there is 1 *ko* 合 left; in the middle one there is 1 *shêng* 升 4 *ko* left;[85] in the right-hand one, there is 1 *ko* remaining. Later the thieves being caught, there are three of them, named *A*, *B*, and *C*. *A* confesses that he picked up a "horse-ladle" [? *ma-shao* 馬杓] at night and filled it several times out of the left-hand bin, putting the contents into his bag; *B* confesses having kicked off his wooden shoe and used it to fill his bag out of the middle bin; *C* says he picked up a bowl and used it to put rice into his bag out of the right-hand bin. They took the rice home to eat, and after this long time they do not remember the quantities. The three vessels were located. The "horse-ladle" is found to contain 1 *shêng* 9 *ko;* the wooden shoe 1 *shêng* 7 *ko*, and the bowl 1 *shêng* 2 *ko*. What is the amount of rice lost, and how much did each take?"[86]

ting-mu	19	17	12
yen-shu	204	228	323
yen-mu		3,876	
chi-shu	14	7	11
ch'êng-lü	15	5	11
yung-shu	3,060	1,140	3,553
remainders	1	14	1
tsung-shu	3,060	15,960	3,553
sum		22,573	
N	$22,573-5\times3,876=3,193$ *ko*$=319.3$ *shih*		

Diagram 86

[85] One *shêng* = 10 *ko*.
[86] Modified from the translation in Wylie (1), p. 184.

The problem is

N≡1 (mod 19)≡14 (mod 17)≡1 (mod 12).[87]

Note that the moduli are relatively prime, and consequently the *yüan-shu* are the *ting-mu* (see Diagram 86).

Problem II, 3

Question: "In the *K'ai-hsi* calendar,[88] the Superior Epoch is 7,848,183 years.[89] We wish to know the method for calculating this period. The computations are to be made for the year 1207; the observations are from 1204, the year of the beginning of the sexagenary cycle."[90]

This problem is very interesting, because
(1) it gives an idea of the way the Chinese calendar makers solved their problems, and
(2) it is rather intricate, even though all possible reductions are applied.[91]

METHOD:

(1) *Jih-fa* 日法 = 16,900. The *jih-fa* 16,900 is in fact the common denominator of the year and month fractions. For example,[92]

$$1 \text{ year} = 365 \tfrac{79}{325} \approx 365 \tfrac{4108}{16900}$$
$$1 \text{ month} = 29 \tfrac{8967}{16900}.$$

According to Ch'in Chiu-shao, this value is derived from Ho Ch'êng-t'ien's 何承天 *Yüan-chia* 元嘉 calendar of 443 A.D.[93]

---

[87] There is a discussion of the problem in Hsü Ch'un-fang (3'), pp. 16 ff.
[88] The *K'ai-hsi* 開禧 calendar was worked out in 1207.
[89] "Superior Epoch," literally "accumulated years" *(chi-nien* 積年*)*.
[90] In fact, in the problem 23 technical terms are to be calculated. These terms are not included here because they only create confusion. Some of them are given in the solution of the problem.
[91] The problem is explained in Ch'ien Pao-tsung (7'), pp. 130 f and (2'), pp. 37 ff; and in Li Yen (9'), pp. 155 ff.
[92] In this sense, Needham in (1), vol. 3, p. 392, translates *jih-fa* as "the number of divisions in a day."
[93] See Yabuuchi (3), p. 454. The source for Ho Ch'êng-t'ien's calendar is *Sung-shu*, ch. 13, p. 1b and p. 3a.

(2) *Shuo-yü* 朔餘 =8,967. According to Ho Ch'êng-t'ien, the "strong ratio" is 339 and the "weak ratio" 17.

If $49x+17y=16,900$, then $x=339$ and $y=17$ is a possible solution. Then[94] $26 \times 339 + 17 \times 9 = 8,967$.

(3) *Shuo-lü* 朔率 =499,067. 1 month= $29\frac{8967}{16900} = \frac{490100}{16900} + \frac{8967}{16900} = \frac{499067}{16900}$

(4) *Tou-fên* 斗分 =4,108. 1 year= $365\frac{79}{325} = 365.2431$

$$\frac{0.2431 \times 16,900}{16,900} = \frac{4,108.39}{16,900} \approx \frac{4,108}{16,900} = \frac{79}{325}.$$

GENERAL PROBLEM[95]

"It was determined on the basis of observations that in the sexagenary year 1 of the *Chia-t'ai* 嘉泰 period[96] (A.D. 1204), astronomical new year and winter solstice took place 11.446154 days after the first day of the sexagenary cycle." Or

$$365\frac{79}{325} \times 60x \equiv 11.446154 \ (\text{mod } 60). \tag{1}$$

"The new moon of the 11th month took place 1.755562 days after the first day of the sexagenary cycle." Or

$$29\frac{399}{752} \equiv 1.755562 \ (\text{mod } 60). \tag{2}$$

Equation (1) can be changed thus:

$$365\frac{79}{325} = \frac{6172608}{16900} \ (6,172,608 \text{ is the } sui\text{-}l\ddot{u} \ 歲率).$$

$$11.446154 = \frac{193,440.26}{16,900}, \text{ or } \frac{193,440}{16,900}.$$

(The figure 193,440 is the *ch'i-ting-ku* 氣定骨.)

Equation (1) becomes

$$\frac{6,172,608}{16,900} \times 60x = \frac{193,440}{16,900} \ (\text{mod } 60)$$

[94] See Chapter 15.
[95] The problem is not clear, but it can be reconstructed from the data in the solution. The version here is taken mainly from Ch'ien Pao-tsung (7'), p. 130.
[96] The period 1201–1205.

$$\frac{6,172,608}{16,900} \times 60x = 60y + \frac{193,440}{16,900}$$

$$6,172,608x = 16,900y + \frac{193,440}{60}$$

$$= 16,900y + 3,224$$

$$= 325 \times 52y + 52 \times 62.$$

$$52\ (365 \times 325 + 79)x = 52\ (325y + 62)$$

$$79x \equiv 62 \ (\text{mod } 325).$$

(The figure 325 is the *pu-lü* 蔀率).

$$a \times 79 \equiv 1 \ (\text{mod } 325) \rightarrow a = 144$$

$$144 \times 79 \times x \equiv 62 \times 144 \ (\text{mod } 325)$$

$$(325 \times 35 + 1)x \equiv 62 \times 144 \ (\text{mod } 325)$$

$$(325 \times 35x) + x \equiv 62 \times 144 \ (\text{mod } 325)$$

$$x \equiv 62 \times 144 \ (\text{mod } 325)$$

$$x \equiv 153 \ (\text{mod } 325)$$

$$60x \equiv 9,180 \ (\text{mod } 19,500).$$

(The figure 9,180 is the *ju-yüan-sui* 入元歲; 19,500 is the *ch'i-yüan-lü* 氣元率.)

From (1) and (2) we get

$$365\tfrac{79}{320} \times 60x \equiv 11.446154 \ (\text{mod } 60)$$

$$\equiv [-1.75562 + 11.446154](\text{mod } 29\tfrac{399}{752}). \tag{3}$$

Or

$$\frac{6,172,608}{16,900} \times 60x \equiv \left[-\frac{29,669}{16,900} + \frac{193,440}{16,900} \right]\left(\text{mod } \frac{499,067}{16,900}\right). \tag{4}$$

$$1 \text{ month} = \frac{499,067}{16,900} \ ; \ 12 \text{ months} = \frac{499,067 \times 12}{16,900} = \frac{5,988,804}{16,900}$$

$$1 \text{ year} - 12 \text{ months} = \frac{6,172,608}{16,900} - \frac{5,988,804}{16,900} = \frac{183,804}{16,900} \ .$$

The figure 183,804 is the *sui-jun* 歲閏 or "annual intercalation factor," the dividend corresponding to the difference, expressed in days, between twelve mean lunar months and one mean tropical year.

$$\frac{1.755562 \times 16,900}{16,900} = \frac{29,668.9978}{16,900} \approx \frac{29,669}{16,900}$$

$$\frac{193,440}{16,900} - \frac{29,669}{16,900} = \frac{163,771}{16,900} \ .$$

(The figure 29,668.9978 is the *shuo-fan-ku* 朔泛骨; 29,669 is the *shuo-ting-ku* 朔定骨; 163,771 is the *jun-fan-ku* 閏泛骨.)

Equation (4) becomes

$$\frac{6,172,608}{16,900} \times 60x \equiv \frac{163,771}{16,900} \left(\mathrm{mod} \frac{499,067}{16,900} \right).$$

As $60x \equiv 9,180 \pmod{19,500}$, we get

$6,172,608(19,500n + 9,180) \equiv 163,771 \pmod{499,067}$
$6,172,608 = 499,067 \times 12 + 183,804$
$183,804(19,500n + 9,180) \equiv 163,771 \pmod{499,067}$ (5)
$183,804(19,500n + 9,180) = 183,804 \times 19,500n + 183,804 \times 9,180$
$183,804 \times 19,500n \equiv 377,873n \pmod{499,067}$
$183,804 \times 9,180 \equiv 474,260 \pmod{499,067}$.

Equation (5) becomes

$377,873n + 474,260 \equiv 163,771 \pmod{499,067}$
$377,873n \equiv (163,771 + 499,067 - 474,260) \pmod{499,067}$
$377,873n \equiv 188,578 \pmod{499,067}$.

(The figure 188,578 is the *jun-suo* 閏縮; 377,873 is the *yüan-jun* 元閏; 499,067 is the *shuo-lü* 朔率). We use the *ta-yen ch'iu-i* method for solving the congruence

$377,873a \equiv 1 \pmod{499,067}$
    $a = 457,999$
$377,873 \times 457,999n \equiv 188,578 \pmod{499,067}$
$1 \pmod{499,067}n \equiv 188,578 \times 457,999 \pmod{499,067}$
$n \equiv 188,578 \times 457,999 \pmod{499,067}$
$n \equiv 402 \pmod{499,067}$.
$60x \equiv 9,180 \pmod{19,500}$,

or

$60x = 19,500n + 9,180$
$60x = 19,500 \times 402 + 9,180 = 7,848,180$.

This is the number of years from the *shang-yüan* 上元 (Superior

Epoch) to A.D. 1204;[97] from the *shang-yüan* to the fourth year of the sexagenary cycle = 7,848,183 years.

[97] The first year of the sexagenary year cycle.

VI

Socioeconomic Information

23

The Shu-shu chiu-chang and Life in Sung China

The *Shu-shu chiu-chang* is interesting not only from the mathematical but also from the socioeconomic point of view. In the first place it gives a good idea of the money and currency in Southern Sung, of the credit system, commercial life, the "harmonious purchase," transportation problems, construction of dykes, architecture, taxes and levies of service, and the military sciences; there are also problems on chronology and meteorology. It has been mentioned that all Ch'in's problems were practical in nature and taken from daily life. Some of these problems are very simple from the mathematical point of view, and all of them reinforce the impression that a mathematician was not a pure scientist in our modern sense but a kind of technician, whose specialty was solving practical problems that required some knowledge of mathematical methods. In European medieval mathematical handbooks the greater part of the problems were imaginary, whereas in China they were derived from situations in daily life. An important corollary is that the numbers used are realistic and represent historically valid information, providing an interesting supplement to the socioeconomic chapters in the dynastic histories. As the integration of this material is the task of scholars who specialize in various fields, the intention here is only to collect the information scattered through the *Shu-shu chiu-chang* and to incorporate it in a more general context, without aiming at a complete description of the socioeconomic situation at that time. This means that from a general point of view this information is very incomplete.

Ch'in Chiu-shao's arithmetical handbook seems to be outstanding in the history of Chinese mathematics, because no other work gives so much contemporary information. Chou Mi[1] says: "By nature he was extremely ingenious: all such things as astronomy, harmonics, mathematics, and even archi-

[1] (1'), C, p. 6b.

tecture . . .he investigated thoroughly." Ch'in's broad education is clearly reflected in his work.

Money and Currency in the Southern Sung

THE TREASURIES AND THE FINANCIAL ADMINISTRATION

The treasuries (*k'u* 庫),[2] where money was stored, are men-

Figure 51. Treasury. From the *San-ts'ai t'u-hui* 三才圖會 (1609), "Kung-shih 宮
室" (Architecture), ch. 2, p. 19a.

[2] See Figure 51.

tioned several times in the problems of the *Shu-shu chiu-chang*. Treasuries of different places made use of local monetary standards (I,4).[3] As financial administrative organs, in VI,7 there are the *hu-pu* 戶部 or Board of Finance, and the *yün-szŭ* 運司 or Salt Comptroller, both traditional institutions; mention is also made of the *tsung-so* 總所, which was established in the Southern Sung and was responsible for the financial affairs of each region, including transport of money to the troops, comparison of increase and decrease of currency and rice, and allotment of punishments and rewards.[4] In the Southern Sung four *tsung-so* were established, in Huai-tung, Huai-hsi, Hu-pei, and Szechwan, the four frontier provinces where troops were encamped.

COINAGE[5]

From antiquity copper coins were the ordinary means of payment in China. The monetary unit was the *wên* 文;[6] and a string of 1,000 *wên* was a *kuan* 貫[7] (Figure 52).

SHORTAGE OF COINS

In I,4 mention is made of the fact that coins were scarce. "The shortage of coins, especially valid coinage, was chronic throughout China."[8] For this reason the number of coins in a string was artificially diminished from time to time.[9] The term *shêng-*

---

[3] Discussed later in this chapter.

[4] Hoshi Ayao (1′), p. 262.

[5] There is a short history in Yang (3), pp. 20–39.

[6] A coin.

[7] There is a picture of a Sung coin with the inscription *Sung-yüan t'ung-pao* 宋元通寶 (current treasure from the beginning of the Sung dynasty) in Morse (1), p. 140; see also Vissering (1), p. 133. There is a general description of Southern Sung coins in Lockhart (1), vol. 1, pp. 35 ff, and pictures in vol. 2, pp. 41 ff.

[8] "Mangel an gemünztem Gelde, besonders an vollwertigem, war chronisch in China." O. Franke (1), vol. 4, pp. 382 f.

[9] O. Franke, in (1), vol. 4, p. 383 speaks of: "the dictatorial, that is, artificial, increase in the value of each piece by the administration, or else the decrease of the number in each string" (die diktatorische, also künstliche Erhöhung des Geltungswertes der Stücke durch die Regierung oder Verringerung ihrer Anzahl in den Schnüren).

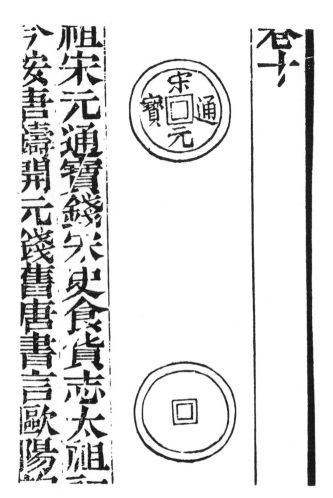

Figure 52. A copper cash of the Sung period, obverse and reverse. From Liang Shih-chêng 梁詩正, *Ch'ien lu* 錢錄 (Numismatics, 1879), ch. 10, p. 1a.

pai 省陌, or short string of cash (I, 4), indicates this diminishing of the real value of a string while the nominal value was maintained.[10] The character *pai* 陌 is a variant of *pai* 百, or one

[10] See Twitchett (1), p. 81; Yang (1), p. 36: "The Sung dynasty accepted 77 cash as '100' and called it *shêng-pai* or simply *shêng*."

hundred,[11] so the literal meaning of *shêng-pai* is "diminished hundred." In 904 "in Lo-yang 80 cash passed current for 100."[12] In the *Shu-shu chiu-chang* the *shêng-pai* is 77/100. This standard must have originated in the Later Han (947–950)[13] at Lo-yang, according to a text of the *Wên-hsien t'ung-k'ao, t'ien-fu k'ao* 文獻通考.田賦考:[14] "In the later Han of the Five Dynasties, Wang Chang, minister of the 'three offices'[15] [*san szǔ* 三司], collected money in a hardhearted manner: when 'old cash' [that is, according to the old standard] was brought in or issued, the 'hundred' was always 80; Wang was the first to decree that [the 'hundred'] of the incoming strings should be 80, and that of the outgoing strings 77; this was called 'diminished hundred' (*shêng-pai*)."

Ch'in also gives the term *kuan-shêng* 官省(official diminution) for 77/100 (I, 4). In I, 4 there is given an amount of 26,950 *kuan* "real value" (*tsu-ch'ien* 足錢, "full money").[16] At a rate of 77/100 this represents a nominal value of 35,000 *kuan*. This nominal value is called *chan-shêng* 展省(literally, increased diminution).[17]

[11] See Shên Kua, who in his *Mêng-ch'i pi-t'an*, ch. 4, p. 4b (in the *Ssǔ-pu ts'ung-k'an*) says that the character *pai* 陌 is used in a figurative sense for an amount of coins less than 100.

[12] Twitchett (1), p. 82; see also p. 83: "It was not an innovation, but seems to have arisen during the southern dynasty of Liang (502–566)." See also Yang (1), p. 36.

[13] See O. Franke (1), vol. 4, p. 73.

[14] Vol. 1, ch. 4, p. 53. On this work, see Têng and Biggerstaff (1), pp. 150 f.

[15] That is, the Finance Board, the Salt and Iron Commission, and the Public Revenue Department. Under the T'ang "the *san-ssǔ* never became a single organisation, a joint co-ordinating ministry of finance, as it did under the Sung. The first single head of the three offices seems to have been appointed under the Later Liang in 906...." Twitchett (1), p. 332.

[16] "The full '100' was called *tsu-pai* 足陌 or simply *tsu* 足." Yang (1), p. 36.

[17] From the calculations it is obvious that the nominal value (N) is $N = x \times 100/77$, where x is the real value. In Okudaira Masahiro (1), 10, pp. 92a–93a, there is a photo of a printing block for printing paper money, with the inscription: "With the exception of Szechwan, this may be circulated in the various provinces and districts to make public and private payments *representing 770 cash per string*." In Yang (1), "The form of the paper note *hui-tzǔ* of the Southern Sung dynasty," pp. 216 ff.

LOCAL STANDARDS

In I,4 the problem mentions several treasuries with their own standards.

In this time of shortage of coins these standards were very low. They are called *shih-pai* 市陌 ("market hundreds"), and in this case are equal to 12, 11, 10, . . . , 6. This means that one *kuan* (a string of 1,000 cash) is represented by strings of 120, 110, . . . , 60 cash.[18]

PAPER MONEY

Bibliographical note: The study of paper money and its devaluation requires a separate monograph in a Western language. The most valuable studies are all in Japanese, namely Wada Sei(1'), Sudô Yoshiyuki (1') and (2'), Katô Shigeshi (1'), and the valuable vocabulary of technical terms by Hoshi Ayao(1'). I have also made use of Vissering (1), H.B. Morse (1), Lien-shêng Yang (2) and (3), E. Stuart Kirby (1), P'êng Hsin-wei (1'), H. Franke (2), the general notes in O. Franke (1), vol.4, and others. On the relationship between paper money and the mathematical works of the Sung and Yüan, there is an article by Yen Tun-chieh (5').[19] On the financial problems in Ch'in Chiu-shao's work there is a short account in Ch'ien Pao-tsung (2'), pp. 102 f.

The paper note *hui-tzŭ* 會子(also referred to as *ch'ao* 鈔) was the typical paper money of the Southern Sung[20] (Figure 53).

---

[18] On this devaluation, see Vissering (1), p. 176, and Morse (1), pp. 152 ff.

[19] This article, published in the "Literary Supplement" of a Chinese newspaper during the war (1943), is very difficult to find. I am much indebted to Joseph Needham for supplying me with a reproduction of his own copy.

[20] "The paper money *par excellence* of the Southern Sung period was *hui-tzŭ*, 'check medium.' Like its predecessor *chiao-tzŭ*, *hui-tzŭ* also originated as a private instrument. In the middle of the twelfth century, notes by this name were issued by private monetary agencies in the area of the capital, Lin-an, the modern Hangchow. In 1160 the government prohibited the circulation of private *hui-tzŭ* and issued their own, which were sometimes designated as *kuan-hui*, 'the official check medium,' in order to distinguish them from the private notes. Denominations were in 1, 2, or 3 strings." Yang (3), p. 55.

Figure 53. A Sung paper note. From Andrew McFarland Davis, *Certain Old Chinese Notes on Chinese Paper Money* (Boston, 1915), facing p. 274.

L. S. Yang provides information on the form of this currency.[21] The value of the note in cash (77/100) was indicated.[22] The

---

An interesting study, written from the economic point of view "to point out certain implications for general monetary theory suggested by the Chinese experience" is Tullock (1).

[21] See Yang (1), pp. 216–224. There are illustrations in Vissering (1), p. 172 and Morse (1), p. 160; and a description in Yang (3), p. 54 and pp. 55 f. See also A. McF. Davis (1), Sung notes on pp. 272 and 274. Vissering (1), p. 165, explains the term *ch'ao* as shortage (*shao* 少) of money (*chin* 金).

[22] Of course this was the nominal value and not the real value.

hui-tzŭ was issued from 1159 on.[23] In 1210 their value was between 300 and 400 *wên*.[24] The process of devaluation accelerated, however,[25] and in Ch'in's time the value was only between 50 and 60 cash (see VI, 1, where the values 53 and 59 are given).[26]

In the *Shu-shu chiu-chang* two different nominal quotations of the *hui-tzŭ* are given: the first of 1,000 *wên* (*chiu-hui* 舊會, old

[23] From 1023 the right to issue paper money was monopolized by the government. See H. Franke (3), p. 194. For the history of paper money in general the reader is referred to the works mentioned in the bibliographical note.

[24] See P'êng Hsin-wei (1'), p. 482. There is an extensive study on the *hui-tzŭ* on pp. 323 ff.

[25] "Under the Southern Sung dynasty the development assumed inflationary characteristics: in 1232, 329 million banknotes were in circulation. In addition to the increasing military expenditures, an unfavorable overseas trade balance was a contributory factor." (Unter der südlichen Sungdynastie nahm die Entwicklung inflationäre Züge an; 1232 waren 329 Millionen Noten in Umlauf. Dazu trug ausser den damals steigenden Kosten für das Militär auch bei, dass sich der Überseehandel stark defizitär gestaltete.) H. Franke (3), p. 194. Yang, in (3), p. 55, deals with this devaluation: in the beginning the value was 770 cash per string. About 1200 "the market price in the capital fluctuated around 700 cash," in the provinces around 600. After 1200 the value constantly diminished. In 1263 "even the official rate sank to 250 cash. Toward the end of the dynasty, it became almost worthless."

[26] Vissering (1), p. 170 gives an example of a value of only 10 cash. In Bi Dung (1), p. 23 there is a description of the devaluation of the *hui-tzŭ* in Szechwan: "A devaluation of bank notes took place when later neither the fixed quota nor the cash redemption was honored. For example in the year 1208 there was a devaluation of 1/60 to 1/90 of the nominal value. In 1137 the value of the paper money in circulation was estimated at about 38 million strings of cash, in 1161 at more than 41 million, in 1178, at 45 million, and in 1204 at about 80 million strings." (Eine Entwertung der Noten trat ein, als später weder das feste Kontingent noch die Bareinlösung innegehalten wurde. So trat zum Beispiel im Jahre 1208 eine Entwertung auf 1/60 bis 1/90 des Nennwertes ein. Man schätzte den Wert des umlaufenden Papiergeldes 1137 auf rund 38 Millionen Strang Cash, 1161 auf über 41 Millionen, 1178 auf 45 Millionen und 1204 auf rund 80 Millionen Strang.) And, "During the Southern Sung dynasty, all notes circulated freely in the capital and in the entire province as unrestricted legal tender, and were not covered by cash." (Während der Zeit der Südsung-Dynastie (1127–1279) zirkulierten alle Noten in der Provinz und in der Hauptstadt als unbeschränkte gesetzliche Zahlungsmittel und hatten keine Bardeckung) (p. 23).

notes), the other of 5,000 *wên* (*hsin-hui* 新會, new notes). However, it is not certain that there were notes of 5,000 *wên*. The meaning seems to be that five notes of the old quotation were exchanged for one note of the new quotation.[27] The *hui-tzŭ* are also mentioned in VI, 2;[28] VI, 3; and VII, 1.

The old and new quotations just mentioned are connected with the periods of convertibility of paper notes, called *chieh* 界 (period, area). "The principle that a bill should be convertible at any time, the Chinese have never known, and from the beginning we see that only at prescribed intervals could the bills be exchanged for specie. A term of three years was generally adopted."[29] According to the *Hsü wên-hsien t'ung-k'ao* 續文獻通考[30], ch. 7 (*Ch'ien pi k'ao* 錢幣考), in the ninth month of the fourth year of the period *Chia-hsi* 嘉熙(1240) it was ordered to arrange the eighteenth *chieh hui-tzŭ* 界會子(conversion of paper notes), to collect and exchange the sixteenth *chieh,* and to use the seventeenth *chieh* with five notes equal to one note of the

---

[27] "It (*hui-tzŭ*) was in four denominations, 500, 300, 200 cash and the usual one-*kuan* note." Wei Wên-pin (1), p. 23.

[28] This quotation, however, is somewhat problematic, because the words "17 *chieh hui-tzŭ*" are not in the text of this problem as preserved in the *Yung-lo ta-tien,* although it is very likely that the copyist did not understand them and omitted them.

[29] Vissering (1), pp. 170 f. See also Yang (3), pp. 53 and 55. "In the years following, new regulations were promulgated, according to which distribution, as in the Northern Sung period, took place in several consecutive issues, each limited to a maximum of 10 million strings, and each having a due date of three years after issue. But later the due date was deferred and the amount increased; this necessarily led to a devaluation of the notes." (In den folgenden Jahren fand eine Neuregelung statt, indem die Emission wie in der Nordsung-Zeit in verschiedenen aufeinanderfolgenden Ausgaben erfolgte, deren Maximalgrenze auf 10 Millionen Strang, und deren Verfallstermin auf je drei Jahre festgesetzt wurde. Aber in späterer Zeit wurde sowohl der Verfallstermin hinausgeschoben als auch der Ausgabebetrag erhöht, was zwangsläufig zu einer Entwertung der Noten führen musste.) Bi Dung (1), p. 23.

[30] An encyclopedia of 1586. See Têng and Biggerstaff (1), p. 151; Hummel (1), p. 123. "It gives a certain amount of new material for the period from 1224 to the end of the Sung dynasty." The quoted text is in vol. 1, ch. 7, p. 2,841; I have translated the summary of the text as given in Yen Tun-chieh (5').

eighteenth *chieh*.[31] From this it is obvious that the seventeenth *chieh* mentioned in the *Shu-shu chiu-chang*, VI, 2[32] and VI, 3, must be this one. Moreover, the old and new quotations, being in the proportion of 1 to 5, are those used in VI, 1. As Ch'in's work was published in 1247, it is beyond all doubt that these are the two *chieh* he is speaking of. The old quotation is that of the seventeenth *chieh*, the new quotation that of the eighteenth *chieh*.[33]

OTHER MEANS OF PAYMENT

Apart from copper coins and paper notes, some other means of payment are mentioned in Ch'in's work.

GOLD AND SILVER[34]

"In Sung times, the uses of precious metals even spread to the lower classes. . . .Gold and silver were also used as indemnity payments, to make loans, to commute taxes, and to redeem paper money."[35] Also, "The use of silver was increasingly extended after the Sung period. On account of the devaluation of bank notes, precious metals, especially gold and silver, were used more and more as tenders of payment of acknowledged value."[36]

In IX, 2 the value of gold is given as 480,000 *wên* 文 (nominal cash value) per *liang* 兩;[37] and in IX, 6 mention is made of a

---

[31] "Under the Southern Sung, the term *chieh* designated a fixed period in which a certain issue of *hui-tzŭ* paper money could validly circulate. At the end of this period, the *hui-tzŭ* had to be exchanged for a new issue. Although the period of circulation varied, it was never longer than six years. The eighteenth *chieh* of *hui-tzŭ* was the last issue of *hui-tzŭ* put out by the Sung government before its downfall." Schurmann (1), p. 189.

[32] See, however, note 28.

[33] This means that there is no doubt about the date of the work and indicates that the information it contains is historically valid.

[34] There is a general description in Kann (1): although mainly modern, it contains some historical notes.

[35] Yang (3), pp. 44 f.

[36] "Der Silbergebrauch gewann seit der Sung-Zeit eine immer grössere Verbreitung. Wegen der Entwertung der Noten wurden im Privatverkehr Edelmetalle, besonders Gold und Silber, immer weitgehender als wertbeständige Zahlungsmittel verwendet." Bi Dung (1), p. 23.

[37] One *liang* = 1/16 *chin* 斤 = 37.5 g.

treasury which held three kinds of gold, namely gold of 80 percent purity, or 400 *kuan/liang*; gold of 75 percent, or 375 *kuan/liang*; and gold of 85 percent, or 425 *kuan/liang*. In all these cases, this gives a value of 500 *kuan/liang* for pure gold. Silver became very important at this time: "During the period Chêng-ta, under the Emperor Ai Tsung (A.D. 1225) silver became a general medium of exchange and this marks the beginning of its full use."[38] In IX, 2 silver is quoted at 50 *kuan/liang*. This must be given in the "old quotation, "because in VI, 1 the value of silver in four prefectures is given as 2,200 *kuan* in full strings; 2,300 *kuan* in full strings; 9,300 *kuan* "new quotation," and 51,000 *kuan* "old quotation." The last agrees with IX, 2.

Silver was much cheaper than gold, and "for the purpose of adulteration, silver was frequently added to gold."[39] Thus "rules against forgery of gold and silver were proclaimed."[40] Problem IX, 6 deals with the refinement of gold. "They wish to smelt [the gold] and to refine it. The price for the food of the workmen, the processing chemicals (*yao* 藥),[41] and the charcoal is 3 *kuan wên* per *liang*. The 'melting' loss is 972.5 *liang* . . . ." Li Ch'iao-p'ing explains the process of refinement: "They [silver and gold] might be separated by beating into thin leaves and cutting into pieces which were then daubed with mud and put into an earthen crucible to be melted with the addition of borax. Silver would be absorbed by the earth, leaving gold in pure state . . . ."[42]

SALT LICENSES

In the Sung, licenses for selling salt were issued. "By this system merchants bought salt licenses . . . from the government, went

---

[38] Precious metals were already used in T'ang times: "To supply the need for a medium of currency suitable for use in large transactions, where copper cash were too bulky and cumbersome, it was the practice in T'ang times to employ either silk cloth or precious metals, in the latter case usually silver." Twitchett (1), p. 70; see also Talcott Williams. (1)

[39] Li Ch'iao-p'ing (1), p. 40.

[40] Yang (3), pp. 44 f.

[41] *Yao* is a general name for all drugs and chemical substances; in this case borax is meant.

[42] (1), pp. 40 f.

to the salt works, obtained the designated amount of salt, and sold it within specified areas."[43] The certificates were called *yen-ch'ao* 鹽鈔. In IX, 2 we find "10 *ch'ao* for 4 bags of salt [each certificate has a value of 4 bags of salt]" and "88 *ch'ao* for 3 bags of salt." There were different kinds of salt licenses: "There were the so-called great and small certificates, which indicated the amount of salt to be obtained."[44]

TU-TIEH 度牒[45]

The *tu-tieh,* or diploma of a Buddhist priest, is mentioned in Ch'in's work IX,2 and IX,3. A *tu-tieh* (literally, "salvation certificate") was an official document given to Buddhist priests and nuns when they went into a monastery; this certificate authorized striking their names off the list of taxpayers. The holder of a *tu-tieh* was thus exempt from all taxes and government services.

From c. 1070 on, it was permissible to sell these certificates at a price of 130,000 copper coins.[46] In the period 1190–1195 this value went up to 800,000 *wên*. In Ch'in's Problem IX, 2 the value 1,500,000 *wên* is given.[47]

In the beginning these certificates were all issued to a specific person whose name was indicated. But the negotiable certificates had to be blank so that the exemption from taxes could be transferred. "The certificates differed from those obtained through regular channels only in the fact that there were blank spaces left for the purchasers to fill in the necessary information."[48] These *tu-tieh* were used in trade as a kind of negotiable paper. "The certificates were traded on the market, and were

[43] Schurmann (1), p. 169; see also Maspero and Balazs (1), p. 291: "...salt licences (*yen-ch'ao*) which were circulating as bank notes" (...de 'bons de sel' . . . qui circulaient comme des billets de banque).

[44] Bi Dung (1), p. 23.

[45] I have relied on Gernet (1), and Yang (3), p. 202; the most extensive study is Yüan Chên (1'); see also Fan Wu (1'), and Eichhorn (1), pp. 23 ff (this last is a translation of the *Ch'ing-yüan t'iao-fa shih-lei* 慶元條法事類, ch. 50 and 51 on "Taoism and Buddhism").

[46] Eichhorn (1), p. 23, note 4.

[47] Yüan Chên (1'), 2nd part, pp. 1 ff, gives all possible sources for the value of a *tu-tieh*. For the year 1227 in Szechwan it amounts to 800,000 *wên*.

[48] Ch'ên (1), p. 313.

not even necessarily used for candidates for the priesthood."[49]

They were issued by the government and were indeed "a curious procedure of public finance."[50] "Scrip certificates were frequently used to obtain funds for public works, and most public offices had a quantity of them on hand."[51]

Problem IX, 3 gives the relative value of these certificates: "They issue *tu-tieh*. They send a man to trade with them. Three *tu-tieh* are worth 13 sacks of salt; 2 sacks of salt are worth 84 rolls of cotton; 15 rolls of cotton are worth 3.5 rolls of silk; 6 rolls of silk are worth 7.2 ounces of silver. The *tu-tieh* fetch 9,172.8 ounces of silver."

SILK

In T'ang times silk was still second to copper as a circulating medium; after that its employment diminished fast, but it remained in use.[52] In all periods it was very important for the payment of taxes.[53] In VI, 1 the value of a standard 40-foot roll (*p'i* 疋) is given as 2,000–2,500 full strings of cash. In this problem taxes are paid in silk and silver.

Credit System[54]

From the financial point of view the life of the lower classes must have been very hard in the period of the Southern Sung.[55] Besides the heavy taxes,[56] there was the usurious interest of the money lenders. "The peasant, compelled by his dire needs to

[49] "Ces certificats se négociaient sur le marché, sans même être nécessairement utilisés pour les candidats à la condition cléricale." Yang (2), p. 50.

[50] "Un curieux procédé de financement public." Ibid.

[51] "Die Blankoordinationsscheine wurden häufig als Geldbeschaffung für öffentliche Arbeiten benutzt und die Ämter hatten deshalb meist einen Vorrat von ihnen." Eichhorn (1), p. 23, note 4.

[52] See Yang (3), pp. 16 ff; see also Chi Yu-tang (1), p. 89.

[53] Yang (3), p. 18.

[54] I have relied mainly on Yang (3).

[55] However, there is not much information available on the lower classes. "We know little about the rural population. No one has concerned himself with providing information on village life and society." (Le peuple des campagnes nous est mal connu. Personne ne s'est soucié de nous renseigner en détail sur sa composition sociale et sur la vie des villages.) Gernet (2), p. 113.

[56] See the section on "Taxes and Levies of Service."

contract debts, sooner or later fell prey to the usurer who deprived him, in the form of interest, of more funds than he could obtain by working."[57]

"In T'ang times, interest was reckoned on a monthly basis in terms of *fên* 分, i.e. percent. Since the Sung period, another word, *li* 釐 has been added to mean one tenth of a *fên*, i.e. per mille. In addition, the word *fên* acquired a new meaning, 'one-tenth, or 10 percent.' Thus a *fên* may be either 10 percent or 1 percent, and a *li* either 1 percent or 1 per mille. Ordinarily the larger *fên* and *li* are used to refer to an annual rate, whereas the smaller *fên* or *li* refer to a monthly rate."[58] Ch'in Chiu-shao is not very consistent in this matter. Although in VI, 8 the monthly interest is given as 6.5 *li*, from the calculations it is obvious that 6.5 *li* means 6.5 percent and not 0.65 percent. Indeed the capital is given as 500,000 *kuan*; after one month the interest is 32,500 *kuan,* and consequently the rate is 6.5 percent.[59] In IX, 7 the rates are 1 *li* (1 percent), 2.5 *li* (2.5 percent), and 3 *li* (3 percent). It is not explicit in the text that these are monthly interest. In IX, 8 the monthly rate is 2 *fên* 2 *li,* which amounts to 2.2 percent. "Two major changes, therefore, have taken place since the Sung period with regard to the technique of charging interest. One is the use of an annual rate in addition to a monthly rate; the other is the introduction of the unit *li* as one-tenth of a *fên*."[60]

These rates were very high. L. S. Yang [(3), p. 95] says that in T'ang times "the ceiling rate was 6 percent per month for private loans, but 7 percent per month for government loans," and in Sung times it was respectively 4 percent and 5 percent. In VI, 8 the interest is at a rate of 6.5 percent per month; the capital lent out by a treasury is 500,000 *kuan*. This has to be paid back in 7 months in repayments of 100,000 *kuan*, but the

[57] "Der Landmann, den der bittere Mangel zum Schuldenmachen zwang, verfiel früher oder später dem Wucherer, der ihm an Zinsen mehr abnahm als er erarbeiten konnte." O. Franke (1), vol. 4, p. 372.
[58] Yang (3), p. 94.
[59] Mikami (1), p. 72, mistakenly says 0.65 percent. Ch'ien Pao-tsung (2'), p. 102, gives the right value, 6.5 percent.
[60] Yang (3), p. 94.

interest of each month is capitalized. In 7 months the interest amounts to about 124,706 *kuan*. This is an interest of about 25 percent. When we take into account the short term and the monthly repayments, this interest is extremely high. " . . . A typical loan in traditional China was on short term, at a high rate of interest, and for consumer expenditure."[61] Problem IX, 8 deals with a loan from a pawnshop.[62] The rate of interest is 2.2 percent monthly. "Rates of interest in pawnbroking were fixed by regulations promulgated by various local governments. In general, the ceiling rate was 3 percent . . . . "[63]

Another common practice of the loan system was to increase the rate as the amount of borrowed capital decreased. "The rate, however, varied with the size of the loan. For instance, the practice in Hu-chou, Chekiang province toward the end of the seventeenth century was to charge 1.5 percent per month if the pledged article was valued at 10 taels or more, 2 percent if it was worth one tael and above, but 3 percent if under one tael."[64] We find the same system as early as Sung times. In IX, 7 we see that the rates for the treasuries were 1 percent for 10,000 *kuan* or more, 2.5 percent for 1,000 *kuan* or more, and 3 percent for 100 *kuan* or more. In VI, 8 the moneylender is a treasury: the rate is 6.5 percent monthly. In IX, 8 the moneylender is a pawnshop: the rate is 2.2 percent monthly. "Official interest" was always higher than private charges.[65]

---

[61] Ibid., p. 92. "A twelfth-century scholar, Yang Shih, stated that the normal annual interest at his time was 50–70 per cent and sometimes as high as 100 per cent and the highest rate was about 8 per cent per month." Ibid., p. 98. In China interest was always very high. For the farm loan system (the so-called "green sprout money") the interest was at least 20 percent and in some cases 40 percent. See Lee Ping-hua (1), pp. 79 f. If we calculate the annual interest in IX,8, we get $2.2 \times 12 = 26.4$ percent (private loan); in VI,8 we get 78 percent (public loan).

[62] On the pawnshops, the oldest credit institution in China, see Yang (3), pp. 71 f.

[63] Yang (3), p. 98.

[64] Ibid.

[65] Ch'ien Pao-tsung (2′), p. 102.

Commercial Life in the Southern Sung

BARTER AND EXCHANGE VALUES

Problem IX, 4: "Three *shêng* 升 of pulse are bartered for 2 *shêng* of wheat; 1 *tou* 斗 5 *ko* 合 of wheat are bartered for 8 *ko* of linseed; 1 *shêng* 2 *ko* of linseed are bartered for 1 *shêng* 8 *ko* of ordinary rice. Now we wish to barter 14 *shih* 石 4 *tou* of pulse for linseed, and 21 *shih* 6 *tou* of wheat for ordinary rice."[66]

If we reduce all the quantities to *ko,* we get:

pulse	wheat	linseed	rice
30	20		
	105	8	
		12	18

By reducing to equivalent prices, we get (approximately):

Pulse		wheat		linseed		rice
10	=	6.5	=	0.5	=	0.75

Problem IX, 5 discusses the barter of cereals: "According to the rates of a granary, for 7 *shih* glutinous millet [*no-ku* 糯穀][67] we give 3 *shih* glutinous rice (*no-mi* 糯米); 1 *tou* of glutinous rice is bartered for 1 *tou* 7 *shêng* of wheat; 5 *shêng* of wheat is changed for 2 *chin* 斤 4 *liang* 兩 of kneaded leaven (*t'a-ch'ü* 踏麴);[68] 11 *chin* kneaded leaven is bartered for 1 *tou* 3 *shêng* "fermented

[66] For more information the reader is referred to "Lectures in Commodities" ("Wheat," p. 1; "Linseed," p. 35) and the bibliographical work by Merrill and Walker (1). On the exchange of commodities at fairs, see Sun and De Francis (1), p. 219. "The periodic fairs of Sung were a pattern of commodity exchange that occurred in an economy of rural self-sufficiency. For the most part the merchandise consisted of things of daily use...." Ibid., p. 221.

[67] This must be a kind of grain rich in proteins. Whether it is a botanical species is not certain.

[68] Note that *chin* and *liang* (1 *liang* = 1/16 *chin*) are measures of weight, but the others are measures of capacity. *T'a-ch'ü* seems to be a kind of malt. *T'a* 踏 (to trample) has perhaps the same meaning as in *t'a-tui* 踏碓, a pestle for hulling rice, worked by the foot. If this holds, *t'a-ch'ü* is ground malt prepared for fermentation, or "kneaded leaven." See Sung Ying-hsing (1), p. 290, and T. L. Davis (1). The term *t'a-ch'ü* is not given in Chia Ssǔ-hsüeh's *Ch'i-min yao-shu* (c. 540 A.D.).

rice" (*yün no-mi* 醞糯米) . . . ."[69] After reducing to the base 100, we get (approximately): 100 *shêng* glutinous grain=43 *shêng* glutinous rice=73 *shêng* wheat=32.9 *chin t'a-ch'ü*=39 *shêng* "fermented rice."[70]

FOREIGN TRADE

"Foreign trade grew tremendously during the Sung dynasty, with Chinese traders reaching the Indian and Persian coasts and Persian and Arab traders coming in large numbers to the leading ports of China."[71] The *Chu-fan chih* 諸蕃志 by Chao Ju-kua 趙汝适 (1225) contains much information on overseas trade in the Southern Sung, principally with Arabian countries.[72] Some of the luxuries that came from Arabia are mentioned in the *Shu-shu chiu-chang*,[73] namely,

1. garu wood (*ch'ên-hsiang* 沈香), packed as bundles (*kuo* 裹) (IX, 1) or measured in *liang* 兩 (IX, 2);[74]

2. tortoiseshell (*tai-mei* 瑇瑁),[75] measured by weight (*chin* 斤) (IX, 1);

3. frankincense (*ju-hsiang* 乳香), coming from south Arabia,[76] packed in cases (*t'ao* 套) (IX, 1);

---

[69] An arrangement for bartering products of the fields is explained in the *Chiu-chang suan shu*, ch. 2, pp. 23 ff; see Vogel (2), p. 17. However, none of the products of IX,5 is mentioned. For a general description of fermentation processes in China, the reader is referred to Li Ch'iao-p'ing (1), pp. 181 ff, and T. L. Davis (1). However, no explanation of the terms *t'a-ch'ü* and *yün-no* appears in their works.

[70] Nothing can be said about weight proportions because it is impossible to determine the specific gravities.

[71] Schurmann (1), p. 222.

[72] See the translation by Hirth and Rockhill (1). Another valuable work is Wheatley (1). See also Kirby (1), p. 147.

[73] This overseas trade was also responsible for the shortage of copper money in Southern Sung. See O. Franke (1), vol. 4, p. 382.

[74] *Aquilaria agallocha* (aloewood); see Chau Ju-kua (1), p. 195; Wheatley (1), p. 47 (and map); Bretschneider (1), vol. 3, p. 459. On the importance of imported wood, see Yang (2), p. 40, and Shafer (1), p. 163.

[75] Chau Ju-kua (1), p. 238; Wheatley (1), p. 81 (and map).

[76] *Boswellia Carteri*; see Chau Ju-kua (1), p. 195; Wheatley (1), p. 47 (and map); Bretschneider (1). vol. 2, p. 303 and vol. 3, p. 460; Shafer (1), pp. 170 f.

4. black pepper (*hu-chiao* 胡椒),[77] packed in parcels (*pao* 包);[78]
5. ivory (*hsiang-ya* 象牙),[79] measured by weight (*ko* 合).

In IX, 2 the seagoing vessels in use are mentioned.

COMMERCIAL AGENTS

There is only one problem in which commercial agents (*ya-jên* 牙人) are mentioned (VI, 6); they buy rice for the "harmonious purchase" (*ho-ti* 和糴;[80] their commission (*ya-ch'ien* 牙錢) is 30 *wên* 文 per *shih* 石 rice, or 1.2 per mille. A small part of this commission (3.2 percent) is converted into rice and made over to the *ch'ien-t'ou* 牽頭 or broker. This part is called *ch'ien-ch'ien* 牽錢.[81]

TRADE AND AGRICULTURE

In I,5 we are told of three farmers who sell their rice. Farmer *A* sells it to an official bureau in his own prefecture; *B* sells it to the people of An-chi,[82] and *C* sells it to a *lan-hu* 攬戶[83] of P'ing-chiang.[84]

Harmonious Purchase (Ho-ti 和糴)

From T'ang times on, the government bought grain from the people at a fair and set price. "During periods of high prices,

[77] Chau Ju-kua (1), p. 222; Wheatley (1), p. 100; Bretschneider (1), vol. 2, p. 323; Laufer (1), pp. 374 ff; Shafer (1), p. 149 ff.
[78] One parcel is equal to 40 *chin* weight.
[79] Chau Ju-kua (1), p. 232; Wheatley (1), p. 111 (and map); Shafer (1), p. 239 ff.
[80] See par. 4.
[81] See Sun and de Francis (2) p. 227: "At one time agents sent out by him (Chu Hsi) from Nan-k'ang-chün with 50,000 odd strings of cash from the official treasury proceeded to other localities that had good harvests that year and bought 23,522 pecks of rice for relief and military provisions." The circumstances seem to be the same as in Ch'in Chiu-shao's time. Chu Hsi lived from 1130 to 1200. We have not found any further information about the organization of the commercial agents nor about the true meaning of *ch'ien-t'ou*.
[82] In Chekiang.
[83] We are not sure of the exact economic function of these *lan-hu*. They were probably commercial agents.
[84] In Chekiang.

Figure 54

grain from the public granaries was thrown on the market, and
during periods of low prices grain was purchased for the public
granaries. This resulted in price stabilization."[85] This was
called the "harmonious purchase." In VI, 3 we read: "Five
official messengers on their way buy grain for the harmonious
purchase [see Figure 54 for place names]. According to mes-

[85] "In Zeiten höher Preise wurde Getreide aus den öffentlichen Speichern
auf den Markt geworfen, in Zeiten niedriger Preise Getreide für die öffentlichen Speicher eingekauft, wodurch eine Preisregulierung ermöglicht
wurde." Bi Dung (1), p. 16; see also Liu Mau-tsai (1), p. 127. This means
that if the supply surpasses the demand and consequently prices fall, the
government buys grain from the people and stores it in the public granaries;

senger *A,* in the prefecture of P'ing-chiang 平江 in Che-hsi
浙西[86] the price for one *shih* 石 is 35 *kuan wên* 貫文. [One *shih*
石 is] 135 *ko* 合;[87] the freight money to Chinkiang 鎮江[88]
is 900 *wên* per *shih*. In the department of An-chi 安吉[89] the price
for one *shih* is 29 *kuan* 500 *wên*; [one *shih* is] 110 *ko*;[90] the freight
money to Chinkiang is 1 *kuan* 200 *wên* per *shih*. In the prefecture
of Lung-hsing[91] 隆興 in Kiangsi 江西 the price per shih is 28
kuan 100 *wên;* [one *shih* is] 115 *ko*; the freight money to Chien-
k'ang 建康[92] is 1 *kuan* 700 *wên* per *shih.* In Chi-chou 吉州[93] the
price per *shih* is 25 *kuan* 850 *wên*; [one *shih* is] 120 *ko*; the
freight money to Chien-k'ang is 2 *kuan* 900 *wên* per *shih*. In
T'an-chou 潭州[94] in Hunan the price per *shih* is 27 *kuan* 300
wên; [one *shih* is] 118 *ko*; the freight money to O-chou 鄂州[95]
is 2 *kuan* 100 *wên* per *shih*. All these prices [are given] in
official *hui-tzŭ* 會子 of the seventeenth *chieh* 界;[96] for the rice,
make use of the grain measure [*hu* 斛] of the *Wên-szŭ Yüan*
文思院[97] to reduce to exact quantities. We wish to know the

---

thus the demand is artificially augmented and the prices rise; if the demand
surpasses the supply, and consequently prices rise, the government sells
grain to the people; thus the supply is augmented and prices fall.

[86] Che-hsi is in the northwest of Chekiang.

[87] See I,5 (Chapter 22).

[88] Or Chên-chiang 鎮江, a port on the Yangtse. See O. Franke (1), vol.
4, p. 222.

[89] In Chekiang.

[90] See I,5 (Chapter 22).

[91] Now Nan-ch'ang 南昌.

[92] In the delta of the Yangtse.

[93] Now Chi-an hsien 吉安縣.

[94] Now Ch'ang-sha 長沙 in Hunan.

[95] Now Wu-ch'ang 武昌 in Hupeh.

[96] See the section on "Money and Currency in the Southern Sung."

[97] The *Wên-szŭ yüan* was the office where the fine workmanship (in gold,
silver, jade, and so on) was managed. It was housed in a building of the
palace. We find it also in I,5, where the official corn-measure is also given
as equal to 83 *shêng*. It seems that this measure was used for all official
purchases (See I,5). See Yang (1), p. 83, note 32.

price per *shih* measured by the official grain measure and the relative cost-price. [The *hu* of the *Wên-szŭ Yüan* is 83 *ko* per *shih*]."[98]

In Problem VI, 5 "Rice of the 'harmonious purchase' is transported to hired granaries where they weigh the stores . . . ." According to VI, 6, "In a 'harmonious purchase' for 3,000,000 *kuan,* they look for a number of *shih* of rice . . . ." From these accounts, we learn that

1. the quantities of grain bought by the government were very large: 3,000,000 *kuan* or 120,000 *shih* (1 *shih* is about 103 liters);

2. there was a kind of "market survey," which took into account local prices, local measures, and transport charges;

3. granaries were hired to house stores of grain before transporting.

Transportation Problems

In the Southern Sung the increase of the population of the towns was very large.[99] On the other hand, the number of officials was high and the troops at the northern frontiers had to be provisioned.[100] "The question of transportation had been a difficult problem in connection with collecting land taxes."[101] For all these reasons the transportation problem was a very important one.

In the *Shu-shu chiu-chang,* VI, 2 we read: "In Kiangsi they transport by water 123,400 *shih* of rice. In the beginning they

[98] The text mistakenly says *tou* instead of *shih.*
[99] Description in Gernet (5).
[100] Moreover, the numerical strength of the troops was huge. See H. Franke (3), p. 199; Bi Dung (1), p. 15. "The demand for rice, grain, and pulse, in the capitals alone, particularly for the use of the military, was tremendous." (Der Bedarf der Hauptstädte allein an Reis, Korn und Hülsenfrüchten, insbesondere für den Unterhalt des Heeres, war gewaltig.) O. Franke (1), vol. 4, p. 376.
[101] "Ein schwieriges Problem bei Abführung der Grundsteuer bildete von jeher die Transportfrage." O. Franke (1), vol. 4, p. 376. This transportation problem was a very old one in China: see Twitchett (1), his interesting chapter on the transportation system in the T'ang (pp. 84–96); see also Chi Yu-tang (1), pp. 182 ff.

From	To	Price/*shih*	Freight/*shih*	%
P'ing-chiang	Chinkiang	35,000	900	2.5
An-chi	Chinkiang	29,500	1,200	4
Lung-hsing	Chien-k'ang	28,100	1,700	6
Chi-chou	Chien-k'ang	28,850	2,900	11
T'an-chou	O-chou	27,300	2,100	8

went to Chinkiang to ship [the rice]. The voyage is calculated at 2,130 *li*. The freight is 1 *kuan* 200 *wên* per *shih* (in *hui-tzŭ* of the seventeenth term, *chieh* 界).[102] Now they are prevented from bringing the load of rice farther and they go to Ch'ih-chou 池州[103] and unload there. From Ch'ih-chou to Chinkiang it is 880 *li*. . . . ." In VI, 3 (see p. 434) we see that transport charges are very important in the purchase of grain. In VI, 1 the prices for porters are given per burden (*tan* 擔) and per *li*. Prefecture *A* supplies 3,200 *liang* of silver and 64,000 rolls of silk. One burden is formed either by 500 *liang* of silver or by 60 rolls of silk. This gives 6.4 burdens of silver and 1066.6 burdens of silk and requires thus 1,073 porters. The value of the silver is 2,200 *wên*/*liang* and the value of the silk 2,000 *wên*/roll. This makes a total value of 135,000 *kuan*. The total porter salary is $6 \times 1,073 \times 1,000 = 6,438$ *kuan*. This is about 5 percent of the total value of the goods.

In VI, 3 the exact distances are not given, but we can calculate the percentage of the freight as shown in the accompanying table, from which it is obvious that over long distances the freight rates were rather high.

Problem VII, 4, which deals with the digging of a junction canal, is discussed in the next section.

Construction of Dykes

It is not necessary to lay stress on the fact that Chinese socioeconomic history is closely related to irrigation and flood

[102] This is not in the text preserved in the *Yung-lo ta-tien*.
[103] Now Kuei-ch'ih 貴池 in Anhwei.

control.[104] In the *Shu-shu chiu-chang* some of the problems mention construction of dykes.[105]

In I, 3 we learn that these dykes were built as a public service by the government. Four prefectures have to supply a number of workmen in proportion to their *wu-li* 物力 or "properties";[106] one man is to be levied for each 770 *kuan* "property."

The "properties" of the prefectures are respectively 138,600, 146,300, 192,500, and 184,800 *kuan;* consequently they have to levy respectively 180, 190, 250, and 240 men. Each workman has to build 60 square feet a day; as the dyke has a length of 19 *li* 235 *pu* 5 *ch'ih*[107] (about 11.300 kilometers) and a width of 20 feet (about 6.40 meters) the daily task is 3 feet lengthwise per man.

In V, 3 we have another problem on the construction of dykes: "Four prefectures together begin to build a dyke with a length of 36.5 *li* (c. 21 kilometers). Prefecture *A* levies 2,780 men; *B*, 1,990 men; *C*, 1,630 men; and *D*, 1,320 men . . . ."

In V, 2 Ch'in mentions the *wei-t'ien* 圍田 or "fields reclaimed from marshlands by dams and levees."[108] The area is 3,021

---

[104] A general study is Chi Ch'ao-ting (1).
[105] On the great extension of water-control activity, see Chi Ch'ao-ting (1), p. 43 and p. 134; "The history of the Sung dynasty is replete with records on water-control developments." Wada Sei (1'): "Irrigation spread rapidly on the south side of the Yangtse River under the Southern Sung."
[106] See the section on "Taxes and Levies of Service."
[107] 1 *li* = 360 *pu*; 1 *pu* = 5.8 feet.
[108] See Wada Sei (1'). The literal meaning is "surrounded fields." "With its new capital established at Hangchow, the unprecedented increase in population stimulated agricultural productivity, which, in turn, led to a demand for more and more land. Under the geographical conditions of South China the solution of this problem was sought in the drainage of the marshes and even of the lakes." Chi Ch'ao-ting (1), p. 134. A technical reason for the enclosure of the fields is given in Chi Yu-tang (1), p. 99: "In the Sung dynasty (960–1127 A.D.) the 'levee fields' were formed in the Kiangsu and Chekiang provinces. It was felt that when fields were surrounded by levees there was little fear of the fields being damaged greatly by heavy rainfall or drought."
On the socioeconomic problem caused by the *wei-t'ien* during the Southern Sung, see Chi Ch'ao-ting (1), pp. 135 ff.

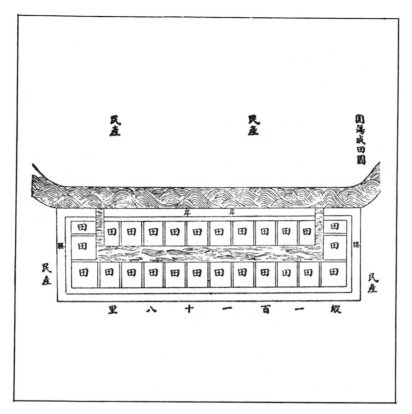

Figure 55. Drainage of marsh lands, Problem III, 9. From the *SSCC*, p. 15.

ch'ing 頃 51 *mou* 畝 15 *pu*.[109] In III, 9 there is a *wei-t'ien* which they wish to drain[110] (Figures 55 and 56). The width is 3 *li* (1.7 kilometers) and the length 180 *li* (c. 103 kilometers). In summer the depth of the water is 2.5 feet. There is a channel with a width of 130 feet and a length of 135 *li* which empties into a lake. They surround the land by dykes and carry off the water by a network of canals.[111]

[109] One *ch'ing* = 100 *mou;* 1 *mou* = 240 square *pu*. If we take 1 *mou* = 1.525 acres, this gives the respectable area of 45,880 acres.
[110] This is not really an irrigation problem [see Needham (1), vol. 3, p. 249], but a drainage problem.
[111] See Chi Ch'ao-ting (1), p. 27.

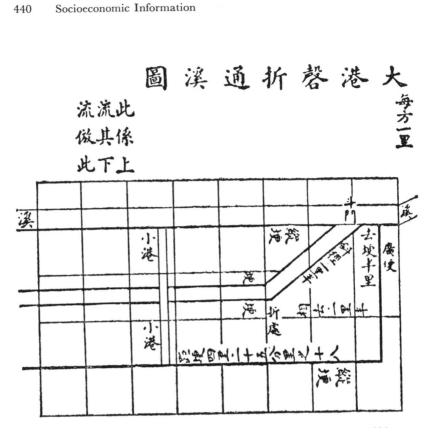

Figure 56. Reconstruction of Ch'in Chiu-shao's Problem III, 9 by Sung Ching-ch'ang. From the *Shu-shu chiu-chang cha-chi,* p. 92.

In VII, 3 we have a description of building a stone embankment:[112] "the length is 300 feet, 'the depth of the water' [the height of the dyke] is 42 feet. The width of the surface has to be 30 feet. The stone slabs have a length of 5 feet, a width of 2 feet, and a thickness of 0.5 feet. They use 10 *chin* 斤[113] of mortar

---

[112] On the construction of stone dykes, see Yang (2), p. 42: "In public works involving conservation and water control as well as transportation, the use of stone appears very early, particularly in the construction of dykes, dams, and bridges of any importance." (Dans les travaux publics intéressant la conservation et le contrôle des eaux ainsi que les transports, l'emploi de la pierre apparaît assez tôt, surtout dans la construction des digues, des barrages et des ponts de quelque importance.)

[113] We are not sure of the literal translation of this sentence; however the sense is clear, because further in the text these workmen are called "those who carry the mortar."

per stone slab. The height of each layer is 2 feet. The difference in width (between two layers) is 1 foot. Each stonemason makes 9 slabs. They use 4 workmen to carry 5 stones. Workmen carry the mortar and at the same time apply it, each workman hand-ling 110 *chin*. For supply [or, as cooks] one servant cares for 60 men; as controller[114] one man is responsible for 120 men . . . ." This gives a good idea of the organization for building a dyke.

Problem VII, 4 is entitled, "Calculations for Digging a Drainage Canal." It says: "They dig a junction canal, and the earth is used for building a dyke. It is ordered that the upper width (of the canal) be 60 feet and the lower width 40 feet. The upper reaches of the canal must have a depth of 8 feet and the lower reaches a depth of 16 feet. The length is 48 *li*. The lower width of the dyke must be 24 feet and the upper width 18 feet. The length is the same as that of the canal. Its height is un-known. The ratio of loose and packed earth is 4 to 3. The autumn standard [applies to] the tasks of men.[115] Each man himself digs out and transports the earth, builds the dyke, and rams the soil. The quantity must be 60 cubic feet daily. They build the dyke halfway and make a ramp to get the earth [on the dyke]. The tasks on the upper part and the lower part differ by 1/5.[116] Within one month [the work] must be finished. . . ."

Taxes and Levies of Service (Fu-i 賦役)

In China there were two major kinds of tax: the land tax (*fu* 賦) and the labor service (*i* 役), which was a tax on the individual or the household. As for installment payments, in the Southern Sung the old two-tax system (*liang-shui fa* 兩稅法)

[114] The controllers are called *pu-ya* 部押, guards from the administration. In Sung times a *ya-ssŭ* 押司 was the name for an underling in a magistrate's office. Further on in the text they are described specifically as *pu-ya hao-chai* 部押濠寨, the official guards of the dykes and strongholds.
[115] See the section on "Taxes and Levies of Service."
[116] For these "tasks," see the section on "Architecture." "Differ by 1/5" means that one task on the lower part of the dyke corresponds to 6/5 task on the upper part.

was followed.[117] "According to the two-tax system, which had replaced the old *tsu yung tiao* 租庸調 system,[118] taxes were paid in two installments, in the summer and in the fall."[119] "Payments during the Sung were made not only in grain but also in copper cash, silver, silk, cotton, various types of cloth, and in such miscellaneous items as salt, tea and honey. On the whole, the fall tax seems to have been paid chiefly in grain and the summer tax in various kinds of textile and yarn."[120]

The *wu-li* 物力 is the productive capacity based on the family property;[121] it was quoted in monetary units (*kuan* 貫). It seems to have originated during the Southern Sung. For country people it was calculated according to the number of *mou* of fields and paddy fields owned by a family; these fields were divided into several classes according to fertility (discussed later), and the *wu-li* was determined according to these classes. For example, in V, 7 the *wu-li* per *mou* of a rice field of a certain class is 1.200 *kuan*, another type of field 0.900 *kuan*. The *wu-li* was also used as base for calculating "harmonious purchase" (*ho-mai* 和買) (V, 7) and for the number of men to be levied for public services (I, 3; V, 8).

Three kinds of land tax are mentioned in the *Shu-shu chiu-chang*:

1. The fall tax, here called *miao-mi* 苗米 (sprout rice), as it was always paid in the form of rice or wheat. In V, 7 there is a fall tax of 3.5 *shêng* per *mou*. In V, 1 the total fall tax of a prefecture is 103,567.8442 *shih*.

2. The summer tax (*hsia-shui* 夏稅) was usually paid in various kinds of textiles.

3. The *ho-mai* was not a tax, but a loan on interest. It originated in the period 976–997 in the beginning of the Sung. In the

[117] "Introduced in the middle of the T'ang after the An Lu-shan rebellion." Schurmann (1), p. 70. There is a thorough treatment in Twitchett (1), pp. 39 ff.
[118] On the *tsu yung tiao* system, see Twitchett (1), p. 40.
[119] Schurmann (1), p. 70.
[120] Schurmann (1), pp. 70 f.
[121] See Liang Fang-chung (1), p. 3.

spring, money from the treasuries was lent to the people. In summer or fall *chüan* 絹 (an ordinary silk fabric) was repaid to the government.[122] In V, 7 for a *wu-li* of 32 *kuan,* the *ho-mai* amounts to 1 roll of cloth.

Rice and also wheat were considered as *pên-se* 本色 or "basic type" in connection with fall tax; for the summer tax and *ho-mai,* the *pên-se* consisted of rolls of cloth; the other goods were considered as *che-se* 折色 or "commutation type," and rates of exchange between them were fixed. The "commutation type" included silver, paper money, copper cash, and silk.

According to V, 7 (translation following this discussion) these land taxes were levied:

1. On paddy fields (*t'ien* 田)

a. fall tax (*miao-mi*), to be paid in basic type (rice)

b. summer tax (*hsia-shui*), to be paid

(1) partly in basic type (rolls of silk)

(2) partly in commutation type, called *che-pai* 折帛; *che-pai* means "converted silk" and was silk of low quality, called *ch'ou-chüan* 紬絹

2. On fields (*ti* 地), there was a tax called *shui ch'ou-chüan* 稅紬絹 (tax in the form of ordinary silk)

a. partly in basic type

b. partly in commutation type

3. *Ho-mai*: the *wu-li* of the paddy fields and other fields is added together; the *ho-mai* was levied on this amount in rolls of cloth

a. partly in basic type.

b. partly in commutation type.

A very important factor was the classification of land. "During the Sung Dynasty (960–1276) a classification of land was made on the basis of land ownership and of the density and organization of the population. The three classes were public land, arable land, and urban land. Under arable land five different grades were established at the beginning and taxes levied

[122] See Katô Shigeshi (1'), vol. 2, p. 403; Sudô Yoshiyuki (1'), p. 347; Hoshi Ayao (1'), p. 424.

accordingly. Later the number of grades was increased to ten."[123] "Both government and private land were classified according to fertility into many grades, such as the 'five classes,' 'the nine classes,' and the 'three classes with nine subclasses.'[124] The rate of tax varied according to fertility."[125]

The whole fifth book of the *Shu-shu chiu-chang* is devoted to tax problems.[126] Problem V, I says: "There are lands of a prefecture that were submerged by the sea; now they have emerged again.[127] After a long time the population again reports and requests the government to survey the land and to place it at the disposal of the six villages. They call the nearest suburb *A* and the farthest *F*. All of them have nine kinds of rice fields." Then these nine kinds of field are given for the six villages; the fields are divided in three categories, and each in three subcategories.[128] "They [the officials] rely on the fall tax [*miao-mi* 苗米] this prefecture collected formerly: 103,567 *shih* 8 *tou* 4 *shêng* 4 *ko* 2 *shuo* [103,567.8442 *shih*]; the *ho-mai* 和買 was 13,498 *p'i* 1 *chang* 7 *ch'ih* 3 *ts'un* 7 *fên* 5 *li*; the summer tax was 9,876 *p'i* 3 *chang* 2 *ch'ih* 6 *ts'un* 5 *fên* 6 *li*.[129] The rice fields of these six villages are of three kinds. *A* is the superior category, *B* and *C* are the next category, *D, E,* and *F* of the last category. We first assume that there are three degrees of land tax in such a way that the proportion of the upper category to the middle one, and of the middle to the lower one, is 11 to 10.[130] Next we

[123] Chi Yu-tang (1), p. 40: see also O. Franke (1), vol. 4, p. 376. This classification was made for the first time in 769: see Twitchett (1), p. 36.
[124] See Problem V,1 (Chapter 22).
[125] Liang Fang-chung (1), p. 3.
[126] On the whole, they are not interesting from the mathematical point of view.
[127] See Lee Ping-hua (1), p. 303: "People were urged to settle on different kinds of fields—such as those confiscated by the government on the extinction of doors (i.e., households), sandy fields left by the Yangtse River and muddy fields due to the receding of the ocean."
[128] The areas are given in *mou* 畝, with subdivision *chüeh* 角 (1 *chüeh* = 1/4 *mou*) and *pu* 步 (1 *mou* = 240 *pu*).
[129] For the metric system, see Chapter 6.
[130] For this proportion, the expression *shih fên wai ch'a i* 十分外差一 ("exclusive discrepancy of 1 part in 10") is used. The meaning is $A/B = (10+1)/10$.

assume that the nine categories [of fields] of each village are calculated according to a relation $(10-a)/10$;[131] the common tax-quotas (*tsu-ô* 租額) are used. The fields of village B are the most fertile, then $D, A, C, F,$ and E follow in that order. . . ."

Problem V, 2 deals with the taxes on a *wei-t'ien* 圍田 or "field reclaimed from marsh lands by dams and levees."[132] These "encompassed fields" are divided into three categories, which respectively bring in as taxes 6 *tou* 斗, 4.5 *tou,* and 4 *tou* [rice] per *mou.*

Problem V, 3, on the levies of service for building a dyke, was explained earlier in this chapter.

According to Problem V, 4, "In a contemplated reduction of the land tax on government land, allowing the use of summer wheat with conversion (*chê-na* 折納) of a fraction, those who cultivate with government oxen get a reduction of 20 percent; those who cultivate with private oxen get a reduction of 40 percent. In addition, for the annual tax grain, one-third is allowed for the summer grain to be commuted into the two types of *mai* 麥: 4/10 *ta-mai* 大麥 (barley), 6/10 *hsiao-mai* 小麥 (wheat); in the commuted categories (*chê-sê* 折色), 3 *shih* barley may be commuted by 2 *shih* wheat, and 2 *shih* wheat by 3.5 *shih* millet.

The old quotas for the land tax of government land are: for 1 *shih* 'government seed' a tax is paid of 5 *shih*; for one *shih* of 'private seed' a tax is paid of 3 *shih*. Now there is [a plot] of government land (*t'un-t'ien* 屯田).[133] Last year it was estimated that the public and private seeds amounted to 9,782 *shih* and the total collected taxes amounted to 39,586 *shih* of cereals . . . ."

The summer wheat was harvested in the fall; the tax treated of here is thus the fall tax. The public oxen were those belonging to the government offices; the point seems to be that they

---

[131] The expression is *shih fên nei ch'a i* 十分內差一 ("inclusive discrepancy of 1 part in 10"). The meaning is $A/B = (10-1)/10$.

[132] See the section on "Construction of Dykes."

[133] Government land belonging to a military colony. See Schurmann (1), p. 29.

were used on public land.[134] We see that the reduction for owners of private land was larger than for those who worked on public land. This was a general rule.[135] We see the same proportion in the tax quota: for one *shih* seed, they had to pay 5 *shih* of cereals from the crops on public land and 3 *shih* on private land. Problem V, 5 says: "Question concerning solicitude [shown for the people] in the provinces. Recently in a certain prefecture, the remainder of the fall tax of the household taxes of the three lowest classes of dutiable families still outstanding was already remitted to the extent of 1,355 *kuan* 706 *wên* in coin, and 5,272 *shih* 1 *tou* 9 *shêng* in grain. The 'properties' [*wu-li* 物力] of the lower [classes] of this prefecture total 37,658 *kuan* 500 *wên*. Now the officials state that in this prefecture there is a majority of families who gladly pay [their taxes] and who have no arrears. If now [the inhabitants] receive the mercy of the remittance of the remainder [literally, the tail] of the tax, this would seem contrariwise to benefit the laggard payers and would be most unreasonable. Consequently it was proposed that the good [tax] payers of the three classes as regards the two taxes of next year should benefit by a decrease according to last year's standards. Upon verification of the records it appears that the 'properties' of the families of the three classes without [tax] debts are 220,815 *kuan* 321 *wên*. We wish to know what would be the fitting decrease in coin and rice per 100 *wên* and what would be the total decrease." (Note: for the calculations they rely on the *wu-li* of the lower classes, and find 3.6 percent coin and 14 percent grain remitted. This standard is applied to the "property" of the families of good payers of all classes.)

Problem V, 6 deals with the payment of taxes in the form of floss silk (*mien* 綿); 11,033 families bring in floss silk for a value of 88,337.6 *liang*; the families are divided into five classes.

---

[134] These public oxen are already mentioned in the *Chin-shu, tsai-chi,* ch. 9, p. 6a.

[135] "Speaking in a general way, the rate of taxation on government land was higher than on private land." Liang Fang-chung (1), p. 3.

Problem V, 7: "Transfer of the household tax [*hu-shui* 戶 稅].[136] In a certain prefecture, *A* states that for the paddy fields and other fields [*ti* 地] of his household he had originally paid a fall tax of 35 *shih* 7 *tou* [rice]. As for the *ho-mai* [he paid] 11 rolls[137] and 29.4875 feet basic type [*pên-se* 本色] and 27 rolls and 0.21375 feet commutation type [*che-pai* 折帛]. As for the *ch'ou-chüan* tax[138] [he paid] 8 rolls 39.739 feet commutation type and 20 rolls 28.939 feet basic type. Farmer *A* made over 407 *mou* of paddy fields to *B* and 516 *mou* of paddy fields to *C*; then he asked to transfer the taxes paid by his household for these rice fields and to incorporate them into the taxes of both households *B* and *C*. They calculate that *B* originally had 375 *mou* of rice fields and *C* had 463 *mou* and that they belonged to this class in this village. The fall tax amounts to 3.5 *shêng* a *mou* and the summer tax to 1.15 feet [silk] a *mou*. The *wu-li* is 1.200 *kuan* [per *mou* rice field]. The *ch'ou-chüan* tax of the fields in this class is 1.34 feet and the *wu-li* 0.900 *kuan* [per *mou*]. On a *wu-li* of 32 *kuan* a *ho-mai* of 1 roll is applied. Three-tenths of the *ho-mai* is collected in 'basic type' and 7/10 in 'commuted silk.' The summer tax is 7/10 basic type and 3/10 commuted silk; the *ch'ou-chüan* is 5/10 basic type and 5/10 commuted silk . . . ."

Problem V,8 deals with the levy of men for transporting rations for the troops to the frontier. The number of men amounts to 12,000; they are levied in proportion to the *wu-li* of the villages and in inverse proportion to the distance to the place where the rations have to be delivered.

Problem V, 9 treats incentives for selling relief grain.

Architecture

INTRODUCTORY NOTE

Any attempt to interpret the few architectural data in Ch'in's work constantly calls to mind the words of Demiéville: "None

---

[136] "This was a money collected from all households at a rate varying in accordance with the household category." Twitchett (1), p. 31.
[137] A roll of textile was 40 feet.
[138] The *ch'ou-chüan* was an ordinary silk fabric; see Liang Fang-chung (1), p. 3.

of the arts of the Chinese is so little known as architecture."[139]
Even the splendid Sung work on architecture, the *Ying-tsao
fa-shih* (1097) cannot give a decisive answer to all our problems.[140] The other works used here are Liang Szŭ-ch'êng (1')
and (2'); Liu Chih-p'ing (1'); Yüeh Chia-tsao (1'); and the
periodical *Chung-kuo ying-tsao hsüeh-shê hui-k'an* (Bulletin of the
Society for Research in Chinese Architecture). The most interesting article in any Western language is Demiéville(1); in addition there are Hsü Ching-chih (1) and Sirén (1).

As stone building material,[141] I, 8 includes "large squares"
(1.3 × 1.3 feet), "small squares" (1.1 × 1.1 feet), "*liu-mên* 六
門"[142] (1 × 0.5 × 0.2 feet), and "city wall bricks" (1.2 × 0.6 ×
0.25 feet). City walls were built of earth packed and rammed
between forms (*terre pisé*) and the foot of the wall was covered
with large stone slabs (5 × 2 × 0.5 feet) (VII, 1).[143] The whole

[139] "Aucun des arts chinois n'est si mal connu que l'architecture." Demiéville (1), p. 213.

[140] "Li Chieh says nothing of porches, nor of door or window frames,
nor of pillars or stone pavements." (Li Kiai ne parle ni de portiques, ni
d'encadrements de portes et de fenêtres, ni de colonnes, ni de pavage en
pierre.) Demiéville (1), p. 244. Li Chieh 李誡 was the author of the *Ying-
tsao fa-shih*.

[141] "Unlike wood, stone was used in Chinese architecture on a very limited
scale." (Contrairement au bois, la pierre a été utilisée sur une échelle
assez restreinte.) Yang (2), p. 40. "The use of stone seems to have been on
an even smaller scale under the Sung than under the Ming and Ch'ing."
(L'utilisation de la pierre aurait été plus réduite encore sous les Song que
sous les Ming et les Ts'ing.) Demiéville (1), p. 244.

[142] Literally, "six gates." No explanation found.

[143] "Bricks were also used as exterior and sometimes interior covering for
earthen city walls. In ancient times, we find this kind of covering in the
walls of the imperial capital and the big cities, but it seems not to have been
in general use before the period of the Ming and the Ch'ing." (La brique
était également utilisée comme revêtement extérieur et parfois intérieur
des murailles en terre de villes. Ce genre de revêtement est attesté à date
ancienne pour les murs de la capitale impériale et des grandes villes, mais
il ne semble pas s'être généralisé avant l'époque des Ming et des Ts'ing.)
Yang (2), p. 43. "The bricks used for constructing city walls, called *ch'êng-
chuan* 城磚, were very big." (Les briques employées pour la construction
des murs de villes, et qu'on appelait *tch'êng-tchouan*, étaient particulièrement
grandes.) Ibid. Stone slabs of the same size are used for the construction of
dykes (VII, 3); see the section of this chapter on that subject. They were also

wall was covered with bricks (*chuan* 甎) in layers above each other. The sizes of the bricks were 1.2 × 0.6 × 0.25 feet.[144] For the walls of a turret, wall bricks were used; their size was 1.6 × 0.6 × 0.2 feet (VII, 2). For flooring a house, tiles of 0.8 × 0.4 × 0.1 feet were used (VII, 7). These tiles also have the name *liu-mên chuan*.

As for wooden materials and the parts used in wooden buildings, the terminological problem is a very difficult one. Nothing is specified in the sources about the nature of the wood.[145] For the construction of a city wall (VII, 1) the following wooden materials (*mu-wu-liao* 木物料) were used:

1. *Yung-ting chu* 永定柱, "eternally fixed" pillars, with a length of 35 feet and a diameter of 1 foot. Twenty of these pillars are used per *chang* (10 feet). In the *Ying-tsao fa-shih* we read: "When the *yung-ting* are placed, their length is in correspondence with the height of the wall. Their diameter is 1 to 1.2 feet."[146] In VII, 1 the height of the wall is 30 feet, from which it is clear that they were driven 5 feet into the ground. The function of these gigantic pillars is not clear; there is no further explanation in the *Ying-tsao fa-shih*.[147] Perhaps they were used as reinforcements of the wall.

2. *P'a-t'ou i-hou mu* 爬頭拽後木 with a length of 20 feet and a diameter of 0.7 foot. Eighty of these beams were used per *chang*.

3. *T'uan-tzŭ mu* 摶子木, with a length of 10 feet and a diameter of 0.3 foot. Three hundred pieces were used per *chang*.

---

used for building watchtowers (VII, 5). See also Needham (1), vol. 4, Part 3, p. 38.

[144] These dimensions agree with those of the *liu-men*, see VII, 5.

[145] Even the *Ying-tsao fa-shih* does not say anything about the nature of the wood. "The nature of these woods has not been determined." (La nature de ces 'bois' n'est pas déterminée.) Demiéville (1), p. 246. We know that a large part of this wood was imported. "In order to supplement indigenous products, Chinese builders also relied on imported wood. This was particularly true under the Sung dynasty. Many beautiful kinds of wood were then imported from Japan." (Pour compléter les produits indigènes, les constructeurs chinois recouraient aussi à du bois importé. Tel fut le cas en particulier sous la dynastie des Song. Beaucoup de beau bois fut alors importé du Japon.) Yang (2), p. 40.

[146] Ch. 3, p. 55.

[147] This was chiefly a manual for architects.

	Length (Feet)	Diameter (Feet)	Number Used
1. *Wo-niu mu* 臥牛本	16	1.1	11 beams
2. *Ta-nao mu* 搭腦本	20	1	11 beams
3. *K'an-hao chu* 看濠柱	16	1.2	11 posts
4. *Fu-hao chu* 副濠柱	15	1.2	11 posts
5. *Kua-chia chu* 掛甲柱	13	1.1	11 posts
6. *Hu-tun chu* 虎蹲柱	7.5	1	11 posts
7. *Yang-huang pan* 仰艎板	10	1.2	45 planks
8. *P'ing-mien pan* 平面板	10	1.2	35 planks
9. *Ch'uan-kua fang* 串掛枋	5	1	73 laths

In VII, 2 all the parts for the construction of a look-out turret are given. As no explanation of this wooden construction could be found, it is possible only to list these parts as a matter for further investigation (see the accompanying table).

In the same problem several kinds of nails used for the wooden constructions are mentioned:[148]

1-foot nails	8 pieces used
8-inch nails	270 pieces used
5-inch nails	100 pieces used
4-inch nails	50 pieces used
Ting-huan 丁環 (annular studs?)[149]	20 pieces used

We have no information about the covering of the roofs. In VII, 2 there are some data on the covering of the ceiling of a lookout turret in order to make a platform. This was done with square tiles, called "four-eight bricks" (*ssŭ-pa-chuan* 四八甎) and built up in three connected layers. Six thousand tiles were used for one platform, and they were fixed by means of mortar, half a *chin* 斤 for one tile.[150]

In VII, 2 the use of paper pulp (*chih-chin* 紙觔) is mentioned, but the application is not specified. On the use of mortar

---

[148] There is a long section on nails in the *Ying-tsao fa-shih*, ch. 28, pp. 104–113.

[149] *Ting* means "a nail"; *huan* means "a ring."

[150] A *chin* is c. 1.3 liter. These tiles can be seen on a picture in the *Wu-ching tsung-yao*, ch. 12, p. 6a.

(*shih-hui* 石灰), we learn in VII, 1 that 10 *chin* were used for a stone slab of $5 \times 2 \times 0.5$ feet (see also VII, 3). In VII,1 Ch'in mentions the *chüeh* 橛 (also called *chüeh-tzŭ mu* 橛子木), which were tenons for fixing the wooden constructions.[151]

Other materials used for building are woven ropes (*jên-so* 紝索) (10 feet long and 0.05 foot in diameter), coarse rush mats (*lu-hsi* 蘆蓆), green reeds (*ch'ing-mao* 青茅), and bamboo, namely, *szŭ-kan kuei-chu* 絲竿筀竹[152] and *jui-tzŭ shui-chu* 芮子水竹.[153] More information on the use of bamboo is given in VII, 8, where bundles of bamboo and rush are described.

The building of a city wall is described in VII, 1, entitled: "Project for the building of a city wall." In the first section the several parts together with their sizes are given: "To build a city wall in Huai-chün 淮郡.[154] The circumference is 1,510 *chang* [15,100 feet]. Outside they build a *yang-ma* wall ["sheep and horse wall"], and dig a moat of the same length as the wall [Figure 57]. The 'body' of the city wall has a height of 3 *chang* [30 feet], an upper width of 3 *chang*, and a lower width of 7.5 *chang*" [Figure 58]. These sizes do not agree with those given in the *Ying-tsao fa-shih*, ch. 3, p. 55. The volume is calculated by means of the usual formula $[(B+b)H/2]L$.[155] "The 'sheep and horse' wall[156] has a height of 10 feet [1 *chang*], an upper width of 5 feet, and a lower width of 10 feet." This wall was

[151] The measures are the same as those given in the *Ying-tsao fa-shih*, vol. 1, p. 55. The length of the tenons was 1 foot and the square profile had a side of 0.1 foot.

[152] *Kuei-chu* is a variety of bamboo from which thin fibers can be made. The meaning of *szŭ-kan kuei-chu* is probably "*kuei*-bamboo with fibrous stems."

[153] *Shui-chu* is water bamboo. The meaning of *jui-tzŭ* is not clear.

[154] In Honan.

[155] Already given in the *Chiu-chang suan-shu*. See Vogel (2), pp. 43 ff. Other data on these city walls are given in the *Wu-pei chih*, ch. 110, p. 19b. There are pictures in *Wu-ching tsung-yao* (see note 158) and *San-ts'ai t'u-hui* (see note 156).

[156] There is a picture in the *San-ts'ai t'u-hui, Kung-shih*, ch. 2. See also *Wu-ching tsung-yao*, ch. 3, p. 16a; *Wu-pei chih*, ch. 109, p. 19a. The sizes were not always the same: for example, the *Wu-ching* gives a height of between 8 and 10 feet.

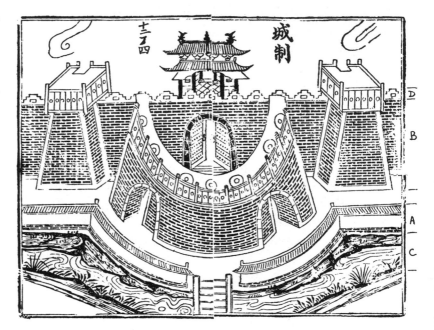

Figure 57. A city wall. A: the "sheep-horse wall." B: the city wall. C: the moat. D: the "women's wall." From *Wu-ching tsung-yao* 武經總要 (Conspectus of Essential Military Techniques, 1044; edition of ca. 1500, reprinted Peking, 1959), ch. 12, pp. 4a-4b.

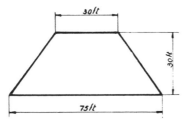

Figure 58. Section of a city wall

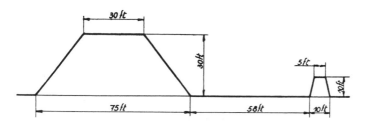

Figure 59. City wall and *yang-ma* wall in the proportion given in the *SSCC* (VII,1)

built outside the city wall and inside the moat; it was a small wall, usually at a distance of 10 paces from the city wall. The general configuration of these fortifications was that shown in Figure 59.[157] "The moat has an upper width of 300 feet and a lower width of 250 feet." The depth is not given, because it must be calculated; in the answer we find it to be 80 feet.

The woman's wall (Figure 60) was built above on the city wall.[158] It was a kind of battlement found also on ships[159] and on carts.[160] In the *Ying-tsao fa-shih*,[161] we read: "We call it woman's wall, because it is as low compared with the city wall as a woman compared with a man."

A projection at the base of the woman's wall was called the "magpie tower" (*ch'üeh-t'ai* 鵲臺) and the upper part the "woman's head" (*nü-t'ou* 女頭). The text says: "The woman's head and the magpie tower have a height of 5.5 feet, a total width of 3.6 feet, and a length of 10 feet. The length of the magpie tower is 10 feet, its height is 0.5 foot, and its width 5.4 feet." The woman's head consists of three different parts, called the base (*tso-tzŭ* 座子),with a length of 10 feet, a height of 2.25 feet, and a width of 3.6 feet; "the shoulder" (*chien-tzŭ*

[157] These walls are clearly represented in the *Wu-ching tsung-yao, ch'ien-chi*, ch. 12, p. 6a. The proportions are about right.
[158] See *Wu-pei chih*, ch. 110, p. 5a; a picture in the *Wu-ching tsung-yao, ch'ien-chi*, ch. 12, p. 19b.
[159] *Wu-ching tsung-yao*, part 5, ch. 11, p. 10a.
[160] Ibid., part 5, ch. 10, p. 27b.
[161] Ch. 1, p. 10.

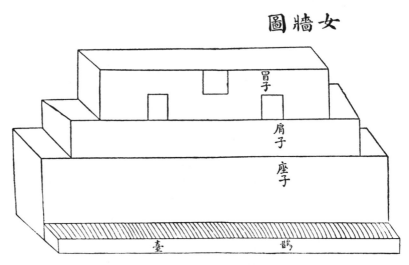

Figure 60. The woman's wall. The projection at the bottom is the magpie tower. From the *SSCC*, p. 328.

肩子), 8.4 × 1.25 × 3.6 feet; and "the cap" (*mao-tzŭ* 帽子), 6.6 × 1.5 × 3.6 feet. In this wall there are three loopholes[162] with a breadth of 0.6 feet and a length of 0.75 feet; the outer hole compared with the inner hole has an incline of 0.3 feet.[163] Another part of the fortifications was the so-called *hu-hsien ch'iang* 護嶮牆, the "wall protecting the vulnerable point." It was 1.2 feet wide and 3 feet high.[164]

Concerning the construction of the city wall, the *Shu-shu chiu-chang* gives these details: "Stone slabs cover the foot of the

[162] For archers.
[163] The text is not very clear. In the solution (p. 330) we have: "The inner and outer holes of the loopholes, although they differ by 0.3 feet and [we have to compute] the 'empty content' by means of the oblique depth, we compensate the surplus by the deficiency, and take it as equal to the perpendicular depth." The volume is given as 0.6 × 0.75 × 3.6. The *Shu-shu chiu-chang cha-chi*, p. 131 says that this is only the volume of one loophole, and that the multiplication by three has been forgotten.
[164] See *Ying-tsao fa-shih*, ch. 16, p. 120.

wall all the way around in three layers.[165] Each slab has a
length of 5 feet, a width of 2 feet, and a thickness of 0.5 foot.
For the whole body [of the wall] one makes use of bricks arranged in layers. For the lower *chang* [10 feet] there are 9
widths, for the middle *chang* 7, and for the upper *chang* 5 widths.
As for the bricks, each plate has a length of 1.2 feet, a width of
0.6 foot, and a thickness of 0.25 foot." The interpretation of
this text seems to be as follows: for covering the foot of the wall,
serving as a solid base for the masonry, large stone slabs are
used, three slabs next to each other, giving a width of $3 \times 2 = 6$
feet. To the height of 10 feet the masonry is made with 9
bricks next to each other, giving a total width of $9 \times 0.6 = 5.4$
feet. Between the heights of 10 and 20 feet the number of bricks
was reduced to 7, and for the highest 10 feet to 5. This is obvious
from the computations in Chin's work, p. 330: a square of 10
square feet is multiplied by 21 layers $(5+7+9)$, giving 2,100
feet of bricks per square *chang* on the average. The area of one
brick is $1.2 \times 0.25 = 0.3$ square foot. By dividing 2,100 by 0.30
we get 7,000 bricks per square *chang*. In the same problem we
see that the proportion of earth dug out to earth rammed was
considered as 4 to 3.[166]

A very interesting point is the organization of the work,
spelled out at great length in the *Ying-tsao fa-shih*, ch. 16–25.
"Work was not measured in units of time, but by tasks."[167] For
instance, the working of ten stone slabs constitutes one task (*kung*
功), the transportation and placing is another task. Carrying
and applying 1,000 *chin* of mortar is also one task.
The following tasks are given:

Placing of the *yung-ting* posts   0.7 task
Chiseling the same   0.3 task
Making of the *p'a-t'ou i-hou* beams   0.3 task
Chiseling the same   0.2 task

[165] See Yang (2), p. 43: note 143. Needham (1), vol. 4, part 3, p. 50.
[166] This was already known in the time of the *Chiu-chang suan-shu*, ch. 5,
p. 65; see Vogel (2), p. 43.
[167] Demiéville (1), p. 263.

Making of the *t'uan-tzŭ* beams	0.2 task
Placing of the same	0.2 task
Making of a tenon	0.07 task
Weaving of a rope	0.09 task
Making 10 stone plates	1 task
Carrying and fixing these plates	1 task
Carrying and using 1,000 *chin* of mortar	1 task
Hacking, digging out, transporting, or ramming 60 cubic feet of earth	1 task
Laying 700 bricks	1 task

Sometimes it is indicated that the performances were calculated according to the "fall standard" (*ch'iu-ch'êng* 秋程) or the "spring standard" (*ch'un-ch'êng* 春程);[168] this means that the length of the day was taken into account.[169] This system of tasks reflects a considerable rationalization of labor.[170]

For digging out the moat and building the earthen wall, four kinds of workmen were used:

hu-shou 钁手, or workmen who hack the earth with a mattock (on the moat).

ch'iao-shou 锹手, or workmen who dig out the earth (on the moat),

tan-t'u 擔土, or workmen who transport the earth,

ch'u-shou 杵手, or workmen who ram down the earth with a pestle (on the wall).

As mentioned before, a task consisted of working 60 cubic feet of earth. As for supplies, one cook took care of 60 workmen. One official (*pu-ya hao-chai* 部押濠寨) was responsible for 120

---

168 For example VII,4 and VII,5 for fall standard; I,3 for spring standard.
169 Only these standards are given; both equal 60 square feet a task. Needham says: "The summer and autumn, times of hoeing and harvest, are never to be used for public works..." [(1), vol. 4, p. 223]; and "Most of the work must be done during the spring agricultural lull when water-levels are low" (ibid.). In Ch'in's work, however, the fall standard is mentioned several times: in VII,4 for digging a canal; in VII,5 for constructing a watch tower (see note 168).
170 However, a daily rate must have been fixed, because the salaries were reckoned per day.

men.[171] The payment was as follows: each workman got 100 *wên* (new *hui-tzŭ*[172]) and 2.5 *shêng* of rice a day.[173]

Problem VII, 2 deals with the construction of 60 lookout turrets on the city wall.[174] The work on one tower is calculated to take 396 man-days. Problems VII, 3 and VII, 4 on the construction of dykes are treated in the preceding chapter.

The construction of a watchtower (Figures 61 and 62) is dealt with in VII, 5: "The perpendicular height is 120 feet. The upper width is 50 feet and the length 70 feet; the lower width is 150 feet and the length 170 feet. The length runs east-west and the width runs north-south[175] [Figure 63]. According

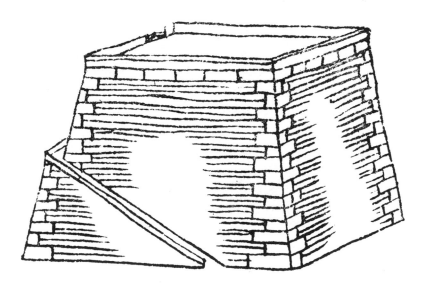

Figure 61. Watchtower. From Yüeh Chia-tsao (1'), fig. 2 (illustrations, p. 1).

[171] See note 114. For cooks and officials the same rule is given in VII,3.
[172] See the section on "Money and Currency in the Southern Sung."
[173] See Yang (2), pp. 49 f.
[174] The number of technical terms in this problem is very great.
[175] The technical term for north-south is *mou* 袤; for east-west it is *kuang* 廣.

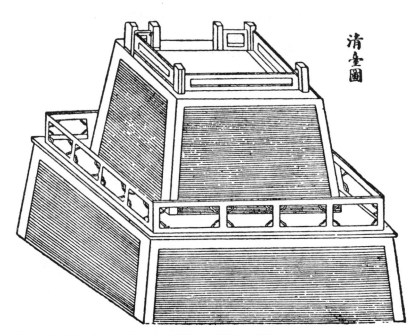

Figure 62. Watchtower, Problem VII, 5. *SSCC*, p. 347.

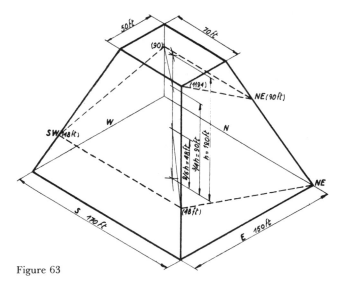

Figure 63

to the fall standard[176] the men daily cover a distance of 60 *li* [a *li* is 360 paces]. Of the men who use the mattock and those who dig each [produces] 200 square feet. Of those who build up the earthen construction each makes 90 square feet. Those who transport the earth carry 1.3 cubic feet; they go 160 paces there and back, including 40 paces up and down on a scaffold ramp. When building has attained one-third [of the projected height], 3 [parts] on the scaffold ramp are equated to 5 [parts] on the flat path;[177] at one-half [of the height] 3 on the ramp equal 7 on the flat; at two-thirds, 2 equal 5.[178] During the time of waiting, to [every] ten [paces] add one.[179] During the time of transporting, for [every] 20 paces deduct one.[180] Now three

[176] See the section on "Taxes and Levies of Service."

[177] The level path is 60 paces, the scaffold ramp 20 paces. The proportion seems to be that of the performances: 5 tasks on the flat path of 120 paces are equal to 3 tasks on the slanting path of 40 paces.

[178] However, I do not understand the application of these proportions made in the computations. Indeed, a general proportion of one unit of the diagonal path to one unit of the level path is computed as follows:

$$\frac{2}{3} = \frac{12}{18}; \quad \frac{1}{2} = \frac{9}{18}; \quad \frac{1}{3} = \frac{6}{18}; \quad \frac{1}{1} = \frac{18}{18}$$

$$\frac{5}{2} = \frac{45}{18}; \quad \frac{7}{3} = \frac{42}{18}; \quad \frac{5}{3} = \frac{30}{18}; \quad \frac{1}{1} = \frac{18}{18}.$$

The first row of fractions gives the heights of the parts of the tower already finished, the second row the proportions between the level and the diagonal part of the way. Now Ch'in tries to calculate a numerical index of task difficulty and performance from these data.

$$\frac{12}{18} \times \frac{45}{18} = \frac{540}{324}; \quad \frac{9}{18} \times \frac{42}{18} = \frac{378}{324}; \quad \frac{6}{18} \times \frac{30}{18} = \frac{180}{324};$$

$$\frac{18}{18} \times \frac{18}{18} = \frac{324}{324}.$$

$$\frac{540 + 378 + 180 + 324}{324 + 324 + 324 + 324} = \frac{1422}{1296} = \frac{79}{72}.$$

Reduced to the actual distance, we get: $79 \times 40 = 3{,}160$ and $72 \times 120 = 8{,}640$. The total performance has the index: $3{,}160 + 8{,}640 = 11{,}800$.

[179] The *ch'ih-ch'u* 踟躕 interval (literally, the period of hesitation) is the time used for loading and discharging; this time is estimated as 1/10 of the transport time.

[180] This seems to mean that when 20 paces have to be covered, for the calculations of performance we have to take 21 paces in order to get real indexes. This gives a strong impression of Taylorism *avant la lettre*.

prefectures, *A, B,* and *C* send people. Prefecture *A* is a neigh-
boring suburb[181] with a tax quota of 133,866; prefecture *B*
is at a distance of 120 *li* from the place of the tower and has a
tax quota of 237,984; prefecture *C* is at a distance of 180 *li*
from the place of the tower and has a tax quota of 312,354.
Everything is equally allotted according to the distances and the
amount of the tax quota.[182] The lower part of the tower
is covered with a stone base in seven layers. First the body of the
tower is covered round with bricks. Next with bricks [put]
layer on layer they raise an encircling ramp, which encircles
the tower in five stages,[183] having a width of 6 feet. One must
have two horizontal paths [facing] the north-south direction
and three sloping paths between them [that is, the horizontal
paths], [facing] the east-west direction. We start from the
northeast corner. On the eastern outmost way we ascend toward the south. From the southeast corner we turn to the west;
we turn around passing through the north and return again to
the east. Next we ascend on the eastern inmost way up to the
southeast corner and mount to the top of the tower. For the
height of the northeast corner of the eastern inmost path and
both corners of the northern horizontal way, and the northeast
corner of the western way, we take 3/4 [of the total height].
For the height of the southwest corner of the western way,
both corners of the southern way, and the southeast corner of
the eastern way, we take 2/5 [of the total height]. The height
of each step of the steep ways is 0.6 feet. For the number of
steps of the inmost way on the east side, we take 1/4;[184] for the
number of steps on the outmost way on the east side, we take
2/5 [as height]; for the number of steps on the western way, we
take 3/4 [as height]. The length of the stones is 5 feet, the width
2 feet, the thickness 0.5 feet, the length of the bricks is 1.2 feet,
the width 0.6 feet, and the thickness 0.25 feet . . . ."

The most interesting part of this problem is the description

[181] This means that the distance was not to be taken into account.
[182] See also the section on "Taxes and Levies of Service."
[183] This access to the top of the tower is clearly illustrated in Yüeh Chia-
tsao (1'), 2, p. 1a.
[184] The meaning is not clear. As there are several mathematical mistakes
in this problem, it is difficult to reconstruct the true meaning.

of the steps by which a height of 120 feet can be reached (Figure 63). Although a picture is given in the text (p. 347), it reflects a literary rather than an architectonic visualization and is not illuminating.[185] But if we reconstruct the true shape by means of the sizes given, we get a good idea of a watchtower at that time. The height of the steps was 0.6 foot, but the length varied according to the steepness (2.29 feet; 2.15 feet, and 2.06 feet).

Problem VII, 6 deals with the laying of a foundation of a large hall 170 by 210 feet. "We first assume that 7 men build and ram 30 cubic feet, and it is estimated that this work requires two days. Now an auspicious day is chosen for setting up the posts. They wish to finish [the work] in 3 days. How many rammers are required per day?"

In VII, 7 some information is given on the division of a house: there is a hall (the middle part of a house) divided into three rooms, each having the dimensions 30×52 feet. There are also a library of 6 rooms, each having the dimensions 15×12 feet; four women's apartments, two inside ones of 10×13 feet and the two next ones of 15×13 feet; ten square arbors with a side of 14 feet.

Military Affairs

Mikami (6') wrote an article devoted to the influences of war and strategy on Chinese (and Japanese) mathematics. This article is mainly a chronological survey based on several authors, and within the scope of this work it was not possible to give a detailed reproduction of problems. As for Ch'in Chiu-shao (pp. 11 f), Mikami recalls the fact that Ch'in himself was concerned with military affairs during his youth, as we learn from Chou Mi's biographical notes;[186] he also mentions the subjects Ch'in deals with in his chapter on war and strategy (Chapter 8). These are listed here.

---

[185] The same is true for other pictures, for instance, in Ch'i Chi-kuang's *Lien-ping shih-chi* (1568), ch. 6, pp. 1b and 2a. The only picture I found that gives a true idea is that mentioned in note 183.
[186] See Chapter 3.

Figure 64. Pontoon bridge. From the *San-ts'ai t'u-hui* 三才圖會 (1609), "Kung-shih 宮室" (Architecture), ch. 2, p. 26b.

FORTIFICATIONS

The building of a city wall (VII, 1) has already been mentioned.

PONTOON BRIDGES

Problem IV, 3 deals with the measurement of the width of a river, in order to find the length of the bamboo hawser used to fasten and support the boats which formed the bridge.[187]

[187] See Figure 64.

MILITARY CAMPS[188]

Problem VIII, 1 says: "One army corps is composed of three divisions, and one division of 33 companies. Toward evening they pitch a camp. One man takes up space equal to a square of 8 feet on a side. It must be arranged so that, between the companies, there is space left for the officers of the companies, who are in the middle. . . . ." In Figure 65 we can see the special arrangement of the soldiers in four groups. The company

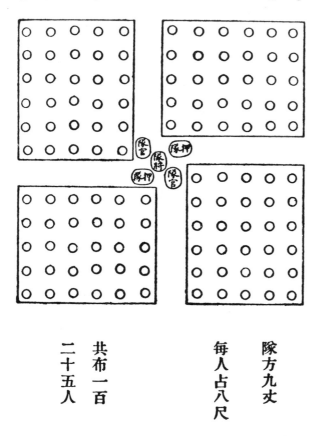

Figure 65. Military camp (see description in text). From the *SSCC*, Problem VIII, 1 (p. 370).

[188] Descriptions of camps are given in the *Wu-pei chih,* ch. 93, p. 10b and following pages, and in the *Wu-ching tsung-yao,* 3, p. 10a and following pages.

commander is the *tui-chiang* 隊將; there are two of each of two types of company officer, *tui-ya* 隊押 and *tui-kuan* 隊官, and 120 soldiers. The part of the camp assigned to a company is a square with a side of 90 feet. The whole camp has a side of 1,710 feet.

Problem VIII, 4 computes the area of a round camp for 12,500 men (an army corps); there are nine circular layers, and each soldier is allotted a circle 6 feet in diameter.

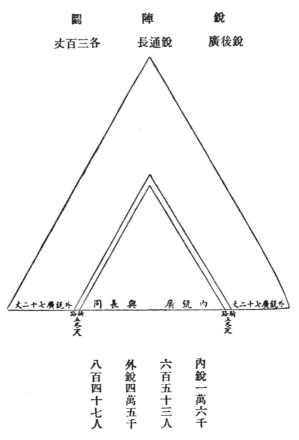

Figure 66. Pointed battle array. From the *SSCC*, Problem VIII, 2 (p. 377).

BATTLE FORMATIONS

Many kinds of battle formations are given in the *Wu-ching tsung-yao* and the *Wu-pei chih*. In the *Shu-shu chiu-chang* there are also two examples.

Problem VIII, 2 says: "Change a square formation into a pointed formation [Figure 66]. There are five corps of infantry; each corps has 12,500 men. They make a square formation in which each man has a square of 8 feet on a side: they wish to change into a pointed formation. For the width of the rear of the array they add the half of the side of the square. Inside the formation there is included a passage for the cavalry of at least 50 feet. . . ."

Problem VIII, 3 deals with the computing of a circular array.[189] The inner diameter must be 720 feet, and each man needs as standing room a circle with a diameter of 9 feet. There are 2,600 soldiers to be drawn up. What is the outer diameter and the number of soldiers on it?[190]

OBSERVATION OF THE ENEMY

Problem IV, 6 deals with the observation of the camp of the enemy and computation of its dimensions by means of gnomons[191] (Figure 67). Problem VIII, 5 concerns the computation of the numerical strength of the enemy by observation and by measurement of his camp.[192]

MANUFACTURE OF WEAPONS

Problem VIII, 8: "Now they wish to manufacture bows and swords [see Figure 68], 10,000 of each, and 1,000,000 arrows, according to following work regulations: 7 men make 9 bows in 8 days, 8 men make 5 swords in 6 days, 3 men make 150 arrows in 2 days. . . ."

[189] See *Wu-pei chih*, ch. 52, p. 7b and following pages.
[190] See Chapter 12.
[191] Formulae are given in Chapter 12.
[192] See Chapter 9.

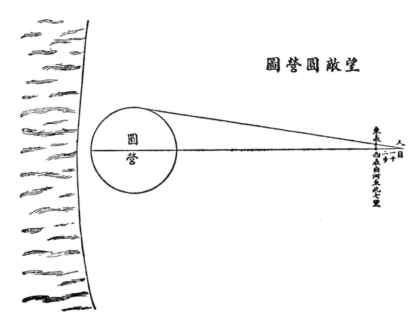

望敵圓營圖

圓營

Figure 67. Observation and measurement of the enemy camp. From the *SSCC*, Problem IV, 6 (p. 190).

MANUFACTURE OF SOLDIERS' UNIFORMS

According to VIII, 9 three kinds of material are used: hemp cloth (*pu* 布), floss silk (*mien* 綿), and "refuse silk" (*hsü* 絮).

Sciences of the Heavens

Chapter 2 is devoted to heavenly phenomena (*t'ien-shih* 天時), and these may represent some of the problems Ch'in learned in the Board of Astronomy.

CHRONOLOGY

Problem II, 1. "Measurements and Examinations of the 'Ways of Heaven' by the Board of Astronomy. In the fourth year of the *Ch'ing-yüan* 慶元 period [A.D. 1198], which was the fifty-fifth year of the sexagenary cycle, the winter solstice was on day 39.9245 [of the sexagenary day cycle, *chia-tzŭ*]. In the third year of the *Shao-ting* 紹定 period [A.D. 1230], which was the

Figure 68. Armaments. From the *SSCC*, Problem IV, 7, reproduced from the *I-chia t'ang ts'ung-shu* 宜稼堂叢書 of 1842, ch. 4, 21b.

twenty-seventh year of the sexagenary year cycle, the winter solstice fell on day 32.9412. We wish to find, for the year 1204 the first year of the sexagenary year cycle in the interval, the *ch'i-ku* 氣骨 [literally, "*ch'i* bone," the time from the winter solstice to the end of the sexagenary day cycle], the *sui-yü* 餘歲, and the *tou-fên* 斗分"[193] (see Figure 69).

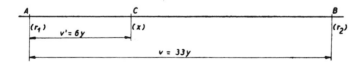

Figure 69

[193] The *sui-yü* or "year remainder" is the difference between the year and 360 days (six-day cycles); the *tou-fên* ("dipper fraction") is the fraction of a day included in the year length constant. The latter is thus the fractional part of the year remainder.

Answer:

The *ch'i-ku* is 11.38208180 days.
The *sui-yü* is 5.24293030 days.
The *tou-fên* is 0.24293030 days.
A=winter solstice of 1198
B=winter solstice of 1230
C=winter solstice of 1204

The values r_1, r_2, and x are the remainders with regard to the sexagenary day cycle.

$$A-B=v$$
$$v=r_2-r_1+n\times 60 \text{ days}$$
$$x=r_1+v'\frac{r_2-r_1+n\times 60}{v}-m\times 60$$
$$j=\frac{(r_2-r_1)+n\times 60}{v}$$

is the difference between one year and a number of sexagenary cycles. Thus

$$5<j<60 \qquad\qquad (1)$$

and $x=r_1+v'j-m\times 60$. From (1),

$$5<\frac{(r_2-r_1)+n\times 60}{v}$$
$$5v<(r_2-r_1)+n\times 60$$
$$\frac{5v-(r_2-r_1)}{60}<n$$
$$\frac{5\times 33-(32.9412-39.9245)}{60}<n$$
$$\frac{165+6.9833}{60}<n$$
$$\frac{171.9833}{60}<n.$$

The minimum value of n is 3.

$$j=\frac{-6.9833+3\times 60}{33}=5.24293030$$

$$x = r_1 + v'j - m \times 60 = 39.9245 + 6 \times 5.24293030 - m \times 60 =$$
$$71.38208180 - 60 = 11.38208180.$$

Problem II, 2 says: "The *K'ai-hsi* 開禧 calendar considers in the fourth year of the *Chia-t'ai* 嘉泰 period [1204], which is sexagenary year 1, the astronomical new year or winter solstice as day 11.446154 [of the *chia-tzŭ*]; the remaining part of the eleventh month [after the beginning of the *chia-tzŭ*] is 1.755562 days. Find the *jun-ku* 閏骨 and the *jun-lü* 閏率."[194] (See Figure 70.)

In the *K'ai-hsi* calendar:

$$1 \text{ year} = \frac{6,172,608}{16,900} \text{ days}$$

$$1 \text{ month} = \frac{499,067}{16,900} \text{ days.}$$

The time between the Superior Epoch and the beginning of sexagenary year 1 is 7,848,180 years.

The *ch'i-ku* 氣骨 or number of days of the year elapsed since the beginning of the last sexagenary day cycle:

$$7,848,180 \times \frac{6,172,608}{16,900} \equiv \frac{193,440}{16,900} \pmod{60}$$

$$\textit{ch'i-ku} = \frac{193,440}{16,900} = 11.446154 \text{ days.}$$

The *jun-ku* (literally, "intercalation bone") or number of days of the year elapsed between the last lunation and the winter solstice:

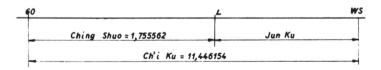

Figure 70

[194] Wylie (1), p. 184 says that the *jun-ku* is the "time to the winter solstice from any previous point in the 11th month." In this problem the previous point is the lunation of the eleventh month.

$$7,848,180 \times \frac{6,172,608}{16,900} \equiv \frac{163,771}{16,900} \left(\mathrm{mod}\,\frac{499,067}{16,900}\right).$$

The *jun-ku* = 163,771/16,900.
The elapsed part of the month at astronomical new year (*t'ien-chêng ching shuo* 天正經朔) is :[195]

$$\frac{193,440}{16,900} - \frac{163,771}{16,900} = \frac{29,669}{16,900} = 1.755562 \text{ days.}$$

The *jun-ku* = 11.446154 − 1.755562 = 9.690592.
The *jun-ku-lü* is the numerator of 163,771/16,900 (the *jun-ku*).
This problem has no mathematical value whatever.
For Problem II, 3, see p. 409. For Problem I, 2, see p. 391.

ASTRONOMY

Problem II, 4 is entitled "To Calculate [the Motion of] a Planet with the 'Technique of Threading' [*chui-shu* 綴術]." "The 'period of invisibility' due to solar conjunction [*ho-fu* 合伏][196] of the planet Jupiter is 16.90 days and [during this period] the planet moves 3.90 degrees.[197] When [the planet] is distant from the sun 13 degrees, it appears [again]. After that its [apparent] forward motion [*shun* 順] takes 113 days. It moves up to 17.83 degrees and is stationary [*liu* 留]. We wish to know the corresponding degrees, the initial speed, the final speed, and the average speed of the phase [*tuan* 段] of invisibility and of the beginning of the accelerated phase [when Jupiter first appears] as a morning star [*ch'ên-chi* 晨疾]."

In this period of invisibility, the planet Jupiter moves 3.90 degrees; in the same period the sun moves 16.90 degrees along the ecliptic; thus, after this period, Jupiter is distant from the sun 13 degrees.[198]

[195] We rely on Yen Tun-chieh (2′), p. 114.
[196] This is the period during which the planet is invisible because of its conjunction with the sun.
[197] Because of its low relative velocity the planet is overtaken by the sun.
[198] See Yabuuchi (2′), p. 116, where a good explanation of the motion of the planet is given. I am much indebted to Professor K. Yabuuchi for his kind assistance in elucidating this problem.

Ch'ien Pao-tsung gives this explanation of the problem:[199] "Suppose the period of invisibility is t_1 days and the motion of the planet during this time is s_1 degrees; the forward motion is t_2 days and the corresponding distance s_2 degrees; v_0 is the initial speed of the phase of invisibility (i.e., the speed during the first day), v_1 is the end speed (i.e., the speed of the last day); v_2 is the end speed of the forward motion, and a is the acceleration (the increase of speed per day). Suppose the speed of the planet is uniformly accelerated; then the speed is a function of the first degree in time, and the distance is a function of the second degree in time. As during the period t_1+t_2, v_0 is the maximum speed and v_2 the minimum speed, then, if we reckon starting from v_2, a has a positive value [see Figure 71]:

$$v_1 = v_2 + t_2.$$

(Here $s_1 = v_1 t_1$; thus s_1 is represented by the rectangle.)

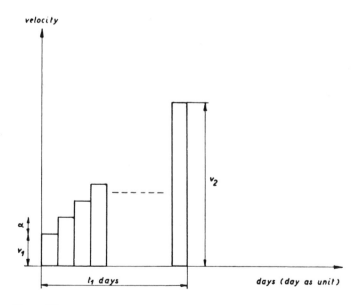

Figure 71

[199] (2'), p. 92.

$$s_1 = v_1 + (v_1 + a) + (v_1 + 2a) + \ldots + [v_1 + (t_2 - 1)a]$$
$$s_1 = v_1 t_1 + a + 2a + \ldots + (t_1 - 1)a.$$
$$= v_1 t_1 + \frac{t_1(t_1 - 1)}{2} a.$$

As $v_1 = v_2 + t_2 a$

$$s_1 = (v_2 + t_2 a)t_1 + \frac{t_1(t_1 - 1)}{2} a. \tag{1}$$

Or $s_1 = v_2 t_2 + t_1\left(t_2 + \frac{t_1 - 1}{2}\right)a. \tag{2}$

For the same reason:

$$s_2 = v_2 t_2 + \frac{t_2(t_2 - 1)}{2} a.$$

The data of the problem are: $t_1 = 16.9$; $s_1 = 3.9$; $t_2 = 113$; $s_2 = 17.83$; find a, v_0, and v_1.
From (1) and (2) we deduce:

$$16.9 v_2 + 2{,}044.055a = 3.9$$
$$113 v_2 + 6{,}328a = 17.83$$

from which: $a = 0.00112366$ degrees."[200]

Problem II, 5: "Successive generations measured the shadows, [but] only the *Ta-yen* calendar of the T'ang is very accurate. According to the *Ch'ung-t'ien* 崇天 calendar[201] of our dynasty, the shadow at winter solstice at Yang-ch'êng 陽城[202] is 12.7150 feet and the summer solstice shadow is 1.4779 feet.[203] This agrees with the *Ta-yen* calendar. At present, according to the *K'ai-hsi* 開禧 calendar,[204] the winter solstice shadow at Lin-an-fu 臨安府[205] is 10.8225 feet and the summer solstice

[200] For further discussion, see Ch'ien Pao-tsung (2'), pp. 93 ff. Li Yen (7'), pp. 58 ff, was of the opinion that the problem was solved by the interpolation formula of the second degree for unequal intervals, but this opinion is rejected by Ch'ien Pao-tsung, mainly because Ch'in Chiu-shao specified as his method simultaneous equations.
[201] A Sung calendar of 1024.
[202] Modern: Kao-ch'êng, with latitude 34°26′ N.
[203] The great precision of these measures shows that they were found by interpolation.
[204] The *k'ai-hsi* period: 1205–1208.
[205] Modern: Hang-chou-fu in Chekiang, with latitude 30°15′ N. During this time Yang-ch'êng was occupied by the Tartar Chin dynasty (1115–1234).

shadow 0.91 feet. Query. Find how many days after the summer solstice at Lin-an-fu the shadow will be the same as on the day of summer solstice at Yang-ch'êng, and how many feet and inches the shadow of the gnomon differs with regard to the *Ta-yen* calendar."

"Answer: (1) On the fifth day at noon after the *ta-shu* 大暑;[206] (2) The shadow length is 1.4885 feet.

Ch'in Chiu-shao solves the problem by means of the formula:

$$S_n = \sqrt{\left[\frac{S_w - S_s}{R + C}\right]^2 M + S_s^2}$$

where:

S_n = the shadow length in a certain point.

S_s = summer solstice shadow at Lin-an-fu

S_w = winter solstice shadow at Lin-an-fu.

R = a right angle (91.3144 degrees)

C = a constant, equal to 11.252750

M = a constant, dependent on the period; for the *hsiao-shu* 小暑[207]: 25; for the *ta-shu* 大暑[208]: 109; for the *li-ch'iu* 立秋[209]: 289.

Ch'in finds the following shadow lengths for Lin-an-fu: *hsiao-shu*: 10.3 inches; *ta-shu*: 13.587 inches; *li-ch'iu*: 18.781 inches.

The summer solstice shadow for Yang-ch'êng is 14.779 inches, or between the *ta-shu* and *li-ch'iu* shadow lengths. By means of a simple linear interpolation, Ch'in obtains the answer: on the fifth day at noon after the beginning of *ta-shu*.

As for the formula used here, the rationale is unclear; we are not able to guess the meaning of the constant C. The formula as a whole gives the impression of being a kind of interpolation formula.[210]

---

[206] The Chinese solar year was divided into 24 solar terms (*chieh-ch'i* 節氣) of 15 days. The *ta-shu* was the second period after summer solstice (approximately 23 July–6 August).

[207] "Slight heat" (7–22 July).

[208] "Great heat" (23 July–6 August).

[209] "Autumn begins" (7 August–23 August).

[210] Although they could not give me further elucidation of this very obscure problem, nevertheless I am very grateful to Dr. S. Nakayama (Tokyo),

METEOROLOGY

"Since it [rain] had its inevitable sequel in the rising of rivers and canals, with the danger of floods, always so serious in China, it would not be surprising to find that the Chinese made use of rain gauges from an early period."[211] As pointed out by Needham the first account of rain gauges is to be found in Ch'in Chiu-shao's work, where we read: "At present in all the capitals of the prefectures and districts there is a rain gauge (*t'ien-ch'ih p'ên* 天池盆) to measure the rainwater." Apart from Ch'in Chiu-shao's mention of rain gauges, the oldest rain-measuring instruments are found in Korea (first half of the fifteenth century).[212] Rain gauges are treated in the *Shu-shu chiu-chang*, Problems II, 6 and II, 7.

There are also two problems, II, 8 and II, 9, concerning the measuring of snow. The first one cannot be said to deal with a snow gauge, but it has something to do with measuring snow, as the first sentence says: "Examine the snow to prognosticate for the coming year."[213]

Problem II, 9 treats the use of snow gauges; these were bamboo cages, and their use was called *chu-ch'i yen hsüeh* 竹器驗雪 (employing bamboo instruments for examining the snow).[214]

---

to Professor N. Sivin (Cambridge, Massachusetts) and to Professor P. Dingens (Ghent) for their kind assistance.

[211] Needham(1), vol. 3, p. 471.
[212] See Wada Yûji (1'); Lyons (1); and Shaw (1), p. 200.
[213] For the mathematical solution, see Chapter 8.
[214] The problem is identical to II,6, but the mathematical solution is very corrupt.

Glossary

Only terms used in the *Shu-shu chiu-chang* are included in this glossary. *Example* refers to a problem in the Chinese text where the term is used; *Reference* refers to the pages in this work where the term is explained.

		Example	Reference
ch'a 差	Difference	II, 6	
chan-wei 展爲	Reduce to (higher metrological units)	II, 6	86
chang 丈	Linear measure for rolls of cloth, 10 feet	V, 7	77
ch'ao 超	To jump over one column on the counting board	III, 1	
ch'ao i wei 超一位	Advance two columns (see also *chin*)		
ch'ao-chin 超進	Advance a number on the counting board	III,1	
ch'ao-pu 超步	See *ch'ao-chin*	III,2	
chê 折	Literally, "break"; used for "divide" when no remainder is expected E.g. 以 4 折之 : divide it by 4	V,6	
chê-pan 折半	To halve	I,2	86
chê-pan ch'a 折半差	To halve repeatedly, so as to form a decreasing geometrical progression with $r = 2$	V,6	
ch'ên 塵	0.0000001 (decimal)		72
chêng 正	Positive coefficient		
chêng-ch'ang 正長	Linear distance	III,6	69, 193
chêng-fu 正負	Literally, "positive-negative"; change the signs	IX,1	154
chêng-kao 正高	Perpendicular height	VII,5	
chêng-shên 正深	Perpendicular depth	VI,4	
chêng-yung-shu 正用數	Technical term in the *ta-yen* rule	I,1	331, 347
ch'êng 乘	To multiply		83

ch'êng-fang 乘方	To square		
ch'êng-lü 乘率	1. Multiplier; factor		84
	2. Congruence factor (*ta-yen*)		340
chi 基	Literally, "the base"; the approximate root of an equation before rounding off	VI,4	
chi 積	1. Area	III,4	
	2. Volume	III,9	
chi-li ch'a 蒺藜差	The "*chi-li* difference," that is, the proportion 1:3:6	IX,7	88
chi 奇	1. Odd number	II,1	87
	2. Remainder		
	3. "Remainder" (*ta-yen*)	I,1	
chi ou pu t'ung lei 奇偶不同類	Mutually indivisible; numbers not having common factors	I,1	333
(*chi ou*) *t'ung lei* [奇偶] 同類	Mutually divisible; numbers having common factors	I,1	
ch'i 弃棄	Cancel (the remainder); cast out (the decimals)	II,1	86
chia 加	To add	II,1	
chia-ling 假令	"Let us now assume ..."		
ch'iang-pan 强牛	Literally, "strong half" (= 3/4)	IV,4	70
chieh 借	To borrow from (another number)	II,2	
chieh-shu 借數	Literally, "borrowed numbers" (*ta-yen*)		349
chieh-yung-fan 借用繁	Literally, "to borrow and use the superfluity"; to compensate	I,1	
chien 減	To subtract		
chien-t'ien 尖田	Literally, "a pointed field"; a quadrangle with the adjacent	III,1	97

	sides equal		
ch'ien 錢	Weight measure		80
chih 置	To place, to set up on the counting board		
ch'ih 尺	Foot	V,7	77
chin 近	To round off; *chin i fên* 近一分 means round off to 1 *fê*n	II,2	
chin 進	Advancing a number on the counting board (see also *ch'ao*)		
chin i wei 進一位	Advance one rank, that is, multiply by ten		
chin 斤	Weight measure		80
ching 徑	Diameter of a circle		
ch'ing 輕	0.00000000001 (decimal)		72
ch'ing 清	0.000000000001 (decimal)		72
ch'ing-fa 頃法	The value of 1 *ch'ing* = 100 *mou*	III,4	77
chiu-chin 就近	Literally, "to approach"; to round off	II,3	86
chiu-wei 就爲	To round off to	VIII,1	
chiu wei ch'üan-shu 就爲全數	To round off to a whole number	VIII,1	
chou 周	Circumference of a circle		
ch'u 除	To subtract; to divide, retaining only the remainder for the next step (in this case the division has the same result as subtracting the divisor as many times as possible).		84
ch'u-lü 除率	Divisor	II,1	
chung-pan 中半	Literally, "middle half" (= 1/2)		70
ch'ung-ch'a 重差	"Double-difference method"; a kind of		122

	prototrigonometry, relying on the properties of similar triangles		
ch'ü 去	1. To remove a number from the counting board		
	2. To subtract	II,1	
chüeh-fa 角法	The value of a *chüeh* = 60 square *pu*	III,6	77
fa 法	Divisor		
fan 泛	1. Provisional numbers	V,2	
	2. Provisional coefficients, before reducing (opposite to *ting,* 定)	IV,6	
fan-chui ch'a 反錐差	"Inverted-wedge difference"; the proportion 3 : 2 : 1	IX,7	87
fan-fa 翻法	Inversion method; the transformation rule by which the constant term in an equation changes from negative into positive		195
fan-yung 泛用	Technical term in *ta-yen*		346
fang 方	1. Square		
	2. Coefficient of x in an equation of higher degree	II,9	181, 192
fang-ch'êng 方程	Literally, "square patterns" (that is, tabulation on the counting board); equations; simultaneous linear equations	IX,1	152, 153
fang-chui ch'a 方錐差	"Squared-wedge difference," or the proportion 1:4:9	IX,7	88
fên 分	1. Fraction		70
	2. 1/10 as fraction		

	3. 0.1 (decimal)		72
	4. 0.01 (centesimal)		73
y *fên chih* 分之 x	The fraction x/y (still used in modern mathematical terminology)		71
fên-li 分釐	Decimal fractions	I,1	82
fu 負	Negative coefficient		193
fu-shu 復數	Multiple of 10		83
hang 行	Column on the counting board	I,1	
hang-lien	See *hsing-lien*		
hao 毫	0.001		72
ho sun i 合損益	Literally, "to join a diminution or an augmention"; to round off	VI,4	86
ho-wên 合問	"It agrees with the [conditions stated in the] problem" (concluding formula)		
hsi 息	Interest	VI,8	
hsia-lien 下廉	Coefficient of x^{n-1} in an equation of the nth degree	II,9	181, 192
hsiang-ch'êng 相乘	Mutual multiplication (of numbers in the same column of the counting board)	I,1	83
hsiang-chien 相減	Mutual subtraction	III,1	182n
hsiang-hsiao 相消	Literally, "mutual canceling"; to make the algebraic sum, used when one coefficient is positive and the other negative (when both have the same sign, *tsêng-ju* or *ju* is used)	III,1	
hsiao-fên 小分	0.000001 (centesimal)		73
hsiao-miao 小秒	0.00000001 (centesimal)		73
hsiao-lien 爻廉	Coefficient of x^{n-3} in an equation of the		192

	nth degree		
hsieh 斜	Oblique line	III,4	
hsien 弦	Hypotenuse	IV,6	
hsing-lien 星廉	Coefficient of x^{n-2} in an equation of the nth degree		192
hsing-lien 行廉	Coefficient of x^{n-4} in an equation of the nth degree		192
hsü 虛	Empty place on the counting board	III,1	193
hsü-fang 虛方	Literally, "empty square"; the coefficient of x is zero		
hsü-ching 虛徑	The inner diameter of an annulus	III,8	
hsü-shang 續商	The second or following digit of a root	III,1	
hu 忽	0.00001 (decimal)		72
hu 斛	Volume measure for grain	I,5	77
hu-ch'êng 互乘	Cross-multiplication	VII,5	83
hu-ch'u 互除	Divide by each other	I,1	
hua 化	Convert (into another metrological unit)	III,3	
hua-tso 化作	Convert into (another metrological unit)	IX,4	
hua-wei 化爲	Same as *hua-tso*	IX,5	
huan-ku 換骨	Literally, "changed bone"; when the given equation is transformed into its depressed equation, it happens that the constant term, which is always negative, changes into positive; this term is called *huan-ku* (changed term)	III,1	195

huan-t'ien 環田	Literally, "ring field"; annulus	III,8	105
i 億	10^8	VII,5	69
i 益	To add; to increase a number in order to round it off to a higher metrological unit	IV,9	
i 益	Negative coefficient in an equation (the literal meaning of *i* in this sense is not clear; however it invariably refers to a negative coefficient)	I,3	181, 193
i a *chien* b 以 a 減 b	Diminish *b* by *a*; subtract *a* from *b*		
i a *ch'u* b 以 a 除 b	Divide *b* by *a* (and retain the remainder). See also *ch'u*		
i ming 異名	Different algebraic signs	IX,1	154
i a *yü* b *hsiang k'o* 以 a 與 b 相課	Literally, "Examine mutually *a* and *b*"; compare *a* with *b*	II,3	
jo-pan 弱半	Literally, "weak half" = 1/4	V,2	70
ju 入	1. Apply a method 2. Add to (= *tsêng-ju*)	I,1 III,1	65
a *ju* b *ch'u chih* a 如 b 除之	Divide *a* by *b*, and retain the remainder. See also *ch'u*	II,1	
ju...êrh i 如...而一	Divide and retain the integral result on the counting board for use in the next step	VI,1	84
ju...nei 入...內	= *tsêng-ju*	III,1	
k'ai chiu-ch'êng fang 開九乘方	Solve an equation of the 10th degree		193
k'ai-fang 開方	Extract the square root		

k'ai li-fang 開立方	Extract the cube root		193
k'ai lien-chih li-fang 開連枝立方	Solve a cubic equation of the type $ax^3 - b = 0$, with $a \neq 1$	VI,4	
k'ai ling-lung (fan-fa) san-ch'êng fang 開玲瓏 [翻法] 三乘方	Solve an equation of the 4th degree having only even powers of x ($ax^4 + bx^2 + c = 0$); biquadratic equation. *Fan-fa* refers to the conversion of the negative constant term into positive	III,1	
k'ai lien-chih san-ch'êng ling-lung fang 開連枝三乘玲瓏方	Solve an equation of the type $-ax^4 + bx^2 - c = 0$ ($a>1; b>1; c>1$)	IV,6	
k'ai ling-lung san-ch'êng fang 開玲瓏三乘方	Solve an equation of the type $-x^4 + bx^2 - c = 0$ ($b>1; c>1$)	IV,6	
k'ai p'ing-fang 開平方	Extract the square root	II,8	193, 209
k'ai san-ch'êng fang 開三乘方	Solve an equation of the 4th degree	II,9	193
k'ai t'ung-t'i lien-chih p'ing-fang 開同體連枝平方	Solve an equation of the type $ax^2 - b = 0$, where a and b are squares ($a^2x^2 - \beta^2 = 0$)	IV,2	194
k'ê 刻	1/100 part of a day	II,1	76
ko 合	Measure of capacity	VII,5	77
kou 句 (modern: 勾)	The base of a right-angled triangle	IV,6	
kou-ku 句股	"Right angles"; the properties of right-angled triangles	IV,1	125
ku 股	The altitude of a right-angled triangle (the perpendicular)	IV,6	
ku-fa 古法	The old value (of π, that is, $\pi = 3$)	IV,5	
ku-lü 古率	Same as *ku-fa*		
kuan 貫	String of cash, monetary unit. One *kuan*		97

	is equal to 1,000 *wên*		
kuang 廣	1. Horizontal line		
	2. Measurement from east to west	VII,5	
k'uei-shu 虧數	Remainder, deficiency	VI,4	
k'uo 闊 .	The width		
k'ung 空	1. An empty place on the counting board	I,1	
	2. Hence: zero	II,1	69
lei-shu 類數	Literally, "numbers of the same class"; numbers having common factors		
li 釐	0.01 (decimal)		72
li-fa 里法	the *li* ratio: 1 *li* = 360 *pu* = 2,160 feet	V,2	77
li-fang 立方	Cube	VI,6	86
liang 兩	Weight measure		
liang-liang lien-huan ch'iu têng 兩兩連環求等	To find the G.C.D. by pairs	I,1	
liang-tu tzŭ-ch'êng 兩度自乘	To cube (literally, to multiply by itself twice)	III,5	86
lieh 列	To arrange on the counting board	II,9	
lien 廉	1. Coefficient of x^2 in an equation of the 3rd degree		192
	2. The name for all the coefficients of the unknowns from x^2 up to x^{n-1} in an equation of the *n*th degree.		
lien-chih 連枝	An equation with the first coefficient $a \neq 1$	III,7	202, 209
ling 零	Remainder	V,1	193, 209
ling-lung 玲瓏	Literally, "harmonious alternating"; equation having only		

	even powers of x		
man x *ch'ü chih* 滿 x 去之	Subtract x as many times as possible (equivalent to dividing by x and retaining the remainder for further use (see also *ch'u*)		84
man ch'ü x 滿去 x	Same as *man* x *ch'ü chih*		
mang 莽	0.0000000001 (decimal)		72
mi 冪	Square (mostly used for geometric measures; for example, the square of the side of a triangle)	IV,2	85
mi-lü 密率	Precise value of π, that is, $\pi = 22/7$	II,7	97
miao 渺	0.000000001 (decimal)		72
miao 秒	0.0001 (centesimal)		73
ming 名	Literally, "sense, name." The algebraic sign (plus or minus)	IX,1	
ming 命	Literally, "to name." Put the remainder of a division on the divisor as denominator (and add as fraction to the quotient) For example, $325:22 = 14$; $r = 17$; result $= 14\ 17/22$	III,7	
mo wei 末位	The last digit of a number	II,3	
mou 袤	Measurement from north to south	VII,5	
mou 畝	Land measure of area; 1 *mou* = 240 square paces	V,1	77
mou-fa 畝法	The *mou* ratio; 1	III,4	77

	mou = 240 square paces		
mu 母	1. Factor 2. Numerator of a fraction.		84 70
mu hu-ch'êng tzŭ 母互乘子	(In adding fractions), multiply each numerator by the denominators of the other fractions (and then multiply all the denominators by one another)		
na-chien 內減	Subtract from it	IV,2	
na-tzŭ 內子	Literally, "include in the numerator." To convert a mixed number into an improper fraction by converting the integer (see also *t'ung-fên nei-tzŭ*)		
ou 偶	Even number	I,1	87
pan 半	To halve		86
pan-ching 半徑	The radius		
p'ao-ch'a 拋差	To diminish gradually (by a constant number). The difference *v* in a diminishing arithmetical progression	V,9	
pei 倍	To double	III,1	86
pei-shu 倍數	A multiple	I,1	
p'i 疋	Linear measure for rolls of cloth	V,7	77
piao 表	A gnomon		
pien 變	Transformation on the counting board. The situation after an operation on the counting board (an equation is solved step by step; each step is called a	III,8	

pien)

ping 併 幷	To add to	II,3	
p'ing-fang 平方	A square		
pu 補	1. To add to	II,2	
	2. To put in the place of	I,1	
pu-chi 不及	Literally, "not extending to." The discrepancy between a constant term and *n* times the given number. For example, the *pu-chi* between 12 months and one year is 10 or 11 days		
shuo pu-chi 朔不及	The lunation discrepancy, defined as above	I,2	
pu-chi ling 不及零	A discrepancy of zero	I,3	
pu-chin 不盡	Literally, "what is not exhausted"; the remainder (in a division)	II,2	
pu-fa 步法	1. Rule for shifting the coefficients of an equation on the counting board	II,9	
	2. The *pu* ratio; 1 *pu* = 5 feet		77
pu-man 不滿	Literally, "the incomplete part"; remainder. See also *man* x *ch'ü chih*		84
san-ch'êng fang 三乘方	Fourth power		86
sha 沙	0.00000001 (decimal)		72
shang (shu) 商 [數]	1. Quotient	II,1	
	2. The root of an equation	II,9	181
shang-lien 上廉	Coefficient of x^2 in an equation of the *n*th degree	II,9	181,192

shao 勺	Measure of capacity	V,1	79
shao-pan 少半	Literally, "Diminished half" (1/3)		70
shêng 升	Measure of capacity	I,5	79
shêng 生	To multiply (only used in equations)	II,9	83
shêng-yü 賸餘	The remainders in the remainder problem		
shih 石	Weight measure	I,5	79
shih 實	1. Literally, "the full." The dividend (the number of full divisors in the original number) 2. The constant term in an equation	II,9	181, 192
shih-chin 適盡	Leaving no remainder (in a division)	III,1	
shih fên nei ch'a i 十分內差一	The proportion $a/b = 10/9$	VII,5	88
shih fên wai ch'a i 十分外差一	The proportion $a/b = 11/10$	V,1	88
shih ju fa ch'u chih 實如法除之	Divide the dividend by the divisor (and retain the remainder for further operations); see also *ch'u*	II,1	
shih ju fa êrh i 實如法而一	Divide the dividend by the divisor (and retain the quotient for further operations). See also *ju... êrh i*	IX,1	
shou-(shu) 收 [數]	Decimal numbers, decimal fractions	II,3	82, 358
shou chi 收棄	The decimals are canceled, that is, rounded off to the lower integer	II,3	
shou shang 收上	The decimals are rounded off to the upper integer	II,2	

shu-ch'u 數處	Places of the numbers on the counting board		
shu yüeh 術曰	Literally, "the method says." The general explanation of the method		
ssŭ 絲	0.0001 (decimal)		72
suan-p'an 算盤	The counting board. It has the form of a chess board, subdivided as follows:		30

(行) 左 次 中 副 右

上
副
中
次
下
(位)

sun 損	To diminish		
sun-ch'ü 損去	To diminish	I,2	
sun ch'i pan-pei 損其半倍	Diminish by half	I,1	86
ta-yen 大衍	Literally, "great extension." Indeterminate analysis	I,1	354
ta-yen-shu 大衍術	*Ta-yen* method, the method for solving indeterminate equations of the first degree	I,1	354
ta-yen ch'iu-i 大衍求一	Solving the congruences in indeterminate equations (linear congruences)	I,1	340
t'ai-pan 太半	Literally, "greater half" (2/3)		70
tê 得	To obtain as a result of solving an operation		
têng-(shu) 等 [數]	Greatest common divisor		86

ti 遞	Transfer (a number to another diagram on the counting board)	I,1	
ti-chien 遞減	Proportional decrease	I,4	
ti-chien-shu 遞減數	The number by which we proportionally decrease	I,4	
t'i 題	Problem, proposition, question		
t'ien-yüan i 天元一	1. The unity symbol (place-indicator)		338
	2. For its special meaning in the *ta-yen* procedure, see chapter 17		345
ting-(lü) 定 [率]	Coefficient after reduction (opposite to *fan* 泛)	IV,6	154
ting-shu 定數	Reduced moduli in the *ta-yen* method		333
to i pei x 多一倍 x	One more than double of x; $2x + 1$	I,2	86
tou 斗	Weight measure; peck	VII,5	79
t'ou-ju 投入	To add to	IV,8	
t'ou-t'ai 投胎	The (negative) constant term in an equation, which after a transformation remains negative but decreases	IV,8	195
tsai tzŭ-ch'êng 再自乘	Cube	III,5	86
ts'ai-lien 才廉	Coefficient of x^{n-6} in an equation of the nth degree		192
ts'ao 草	Detailed workings		64
tsêng-ju 增入	To add to	II,3	
ts'un 寸	Inch	V,7	77
tsung 縱	Perpendicular line	III,4	
tsung 總	Technical term in *ta-yen*		349

tsung-têng 總等	Greatest common divisor	I,2	86
ts'ung 從	Positive coefficient		181,193
tu 度	Degree	II,4	76
t'u 圖	Disposition on the counting board		
kan-t'u 干圖	Terms used to dis-	IX,1	91 ff
kung-t'u 宮圖	tinguish individual	IX,1	
chih-t'u 支圖	counting boards, when	IX,1	
jun-t'u 閏圖	several boards are	IX,1	
ho-t'u 合圖	used together for solving simultaneous equations	IX,5	
tui-ch'êng 對乘	Multiply by the corresponding coefficients in another equation (that is, by the numbers opposite on the counting board)	I,2	
tui-wei 對位	Corresponding places on the count- ing board		
t'ui 退	To "retire" a number on the counting board, that is, to divide by 10 (一退), by 100 (再退), by 1000 (三 退), and so on.	V,4	
t'ui x *wei* 退 x 位	To shift backwards x columns	III,8	
t'ung 同	Common denomi- nator		
t'ung 通	Convert an integer into a fraction with a given denomina- tor. For example, 以 29 通日法 (convert 29 days by use of the day con- stant, 16,900): 29 = (29 × 16,900)/ 16,900	II,3	
t'ung-fên na-tzu 通分內子	In a mixed number, to multiply the	II,2	

	integer by the denominator *(t'ung-fên)* and add the numerator *(na-tzŭ)* in order to convert to an improper fraction. E.g, $365\ 1/4 = 1{,}461/4$ $365 \times 4 = 1{,}460$ *(t'ung-fên)* $1{,}460 + 1 = 1{,}461$ *(na-tzŭ)*		
t'ung-ming 同名	Terms having the same algebraic sign	IX,1	154
t'ung-ming hsiang-ch'u 同名相除	1. Subtraction of terms of the same sign 2. Canceling of identical terms in both terms of an equation		
t'ung-shu 通數	Fractions	I,1	83, 353
t'ung-wei 通爲	Reduce a number to (a lower metrological unit)	II,8	86
tzŭ 子	1. Factor 2. Numerator of a fraction		84
tzŭ-ch'êng 自乘	Literally, "to multiply by itself"; to square		
tzŭ-chih 自之	Literally, "to self it"; same as *tzŭ-ch'êng*	II,8	85
tz'ŭ-lien 次廉	Coefficient of x^{n-7} in an equation of the nth degree		192
wan 萬	10^4		69
wei 位	1. A place on the counting board 2. A number which occupies it 3. A "rank" of the counting board. See *suan-p'an*.		

wei 微	0.000001 (decimal)		72
wei-fên 微分	0.0000000001 (centesimal)		73
wei-lien 維廉	Coefficient of x^{n-5} in an equation of the *n*th degree		192
wei-miao 微秒	0.000000000001 (centesimal)		73
wei-shu 尾數	Literally, "the tail number." The last non-zero digit in a number. E.g., in 225,600 the *wei-shu* is 6.	I,2	
wei-wei 尾位	Literally, "the tail positions." See *wei-shu*	I,1	71
wên 文	Cash; monetary unit		86
wên-shu 問數	The problem numbers; the data of a problem		328
wu 無	"Zero"	VI,8	
wu-t'ung lei 五同類	Fivefold	I,2	334
yen 煙	0.0000000000001 (decimal)		72
yen-ch'ih 雁翅	Literally, "wings of wild geese," that is, the form of the long wing of a goose. A special disposition of proportional numbers on the counting board	IX,3	89
yen-mu 衍母	Technical term of *ta-yen* rule	I,1	337
yen-shu 衍數	Technical term of *ta-yen* rule	I,1	337
yin 因	To multiply	I,6	83
yin-lü 因率	Factor; multiplier		84
ying-fei 盈朒	Rule of false position; "determinants"	VIII,9	167
yung-shu 用數	Technical term of *ta-yen* rule	I,1	

yü 餘	Remainder of a subtraction		
yü 隅	Coefficient of x^n in an equation of the nth degree	II,8	181, 192
yü a *yen* b 與 a 驗 b	Literally, "investigate b with regard to a"; compare a with b		
yüan (t'ien) 圓 [田]	Circle		
yüan-chou lü 圓周率	π		
yüan-fa shu 元法數	Technical term of *ta-yen* rule	I,1	
yüan-pien 圓邊	Circumference of a circle	VIII,3	
yüan-shu 元數	Whole number	I,1	82
yüeh 約	To reduce	II,2	86
yüeh-fên 約分	To reduce fractions to their simplest form (by dividing by the G.C.D. of numerator and denominator)		

Bibliography

Chinese and Japanese Books and Articles

Collectanea (*ts'ung-shu*), Encyclopedias (*lei-shu*), and Periodicals: Abbreviations

BNLP
Bulletin of the National Library of Peiping
Kuo-li Pei-p'ing t'u-shu-kuan k'an 國立北平圖書館刊

CHT
Ch'un-hui t'ang ts'ung-shu 春暉堂叢書 [Ch'un-hui hall collection], 1841

CKSHCCSCK
Chung-kuo she-hui ching-chi shih chi-k'an 中國社會經濟史集刊 [Chinese social and economic history review (Institute of Social Research, Academia Sinica)]

CLTC
Chên-li tsa-chih. 眞理雜誌 [Truth miscellany].

CPTC
Chih-pu-tsu chai ts'ung-shu, 知不足齋叢書 [Collection of the Chih-pu-tsu library], 1776–1798

HCJP
Hsi-ching jih-pao 西京日報 [Hsi-ching daily]

HCTY
Hsüeh-chin t'ao-yüan 學津討源 [Hsüeh-chin t'ao-yüan collection], 1806. Photolithographic reproduction, Commercial Press, 1935.

HHLP
Hsüeh-hai lei-pien 學海類編 [Classified anthology from the Ocean of Learning], 1831

HI
Hsüeh i 學藝 [Art and science]

HITH
Hsüeh-i t'ung-hsün 學藝通訊 [Science and art correspondent]

ICT
I-chia t'ang ts'ung-shu 宜稼堂叢書 [I-chia hall collection], 1842

ISP/WSFK
I-shih pao; Wên-shih fu-k'an 益世報. 文史副刊 [Literary supplement of the People's Betterment Daily]

JMJ
Japanese Meteorological Journal

KHCP
Kuo-hsüeh chi-pên ts'ung-shu 國學基本叢書 [Basic Sinological Series], 1936

KH
K'o-hsüeh 科學 [Science]

KHSCK

K'o-hsüeh shih chi-k'an 科學史集刊 [Journal of the history of science]

KK

Kagakushi kenkyû 科學史研究 [Researches on the history of science]

KTHP

Kuo-ts'ui hsüeh-pao 國粹學報 [Journal of studies in the national heritage]

KTIS

K'ung Tai i shu 孔戴遺書 [Lost books (rediscovered) by K'ung (Chi-han 孔繼涵) and Tai (Chên 戴震)], Ch'ing

LYI

Lien-yün-i ts'ung-shu 連筠簃叢書 [Collection from the Lien-yün-i library (owned by Yang Shang-wên)], 1848

MSKGK

Meiji Seitoku Kinen gakkai kiyô 明治聖德記念學會紀要 [Reports of the Meiji Memorial Society]

PCTHYK

Pei-ching ta-hsüeh yüeh-k'an 北京大學月刊 [Peking university monthly]

PFTSH

Pai-fu t'ang suan-hsüeh ts'ung-shu 白芙堂算學叢書 [Mathematical collection of the Pai-fu hall], 1875

Po-na

Po-na pên êrh-shih-ssŭ shih 百衲本二十四史 [*Po-na* edition of the 24(dynastic) histories]

QBCB

Quarterly Bulletin of Chinese Bibliography

T'u-shu ch'i-k'an 圖書集刊

SHS

Su-hsiang shih ts'ung-shu 粟香室叢書 [Collection from the Su-hsiang house], 1886–1887

SHT

Shih-hsün t'ang ts'ung-shu 式訓堂叢書 [Collection from the Shih-hsün hall], 1880

SK

Sanjutsu kyôiku 算術教育 [Mathematical education]

SKCSCP

Ssŭ-k'u ch'üan-shu chên-pên 四庫全書珍本 [Rare books from the quadripartite imperial library], 1935. See Hummel (1), 121.

SPTK

Ssŭ-pu ts'ung-k'an 四部叢刊 [Collected books from the four branches (of literature)], 1936

SSK

Shou-shan ko ts'ung-shu 守山閣叢書 [Collection from the Shou-shan Pavilion], 1862?

SY

Shih-yüan ts'ung-shu 適園叢書 [Shih-yüan collection], 1910.

TBGZ
Tôkyô Butsuri Gakko Zasshi 東京物理學校雜誌 [Journal of the Tokyo College of Physics]

TFTC
Tung-fang tsa-chih 東方雜誌 [Eastern miscellany]

TG
Tôyô Gakuhô 東洋學報 [Reports of the Oriental Society (of Tokyo)]

TLLL
T'ien-lu Lin-lang ts'ung-shu 天祿琳琅叢書 [T'ien-lu Lin-lang collection] Photographic reprint, 1931

TMJ
Tôhoku Mathematical Journal

TNK
Tai-nan ko ts'ung-shu 岱南閣叢書 [Collection from the Tai-nan Pavilion]

TSCC
Ts'ung-shu chi-ch'êng 叢書集成 [Collection of collectanea], 1936

WSTC
Wên shih tsa-chih 文史雜誌 [Journal for literature and history]

WYTCCP
Wu-ying tien chü-chên-pan ts'ung-shu 武英殿聚珍版叢書 [Collection of rare books from the Wu-ying Palace]. Recut block-print edition of 1874.

YCHP
Yen-ching hsüeh-pao 燕京學報 [Yenching journal (of Chinese studies)]

Reference Works

H
Hummel (1)

LY
Li Yen (10')

SK
Ssŭ-k'u ch'üan-shu tsung-mu t'i-yao (references follow the pagination of the Taiwan reprint, 1968)

TB
Têng and Biggerstaff (1)

TFP
Ting Fu-pao (1') (Page references are to the first part, numbers to the appendices)

Books and Articles

Chang Chin-wu (1') 張金吾
Ai-jih-ching-lu ts'ang-shu chih 愛日精廬藏書志 [Annotated catalogue of Ai-jih-ching-lu studio], 1826. *Wu-hsien Ling-fên ko hsü shih* 吳縣靈芬閣徐氏. Block-print edition of 1887. Ref.: TB, 44; H, 33.

Chang Ch'iu-chien (1') 張邱建
Chang Ch'iu-chien suan-ching 張邱建算經 [Mathematical classic of Chang Ch'iu-chien]. Sui, late sixth century. *KTIS*, vol 33; *TSCC*, 1,267; *TLLL*, 19. Ref.: TFP, 568b; *SK*, 2,204.

Chang P'êng-fei and Hsü T'ien-yu (1') 張鵬飛, 徐天游
Shu-hsüeh fa-ta shih 數學發達史 [History of the development of mathematics]. Taipei, 1962.

Chang Tun-jên (1') 張敦仁
Ch'iu-i suan-shu 求一算術 [Mathematical method for seeking unity (i.e., indeterminate analysis)], 1803. *TNK*, 46. Ref.: H, 417; LY, 289.

Chang Yung-li (1') 張永立
"Shêng-ching chung chih yüan-chou-lü" 聖經中之圓周率 [π in ancient Chinese literature]. *CLTC*, 1943, *1*, 267.

Chao Jan-ning (1') 趙然凝
"Yen-lü ming t'an" 衍率冥探 [Investigation of the secrets of the "extension numbers"]. *TFTC*, 1944, *41*, (23), 55.

Chao Liao (1') 趙繚
Shu-hsüeh tz'ŭ-tien 數學辭典 [Mathematical dictionary], Shanghai, 1923.

Ch'ên Chên-sun (1') 陳振孫
Chih-chai shu-lu chieh-t'i 直齋書錄解題 [Catalogue of the books of Chih-chai (i.e.,Ch'ên Chên-sun)and explanation of their contents], c.1250. *WYTCCP*, (ed. 1874), 49–56. Ref.: TB, 21; *SK*, 1780.

Ch'êng Hung-chao (1') 程鴻詔
Yu-hêng Hsin-chai wên 有恒心齋文 [Collection of essays by Yu-hêng Hsin-chai (i.e., Ch'êng Hung-chao)], 1865. (Ch'iu-i shu chi, 求一術記, Note on the Ch'iu-i method. In vol. 3, pp. 5a–5b.)

Ch'êng Ta-wei (1') 程大位
Suan-fa t'ung tsung 算法統宗 [Systematic treatise on arithmetic], 1593. Sankeidô 三桂堂 (Jap.) ed. Ref.: TFP, 585a; *SK*, 2,225. See Mei Ku-chêng (1').

Ch'i Chi-kuang (1') 戚繼光
Lien ping shih-chi 練兵實紀 [Authentic record on training troops], 1568. *SSK*, vols. 80–83 (also *TSCC*, vols. 948–950). Ref.: *SK*, 2,045.

Chiao Hsün (1') 焦循
T'ien-yüan-i shih 天元一釋 [Explanation of the *t'ien-yüan* algebra], 1799. In *Li t'ang hsüeh-suan chi*, 里堂學算記. Published in *Chiao shih i-shu*, 焦氏遺書, 1896, vol. 35. Ref.: H, 144; TFP, 592b; LY, 323.

Ch'ien-ch'ing t'ang shu-mu
See Huang Yü-chi (1').

Ch'ien Mu(1') 錢穆
"Liu Hsiang Hsin fu tzŭ nien-p'u" 劉向歆父子年譜 [Chronological biography of Liu Hsiang (77–6 B.C.) and Liu Hsin (c. 53 B.C.–23 A.D.) father and son]. *YCHP*, 1930, *7*, 1,189–1,318.

Ch'ien Mu (2')

"Chou-kuan chu-tso shih-tai k'ao" 周官著作時代考 [Study about the period of compilation of the *Chou-kuan* (i.e. *Chou-li*]). *YCHP*, 1932, 11, 2,191–2,300.

Ch'ien Pao-tsung (1′) 錢寶琮
"Pai-chi shu yüan-liu k'ao" 百鷄術源流考 [On the origin and development of the method of the "hundred fowls"]. *HI*, 1921, *3*, (no. 3), 1–6. Repr. in (4′), pp. 37–44.

Ch'ien Pao-tsung (2′)
Sung Yüan shu-hsüeh shih lun-wên-chi 宋元數學史論文集 [Collection of essays on the history of Sung and Yüan mathematics], Peking, 1966.

Ch'ien Pao-tsung (3′)
"Fang-ch'êng suan-fa yüan-liu k'ao" 方程算法源流考 [Study of the origin and development of the mathematical method for solving equations]. *HI*, 1921, *3* (no. 2), 1–12.

Ch'ien Pao-tsung (4′)
Ku-suan k'ao yüan 古算考源 [On the origin of Chinese mathematics]. Shanghai, 1930.

Ch'ien Pao-tsung (5′)
"Wang Hsiao-t'ung *Ch'i ku suan-shu* ti êrh t'i, ti san t'i shu wên shu-chêng" 王孝通緝古算術第二題第三題術文疏證 [Commentary on the method of problems 2 and 3 of Wang Hsiao-t'ung's *Ch'i ku suan-shu*]. *KHSCK*, 1966, *9*, 31–52.

Ch'ien Pao-tsung (6′)
"Tsêng-ch'êng k'ai-fang fa ti li-shih fa-chan" 增乘開方法的歷史發展 [Historical development of Horner's method]. *KHSCK*, 1959, *2*, 126–143. Repr. in (2′), pp. 36–59.

Ch'ien Pao-tsung (7′)
Chung-kuo suan-hsüeh shih 中國算學史 [A history of Chinese mathematics]. Peking, 1932. (National Research Institute of History and Philology Monographs, Ser. A, No. 6; Academia Sinica.)

Ch'ien Pao-tsung (8′)
Chung-kuo shu-hsüeh shih 中國數學史 [A history of Chinese mathematics] Peking, 1964.

Ch'ien Pao-tsung (9′)
Chung-kuo shu-hsüeh shih hua 中國數學史話 [On the history of Chinese mathematics]. Peking, 1957.

Ch'ien Pao-tsung (10′)
"Ch'iu-i shu yüan-liu k'ao" 求一術源流考 [Investigation of the origin and development of indeterminate analysis]. *HI*, 1921, *3*, (no. 4). Repr. in (4′), p. 45.

Ch'ien Pao-tsung (11′)
"*Chiu-chang suan-shu* ying-pu-tsu shu liu-ch'uan Ou-chou k'ao" 九章算術盈不足術流傳歐洲考 [On the transmission of the rule of "too much and not enough" (algebraic rule of false position) of the *Chiu-chang suan-shu* to

Europe]. *KH*, 1927, *12* (no. 4), 701–714.

Ch'ien Pao-tsung (12′) (ed.)
Suan-ching shih shu 算經十書 [Ten mathematical manuals]. Peking, 1963.

Ch'ien Ta-hsin (1′) 錢大昕
Ch'ien-yen t'ang wên-chi 潛研堂文集 [Collection of studies from Ch'ien-yen studio], 1806. *SPTK*, 1839–1854. Ref.: H, 155.

Ch'ien Ta-hsin (2′)
Shih-chia chai yang-hsin lu 十駕齊養新錄 [Record of cultivating self-renewal from the Shih-chia studio], c. 1790. Commercial Press, 1957, Ref.: H, 154.

Ch'ien Ta-hsin (3′)
Chu-t'ing hsien-shêng jih-chi ch'ao 竹汀先生日記鈔 [Diary of Mr. Chu-t'ing (i.e. Ch'ien Ta-hsin)], c. 1800. Ed. Ho Yüan-hsi 何元錫.

Ch'ien T'ai-chi (1′) 錢泰吉
P'u-shu tsa-chi 曝書雜記 [Miscellaneous notes on sun-dried books], 1838. *SHT*, vol. 10. Ref.: H, 155.

Ch'ien Tsêng (1′) 錢曾
Ch'ien Tsun-wang, "Tu-shu min-ch'iu chi" chiao-chêng 錢遵王讀書敏求記校證 [Critical edition of Ch'ien Tsêng's *Tu-shu min-ch'iu chi*], second half of the seventeenth century. Ed. Chang Yü 章鈺, 1926. Ref.: H, 157; TB, 42.

Chin-shu 晉書
[History of the Chin dynasty (265–419)]. Compiled by Fang Hsüan-ling, 房玄齡. *Po-Na* edition.

Ch'in Chiu-shao(1′) 秦九詔
Shu-shu chiu-chang 數書九章 [Mathematical treatise in nine sections], 1247. *ICT*, vols. 33–36; repr. *TSCC*, vols. 1,269–1,273.

Ching-ting Chien-k'ang chih 景定建康志
[Gazetteer of Nanking in the Ching-ting period (1260–1265)]. Ed. of 1802. See *SK*, 1,468.

Ch'ing-shih lieh-chuan 清史列傳
[Biographies from the Ch'ing history]. Taipei, 1962.

Chiu-chang suan-shu 九章算術
[Nine chapters on mathematical techniques]. *TSCC*, vol. 1,263. Ref.: Needham (1), vol. 3, 24; K. Vogel(2).

Chiu-chang suan-shu yin-i 九章算術音義
[Phonological glosses on the *Chiu-chang suan-shu*]. Sung. *TSCC*, 1936. Ref.: TFP, 563b.

Chou Ch'ing-shu (1′) 周清澍
"Wo-kuo ku-tai wei-ta ti k'o-hsüeh chia; Tsu Ch'ung-chih" 我國古代偉大的科學家, 祖冲之 [A great scientist of ancient China: Tsu Ch'ung-chih]. Essay in Li Kuang-pi 李光璧 and Ch'ien Chün-hua 錢君曄, *Chung-kuo k'o-hsüeh chi-shu fa-ming ho k'o-hsüeh chi-shu jên-wu lun-chi* 中國科學技術發明和科學技術人物論集 [Essays on Chinese discoveries and inventions in science and technology, and on the men who made them]. Peking, 1955, pp. 270–287.

Chou Chung-fu (1′) 周中孚
Chêng t'ang tu-shu chi 鄭堂讀書記 [Studies from the Chêng hall], 1921.
KHCP. Ref.: TB, 32.

Chou Mi(1′) 周密
Kuei-hsin tsa-chih, hsü-chi 癸辛雜識續集 [Miscellaneous notes from Kueihsin street, first addendum], c. 1290. *HCTY*, vols. 184–186. Photolithographic reproduction of the edition of 1806, Commercial Press, 1935. Ref.: *SK*, 2,930; Des Rotours (1), cxii.

Chou Mi (2′)
Chih-ya t'ang tsa-ch'ao 志雅堂雜鈔 [Miscellaneous notes from Chih-ya hall], Sung (thirteenth century). *HHLP*, vols. 77–78. Reprint, Commercial Press, 1920. Ref.: H, 740; *SK*, 2,665.

Chou-pei suan-ching 周髀算經
[Arithmetical classic of the Chou gnomon], Chou, Ch'in, and Han. *TSCC*, vol. 1,262. Ref.: TFP, 560a.

Chou Yü-chin (1′) 周玉津
Shang-yeh shih 商業史 [History of trade]. Taipei, 1963.

Chung-kuo ti-ming tz'ŭ-tien 中國地名辭典
[Dictionary of Chinese place names], n.d.

Chung-kuo ying-tsao hsüeh-shê hui-k'an 中國營造學社彙利
[Bulletin of the Society for Research in Chinese Architecture], 1930–.

Ch'ü Yung, (1′) 瞿鏞
T'ieh-ch'in-t'ung-chien lou ts'ang-shu mu-lu 鐵琴銅劍樓藏書目錄 [Catalogue of the T'ieh-ch'in-t'ung-chien lou library]. Tung K'ang 董康 Sung Fên Shih 誦芬室 block-print edition of 1897. Ref.: H, 34; TB, 45.

Fan Wu (1′) 范午
"Sung-tai tu-tieh shuo" 宋代度牒說 [On the *tu-tieh* in the Sung period]. *WSTC*, 1942, *2* (no. 4), 45–52.

Fang Kuo-yü (1′) 方國瑜
"Shu-ming yuan-shih" 數名原始 [Origin of the names of numbers]. *TFTC*, 1931, *28*, (no. 10), 83–88.

Fujiwara, M. (1′) 藤原松三郎
"Shina sugaku no kenkyû" 支那數學史の研究 [Researches on the history of Chinese mathematics]. *TMJ*, 1939, *46*, 284; 1940, *47*, 35, 309; 1941, *48*, 78.

Hoshi Ayao (1′) 星斌夫
Chûgoku shakai keizai shi goi 中國社會經濟史語彙 [Glossary of Chinese socioeconomic history]. Tokyo, 1966.

Hsü Chên-ch'ih (1′) 徐震池
"Shang-yü ch'iu yüan fa" 商餘求原法 [The remainder problem]. *KHS*, 1925, *10*, (2), 174.

Hsü Ch'un-fang (1′) 許蒪舫
Chung-kuo suan-shu ku-shih 中國算術故事 [The story of Chinese mathematics]. Peking, 1965.

Hsü Ch'un-fang (2′)
Chung suan-chia ti chi-ho-hsüeh yen-chiu 中算家的幾何學研究 [The study of geometry by Chinese mathematicians]. Peking, 1954.

Hsü Ch'un-fang (3′)
Chung suan-chia ti tai-shu-hsüeh yen-chiu 中算家的代數學研究 [The study of algebra by Chinese mathematicians]. Peking, 1954.

Hsü Ch'un-fang (4′)
Ku-suan ch'ü-wei 古算趣味 [Mathematical recreations in old China]. Peking, 1956.

Hsü Ch'un-fang (5′)
Ku suan-fa chih hsin yen-chiu 古算法之新研究 [New researches on old (Chinese) mathematics]. Shanghai, 1935.

Hsü Ch'un-fang (6′)
Ku suan-fa chih hsin yen-chiu hsü-pien 續編 [New researches on old (Chinese) mathematics, Supplement]. Shanghai, 1945.

Hsü Yüeh(1′) 徐岳
Shu-shu chi-i 數術記遺 [Record of mathematical methods], H/Han, c. 190. With commentary by Chên Luan 甄鸞 (sixth century). *TSCC*, vol. 1,266. Ref.: TFP, 568a.

Hu Yü-chin (1′) 胡玉縉, completed by Wang Hsin-fu 王欣夫.
Ssŭ-k'u t'i-yao pu-chêng 四庫提要補正 [Supplements to the *Ssŭ-k'u (ch'üan-shu tsung-mu) t'i-yao*]. Taipei, 1964¹, 1967².

Hua Lo-kêng (1′) 華羅庚
Ts'ung Tsu Ch'ung-chih ti yüan-chou-lü t'an-ch'i 從祖冲之的圓周率談起 [A discussion beginning with Tsu Ch'ung-chih's determination of π]. Hongkong, 1967.

Huang Chieh (1′) 黃節
"Ku suan-ming yüan" 古算名原 [Origin of old mathematical terminology]. *KTHP*, 1908, (3) *46*, 1–6; *49*, 1–9.

Huang Tsung-hsien (1′) 黃宗憲
Ch'iu-i shu t'ung-chieh 求一術通解 [Complete explanation of the method for finding unity], 1875. *PFTSH*, vol. 10. Ref.: Van Hee (5); TFP, no. 322; LY, 342.

Huang Yü-chi (1′) 黃虞稷
Ch'ien-ch'ing t'ang shu-mu 千頃堂書目 [Bibliography from the Ch'ien-ch'ing hall], second half of the seventeenth century. SY, vols. 13–24. Ref.: TB, 25.

Juan Yüan (1′) 阮元
Ch'ou-jên chuan 疇人傳 [Biographies of mathematicians and astronomers], 1799. Repr. Commercial Press, 1955. Ref.: W. Franke (1); H, 402; TFP, no. 536.

Kao Hêng (1′) 高亨
Mo-ching chiao-ch'üan 墨經校詮 [Critical study of the *Mo-ching*]. Peking, 1962.

Katô Shigeshi (1′) 加藤繁

Shina keizaishi kôshô 支那經濟史考證 [Philological studies in Chinese economic history]. 2 vols., Tokyo, 1952–1953. (Tôyô Bunko Pub., Series A, No. 34; English summary.)

Ku chin t'u-shu chi-ch'êng 古今圖書集成
[Encyclopedia–old and new], 1725. Chung-hua shu-chü 中華書局, photo-lithographic reproduction of the palace edition, 1934.

Ku Kuan-kuang (1′) 顧觀光
Wu-ling shan jên i-shu 武陵山人遺書 [Posthumous works of the man of the Wu-ling mountain], 1883. Ref.: LY, 400; TFP, no. 97.

Ku Kuang-ch'i (1′) 顧廣圻
Ssŭ-shih chai chi 思適齋集 [Collected works from Ssŭ-shih's (that is, Ku Kuang-ch'i's) studio], 1894. *CHT*, vol. 13. Ref.: H, 419.

Kuo Pai-kung (1′) 郭伯恭
Ssŭ-k'u ch'üan-shu tsuan-hsiu k'ao 四庫全書纂修考 [Study on the compilation of the *Ssŭ-k'u ch'üan-shu*]. Commercial Press, 1937. Ref.: TB, 29, note 1.

Kuo Pai-kung (2′)
Yung-lo ta-tien k'ao 永樂大典考 [Study of the Yung-lo encyclopedia], 1962².

Lao Nai-hsüan (1′) 勞乃宣
Ku ch'ou-suan k'ao-shih 古籌算考釋 [Investigation of the ancient computing rods], 1886.

Li chi 禮記
[Book of rites], Early Han. Shanghai, 1912 (reprint of the *Ch'ung-k'an Sung pên, shih-san ching,* Reprinted Sung editions of the thirteen classics, 1815).

Li Chieh (1′) 李誡
Ying-tsao fa-shih 營造法式 [Architectural standards], 1103–1106. Ed. 1919. Ref.: Demiéville (1).

Li Jui (1′) 李銳
Jih-fa shuo-yü ch'iang jo k'ao 日法朔餘强弱考 [Investigation of the "day ratio," the "month remainder," and the "strong" and "weak" factors], 1799. In *Li shih i-shu* 李氏遺書 [Posthumous works of Mr. Li], 1823. Ref.: TFP, no. 93; LY, 251.

Li Liu (1′) 李劉
Mei-t'ing hsien-shêng ssŭ-liu piao-chun 梅亭先生四六標準 [Standards for the "four-six" (prose) of Mr. (Li) Mei-t'ing (Li Liu)], Sung. *SPTK*, 421–428. Ref.: *SKCS*, vol. 5, 3,402.

Li Ts'êng-po (1′) 李曾伯
K'o chai hsü-kao hou 可齋續稿後 [Supplementary manuscripts of K'o Chai (i.e. Li Ts'êng-po), second part], Sung. *SKCSCP*, vol. 4. Ref.: *SK*, 3,413.

Li Yeh (1′) 李冶
Ts'ê-yüan hai-ching 測圓海鏡 [Sea mirror of circle measurement], 1248. *TSCC*, 1,277–1,278. Ref.: Van Hee (1); TFP, no. 9.

Li Yen (1′) 李儼
"Tung-fang T'u-shu-kuan shan-pên suan-shu chieh-t'i" 東方圖書館善本

算書解題 [Descriptive notes on rare Chinese mathematical books in the Oriental Library of the Commercial Press] *BNLP*, 1933, *7*, no. 1.

Li Yen (2')
"Êrh-shih nien lai Chung-suan shih lun-wên mu-lu" 二十年來中算史論文目錄 [A bibliography of articles on Chinese mathematics published in the last twenty years] *BNLP*, 1932, *6*, 197–205.

Li Yen (3')
Shih-san shih-ssŭ shih-chi Chung-kuo min-chien shu-hsüeh 十三十四世紀中國民間數學 [Chinese mathematics in the thirteenth and fourteenth centuries], Peking, 1957.

Li Yen (4')
"Êrh-shih-pa nien lai Chung-suan shih lun-wên mu-lu" 二十八年來中算史論文目錄 [Bibliography of papers on the history of Chinese mathematics during the past 28 years]. *QBCB*, 1940, *2*, 372. Addenda: *QBCB*, 1944, *5*, (no. 4), 55.

Li Yen (5')
"San-shih nien lai chih Chung-kuo suan-hsüeh shih" 三十年來之中國算學史 [(Researches into) the history of Chinese mathematics during the last thirty years]. *KH*, 1947, *29*, 101. Repr. in Li Yen (6'), 146.

Li Yen (6')
Chung-suan shih lun-ts'ung 中算史論叢 [Collected essays on Chinese mathematics]. Shanghai, 1933–1947. 3 vols., 1933–35; 4th vol. (in 2 parts), 1947. Commercial Press, Shanghai. New edition in 5 vols., Peking, 1954–1955.

Li Yen (7')
Chung suan-chia ti nei-ch'a fa yen-chiu 中算家的內插法研究 [Researches on the interpolation formulae of Chinese mathematicians]. Peking, 1957.

Li Yen (8')
Chung-kuo shu-hsüeh ta-kang 中國數學大綱 [Outline of Chinese mathematics]. Shanghai, 1931. Repr.: Peking, 1958.

Li Yen (9')
"Ta-yen ch'iu-i shu chih kuo-ch'ü yü wei-lai" 大衍求一術之過去與未來 [The past and future of indeterminate analysis]. *HI*, 1925, *7*, (no. 2), 1–45. Repr. in Li Yen (6'), vol. 1, 122–174.

Li Yen (10')
Chung-kuo suan-hsüeh shih lun-ts'ung 中國算學史論叢 [Collected essays on the history of Chinese mathematics]. Taipei, 1954.

Li Yen (11')
"Li Yen so ts'ang Chung-kuo suan-hsüeh shu mu-lu" 李儼所藏中國算學書目錄 [Catalogue of the Chinese mathematical books in the library of Li Yen]. *KHS*, 1920, *5* (no. 4), 418; (no. 5), 525; 1925, *10* (no. 4), 548; 1926, *11* (no. 6), 817; 1927, *12* (no. 12), 1,825; 1929, *13* (no. 8), 1,134; 1930, *15* (no. 1), 158; 1932, *16* (no. 5), 856, (no. 11), 1,710; 1933, 17 (no. 6), 1,005; 1934, *18* (no. 11), 1,547.

Li Yen (12')
"Chung-kuo suan-hsüeh shih yü lu" 中國算學史餘錄 [Additional remarks

on the history of mathematics in China]. *KH*, 1917, *2* (no.2), 238.

Li Yen (13′)
Chung-kuo ku-tai shu-hsüeh chien-shih 中國古代數學簡史 [Brief history of old Chinese mathematics]. 2 vols., Peking, 1963–1964.

Li Yen (14′)
"Chung-kuo suan-hsüeh lüeh-shuo" 中國算學略說 [Summary of Chinese mathematics]. *KH*, 1934, *18*, 1,135. Repr. in Li Yen (10′), pp. 1 ff.

Li Yen (15′)
Chung-kuo ku-tai shu-hsüeh shih-liao 中國古代數學史料 [Materials for the historical study of ancient Chinese mathematics]. Shanghai, 1954.

Li Yen (16′)
Chung-kuo suan-hsüeh shih 中國算學史 [A history of Chinese mathematics]. Shanghai, 1937.

Li Yen (17′)
Chung-kuo suan-hsüeh hsiao-shih 中國算學小史 [Brief history of Chinese mathematics]. Shanghai, 1930.

Li Yen and Yen Tun-chieh (18′)
"K'ang-chan i lai Chung-suan shih lun-wên mu-lu" 抗戰以來中算史論文目錄 [Bibliography of papers on the history of Chinese mathematics during the period of the Second World War]. *QBCB*, 1944, *5* (no. 4), 51.

Li Yen (19′)
"Chung-suan shih chih kung-tso" 中算史之工作 [Work on the history of Chinese mathematics]. *KH*, 1928, *13* (no.6), 785–809. Repr. in Li Yen (6′), vol. 5, pp. 93 ff.

Li Yen (20′)
"Chu-suan chih-tu k'ao" 珠算制度考 [Investigation of the system of abacus reckoning]. *YCHP*, 1931, *1* (no.10), 2,123.

Li Yen (21′)
"Chung-kuo shu-hsüeh yüan-liu k'ao-lüeh" 中國數學源流考略 [Outline of the origin and evolution of Chinese mathematics]. *PCTHYK*, 1919, *1* (no.4), 1; 1919, *1* (no.5), 59; 1920, *1* (no.6), 65.

Li Yen (22′)
"Chung suan-chia chih chi-shu lun" 中算家之級數論 [On the treatment of series by Chinese mathematicians]. *KHS*, 1929, *13* (no.9), 1,139; 1929, *13* (no.10), 1,349. Repr.: Li Yen (6′), vol.1, pp. 315 ff.

Li Yen (23′)
"Hsien-shou Chung suan-shu mu-lu" 現售中算書目錄 [Catalogue of books in print on Chinese mathematics]. *HCJP*, 1935, *7*. Repr. in Li Yen (10′), pp. 18 ff.

Li Yen (24′)
"Tsui-chin shih nien lai Chung-suan shih lun-wên mu-lu" 最近十年來中算史論文目錄 [Bibliography of papers on the history of Chinese mathematics during the last ten years]. *HITH*, 1948, *15*, 16.

Liang Ssŭ-ch'êng (1′) 梁思成

Ch'ing-shih ying-tsao tsê-li 清式營造則例 [Architectural rules of the Ch'ing style]. 1934. Ed. by 中國營造學社 [Society for Research in Chinese Architecture].

Liang Ssǔ-ch'êng (2′)
Ying-tsao suan-li 營造算例 [Mathematical laws in architecture], 1934. Ed. by 中國營造學社 [Society for Research in Chinese Architecture].

Liu Chih-p'ing (1′) 劉致平
Chung-kuo chien-chu lei-hsing chi chieh-kou 中國建築類型及結構 [Chinese architecture: characteristics and constructions]. Peking, 1957.

Liu Hsiao-sun (1′) 劉孝孫
Chang Ch'iu-chien suan-ching hsi-ts'ao 張邱建算經細草 [Detailed solutions of the problems in the *Chang Ch'iu-chien suan-ching*], late sixth century. *Suan-ching shih shu*, vols. 27–28.

Liu K'ê-chuang (1′) 劉克莊
Hou Ts'un hsien-shêng ta-ch'üan-chi 後村先生大全集 [Collected works of Mr. Hou Ts'un (Liu K'ê-chuang)], Sung. *CP*, 1929. Ref.: *SK*, 3,413.

Liu Shêng-mu (1′) 劉聲木
T'ung-ch'êng wên-hsüeh yüan yüan k'ao 桐城文學淵源考 [Investigation of T'ung-ch'êng literary sources], 1909. *CCT*, vol. 3. Ref.: H, 65.

Liu To (1′) 劉鐸
Jo-shui chai ku-chin suan-hsüeh shu-lu 若水齋古今算學書錄 [Jo-shui studio bibliography of mathematical books]. Shanghai, 1898. Included in TFP fol. 54–93 with notes.

Lo T'êng-fêng (1′) 駱騰鳳
I-yu lu 藝游錄 [The pleasant game of mathematical art], 1815. Ho Chin 何錦 edition of 1843. Ref.: LY, 384; TFP, 522.

Lu Hsin-yüan (1′) 陸心源
I-ku t'ang t'i-pa 儀顧堂題跋 [Colophons from the I-ku hall], 1890–1892. Suppl. vol. *I-ku t'ang t'i-hsü* (續) *pa*. Ref.: TB, 47; H, 546; TFP, 3c.

Ma Tuan-lin (1′) 馬端臨
Wên-hsien t'ung k'ao 文獻通考 [Encyclopedia of documentary researches], Sung. Wan Yu Wên Ku, 萬有文庫, Commercial Press, Shanghai, 1936. Ref.: TB, 150.

Mao I-shêng (1′) 茅以昇
"Chung-kuo yüan-chou-lü lüeh-shih" 中國圓周率略史 [Outline history of π in China]. *KH*, 1917, *3* (no.4), 411.

Mao Yüan-i (1′) 茅元儀
Wu-pei chih 武備志 [Treatise on armaments], 1628. Repr. Canton, late Ch'ing. Ref.: *SK*, 2,060.

Mateer, C. W. (1′), tr. Ti K'ao-wên 狄考文 and Tsou Li-wên 鄒立文
Pi-suan shu-hsüeh 筆算數學 [(A handbook of) "written" (European) arithmetic]. 3 vols., 1892.

Mei Ku-ch'êng (1′) 梅瑴成
Tsêng-shan suan-fa t'ung-tsung 增刪算法統宗 [Emended edition of the *Suan-*

fa t'ung tsung]. Shanghai, 1760. Ref.: H, 569; LY, 303.

Mei Wên-ting (1′) 梅文鼎
Li-suan shu-mu 曆算書目 [Catalogue of works on calendar and mathematics], 1723. *CPTC*, vol. 149 (also *TSCC*). Ref.: H, 570; LY, 299.

Miao Ch'üan-sun and Ch'ên Ssŭ(1′) 繆荃孫, 陳思
Chiang-yin hsien hsü-chih 江陰縣續志 [Supplement to the Gazetteer of Chang-yin district]. 28 ch. (12 vols.), 1920.

Mikami Yoshio (1′) 三上義夫
"Loria hakushi no Shina Sûgaku ron" Loria 博士の支那數學論 [Dr. Loria on Chinese mathematics] *TG*, 1923, *12* (no. 4), 500.

Mikami Yoshio (2′)
"Dickson shi no seisûron rekishijô ni okeru Wa-Kan-jin no kôken" Dickson 氏ノ整數論歷史上ニ於ケル和漢人ノ貢獻 [The contributions of Chinese and Japanese (mathematicians) in Dickson's *History of the Theory of Numbers*]. *TBGZ*, 1921, no. 354, 192; no. 356, 274, no. 362, 11.

Mikami Yoshio (3′)
Chung-kuo suan-hsüeh chih t'ê-sê 中國算學之特色 [Special characteristics of Chinese mathematics]. Orig. pub. in Japanese, *TG*, 1929, *15* (no.4); *16* (no.1). Tr. into Chinese by Lin K'o-t'ang 林科棠, Shanghai, 1934.

Mikami Yoshio (4′)
"Shina kodai no Sûgaku" 支那古代の數學 [Ancient Chinese mathematics]. *MSKGK*, 1932, *37*, 19.

Mikami Yoshio (5′)
"So Gen sûgakujô ni okeru endan oyobi shakusa no igi" 宋元數學上に於ける演段及び釋鎖の意義 [The significance of the *yen-tuan* and *shih-so* methods of solving equations in the Sung and Yüan dynasties]. *SK*, 1937, *179*, 1.

Mikami Yoshio (6′)
"Wakan sûgaku-shijô ni okeru senran oyobi gunji no kankei" 和漢數學史上に於ける戰亂及び軍事の關係 [Matters relating to war and military affairs in the history of mathematics in China and Japan]. *SK*, 1937, *182*, 1.

Min Êrh-ch'ang (1′) 閔爾昌
Pei-chuan chi pu 碑傳集補 [Additions to the '*Pei-chuan chi*' (by Ch'ien I-ch'i)]. Peking, 1932. Ref.: H, 151; TB, 243.

Ming-jên chuan-chi tzŭ-liao so-yin 明人傳記資料索引
[Index to documents on Ming biographies]. 2 vols., Taipei, 1965. Ed. National Central Library, Taipei 國立中央圖書館

Nagasawa Kamenosuke (1′) 長澤龜之助
Suan-shu tzŭ-tien 算術辭典 [Mathematical dictionary]. Shanghai, 1959. Tr. from Japanese.

Nan-Sung kuan-ko hsü-lu 南宋館閣續錄
[Additional palace records of the Southern Sung]. Ref.: *SK*, 1671.

Ogura Kinnosuke and Ôya Shinichi (1′) 小倉金之助, 大矢眞一
"Mikami Yoshio hakushi to sono gyôseki Mikami Yoshio sensei chosaku rombun mokuroku" 三上義夫博士とその業績・三上義夫先生著作論文目錄

[Life and work of Dr. Yoshio Mikami with bibliography of his publications]. *K,* 1951, (no. 18), 1, 9.

Pai Shang-shu (1′) 白尚恕
"Ch'in Chiu-shao ts'ê-wang chiu wên tsao shu chih t'an-t'ao" 秦九韶測望九問造術之探討 [Investigation into the structure and method of the nine "trigonometrical" problems of Ch'in Chiu-shao]. In Ch'ien Pao-tsung (2′), 290.

P'êng Hsin-wei (1′) 彭信威
Chung-kuo huo-pi shih 中國貨幣史 [History of Chinese currency].Shanghai, 1958.

San-ts'ai t'u-hui 三才圖會
Universal encyclopedia (literally, Encyclopedia of the "Three Powers," that is, heaven, earth and man), 1607. Compiled by Wang Ch'i 王圻 and Wang Ssŭ-i 王思義. Ref.: TB, 144.

Ssŭ-k'u ch'üan-shu chien-ming mu-lu 四庫全書簡明目錄
[Simplified catalogue of the "Complete Library in Four Branches"], 1782. Compiled by Chi Yün 紀昀 and others. Kuang-chou Ching-yün Lou 廣州經韻樓 edition, 1868. Ref.: TB, 32; H, 121.

Ssŭ-k'u ch'üan-shu tsung-mu t'i-yao 四庫全書總目提要
[Annotated catalogue of the "Complete Library in Four Branches"], completed in 1782. Shanghai, Ta-tung Book Co., 1930. Repr. in 6 vols., Commercial Press, Taipei, 1968. Ref.: TB, 27; H, 121.

Suan-ching shih-shu 算經十書
[Ten mathematical manuals], collected and printed in 1084. *KTIS.* Ref.: H, 697; see also Ch'ien Pao-tsung (12′).

Sudô Yoshiyuki (1′) 周藤吉之
Tô Sô shakaikeizaishi kenkyû 唐宋社會經済史研究 [Studies on the socioeconomic history of the T'ang and Sung]. Tokyo, 1965.

Sudô Yoshiyuki (2′)
Sôdai keizaishi kenkyû 宋代經済史研究 [Studies on the economic history of the Sung dynasty]. Tokyo, 1962.

Sui-shu 隋書
[History of the Sui dynasty (581–617)], T'ang, 636–656. *Po-Na* ed., CP, 1935.

Sun Tzŭ suan-ching 孫子算經
[Mathematical classic of Sun Tzŭ], San Kuo, Chin or L/Sung. *TLLL, ts'ê* 15 (photographic repr., 1931). Also *TSCC,* vol. 1,266). Ref.: TFP, 566.

Sung Ching-ch'ang (1′) 宋景昌
Shu-shu chiu-chang cha-chi 數書九章札記
[Notes on the *Shu-shu chiu-chang*], 1842. *ICT,* 49–50; also *TSCC,* 1,274–1,275). Ref.: TFP, no. 469.

Sung-shih 宋史
[History of the Sung Dynasty], c. 1345. *SPPY,* 745–824.

Takeda Kusuo (1') 武田楠雄
"Tengenjutsu sôshitsu no shosô; Mindai sûgaku no tokushitsu III" 天元術喪失の諸相（明代數學の特質Ⅲ）[Various aspects of the loss of the *t'ien-yüan* algebra. The character of Chinese mathematics in the Ming dynasty, part III]. *KK, 34,* 1955, pp, 12–22.

Têng Yen-lin and Li Yen (1') 鄧衍林, 李儼
Pei-p'ing ko t'u-shu-kuan so ts'ang Chung-kuo suan-hsüeh shu lien-ho mu-lu 北平各圖書館所藏中國算學書聯合目錄 [Union catalogue of Chinese mathematical books in the libraries of Peiping]. Peking, 1936. Ref.: Hummel (2).

Ting Ch'ü-chung (ed.) (1') 丁取忠
Pai-fu t'ang suan-hsüeh ts'ung-shu 白芙堂算學叢書 [Collection of mathematical books from the Pai-fu hall], 1875. Ref.: Van Hee (5); TFP, no. 81.

Ting Ch'ü-chung (2')
Shu-hsüeh shih-i 數學拾遺 [Mathematical gleanings]. *PFTSH*, no. 12. Ref.: Van Hee (5), 151; TFP, no. 523.

Ting Fu-pao and Chou Yün-ch'ing (1') 丁福保, 周雲靑
Ssŭ-pu tsung-lu, suan-fa pien, 四部總錄算法編 [General catalogue of the four departments of literature; section "mathematics"]. Shanghai, 1957. This compilation is derived from

1. *Tzŭ-pu suan-fa pien* 子部算法編 [Miscellaneous notes—mathematical section] (by Ting Fu-pao), 1899.

2. *Pu-i* 補遺 [Appendix] (by Chou Yün-ch'ing), 1956.

3. *a). Ku-chin suan-hsüeh shu-lu,* 古今算學書錄 [Bibliography of old and new mathematical books] (by Liu To 劉鐸), 1898.

b). Suan-hsüeh k'ao ch'u-pien 算學考初編 [First compilation of studies on mathematics] (by Fêng Chêng 馮澂), 1897.

Tsou An-ch'ang (1') 鄒安鬯
K'ai-fang chih fên huan-yüan shu 開方之分還原術 [Treatise on root-extraction]. 1841. *SHS*, vol. 19.

Tu-shu min-ch'iu chi chiao-chêng
See Ch'ien Tsêng.

Tuan Yü-hua and Chou Yüan-jui (1') 段玉華, 周元瑞
Suan-hsüeh ta-tz'ŭ-tien 算學大辭典 [A complete dictionary of mathematical terms]. Hong Kong, 1957.

Wada Sei (1') 和田清
Sôshi shokukashi yakuchû 宋史食貨志譯註 [Treatise on commodities of the Sung]. Tokyo, 1960. (Tôyô Bunko Pub., Series A, no. 44).

Wada Yûji (1') 和田雄治
"Seisô Ei-so Ryôchô no soku-u-ki" 世宗英祖兩朝之測雨器 [Korean rain gauges from the reigns of Sejong (1419–1450) and Yongjo (1725-1776)]. *JMJ*, 1911, no. 3. Chinese résumé: *KH*, 1916, *2*, 582.

Wang Ch'i
See *San-ts'ai t'u-hui.*

Wei Liao-wêng (1') 魏了翁
(Pai)-ho shan hsien-shêng ta-ch'üan wên-chi (白)鶴山先生大全文集 [Complete literary works of the master of the Pai-ho mountain], Sung. *SPTK*, vols. 1,240–1,263. Ref.: *SK*, 3391.

Wu Ch'êng-lo (1') 吳承洛
Chung-kuo tu-liang-hêng shih 中國度量衡史 [History of Chinese metrology (weights and measures)]. Shanghai, 1937.

Wu-ching tsung-yao 武經總要
[Conspectus of essential military techniques], Sung, 1040. Compiled by Tsêng Kung-liang 曾公亮 and Ting Tu, 丁度. *SKCSCP*, vols. 807–836. Ref.: *SK*, 2,041.

Wu-pei chih
See Mao Yüan-i (1').

Yabuuchi Kiyoshi, ed. (1') 藪內清
Shina Sûgakushi gaisetsu 支那數學史概説 [Outline of Chinese mathematics]. Kyoto, 1944.

Yabuuchi Kiyoshi, ed. (2')
Sô Gen jidai no kagaku gijutsu shi 宋元時代の科學技術史 [A history of science and technology in the Sung and Yüan periods]. Kyoto, 1967.

Yabuuchi Kiyoshi,ed. (3')
Chûgoku chûsei kagaku gijutsu shi no kenkyû 中國中世科學技術史の研究 [Studies on the history of Chinese medieval science and technology], Kyoto, 1963.

Yang Hui (1') 楊輝
Hsü-ku chai-ch'i suan-fa 續古摘奇算法 [Continuation of ancient and curious mathematical methods], Sung, 1275. *CPTC*, vol. 103.

Yang Hui (2')
Hsiang-chieh chiu-chang suan-fa tsuan lei 詳解九章算法纂類 [Compendium of analyzed mathematical methods in the Nine Chapters], 1261. *TSCC*. Ref.: TFP, 3, 79a.

Yang K'uan (1') 楊寬
Chung-kuo li-tai chih-tu k'ao 中國歷代尺度考 [Study on the foot measure in the Chinese historical periods]. Shanghai, 1955.

Yao Chin-yüan (1') 姚覲元
Fu-chou Shih-yü wên-tzŭ so-chien lu 涪州石魚文字所見錄 [Records from the inscriptions of the "Stone Fish" at Fu-chou], 1912. Ku-hsüeh Hui-k'an, 古學彙刊. Photostatic repr., Taipei, 1964. Vol. 2, pp. 1207–1271.

Yen Tun-chieh (1') 嚴敦傑
"Shanghai suan-hsüeh wên-hsien shu-lüeh" 上海算學文獻述略 [Outline of mathematical documents in Shanghai]. *KH*, 1939, *23*, 72.

Yen Tun-chieh (2')
"Sung Yüan suan-hsüeh ts'ung k'ao" 宋元算學叢考 [Collected investigations on Sung and Yüan mathematics]. *KH*, 1947, *29*, 109.

Yen Tun-chieh (3')

Chung-kuo ku-tai shu-hsüeh ti ch'êng-chiu 中國古代數學的成就 [Contributions of ancient Chinese mathematics]. Peking, 1956.

Yen Tun-chieh (4′)
"Ch'ou-suan suan-p'an lun" 籌算算盤論 [On the abacus and the counting board]. *TFTC*, 1944, *41*, 33.

Yen Tun-chieh (5′)
"Sung Yüan suan-shu yü hsin-yung huo-pi shih liao" 宋元算書與信用貨幣史料 [Historical data concerning the mathematical books of the Sung and Yüan and the paper money]. *ISP/WSFK*, 1943, no. 38.

Ying-tsao fa-shih
See Li Chieh (1′)

Yung-lo ta-tien 永樂大典
[Yung-lo encyclopaedia], 1407. Chung-hua Book Co. (中華書局), Photographic reprint, 1960. Ref.: Li Yen (6′), vol. 2, p. 83; TFP, no. 164; Yüan T'ung-li (1′).

Yung-lo ta-tien mu-lu 永樂大典目錄
[Catalogue of the Yung-lo encyclopaedia]. LYI, ch. 42.

Yü-hai 玉海
[Ocean of jade (encyclopaedia)], 1247. Compiled by Wang Ying-lin 王應麟. Chekiang Book Co. 浙江書局, 1883. Ref.: TB, 122.

Yüan Chên (1′) 袁震
"Liang-Sung Tu-tieh k'ao" 兩宋度牒考 [The *tu-tieh* in the Sung dynasty]. *CKSHCCSCK*, 1944, *7* (no. 1), 42; 1946, *7* (no. 2), 1.

Yüan-shih 元史
[History of the Yüan dynasty], c. 1370, *SPPY*, 853–881.

Yüan T'ung-li (1′) 袁同禮
"*Yung-lo ta-tien* hsien-ts'un chüan mu-piao" 永樂大典現存卷目表 [Bibliography of extant chapters of the *Yung-lo ta-tien*]. *BNLP*, 1933, *1*, 103.

Yüeh Chia-tsao (1′) 樂嘉藻
Chung-kuo chien-chu shih 中國建築史 [A history of Chinese architecture]. Shanghai, 1933.

Books and Articles in Western Languages

Western Periodicals: Abbreviations

AGM
Abhandlungen zur Geschichte der Mathematik

AGMW
Abhandlungen zur Geschichte der Mathematischen Wissenschaften

AHES
Archive for History of Exact Sciences

AM
Asia Major

AMM
American Mathematical Monthly

AMN
Archiv für Math. og Naturvidenskab

AMP
Archiv für Mathematik und Physik

ARCHEION
Archives internationales d'histoire des sciences

ASPN
Archives des sciences physiques et naturelles

B
Boethius; Texte und Abhandlungen zur Geschichte der exakten Wissenschaften

BBSSMF
Bolletino di bibliografia e di storia della scienze matematiche e fisiche (Boncompagni's)

BCMS
Bulletin of the Calcutta Mathematical Society

BDM
Bolletino di matematica

BEFEO
Bulletin de l'Ecole française d'Extrême-Orient (Hanoi)

BG
Bulletin de géographie

BM
Bibliotheca mathematica

BSM
Bulletin de la société mathématique (de France)

BSOAS
Bulletin of the School of Oriental and African Studies

CHM
Cahiers d'histoire mondiale (Journal of World History)

CJSA
China Journal of Sciences and Arts

CRAIBL
Comptes rendus de l'Académie des inscriptions et belles lettres (Paris)

CRAS
Comptes rendus de l'Académie des sciences (Paris)

E
Euclides

EHR
Economic Historical Review

ER
Edinburgh Review

FA
France-Asie

FF
Forschungen und Fortschritte

HJAS
Harvard Journal of Asiatic Studies

HTR
Harvard Theological Review

IA
Indian Antiquary

IMI
Istoriko-matematičeskie issledovaniya

JA
Journal asiatique

JACM
Journal of the Association for Computing Machinery

JAOS
Journal of the American Oriental Society

JAS
Journal of Asian Studies

JDMV
Jahresberichte der deutschen Mathematiker-Vereinigung

JIMC
The Journal of the Indian Mathematical Club (Madras)

JIMS
The Journal of the Indian Mathematical Society (Madras)

JPASB
Journal and Proceedings of the Asiatic Society of Bengal

JRAM
Journal für reine und angewandte Mathematik (Crelle's)

JRAS/MB
Journal of the Royal Asiatic Society/Malayan Branch

JRAS/NCB
Journal of the Royal Asiatic Society/North China Branch

JS
Journal des savants

JSHS
Japanese Studies in the History of Science

KDBS/MFM
Kgl. Danske Videnskabernes Selskab (Math.-fysiske Medd.)

LM
Les mondes (Revue hebdomadaire des sciences et arts)

MASIB
Memorie della Accademia della Scienze dell' Istituto di Bologna

MCB
Mélanges chinois et bouddhiques

MDGNVO
Mitteilungen der deutschen Gesellschaft für Natur- und Volkskunde Ostasiens

MKAWB
Monatsberichte der königlichen Akademie der Wissenschaften zu Berlin

MMG
Mitteilungen der Mathematischen Gesellschaft (Hamburg)

MN
Mathematische Nachrichten

MS
Monumenta Serica

MT
The Mathematics Teacher

MZ
Meteorologische Zeitschrift

NAM
Nouvelles annales mathématiques

NAW
Nieuw Archief voor Wiskunde

NC
Numismatic Chronicle

O
Observatory

OE
Oriens Extremus

PAAAS
Proceedings of the American Academy of Arts and Sciences

PC
People's China

PSM
The Popular Science Monthly (see SM)

PTMS
Proceedings of the Tokyo Mathematical Society

PTRS
Philosophical Transactions of the Royal Society

QJRMSL
Quarterly Journal of the Royal Meteorological Society of London

QSGM/B
Quellen und Studien zur Geschichte der Mathematik (Part B: Astronomie und Physik).

RHS
Revue d'histoire des sciences
RQS
Revue des questions scientifiques (Brussels)
RSO
Revista degli studi orientali
S
Sinica (originally *Chinesische Blätter für Wissenschaft und Kunst*)
SH
Scientiarum Historia
SM
Scientific Monthly (formerly *PSM*)
SMGBO
Studien und Mitteilungen zur Geschichte des Benediktiner-Ordens
SPMSE
Sitzungsberichte Physikalisch-Medizinische Sozietät Erlangen
SS
Science and Society
TASJ
Transactions of the Asiatic Society of Japan
TP
T'oung Pao
UMN
Unterrichtsblätter für Mathematik und Naturwissenschaften
VNGZ
Vierteljahrschrift der Naturforschenden Gesellschaft in Zürich
ZDMG
Zeitschrift der Deutschen Morgenländischen Gesellschaft
ZMNWU
Zeitschrift für Mathematischen und Naturwissenschaftlichen Unterricht
ZMP
Zeitschrift für Mathematik und Physik

Books and Articles

Aiyar, P. V. Seshu (1)
"A historical Note on Indeterminate Equations." *JIMC*, 1910, *2* (no. 5), 216–219.

Amiot and Bourgeois et al. (1)
Mémoires concernant l'histoire, les sciences, les arts, les moeurs, les usages, etc. par les missionnaires de Pe-kin. 15 vols., Paris, 1776–1791.

Ang Tian Se (1)
A Study of the Mathematical Manual of Chang Ch'iu-chien. M.A. thesis. Univer-

sity of Malaya, Kuala Lumpur, 1969.

Archibald, R. C. (1)
"The Cattle Problem of Archimedes." *AMM*, 1918, *25*, 411–414.

Archibald, R. C. (2)
"Outline of the History of Mathematics." *AMM*, 1949, *56*, 1–114 (6th ed.).
First ed. 1932.

Ayyangar, A. A. Krishnaswami (1)
"Bhaskara and Samçlishta Kuttaka." *JIMS*, Notes and Questions, 1929,
18 (no. 1), 1–7.

Bachet, C. G., Sr. de Méziriac (1)
Problèmes plaisans et delectables, qui se font par les nombres. Lyon, 1612.

Bagchi, P. C. (1)
India and China; a Thousand Years of Sino-Indian Cultural Relations. Bombay,
1944¹, 1950².

Balazs, E. (1)
La bureaucratie céleste. Paris, 1968.

Ball, W. Rouse (1)
A Short Account of the History of Mathematics. London, 1915⁶.

Barde, R. (1)
"Recherches sur les origines arithmétiques du Yi-king." *ARCHEION*,
1952, *31* (n.s.), 234–281.

Bauschinger, J. (1)
"Interpolation." *Encyklopädie der Mathematischen Wissenschaften*, vol. 1, part 1
(1898–1904), 799.

Becker, O. (1)
Grundlagen der Mathematik in geschichtlicher Entwicklung. Orbis Academicus, vol.
2/6. Freiburg-Munich, 1954.

Becker, O., and Hofmann, J. E. (1)
Geschichte der Mathematik. 3 vols. Bonn, 1951. Review by O. Neugebauer,
Centaurus, 1953, *2*, 364.

Bell, E. T. (1)
The Development of Mathematics. New York, London, 1945².

Berezkina, E. I. (1)
"Drevnekitajskij Traktat *Matematika v devjati Knigach*" [The ancient Chinese
work *Nine Chapters on the Mathematical Art*]. *IMI*, 1957, *10*, 423–584.

Bernard-Maître, H. (1)
"Un Historien des mathématiques en Europe et en Chine. Le Père Henri
Bosmans, S.J. (1852–1928)." *Archeion*, 1950, *29* (n.s.), 619–656.

Bertrand, J. (1)
"Les mathématiques en Chine." *JS*, 1869, 317–464. Translation of Biernatzki (1).

Beveridge, W. (1)
Institutionum chronologicarum libri II. Londini, 1669.

Bi Dung (1)
Eine Studie über die Chinesische Finanzwissenschaft. Diss. Hamburg, 1934.

Biernatzki, K. L. (1)
"Die Arithmetik der Chinesen." *JRAM,* 1856, *52,* 59.

Biot, E. (1)
("On Pascal's Triangle in the *Suan-fa t'ung-tsung*"). *JS,* 1835 (May), 270–273.

Biot, E. (2)
"Note sur la connaissance que les Chinois ont eue de la valeur de position des chiffres." *JA,* 1839 (3rd series), *8,* 497–502.

Biot, E. (3)
"Table générale d'un ouvrage chinois intitulé... *Souan-Fa Tong-Tsong,* ou Traité Complet de l'Art de Compter." *JA,* 1839 (3rd series), *7,* 193–217.

Bischoff, B. (1)
"Studien zur Geschichte des Klosters St. Emmeran in Spätmittelalter (1324–1525)." *SMGBO,* 1953/54, *65,* (nos. 1–2).

Bochenski, I. M. (1)
Die zeitgenössischen Denkmethoden. Bern, 1959².

Bodde, D. (1)
"The Chinese Cosmic Magic Known as Watching for the Ethers." *Studia Serica Bernhard Karlgren dedicata.* Copenhagen, 1959, pp. 14–35.

Bortolotti, E. (1)
"Le Fonti Arabe di Leonardo Pisano." *MASIB,* Classe di Scienze Fisiche, Series 8, volume 7 (1929–1930), 91.

Breitschneider, M. (1)
"Mathematik bei den Juden (1505–1550)." *AGM,* 1899, *9,* 473–483.

Bretschneider, E. (1)
Botanicon Sinicum; Notes on Chinese Botany from Native and Western Sources. 3 vols. London, 1882. Reprinted from *JRAS/NCB,* 1881, *16.*

Brockelmann, C. (1)
Geschichte der arabischen Literatur. 2 vols. Suppl. vols. 1–3. Weimar, 1898–1902; reprint Leiden, 1943–1949.

Burgess, E. (1)
"(Translation of) the *Sûryasiddhânta.*" *JAOS,* 1860, *6,* 141–498.

Burk, A. (1)
"Das Apastamba-Sulba-Sutra." *ZDMG,* 1901, *55,* 543–591; 1902, *56,* 327–391.

Cajori, F. (1)
A History of Mathematics. New York, 1919². (The first edition does not contain the section on Chinese mathematics.)

Cajori, F. (2)
A History of Mathematical Notations. 2 vols. Chicago, 1928–1929.

Cajori, F. (3)

A History of Elementary Mathematics. New York, 1896[1], 1917[2], 1930[3].

Camman, S. (1)
"The Evolution of Magic Squares in China." *JAOS,* 1960, *80,* 116–124.

Cantor, M. (1)
"Zur Geschichte der Zahlzeichen." *ZMP,* 1858, *3,* 325–341. Critique by Matthiessen (2).

Cantor, M. (2)
Vorlesungen über die Geschichte der Mathematik. 4 vols. Leipzig, 1880–1908. Critique of Chapter 31, "Die Mathematik der Chinesen" in Mikami (5).

Carmichael, R. D. (1)
Theory of Numbers. Mathematical monographs, no. 13. New York, 1914.

Cassien-Bernard, F. (1)
"Le pa-koua et l'origine des nombres." *FA,* 1950, *6,* 848–856.

Chalfant, F.H. (1)
"Early Chinese writing." *Memoirs of the Carnegie Museum,* vol. 4, no. 1 (1906).

Chang Fu-jui (1)
Les fonctionnaires des Song. Index des titres. Paris, 1962.

Chatley, H. (1)
"The Cycles of Cathay." *JRAS/NCB,* 1934, *65,* 36–54. Addendum: *0,* 1917, *60,* 171.

Chatley, H. (2)
"The Sixty-year and Other Cycles." *CJ,* 1934, *20,* 129.

Chau Ju-kua (1)
On the Chinese and Arab Trade [*Chu-fan chih*]. Edited and translated by F. Hirth and W.W. Rockhill as "Chau Ju-Kua; His Work on the Chinese and Arab Trade in the Twelfth and Thirteenth Centuries, Entitled Chu-Fan-Chi." Amsterdam, 1966 (first ed. St. Petersburg, 1911).

Chêng Chin-te, D. (1)
"The Use of Computing Rods in China." *AMM,* 1925, *32,* 492–499.

Chêng Chin-te, D. (2)
"On the Mathematical Significance of the Chinese *Ho T'u* and *Lo Shu.*" *AMM,* 1925, *32,* 499.

Ch'ên, Kenneth (1)
"The Sale of Monk Certificates during the Sung Dynasty, a Factor in the Decline of Buddhism in China." *HTR,* 1956, *49* (no. 4), 307–327.

Chi Ch'ao-ting (1)
Key Economic Areas in Chinese History, as Revealed in the Development of Public Works for Water Control. London, 1936.

Chi Yu-tang (1)
An Economic Study of Chinese Agriculture. Ithaca, 1924.

Ch'ien Pao-tsung (1)
"On the Shou-shih Calendar System of Kuo Shou Ching." *Actes du 8e Congrès International d'Histoire des Sciences.* Florence, 1956, 430.

Clark, W. E. (1)
The Aryabhatya of Aryabhata. Chicago, 1930.

Cochini, C. and Seidel, A. (1)
Chronique de la Dynastie des Sung (960–1279) (Extraite et traduite du *Chung-wai li-shih nien-piao*). Matériaux pour le Manuel de l'Histoire des Sung (Sung project). Munich, 1968.

Colebrooke, H. T. (1)
Algebra, with Arithmetic and Mensuration, from the Sanskrit of Brahmagupta and Bhaskara. London, 1817. Rev. ed. by H.C. Banerji, Calcutta, 1927.

Curtze, M. (1).
"Ein Beitrag zur Geschichte der Algebra in Deutschland im fünfzehnten Jahrhundert." *AGM*, 1895, *7*, 31–74.

Curtze, M. (2)
Urkunden zur Geschichte der Mathematik im Mittelalter. 2 vols in 1 (Abhandlungen zur Geschichte der Mathematik, 12, 13). Leipzig, 1902.

Curtze, M. (3)
"Über die sogenannte Regel Ta Yen in Europa." *ZMP*, 1896, *41*, Hist.-Lit. Abt., 81–82.

Datta, B. (1)
"Elder Aryabhaṭa's Rule for the Solution of Indeterminate Equations of the First Degree." *BCMS*, 1932, *24*, 19–36.

Datta, B. (2)
"The Origin of Hindu Indeterminate Analysis." *ARCHEION*, 1931, *13*, 401–407.

Datta, B. (3)
The Science of the Sulba. Calcutta, 1932.

Datta, B. (4)
"On the Supposed Indebtedness of Brahmagupta to *Chiu-chang Suan-shu*." *BCMS*, 1930, *22* (no. 1), 39–51.

Datta, B. (5)
"The Bakhshali Mathematics." *BCMS*, 1929, *21* (no. 1), 1–60.

Datta, B. (6)
"Early Literary Evidence of the Use of the Zero in India." *AMM*, 1929, *33*, 449–454.

Datta, B. and Singh, A. N. (1)
History of Hindu Mathematics. 2 vols. Lahore, 1935, 1938.

Davis, A. McF. (1)
Certain Old Chinese Notes or Chinese Paper Money. Proceedings of the American Academy of Arts and Sciences, June 1915, *50* (no. 11) Boston, 1915.

Davis, T. L. et al. (1)
"The Preparation of Ferments and Wine." *HJAS*, 1945–1946, *9*, 24–44.

De Lacouperie, Terrien (1)
"The Old Numerals, the Counting Rods, and the Swan-Pan in China." *NS*, 1883 (3rd series), *3*, 297.

Demiéville, P. (1)
"(Review of) *Che-yin Song Li Ming-tchong Ying-tsao fa che*, 'Edition photolith-ographique de la Méthode d'architecture de Li Ming-tchong des Song.'"
BEFEO, 1925, *25*, 213–264. Reprint in *Chung-kuo ying-tsao hsüeh-shê hui-k'an* [Bulletin of the Society for Research in Chinese Architecture], 1931, *2*, 1–36.

de Rocquigny, G. (1)
"Solution du problème des restes." *LM*, 1881, *54*, 304.

Des Rotours, R. (1)
Traité des fonctionnaires et traité de l'armée, traduits de la nouvelle histoire des T'ang (ch. 46–50). 2 vols., Leiden, 1948 (Bibl. de l'Inst. des Hautes Etudes Chinoises, no. 6).

Des Rotours, R. (2)
Le traité des examens, traduit de la nouvelle histoire des T'ang (ch. 44–45). Paris, 1932 (Bibl. de l'Inst. des Hautes Etudes Chinoises, no. 2)

de Saussure, L. (1)
"Le système astronomique des Chinois." *ASPN*, 1919, *124* (5th series: 1), 186, 561.; 1920, *125*, (5th series: 2), 214, 325.

Dickson, L. E. (1)
History of the Theory of Numbers. 1st ed. Carnegie Institution, Washington, 1919–1920 (Carn. Inst. Publ. no. 256), 3 vols., 2nd ed. New York, 1934.

Dirichlet, P. C. Lejeune (1)
Vorlesungen über Zahlentheorie. (Ed. R. Dedekind.) Braunschweig, 1879[3].

Domingues, A. (1)
"Solution du problème chinois des restes." *LM*, 1881, *55*, 62.

Doolittle, J. (1)
A Vocabulary and Handbook of the Chinese Language. Fuchow, 1872. (Vol. 2, pp. 354–364: Wylie, A. "(Glossary of Chinese) Mathematical and Astronomical Terms)."

Dostor, C. (1)
"Propriétés élémentaires des nombres." *AMP*, 1879, *63*, Miscellen. 3, 324.

Edmunds, C. K. (1)
"Science among the Chinese." *PSM*, 1911, *79* (no. 6), 521–531; 1912, *80* (no. 1), 22–35.

Eichhorn, W. (1)
Beitrag zur rechtlichen Stellung des Buddhismus und Taoismus im Sung-Staat. (Translation of the "Taoism und Buddhism" section of the *Ch'ing-Yüan T'iao-Fa Shih-Lei*, classified legal documents of the Ch'ing-Yüan period (1195–1200), ch. 50 and 51). Monographies du T'oung Pao, vol. 7. Leiden 1968.

Eneström, G. (1)
"(Notiz über die 'Regula cecis')." *BM*, 1904, *55*, no. 3, 201–202.

Eneström, G. (2)
"(Indeterminate Equations in the 'Triparty' of Chuquet—Note)." *BM*,

1905, *66,* no. 3, 399.

Eneström, G. (3)
"Über eine alte Scherzfrage, die der Lösung einer unbestimmten Gleichung ersten Grades entspricht." *BM,* 1908, *8,* no. 3, 311.

Eneström, G. (4)
"(Note on Alcuin's Problem 34)." *BM,* 1908, *8,* no. 4, 415.

Eneström, G. (5)
"Über die Bedeutung historischer Hypothesen für die Mathematische Geschichtsschreibung." *BM,* 1905, *6,* no. 1, 1–8.

Ens, Casper (1).
Thaumaturgus mathematicus, id est, admirabilium effectorum et mathematicarum disciplinarum fontibus profluentium sylloge. Munich, 1636.

Euler, L. (1)
"Solutio problematis arithmetici de inveniendo numero qui per datos numeris divisus relinquat data residua." *Commentarii academiae scientiarum Petropolitanae* 7 (1734/5), 1740, 46–66. Repr. in *Commentationes arithmeticae collectae,* 1849, 11–20.

Fabricius, J. A. (1)
*Bibliotheca graeca...*Edition of G.C. Harles, 12 vols. Hamburg, 1808.

Feldhaus, F. M. (1)
Die Technik der Vorzeit, der geschichtlichen Zeit, und der Naturvölker (encyclopaedia). Leipzig and Berlin, 1914.

Fibonacci
See Leonardo Pisano

Forke, A. (1)
Geschichte der neueren chinesischen Philosophie. Hamburg. 1938.

Franke, H. (1)
Sinologie. Orientalistik, Part 1. Wissenschaftliche Forschungsberichte, Geisteswissenschaftliche Reihe. (Ed. K. Hönn), vol. 19. Bern, 1953.

Franke, H. (2)
Geld und Wirtschaft in China unter der Mongolen-Herrschaft. Leipzig, 1949.

Franke, H. (3)
Das Chinesische Kaiserreich. Fischer Weltgeschichte, vol. 19. Frankfurt a. M., 1968.

Franke, O. (1)
Geschichte des chinesischen Reiches. 3 vols. Berlin, 1930–1937.

Franke, W. (1)
"Juan Yüan." *MS,* 1944, *9,* 53.

Fung Yu-lan (1)
A History of Chinese Philosophy (tr. D. Bodde), 2 vols. Princeton, 1952–1953.

Gandz, S. (1)
"On the Origin of the Term Root." *AMM,* 1926, *33,* 261–265; 1928, *35,* 67–75.

Ganguli, S. (1)
"India's Contribution to the Theory of Indeterminate Equations of the First Degree." *JIMS*, 1931, *19*, 110–169.

Ganguli, S. (2)
"Notes on Indian Mathematics. A Criticism of G. R. Kaye's Interpretation." *Isis*, 1929, *12*, 132–145.

Ganguli, S. (3)
"Bhaskaracharya and Simultaneous Indeterminate Equations of the First Degree." *BCMS*, 1926, *17*, 89–98.

Ganguli, S. (4)
"The Source of the Indian Solution of the So-called Pellian Equation." *BCMS*, 1928, *19*, 151–176.

Gauchet, L. (1)
"Note sur la trigonométrie sphérique de Kouo Cheou-King [Kuo Shou-ching]." *TP*, 1917, *18*, 151–174.

Gauchet, L. (2)
"Note sur la généralisation de l'extraction de la racine carrée chez les anciens auteurs chinois, et quelques problèmes du *Chiu Chang Suan Shu*." *TP*, 1914, *15*, 531.

Gauss, C. F. (1)
Disquisitiones arithmeticae. Carl Friedrich Gauss Werke, vol. 1. Göttingen, 1863. (1st ed. Leipzig, 1801.)

Gerhardt (first name unknown) (1)
"Zur Geschichte der Algebra in Deutschland." *MKAWB*, 1870, 141.

Gernet, J. (1)
Les Aspects Economiques du Bouddhisme. Publications de l'Ecole Française d'Extrême-Orient, vol. 39. Saigon, 1956. (Critiques by A.F. Wright, *JAS*, 1956–1957, *16*, 408–414; D.C. Twitchett, *BSOAS*, 1957, *19*, 526–549.)

Gernet, J. (2)
La vie quotidienne en Chine à la veille de l'invasion mongole 1250–1276. Paris, 1959.

Giesing, J. (1)
Leben und Schriften Leonardos da Pisa. Progr. Döbeln, 1886.

Ginzel, F. K. (1)
Handbuch der Mathematischen und Technischen Chronologie. 3 vols. Berlin, 1906, 1911, 1914.

Glathe, A. (1)
"Die chinesischen Zahlen." *MDGNVO*, 1932, *26*, B, 1.

Goodrich, L. C. (1)
Introduction to Chinese History, and Scientific Developments in China. Sino-Indian Pamphlets, no. 21. Santiniketan, 1954.

Goodrich, L. C. (2)
"Measurements of the Circle in Ancient China." *Isis*, 1948, *39*, 64–65.

Goodrich, L. C. (3)

"The Abacus in China." *Isis*, 1948, *39*, 239.

Goschkevitch, J. (1)
"Über das chinesische Rechenbrett." In *Arbeiten der kaiserlichen Russischen Gesandtschaft zu Peking*, vol. 1, 295. Berlin, 1858.

Grosswald, E. (1)
Topics from the Theory of Numbers. New York, 1966.

Gurjar, L. V. (1)
Ancient Indian Mathematics and Vedha. Poona, 1947.

Hankel, H. (1)
Geschichte der Mathematik im Altertum und Mittelalter. Leipzig, 1874.

Hartner, W. (1)
"Chinesische Kalenderwissenschaft." *S*, 1930, *5*, 237.

Hartner, W. (2)
"Al-Djabr." In *Encyclopaedia of Islam*, vol. 2, 360–362.

Haskins, C. H. (1)
Arabic Science in Western Europe. Brussels, 1925 (Abstract from *Isis*, 1925, no. 23, vol. 7, no. 3).

Haskins, C. H. (2)
Studies in the History of Mediaeval Science. Cambridge, Mass., 1924.

Hayashi, Tsuruichi (1)
"Brief History of Japanese Mathematics." *NAW*, 1905, *6*, 296; 1907, *7*, 105. (Critique by Mikami, Y., *NAW*, 1911, *9*, 373.)

Hayashi, Tsuruichi (2)
"The Fukudai and Determinants in Japanese Mathematics." *PTMS*, 1910, *5*, 254.

Heiberg, J. (1)
"Byzantinische Analekten." *AGM*, 1899, *9*, 161.

Herrmann, A. (1)
An Historical and Commercial Atlas of China. Cambridge, Mass., 1935. Repr. Edinburgh, 1966.

Hightower, J. R. (1)
"Some Characteristics of Parallel Prose." *Studia Serica Bernhard Karlgren dedicata.* Copenhagen, 1959, pp. 60–91.

Hirth, F. and Rockhill, W.W. (1)
See Chau Ju-kua (1).

Ho Pêng-yoke (1)
"Ch'in Chiu-shao." In *Dictionary of Scientific Biography* (in press).

Ho Pêng-yoke (2)
"Yang Hui." In *Dictionary of Scientific Biography* (in press).

Ho Pêng-yoke (3)
"Li Chih." In *Dictionary of Scientific Biography* (in press).

Ho Pêng-yoke (4)
"Chu Shih-chieh." In *Dictionary of Scientific Biography* (in press).

Ho Pêng-yoke (5)
"Liu Hui." In *Dictionary of Scientific Biography* (in press).

Ho Pêng-yoke (6)
"The Lost Problems of the *Chang Ch'iu-chien Suan Ching*, a 5th-century Mathematical Manual." *OE*, 1965, *12*, 37.

Hoche, R. (ed.) (1)
The Eisagogè Arithmetikè εισαγωγη αριθμητικη of Nichomachus of Gerasa. Leipzig, 1866.

Hoernle, R. (1)
"The Bakhshali Manuscript." *IA*, 1888, *17*, 33, 275.

Hofmann, J. E. (1)
Geschichte der Mathematik, 3 vols., Berlin, 1953–1957.

Hofmann, J.E. (2)
"Frans Van Schooten der Jüngere." *B*, 1962, *2*.

Hopkins, L.C. (1)
"The Chinese Numerals and Their Notational Systems." *JRAS*, 1916, *35*, 737.

Hoppe, E. (1)
"Michael Stifels handschriftlicher Nachlass." *MMG*, 1900, *3* (no. 10), 411.

Horner, W. G. (1)
"A New Method of Solving Numerical Equations of all Orders, by Continuous Approximation." *PTRS*, 1819, *1*, 308.

Hsü Ching-chih (Su Gin-djih) (1)
Chinese Architecture—Past and Contemporary. Hong Kong, 1964.

Hucker, C. O. (1)
China, A Critical Bibliography. Tucson, 1962.

Hulsewé, A. F. P. (1)
Remnants of Han Law. Leiden, 1955.

Hummel. A. W. (1)
Eminent Chinese of the Ch'ing Period. 2 vols., Washington, 1944.

Hummel, A. W. (2)
"Chinese Mathematics", being a review of Têng Yen-lin (1'). *Annual Reports of the Librarian of Congress (Division of Orientalia)*, 1939, 179.

Hung, William (1)
"Preface to an Index to *Ssŭ-k'u ch'üan-shu tsung-mu*." *HJAS*, 1939, *4*, 47.

Hunger, F. (1)
"Das älteste deutsche Rechenbuch." *ZMP*/ Hist. Lit. Abt., 1888, *33*, 125.

Hunger, H. and Vogel, K. (1)
Ein Byzantinisches Rechenbuch des 15. Jahrhunderts. Österreichische Akademie der Wissenschaften, Phil. Hist. Klasse. Denkschriften, vol. 76, part 2. Vienna, 1957.

Hunter, J. (1)
Number Theory. New York, 1964

Itard, J. (1)
"Claude-Gaspar Bachet de Méziriac." *RHS,* 1947, *1* (no. 1), 35.

Kamalamma, K. N. (1)
"On the Solution of a Problem on Indeterminate Equations Occurring in the Sulva-Sutras." *BCMS,* 1948, *40,* 140.

Kann, E. (1)
The Currencies of China. Shanghai, 1927.

Karlgren, B. (1)
"On the Early History of the *Tso-chuan* and *Chou-li* texts." *BMFEA,* 1936, *3,* 1.

Karpinski, L. C. (1)
"The Algebra of Abu Kamil Shoja' ben Aslam." *BM,* 1912, *12,* 3rd series, 40.

Karpinski, L. C. (2)
"The Algebra of Abu Kamil." *AMM,* 1914, *21,* (no. 2), 37.

Karpinski, L. C. (3)
History of Arithmetic. New York, 1925.

Kaye, G. R. (1)
"Notes on Indian Mathematics. No. 2—Aryabhaṭa." *JPASB,* 1908, *4* (n.s.), 111.

Kaye, G. R. (2)
"Indian Mathematics," *Isis,* 1914, *2,* 326,.

Kaye, G. R. (3)
Indian Mathematics. Calcutta-Simla, 1915.

Kaye, G. R. (4)
"Some notes on Hindu Mathematical Methods." *BM,* 1910, *11,* no. 4, 289.

Kaye, G. R. (5)
"Notes on Indian Mathematics—Arithmetical Notations." *JPASB,* 1907, *3* (no. 7), 475.

Kaye, G. R. (6)
The Bakhshali Manuscript. 3 parts, Calcutta, 1927–1932.

Keith, A. B. (1)
The Religion and Philosophy of the Veda and Upanishads. Cambridge, Mass., 1925.

Kimm Chung-se (1)
"Kuei Ku Tzŭ." *AM,* 1927, *4,* 108.

Kirby, E. Stuart (1)
Introduction to the Economic History of China. London, 1954.

Klügel, G. S. (1)
Mathematisches Wörterbuch. 3 vols., Leipzig, 1808.

Knott, C. G. (1)
"The Abacus in Its Historic and Scientific Aspects." *TASJ,* 1886, *14,* 18.

Konantz, E. L. (1)

"The Precious Mirror of the Four Elements." *CJSA*, 1924, *2*, 304.

Krause, M. (1)
"Stambuler Handschriften islamischen Mathematiker." *QSGM/B*(Studien), 1936, *3*, 437.

Kuo Ping-wên (1)
The Chinese System of Public Education. New York, 1914.

Laforêt, J.B. (1)
Alcuin restaurateur des sciences. Louvain, 1851.

Lagrange, J. L. (1)
"Sur la solution des problèmes indéterminés." *Mémoires de l'Académie Royale des Sciences, Berlin* 1769, *23*; 1770, *24*, 222. Also in *Oeuvres,* vol. 2, 519, 698.

Lam Lay Yong (1)
"The Geometrical Basis of the Ancient Chinese Square-root Method" (in press).

Lam Lay Yong (2)
"The Yang Hui Suan Fa: A Thirteenth-Century Chinese Mathematical Classic," *The Chung Chi Journal,* 1968, *7,* (no. 2), 171–176.

Lam Lay Yong (3)
"On the Existing Fragments of Yang Hui's *Hsiang Chieh Suan Fa,*" *AHES,* 1969, *6,* (no 1), 82–88.

Laufer, Berthold (1)
Sino-Iranica: Chinese Contributions to the History of Civilisation in Ancient Iran, with Special Reference to the History of Cultivated Plants and Products. Chicago, 1919. (Field Museum of National History, pub. 201; Anthropological Series, vol. 15, no. 3)

Leavens, D. H. (1)
"The Chinese Suan P'an," *AMM*, 1920, *27*, 180.

Lebèsgue, V. A. (1)
Exercices d'analyse numérique. Paris, 1859.
Lectures in Commodities. Vegetable Products. Economic Studies, no. 3. Hautes Etudes Industrielles et Commerciales. Tientsin/ Shanghai, 1936.

Lee Ching-ying, M. (1)
Index des noms propres dans les Annales principales de l'histoire des Song. Paris, 1966.

Lee Ping-hua, M. (1)
The Economic History of China. With Special Reference to Agriculture. New York, 1921.

Leonardo Pisano (Fibonacci) (1)
Liber Abbaci (1202). Rome, 1857. (Part 1 of *Scritti di Leonardo Pisano.*)

Levey, M. (1)
The Algebra of Abu Kamil. Madison, Wisconsin, 1966.

Li Ch'iao-ping (1)
The Chemical Arts of China. Easton, 1948.

Li Shu-t'ien (1)
"Origin and Development of the Chinese Abacus." *JACM*, 1959, *6*, 102.

Li Yen (1)
"The Interpolation Formulae of Early Chinese Mathematicians." *Proc. 8th International Congress of the History of Science*. Florence, 1956, pp. 70–72.

Li Yen (2)
"Tsu Ch'ung-chih, Great Mathematician of Ancient China." *PC*, 1956, *24*, 34.

Liang Fang-chung (1)
The Single-whip Method of Taxation in China. Cambridge, 1956.

Libri, G. (1)
Histoire des sciences mathématiques en Italie. 4 vols., Paris, 1838–1840.

Liu Mau-tsai (1)
"Soziale Einrichtungen in der Hauptstadt der Süd-Sung (1127–1279)." *OE*, 1968, *15* (no. 2), 123.

Lockhart, J.H. Stewart (1)
Currency of the Farther East. 3 vols., Hongkong, 1895.

Loria, G. (1)
"Documenti relativi all'antica matematica dei Cinesi." *Archeion*, 1922, *3*, 141.

Loria, G. (2) and McClenon, R.B.
"The Debt of Mathematics to the Chinese People." *SM*, 1921, *12*, 517. Critique in Mikami (1').

Loria, G. (3)
Storia delle Matematiche dall' alba della civiltà al secolo XIX. 3 vols., Turin, 1929¹, Milan, 1950². ("L'Enigma Cinese" is in vol. 1, Chapter 9, p. 145.)

Loria, G. (4)
"Che cosa debbono le matematiche ai Cinesi?" *Bollettino della mathesis*, 1920, *12*, 63.

Luckey, P. (1)
"Zur islamischen Rechenkunst und Algebra des Mittelalters." *FF*, 1948, *24*, 17.

Lyons, H.G. (1)
"An Early Korean Rain-Gauge." *QJRMSL*, 1924, *50* (no. 209), 26.

Ma, C.C. (1)
"The Origin of the Term 'Root' in Chinese Mathematics." *AMM*, 1928, *35*, 29.

Mahler, K. (1)
"On the Chinese Remainder Theorem." *MN*, 1958, *18*, 120.

Majumdar, R.C. (1)
The History and Culture of the Indian People. The Vedic Age. London, 1951.

Marakuev, A.V. (1)
"Istorija razvitija matematiki v Kitae, a takže v Japonii" [History of the

development of mathematics in China and Japan]. In *Otčet o dejatel'-nosti matematičeskoj konferencii* [Report on the activity of the conference on mathematics], Jan.–Feb. 1930, pp. 47–60. Vladivostok, 1931.

Marchand, D. (1)
"Problème des restes." *LM*, 1881, *19* (no. 54), 437.

Maspero, H. (1)
"Les instruments astronomiques des Chinois au temps des Han." *MCB*, 1939, *6*, 183.

Maspero, H. and Balazs, E. (1)
Histoire et institutions de la Chine ancienne. Paris, 1967.

Matthiessen, L. (1)
"Vergleichung der indischen Cuttaca und der chinesischen Tayen-Regel." In *Sitzungsberichte der mathematisch-naturwissenschaftlichen Section in der 30. Versammlung deutscher Philologen und Schulmänner in Rostock, 1875,* p. 75. Leipzig, 1876.

Matthiessen, L. (2)
"Zur Algebra der Chinesen. Auszug aus einem Briefe an M. Cantor." *ZMP*, 1876, *19*, 270; *ZMNWU*, 1876, *7*, 73.

Matthiessen, L. (3)
"Über das sogenannte Restproblem in den Chinesischen Werken Swan-King von Sun-Tsze und Tayen Lei Shu von Yih-Hing." *JRAM*, 1881, *91*, 254.

Matthiessen, L. (4)
"Die Methode Ta jàn (Ta Yen) im Suan-King von Sun-tsè und ihre Verallgemeinerung durch Yih-Hing im I. Abschnitte des Ta jàn li shu." *ZMP*, 1881, *26*, 2 (Hist. Lit. Abt.), 33.

Matthiessen, L. (5)
"Über eine antike Auflösung des sogenannten Restproblems in moderner Darstellung." *ZMNU*, 1879, *10*, 106.

Matthiessen, L. (6)
[Same title as (5).] *ZMNU*, 1882, *13*, 187.

Mazumdar, N. K. (1)
"Aryyabhatta's Rule in Relation to Indeterminate Equations of the First Degree." *BCMS*, 1911–1912, *3*, 11.

Mazumdar, N. K. (2)
"On Chinese Indeterminate Analysis." *BCMS*, 1913–1914, *5*, 9.

Mc Clenon, R. B.
See Loria, G. (2).

Melchers, B. (1)
China, Der Tempelbau (Schriftenreihe: Geist, Kunst und Leben Asiens). Hagen, 1921.

Merrill, E. D., and Walker, E. H. (1)
A Bibliography of Eastern Asiatic Botany. Cambridge, Mass., 1938.

Mieli, A. (1)
La science arabe et son rôle dans l'évolution scientifique mondiale. Leiden, 1938.

Migne, I. F. (1)
Propositiones Alcuini doctoris Carolomagni Imperatoris ad acuendos juvenes. In *Alcuini opera omnia* (Patrologiae cursus completus, Tomus 101), vol. 3. Paris, 1851.

Mikami, Y. (1)
The Development of Mathematics in China and Japan (Abhandlungen zur Geschichte der mathematischen Wissenschaften, no. 30). Leipzig, 1912. (Review by H. Bosmans, *RQS*, 1913, *74*, 641).

Mikami, Y. (2)
"The Influence of Abaci on the Chinese and Japanese Mathematics." *JDMV*, 1911, *20*, 380.

Mikami, Y. (3)
"Arithmetic with Fractions in Old China." *AMN*, 1911, *32* (no. 3), 3.

Mikami, Y. (4)
"The Circle-Squaring of the Chinese." *BM*, 1910, *10* (4th series), 193.

Mikami, Y. (5)
"A Remark on the Chinese Mathematics in Cantor's *Geschichte der Mathematik.*" *AMP*, 1909, *15*, 68; 1911, *18*, 209.

Mikami, Y. (6)
"The Ch'ou-Jên Chuan of Yüan Yüan (Juan Yüan). [Reply to Van Hee (3)]. *Isis*, 1928, *11*, 123.

Mikami, Y. (7)
"Mathematics in China and Japan." In *Scientific Japan, Past and Present.* 3rd Pan-Pacific Science Congress Volume, p. 177. Tokyo, 1926.

Mikami, Y. (8)
A History of Japanese Mathematics. Chicago, 1914.

Mikami, Y. (9)
"On the Japanese Theory of Determinants." *Isis*, 1914, *2*, 9.

Montucla, J.E. (1)
Histoire des mathématiques. 4 vols., Paris, 1758–1802.

Morse, H.B. (1)
The Trade and Administration of China. London, 1920[3].

Muir, Sir Thomas (1)
Theory of Determinants in the Historical Order of Development. 4 vols., London, 1890–1919; 2nd. ed. 1906–1923.

Murr, C.T. de (1)
Memorabilia bibliothecarum publicarum Norimbergensium et Universitatis Altdorfinae, Pars I, Nürnberg, 1786.

Nakayama, S. (1)
A History of Japanese Astronomy (Harvard-Yenching Institute Monograph Series, vol. 18). Cambridge, Mass., 1969.

Nakayama, S. (2)
"Accuracy of Pre-Modern Determinations of Tropical Year Length."
JSHS, 1963, *2*, 101.

Needham, J. (1)
Science and Civilisation in China. 7 vols., Cambridge, 1954–

Needham, J. (2)
"Mathematics and Science in China and the West." *SS*, 1956, *20*, 320.

Nesselmann, G.H.F. (1)
Die Algebra der Griechen. Berlin, 1842.

Nichomachus
See Hoche (1)

Pannekoek, A. (1)
De Groei van ons Wereldbeeld. Amsterdam, 1951.

Pearson, J. D. (1)
Index islamicus. Cambridge, 1958. Suppl. 1, 1962; Suppl. 2, 1968.

Petrucci, R. (1)
"Sur l'algèbre chinoise." *TP*, 1912, *13*, 559.

Rademacher, H. (1)
Lectures on Elementary Number Theory. New York, 1964.

Rangacarya, M. (1)
The Ganita-Sara-Sangraha of Mahaviracarya, Sanskrit and English. Madras, 1912.

Renaud, H.P.J. (1)
"Additions et Corrections à Suter, 'Die Mathematiker und Astronomen der Araber.'" *Isis, 1932, 18,* 166.

Review of "Algebra with Arithmetic and Mensuration, from the Sanskrit of Brahmegupta and Bhascara. Translated by Henry Thomas Colebrooke, Esq. F.R.S., etc. London, 1817." *ER*, 1817, *57* (vol. 29), 141.

Rodet, L. (1)
"Leçons de calcul d'Aryabhaṭa." *JA*, 1879, *13*, (7th series) 393.

Rodet, L. (2)
"Le Souan-Pan et la banque des argentiers." *BSM*, 1880, *8*, 158.

Rohrberg, A. (1)
"Das Rechnen auf dem chinesischen Rechenbrett." *UMN*, 1936. *42*, 34.

Ruska, J. (1)
"Zur ältesten arabischen Algebra und Rechenkunst." *Sitzungsberichte der Heidelberger Akad. d. Wissenschaften*, Phil.-hist. Kl., 1917.

Rybnikov, K. A. (1)
"On the Role of Algorythms in the History of Mathematical Analysis." *Actes du VIIIe Congrès International d'Historire des Sciences*, p. 142.

Sanford, V. (1)
The History and Significance of Certain Standard Problems in Algebra. New York,

1927 (Contributions to Education, Teachers College, Columbia University, no. 251).

Sanford, V. (2)
A Short History of Mathematics. Boston, 1930.

Sarton, G. (1)
Introduction to the History of Science. 3 vols., Baltimore, 1927–1947 (Carnegie Institution Publ. no. 376).

Sarton, G. (2)
A Guide to the History of Science. Waltham, Mass., 1952.

Saunderson, N. (1)
The Elements of Algebra. Cambridge, 1740. French translation, *Elemens d'algèbre*, 1756.

Schäfer, J.C. (1)
Die Wunder der Rechenkunst. Weimar, 1831,[1] 1842[2].

Schram, R. (1)
Kalendariographische und Chronologische Tafeln. Leipzig, 1908.

Schurmann, H.F. (1)
Economic Structure of the Yüan Dynasty (Translation of ch. 93 and 94 of the *Yüan Shih*). Cambridge, Mass., 1956 (Harvard-Yenching Institute Monograph Series, vol. 16).

Scott, J. F. (1)
A History of Mathematics from Antiquity to the Beginning of the Nineteenth Century. London, 1958[1], 1960[2].

Sédillot, L. (1)
Courtes observations sur quelques points de l'histoire de l'astronomie et des mathématiques chez les orientaux. Paris, 1863.

Sédillot, L. (2)
Matériaux pour servir à l'histoire comparée des sciences mathématiques chez les Grecs et les Orientaux. 2 vols., Paris, 1845–1849.

Sédillot, L. (3)
"De l'Astronomie et des mathématiques chez les Chinois, lettre de M. L. Sédillot à D. B. Boncompagni." *BBSSMF*, 1868, *1*, 161.

Sen, S.N. (1)
"Study of Indeterminate Analysis in Ancient India." *Proc. 10th International Congress of the History of Science* (Ithaca, 1962), 1964, 493.

Sengupta, P.C. (1)
"The Aryabhatiyam." *Journal of the Department of Letters, Calcutta University,* vol. 16, 1927.

Shafer, E.H. (1)
The Golden Peaches of Samarkand, A Study of T'ang Exotics. Berkeley, 1963.

Shaw, W. (1)
Manual of Meteorology. 4 vols., Cambridge, 1926.

Singh, A.N.
"A Review of Hindu Mathematics up to the XIIth century. "*ARCHEION*, 1936, *18*, 43.

Siren, Oswald. (1)
The Walls and Gates of Peking. London, 1924.

Sivin, N. (1)
An Introduction to the Bibliography of Science and Natural Philosophy in the Chinese Tradition (Draft). 1965.

Sivin, N. (2)
"On the Reconstruction of Chinese Alchemy." *JSHS*, 1967, *6*, 60.

Sivin, N. (3)
Cosmos and Computation in Early Chinese Mathematical Astronomy. Leiden, 1969. Reprint from *TP*, *55*.

Sivin, N. (4)
Chinese Alchemy: Preliminary Studies. Cambridge, Mass., 1968 (Harvard Monographs in the History of Science, vol. 1).

Smith, D.E. (1)
History of Mathematics. 2 vols., New York, 1923–1925.

Smith, D.E. (2)
"Chinese Mathematics." *PSM*, 1912, *80*, 597.

Smith, D.E. (3)
"Unsettled Questions concerning the Mathematics of China." *SM*, 1931, *33*, 244.

Smith, D.E. (4)
"The Ganita-Sara-Sangraha of Mahaviracarya." *BM*, 1908/9, *9*, no. 2, 106.

Smith, D.E. (5)
A Source Book in Mathematics. New York and London, 1929.

Smith, D.E. and Mikami, Y. (1)
A History of Japanese Mathematics. Chicago, 1914.

Solomon, B.S. (1)
"One Is No Number in China and the West." *HJAS*, 1954, *17*, 253.

Srinivasiengar, C.N. (1)
The History of Ancient Indian Mathematics. Calcutta, 1967.

Steinschneider, M. (1)
"Mathematik bei den Juden." *AGM*, 1880, *3*, 57.

Stieltjes, R.J. (1)
Oeuvres complètes. 2 vols., Groningen, 1918.

Stifel, Michael (1)
Arithmetica integra. Nürnberg, 1545.

Stifel, Michael (2)
Die Coss Christoffs Rudolfs. Königsberg, 1553.

Storey, C.A. (1)
Persian Literature: a Bibliographical Survey, vol. 2, part. 1. London, 1958.

Struik, D.J. (1)

"On ancient Chinese Mathematics." *MT*, 1963, *56*, 424. Reprint in *E*, 1964/5, *40*, 65.

Struik, D.J. (2)
A Concise History of Mathematics. 2 vols., New York, 1948.

Su Gin Djih
see Hsü Ching-chih.

Sun, E.T.Z., and de Francis, J. (1)
Bibliography on Chinese Social History. New Haven, 1952.

Sun, E.T.Z., and de Francis, J. (2)
Chinese Social History. Washington, 1956.

Sung Ying-hsing (1)
T'ien-kung K'ai-wu. Chinese Technology in the 17th Century. University Park, Pennsylvania, 1966.

Sung, Z.D. (1)
The Text of Yi King (and Its Appendices). Chinese original, with English translation. 2 vols., Shanghai, 1935.

Suter, H. (1)
"Das Buch der Seltenheiten der Rechenkunst von Abu Kamil el-Misri." *BM*, 1910-11, *11* (3rd series), 100.

Suter, H. (2)
"Die Mathematiker und Astronomen der Araber und ihre Werke." *AGMW*, 1900, *10*, 43.

Suter, H. (3)
"Über die Bedeutung des Ausdruckes 'regula coeci.' " *BM*, 1905, *6*, (3rd series), 112.

Swann, Nancy Lee (1)
Food and Money in Ancient China. Princeton, 1950.

Taranzano, C. (1)
Vocabulaire des sciences mathématiques, physiques et naturelles. Vocabulaire français-chinois, Sien-hsien, 1914; vocabulaire chinois-français, Sien-hsien, 1921. (Reviewed by P. Pelliot in *TP*, 1914, *13*, 304; *TP*, 1923/24, *22–23*, 40).

Taylor, J. (1)
Lilawati, or a Treatise on Arithmetic and Geometry, by Bhascara Acharya, Translated from the Original Sanskrit. Bombay, 1816.

Têng Ssŭ-yü
See also Yen Chih-t'ui.

Têng Ssŭ-yü and Knight Biggerstaff (1)
An Annotated Bibliography of Selected Chinese Reference Works. Revised edition. Harvard-Yenching Institute Monograph Series, vol. 2. Cambridge, Mass., 1950. (First edition, Peking, 1936).

Terquem, O. (1)
" 'Arithmétique et algèbre des Chinois' par M.K.L. Biernatzki, Docteur à Berlin." *NAM*, 1862, *1*, 35; 1862, *2*, 529.

The Tjoe Tie (1)
"De Oorsprong van het tientallig stelsel." *SH*, 1962, *4*, 24.

Thibaut, G. (1)
Astronomie, Astrologie und Mathematik (der Inder). In *Grundriss der Indo-Arischen Philologie und Altertumskunde*, vol. 3, no. 9, 1899.

Thibaut, G. and Dvivedi, S. (1)
Pañca-siddhântikâ of Varaha-Mihira. Benares, 1889; Lahore, 1930².

Thureau-Dangin, F. (1)
"L'origine de l'algèbre." *CRAIBL*, 1940, 292.

Treutlein, P. (1)
"Das Rechnen in 16. Jahrhundert." *AGM*, 1877, *1*, 3.

Tropfke, J. (1)
Geschichte der Elementarmathematik. 7 vols. Leipzig, 1921–1924².

Tullock, G. (1)
"Paper Money—a Cycle in Cathay." *EHR*, 1956/57, *9*, 393.

Twitchett, D. C. (1)
Financial Administration under the T'ang Dynasty. Cambridge, England, 1963 (Cambridge University Oriental Publications, no. 8).

Vacca, G. (1)
"(Remark on Biernatzki and Wylie)." *BM*, 1901, *2*, (3rd series) 143.

Vacca, G. (2)
"Note Cinesi. "*RSO*, 1914/15, *6*, 131.

Vacca, G. (3)
"Sulla matematica degli antichi Cinesi." *BBSSMF*, 1905, *8*, 1.

van der Loon, P. (1)
"On the Transmission of *Kuan-tzŭ*." *TP*, 1952, *41*, 357.

Van Der Waerden, B. L. (1)
"Diophantische Gleichungen und Planetenperioden in der indischen Astronomie." *VNGZ*, 1955, *100* (no. 3), 153.

Van Gulik, R. H. (1)
"De mathematische conceptie bij de oude Chinezen." *E*, 1927/28, *5*, 104–109.

Van Hee, L. (1)
"Le Hai-Tao Souan-King de Lieou" (Sea Island Mathematical Manual). *TP*, 1920, *20*, 51.

Van Hee, L. (2)
"The Arithmetic Classic of Hsia-hou Yang". *AMM*, 1924, *31*, 235.

Van Hee, L. (3)
"The Ch'ou-jên Chuan of Yüan Yüan [Juan Yüan]." *ISIS*, 1926, *8*, 103.

Van Hee, L (4)
"Le Zéro en Chine." *TP*, 1914, *15*, 182.

Van Hee, L. (5)

"Bibliotheca mathematica sinensis Pé-Fou." *TP*, 1914, *15*, 111–164.

Van Hee, L. (6)
"Le classique de l'ile maritime, ouvrage chinois du IIIe siècle." *QSGM/B*, 1932, *2*, 255.

Van Hee, L. (7)
"Le grand thésaurus des mathématiques chinoises et européennes." *ARCHEION*, 1926, *7*, 18.

Van Hee, L. (8)
"Les séries en extrême-orient." *ARCHEION*, 1930, *12*, 117.

Van Hee, L. (9)
"Algèbre chinoise." *TP*, 1912, *13*, 291. (Comment by H. Bosmans, *RQS*, 1912, *72*, 654.)

Van Hee, L. (10)
"Problèmes chinois du second degré." *TP*, 1911, *12*, 559.

Van Hee, L. (11)
"Li-Yé, mathématicien chinois du XIIIe siècle." *TP*, 1913, *14*, 537.

Van Hee, L. (12)
"Les Cent Volailles, ou l'analyse indéterminée en Chine." *TP*, 1913, *14*, 203, 435.

Van Hee, L. (13)
"The Great Treasure House of Chinese and European Mathematics." *AMM*, 1926, *33*, 502.

Van Hee, L. (14)
"La notation algébrique en Chine au XIIIe siècle." *RQS*, 1913 (3rd series), *24*, 574.

Van Hee, L. (15)
"Le précieux miroir des quatre éléments." *AM*, 1932, *7*, 242. (Review by P. Pelliot in *TP*, 1932, *29*, 255.)

Van Hee, L. (16)
"Léopold de Saussure et l'astronomie chinoise." *Archeion*, 1931, *13*, 43.

Van Name, A. (1)
"On the Abacus of China and Japan." *JAOS*, 1880, *10*, cx. (Proceedings at Boston, May, 1875.)

Van Schooten, Frans (1)
Exercitationum mathematicarum liber primus. Lugd. Batav., 1657. Dutch translation *Eerste Bouck der Mathematische Oeffeningen*, Amsterdam, 1660.

Veque, W. J. Le (1)
Topics in Number Theory. London, 1965³.

Vissering, W. (1)
On Chinese Currency. Leiden, 1877.

Vissière, A. (1)
"Biographie de Jouan Yüan" *TP*, 1904, 561.

Vissière, A. (2)

"Recherches sur l'origine de l'abaque Chinois et sur sa dérivation des anciennes fiches à calcul." *BG*, 1892, 54.

Vogel, K. (1)
"The Role of Byzantium as an Intermediary in the Transmission of Ancient and Arabic Mathematics to the West," *Proc. 10th International Congress of the History of Science* (Ithaca, 1962), 1964, p. 537.

Vogel, K. (2)
Neun Bücher Arithmetischer Technik. Braunschweig, 1968.

Vogel, K. (3)
Die Practica des Algorismus Ratisbonensis. Ein Rechenbuch des Benediktinerklosters St. Emmeran aus der Mitte des 15. Jahrhunderts. München, 1954 (Schriftenreihe zur bayerischen Landesgeschichte, vol. 50).

Wada, Y. (1)
"A Korean Rain-Gauge of the 15th century." *QJRMSL*, 1911, *37*, 83. (Also *MZ*, 1911, 232).

Wallis, J. (1)
Opera. 3 vols., Oxford, 1693.

Wang Chu-chi (1)
English-Chinese Dictionary of Mathematical Terms. Hongkong, 1964.

Wang Ling (1)
"The Chinese Origin of the Decimal Place-Value System in the Notation of Numbers." Communication to the 23rd International Congress of Orientalists. Cambridge, 1954.

Wang Ling and Needham, J. (2)
"Horner's Method in Chinese Mathematics; Its Origins in the Root-Extraction Procedures of the Han Dynasty." *TP*, 1955, *43*, 345.

Wang Ling (3)
"The Development of Decimal Fractions in China." *Proc. 8th International Congress of the History of Science,* Florence, 1956, vol. 1, pp. 13–17.

Wang Ling (4)
"The Date of the *Sun Tzŭ Suan Ching* and the Chinese Remainder Problem." *Proc. 10th International Congress of the History of Science* (Ithaca, 1962), 1964, vol. 1, pp. 489–492.

Wang Ling (5)
The 'Chiu Chang Suan Shu' and the History of Chinese Mathematics during the Han Dynasty. Inaug. Diss., Cambridge, 1956.

Wang Ping (1)
"Alexander Wylie's Influence on Chinese Mathematics." International Association of Historians of Asia, *Second Biennial Conference Proceedings,* Taipei, Oct. 6–9, 1962, 777.

Wei Wên-pin (1)
The Currency Problem in China. New York, 1914.

Weinberg, J. (1)
Die Algebra des Abu Kamil Soja ben Aslam. Munich, 1935.

Wêng T'ung-Wên (1)
Répertoire des dates des hommes célèbres des Song (Matériaux pour le Manuel de l'Histoire des Song). Paris, The Hague, 1962.

Werner, E.T.C. (1)
Descriptive Sociology; Chinese. (ed. H. Spencer). London, 1910.

Wertheim, G. (1)
Die Arithmetik des E. Misrachi. Frankfurt, 1893.

Westphal, A. (1)
"Über die Chinesisch-Japanische Rechenmaschine." *MDGNVO,* 1875, *8,* 27.

Westphal, A. (2)
"Über die chinesische Swan-Pan." *MDGNVO,* 1876, *9,* 43.

Wheatley, P. (1)
"Geographical Notes on Some Commodities involved in Sung Maritime Trade." In *Journal of the Malayan Branch of the Royal Asiatic Society,* vol. 32, part 2 (No. 186).

Wiedemann, E. (1)
"Notiz über ein von Ibn al Haitam gelöstes arithmetisches Problem." *SPMSE,* 1892, *24,* 83.

Wilhelm, R. (1)
I Ging; Das Buch der Wandlungen. 2 vols., Jena, 1924.

Williams, Talcott (1)
Silver in China. Philadelphia, 1897.

Winter, H.J.J. (1)
Eastern Science. London, 1952.

Woepcke, F. (1)
Extrait du Fakhri, précédé d'un mémoire sur l'algèbre indéterminée chez les Arabes. Paris, 1853.

Wong Ming (1)
"Le Professeur Li Yen (1892–1963)." *BEFEO,* 1964, *52,* (no. 1), 310.

Wylie, A. (1)
"Jottings on the Science of the Chinese: Arithmetic." *North China Herald,* Aug.–Nov. 1852, nos. 108, 111, 112, 113, 116, 117, 119, 120, 121. Repr. *Copernicus,* 1882, *2,* 169, 183. (For other reprints, see Needham's *Science and Civilisation in China.* vol. 3, p. 800).

Wylie, A. (2)
Notes on Chinese Literature. Shanghai, 1867. (Repr. Peiping, 1939).

Wylie, A. (3)
see J. Doolittle (1)

Yabuuchi, K. (1)
"The Development of the Sciences in China from the 4th to the end of the 12th century." *CHM,* 1958, *4,* 330.

Yabuuchi, K. (2)

"Indian and Arabian Astronomy in China." In *Silver Jubilee Volume of the Zinbun Kagaku Kenkyusyo*, Kyoto University, Kyoto, 1954, pp. 585–603.

Yabuuchi, K. (3)
"Astronomical Tables in China, from the Han to the T'ang Dynasties." In Yabucchi (3').

Yabuuchi, K. (4)
"Indian Astronomy in China." *Proc. 10th International Congress of the History of Science*, Ithaca, 1962 (pub. 1964), p. 555.

Yajima, S. (1)
"Bibliographie du Dr. Mikami Yoshio; Notice biographique." In *Actes de VIIe Congrès Internationale d'Histoire des Sciences*, Jerusalem, 1953, pp. 646–658.

Yang, Lien-shêng (1)
Studies in Chinese Institutional History. Cambridge, Mass, 1961. (Harvard-Yenching Institute Monograph Studies, vol. 20.)

Yang, Lien-shêng (2)
Les aspects économiques des travaux publics dans la Chine impériale. Paris, 1964.

Yang, Lien-shêng (3)
Money and Credit in China, a Short History. Cambridge, Mass., 1952.

Yang, Lien-shêng (4)
"Buddhist Monasteries and Four Money-raising Institutions in Chinese History." *HJAS*, 1950, *13*, 174.

Yen Chih-t'ui (1)
Family Instructions for the Yen Clan (Yen-shih chia-hsün). Leiden, 1968. (Monographies du T'oung Pao, vol. 4.)

Yoshino, Y. (1)
The Japanese Abacus Explained. Tokyo, 1938. Repr. New York, 1963.

Yushkevitch, A.P. (1)
"O dostiženijax kitajskix učenyx v oblasti matematiki," [On the achievements of Chinese scholars in the field of mathematics]. In *Iz Istorii Nauki i Texniki Kitaja* [Essays in the History of science and technology in China], p. 130. Moscow, 1955.

Yushkevitch, A.P. (2)
[Same title as (1)] In *Istoriko-matematičeskie issledovanija*, 1955, *8*, 539.

Yushkevitch, A.P. and Rosenfeld, B.A. (3)
Die mathematik der Länder des Ostens im Mittelalter. (Sovjetische Beiträge zur Geschichte der Naturwissenschaften), Berlin, 1960, pp. 62–160.

Yushkevitch, A.P. (4)
Geschichte der Mathematik im Mittelalter (tr. from Russian). Leipzig, 1964. English translation, *History of Mathematics in the Middle Ages.* Cambridge, Mass., and London, in press.

Yushkevitch, A.P. (5)
"Sur certaines particularités du développement des mathématiques arabes." *Actes du VIIIe Congrès International d' Histoire des Sciences*, Florence, 1956, p. 156.

Zeuthen, H.G. (1)
Geschichte der Mathematik im Altertum und Mittelalter. Copenhagen, 1896[1];
Berlin, 1912[2].

Zeuthen, H.G. (2)
"Sur l'origine de l'algèbre." *KDVS/MFM,* 1919, *2,* no. 4.

Ziegler, A. (1)
Regiomontanus. Dresden, 1874.

Zinner, E. (1)
Leben und Wirken des Johannes Müller von Köningsberg, genannt Regiomontanus.
Munich, 1938. (Schriftenreihe zur bayerischen Landesgeschichte, vol. 31).

Zurmühl, R. (1)
Praktische Mathematik für Ingenieure und Physiker. Berlin, Göttingen, Heidel-
berg, 1963[4].

Index

Abû Bakr al-Farajî (fl. c. 1000), 216, 235
Abû Kâmil al-Misrî (c. 850–930), 216, 234, 240
Abu'l Wafâ, 235 n.84
Agra (Indian technical term), 221–223, 224 n.33, 225 n.41
Agriculture, 433
Ai-jih-ching-lu ts'ang-shu-chih, 48
Ai Tsung (Emperor, A.D. 1225), 426
Alcuin (730?–804), 236
Algebra, 2, 10, 12, 13, 14, 43, 64, 153, 163, 171, 177, 295, 376
European. *See* European mathematics
Indian. *See* Indian mathematics
limited by the possibilities of the counting-board, 207
rhetorical, 27 n.33
Algorithms, 10, 12, 13, 178–180, 204, 362, 377–379
Euclidean, 87 n.19, 223, 225, 257, 365–366, 371
An-yüeh, 24, 26
Analogy, taste for, 15
Ang Tian Se, 278 n.56
Angles, measurement of, 123
Annulus, area of, 105–106
Approximation formulae, 196–201
Arabia, trade with China, 432
Arabian mathematics, 100, 196, 197–198, 208, 214, 233 n.74, 236 n.87, 240–241
Arc-sagitta method, 17
Archimedes, 218
Architecture, 7, 8, 416, 447–461
Area, 16, 97–109
measures, 77
of an annulus, 105–106
of a quadrangle, 100
of a triangle, 99, 105
Argyros, Isaac (1318–1372), 216, 219 n.13, 241–242, 316, 380, 381
Arithmetic, 10, 43 n.46, 55, 82–95, 359 n.1
in Europe. *See* European mathematics

mental, 249
"official," 15
oldest textbook on, 3
terminology, 82–89
Arithmetica (Symon Jacobs of Coburg), 263
Arithmeticae Liber I (Stifel), 251
Arithmetical progressions. *See* Progressions
Armaments. *See* Weapons
Artisans, 7
Âryabhaṭa I (475–550), 197, 214, 216, 220, 221–229, 276, 361, 364, 365
Âryabhaṭa II (fl. A.D. 950), 230
Âryabhaṭiya, 220 229
Astrology, 58
Astronomy, 4, 7, 44, 58, 219, 360, 466, 467, 470–474
Ayao, Hoshi, 421

Babylonian mathematics, 3
Bachet de Mèziriac, C. G., (fl. 1612), 256–260, 360, 365, 371 n.11
Bamboo, 451
Barter, 15, 431–432
Battle formations, 465
Becker, O. and Hofmann, J. E., 325
Bertrand, J., 314–315
Beveridge, William, 216, 263–265, 365, 370, 380
Bhâskara I (fl. A.D. 525), 214, 216, 221, 227, 229–230, 232, 234, 239 n.97, 240 n.99, 250 n.143, 259 n. 170, 265
Bhâskara II (fl. twelfth century), 215, 216, 230–231, 234
Bianchini (fl. fifteenth century), 247–249
Biernatzki, K. L., 177, 311–315, 359 n.1
Biot, E., 291, 310, 315
Board of Astronomy, 5, 7, 20, 26, 62, 65, 369, 466, 467
Board of Finance *(hu-pu)*, 418
Bochenski, I. M., 378
Book of Changes. *See* I-ching
Botanicon Sinicum (Bretschneider), 88